AF673727

Handbook of Rubber Technology

Volume 2

Processing, Compounding, Manufacturing and Uses of Rubber

Handbook of Rubber Technology

Volume 2

Processing, Compounding, Manufacturing and Uses of Rubber

J.M. Martin
M.Sc., Ph.D., C.Eng.

W.K. Smith
C.Eng.

Edited by
S.C. Bhatia
B.E. (Chemical), M.B.A.

CBS

CBS PUBLISHERS & DISTRIBUTORS
NEW DELHI • BANGALORE

ISBN : 81-239-1054-1

First Edition: 2004
Reprint : 2007

Publishing Director : Vinod K. Jain

Published by :
Satish Kumar Jain for CBS Publishers & Distributors,
4596/1-A, 11 Darya Ganj, New Delhi - 110 002 (India)
E-mail : cbspubs@del3.vsnl.net.in
Website : http://www.cbspd.com

Branch Office :
Seema House, 2975, 17th Cross, K.R. Road,
Bansankari 2nd Stage, Bangalore - 560070
Fax : 080-6771680 • E-mail : cbsbng@vsnl.net

Printed at :
Asia Printograph, Shahdara, Delhi - 110 032

Preface

This volume dealing with various types of processing, compounding, manufacturing techniques and applications of natural and synthetic rubbers is the second volume of 'Handbook of Rubber Technology'. This useful handbook endeavours to present a comprehensive, accurate and the most recent account of rubber technology.

Every industry uses rubber for conveyor belts, hoses, tubings or as insulation in heavy cables, every vehicle, such as motor car, bus and aeroplane moves on rubber tyres and consists of other parts made out of rubber such as anti-vibration mountings, window and door seals, radiation hoses or fan belts, and every home uses rubber-based household equipments.

Recognising the great importance of rubber, this reference text book is intended for the practising rubber and polymer technologists, chemical and mechanical engineers, industrial chemists and research students. This book also provides an upto date information for the users of rubber products in other industries.

Emphasis has been placed on the main scientific facts, technological methods, and manufacturing techniques with comprehensive coverage of both raw materials and testing methods. For those new to the industry, attention has been paid to nomenclature and to the terms and abbreviations. Thus this book can also be used as reference manual offering complete information, not only in a panoramic survey, but in-depth details, giving a clear perspective of rubber technology.

This volume is divided into five sections. Section I explains the processing and manufacturing technology of rubber, to make them usable, number of operations are carried out that alter their physical shape or chemical composition such as mixing, masticating, extruding, calendering, coating of textile products and finally moulding and finishing of rubber components. Chapter 1 deals with various mixing machineries and equipments used for processing rubber. Masticating is the breaking down of polymers to make them soft and to reduce their nerves to a point where other mixing ingredients may be added without any difficulty, so chapter 2 is devoted to masticating and mixing. Chapter 3 focuses on extruding of rubber products, which explains screw extruders, ram extruders and vulcanisation of extruders. Chapter 4 discusses calendering which is a process used for the continuous manufacture of sheet or film. Fabrics coated with rubber produced either by impregnation or by surface coating or lamination are called rubberised fabrics, or alternatively proofed fabrics. Hence, chapter 5 is devoted to coating of textile products. Chapter 6 deals with various moulding and finishing aspects of rubber components.

Section II deals with the principles of compounding. Compounding is the mixing process which is used to produce a homogeneous mixture for use in plastics or rubber composition. Thus the compounding describes the selection of additives and their incorporation into polymer so as to give a homogeneous mixture ready for subsequent process. Chapter 7 focuses on compounding ingredients and compound design. Selection of ingredients, formulation of a mix, solubility parameter concepts and designing of compounds are discussed in detail. Chapter 8 concentrates on compounding to meet processing requirements and vulcanisate properties and chapter 9 is devoted to compounding aspects of metal to rubber bonding.

Section III describes various manufacturing techniques of rubber. Recently significant changes have taken place in various manufacturing techniques of rubber related products such as-tyres, footwear, hoses, V-belts, fan belts, sport goods, cables and sealing technology. For example in tyre industry, there is a growing interest in tyres offering safety in the event of a sudden deflation and tyres which, if deflated, enable the driver to continue journey until it becomes convenient to change a wheel. Tyres are also designed to meet a wide spectrum of service and environmental conditions. Chapter 10 is devoted to tyre technology, its properties, mixing techniques, compounding aspects, compound cure systems, processing aids, tyre components, tyre design and tyre performance analysis. Chapter 11 focuses on footwear technology that includes rubber compounds in their make up. This can be classified as-heavy industrial boots, fashion boots, sport shoes and casual with uppers, slippers and rubber soled leather footwear. Also discussed are the typical formulations and trouble shooting guide for foot wear, highlighting the problems, possible causes and possible solution. Chapter 12 concentrates on conveyor belt technology. Conveyor belts are constructed of a reinforcement, either textile or steel cord that is protected from mechanical damage arising from impact and the ingress of moisture by a rubber or plastics cover. The rubber conveyor belt is usually manufactured by laminating a number of plies of rubberised textile reinforcement and consolidating and vulcanising them in a special press or rotary curing machines. The reinforcement aspects, materials used, manufacturing techniques and belt selection are discussed in detail. Chapter 13 focuses on V-belts and fan belt manufacturing technology and chapter 14 is devoted to hose technology. Chapter 15 concentrates on rubber sports goods manufacturing technology. Chapter 16 deals with cable technology, and explains the various types of cables, their properties, manufacturing techniques and uses. In order to use rubber effectively as an engineering material, it is necessary in most cases to bond it securely to metals, usually during the vulcanisation process. Keeping this in mind chapter 17 focuses on rubber to metal bonding components. Various bonding agents, cleaning of metals or other substrate and compounding of rubber for bonding are discussed in detail. Chapter 18 is devoted to vibration isolators and mounts where as chapter 19 concentrates on rubber covered rolls. Chapter 20 is devoted to sealing technology and highlights physical and chemical considerations of elastomeric seals, sealing configurations and environmental considerations. There are large number and variety of other products consuming little rubber because of their small size, their small quantity usage or their low content of new polymer such as, aircraft, electrical and electronic industries. Because of their variety, it is difficult to group or classify them. Keeping this in mind chapter 21 focuses on mechanical and miscellaneous products. Chapter 22 is devoted to computer aided manufacturing. To help the operator maintain tight control, key variable of calendering and other processes can be simultaneously measured and controlled by accomplishing such integration and decision making through computer aided manufacturing (CAM). The introduction of computer to production planning and design will have the effect of greatly increasing efficiency.

Section IV explains the various uses of rubber products in automobiles, aerospace, railways, cables, bearing and bridges, piers and fenders, machinery and equipments, and packaging. The automobile combines many different materials that must be integrated with the ever changing design and manufacturing processes. Without rubber components, the automobile in its current state of technology cannot exist. Keeping this in mind chapter 23 is devoted to rubber in automotive applications. Chapter 24 concentrates on thermoplastic elastomers for automotive applications where as chapter 25 focuses on high tension ignition cables for special automotive applications. Chapter 26 describes in detail the uses of rubber and composite in aerospace, as typical uses in dry bearings and structural composites. Shock and vibrations in rail vehicles are much more severe due to rail-wheel interactions, track irregularities, crossings and rail joints, firing impulses of engine. Keeping this is mind chapter 27 focuses on some novel applications of rubber in railway engineering. Chapter 28 concentrates on various special applications of rubber. Chapter 29 explains the antistatic and conductive uses of rubber in antistatic vehicle tyres, belts, hoses, hospital equipments and cables. Chapter 30 is devoted to uses of rubber in bearing, piers and fenders where as chapter 31 describes some typical uses of nitrile rubber in construction industry.

Section V describes some of the special topics which are of great importantce to rubber technologists. Chapter 32 focuses on chemical modification of polymers with special emhasis on esterification, hydrolysis and hydrogenation of polymers. Chapter 33 deals with control of biological degradation of polymers by means of biocides. Rheology is the science that deals with the way materials deform when forces are applied to them. The term is commonly applied to the study of liquids and liquids like material such as polymer solutions, molten plastics etc. Keeping this in mind chapter 34 is devoted to basic concepts of rehology. Chapter 35 discusses the rubber resins pressure sensitive adhesive tape technology and its manufacturing aspects.

The text in this book is throughout supplemented with diagrams, and tables. The treatment of all topics is in a cogent, lucid style aimed at enabling the reader to grasp the information quickly and easily. The references and index are also given at the end of the book.

While painstaking care has gone into producing a useful and exhaustive reference textbook, the authors would welcome any constructive criticism and creative feedbacks from the students, teachers and professionals in the industry.

We wish to thank all those colleagues who provided figures and photographs and offered important advice during its preparation.

J. M. Martin
W. K. Smith

Contents at a Glance

SECTION I

SECTION II

SECTION III

SECTION IV

SECTION V

Contents

SECTION V

SECTION I

Processing/Manufacturing Technology

CHAPTER 1

Mixing Machinery and Equipments for Rubber

INTRODUCTION

The raw polymers cannot be used as such. To make them usable a number of operations are carried out on them and which alter their physical shape or chemical composition. Mixing is central to rubber technology. If the base compound is inadequately mixed problems cascade down through the subsequent processes of shaping and curing into the end product. It is therefore of value to understand not only the procedures and practice of mixing but also physical processes involved in mixing.

Wide variety of materials are added to the polymer to obtain desired processing and vulcanisate properties. When rubber contains all the ingredients needed it is known as a compound. The purpose of the mixing rubber compound is to blend the ingredients of the formulation into a uniform, well dispersed stock. The mixing of rubber compound is generally done in batch process using equipment such as: (i) two roll mills; and (ii) internal mixers.

MILLS

Just over a century ago commercial processing of rubber was started using two roll mills. The roll axes were parallel and distance (nip) between the two rolls was made adjustable by having the bearing blocks of the front roll resting against adjusting screws. The mills remain substantially unchanged to the present day, the main difference being only modification and refinement of the original design.

Originally mills were double geared and driven by steam engine, now electric motor is used and mills are single geared as they are cheaper to manufacture. Some industry use partially or fully corrugated back roll for quick pre-warming of the rubber or rubber compound.

Friction Ration of Mills

The speeds of the two rolls are often different, the friction ratio depending on the mills use. For natural rubber mixing a ratio of 1:1.25 for the front-to-back roll is common. High friction ratios are used for refining compounds. For mixing of some of the synthetic rubbers (say butyl) a near-even speed is best. Most of the synthetic rubbers are very difficult to mix on mills so internal mixers are used for them.

Cooling of Mills

Mixing of rubber compounds are generally carried out at around 70°C. But during mixing heat is generated and temperature of mills goes higher then this limit. So cooling of mills is necessary. Two types of cooling are used. One employs cored rolls i.e., water is sprayed on to the outside of an, axially drilled central core. Other type is peripherally drilled rolls, the water is circulated through a labyrinth of passages about 50mm under the roll surface.

Safety of Mills

Mills are usually fitted with various safety devices, some to protect the operator, the others to protect the mills.

The operator is protected by the Lurin safety bar. This bar is brest high bar and it stops the mill if pressure is applied. If, by some mischance, the operator is caught up and dragged towards the nip, the pressure on the Lurin bar will stop the mill.

The mill is protected by fitting cast iron breaker plates between the roll nip adjusting screws and the roll end bearing housing. Now-a-days cast-iron breaker plates are replaced by a hydraulic relief system which allows the nip to open in the even of overload.

Ancilliary Attachments

Mills are fitted with a metal tray under the rolls to collect dropping from the mill and with guides which are fitted to the ends of rolls to prevent rubber being contaminated with grease, etc.

Other attachments

1. Stock Blender: It is a device for blending the rubber compound mechanically instead of depending upon the operator.
2. Cutting knives mounted against the roll to produce strips for feeding other equipment (say cold feed extruder).
3. Scraper blades to remove soft and sticky compounds.

INTERNAL MIXERS (THEORY OF MIXING)

As rubber industry grew, mills were replaced by internal mixer because of the versatility rapid mixing and high productivity. An internal mixer can simply be defined in the following way. The closed mixing chamber has an outer cooling or heating jacket, and is in the form of a divided, double cylinder in which two hollow spiral rotors rotate in opposite directions. The rubber is kneaded and mixed between the chamber walls and the rotors and between the rotors themselves. The compound is kept under pressure by means of pneumatic ram.

Mixing in internal mixer can be better understood in the following manner. Mixing is accompanied by four different actions (Fig. 1.1): (i) milling; (ii) kneading; (iii) longitudinal cut-back; and (iv) lateral overlap. Milling is the high shear action smearing material around the periphery of the chamber.

Due to the helix (spiral) angles of the rotor blades material is moved towards the centre of the mixing chamber, thereby causing longitudinal cut-back. Because of the different rotor speeds an intensive kneading action takes place between the rotors and lateral overlap from one side of the chamber to the others. These four actions, together with correct temperatures control of the mixing chamber result in high quality compounds with relatively short mixing cycle times.

Types of Internal Mixer

Internal mixers are generally of two types. One is 'Banbury Mixer' which has a friction ratio 1:1.11 between the rotors. These rotors are non-interlocking and work required to incorporate the powders is carried out between rotors and the chamber. Other is 'Intermix' in which the rotors run at even speed and the wings are designed to produce a friction ratio between the rotor. The work is done between the rotors rather then between the rotors and the chamber. These rotors are inter-locking type. Both types of rotors are shown in Fig. 1.2.

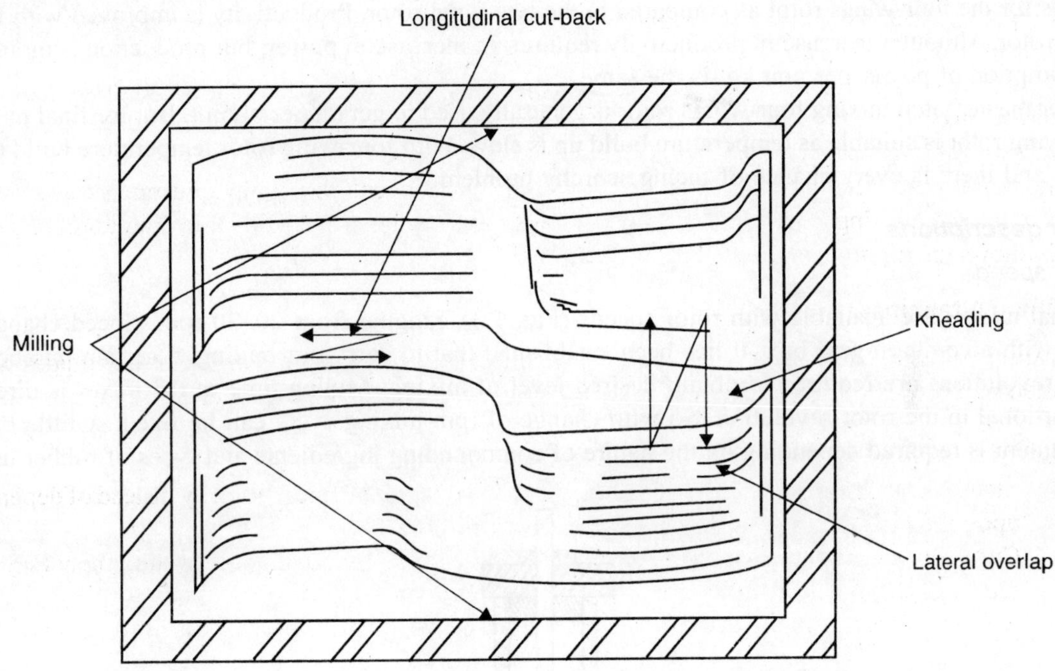

Fig. 1.1. Lateral overlap.

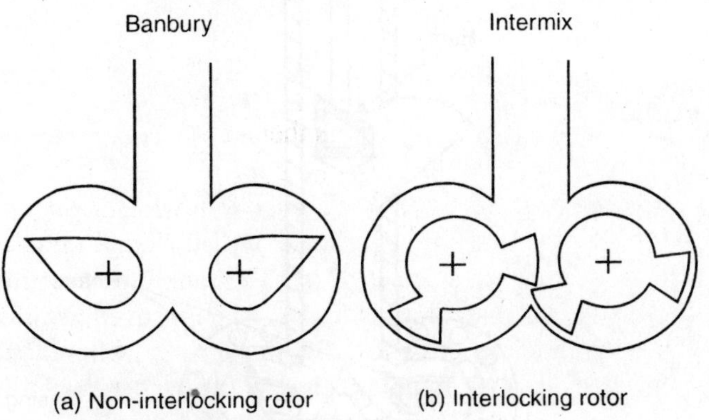

Fig. 1.2. Rotor design.

With inter-locking type rotors materials do not pass easily through the rotors. Mixing takes place between the nip of the rotors. The break down of rubber is better, viscosity is lower, dispersion is better then mixer with non-interlocking type rotors. This is ideal mixer for compounds with high loading and with less oil or no oil. But output is low. With mixers with non-interlocking rotors material easily pass through the rotors, so output is high.

Rotor design

Lot of changes have taken place in rotor design for internal batch mixer. The ultimate rotor design is the four-wing rotor, in comparison to the two-wing rotor. The four-wing rotor is a superior design to the two-wing in actual performance. Reduction of mixing times can be obtained at various mixer rotor speeds for the four-wings rotor as compared to the two-wing rotor. Productivity is improved with four-wing rotor. Although increase in productivity requires an increase in power, but production is more. So consumption of power per unit kg. is the same.

For masterbatch mixing four-wings rotor is generally used to get higher output. But for final mixing two-wing rotor is suitable as temperature build up is slow. With four-wing rotor temperature build up is faster and there is every chance of facing scorchy problem.

Mixer descriptions

Rotor speed

Internal mixers are available with rotor speeds (Fig. 1.3), ranging from 20–30 rpm. Speed change is done with a common gear box. It has been established that to carry out mixing, a certain number of rotor revolutions are required to obtain desired level of mixing. Mixing time in the mixer is directly proportional to the rotor revolution. So with change of rpm mixing cycle can be fixed. A little bit of adjustment is required depending on the nature of compounding ingredients and types of rubber used.

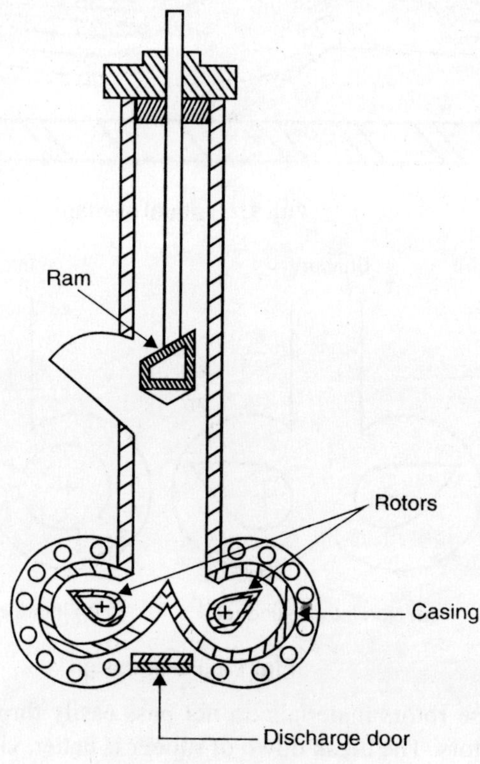

Fig. 1.3. Diagrammatic section of a Banbury mixer.

Ram pressure

The thrust applied to the ram affect the output. Increase in ram pressure reduces the voids in the mixer and increases the speed of engagement of materials in the mixers by rotors and hence reduces mixing cycles. Increase in raw pressure also increases contact between the material in the mixer and so gives more rapid ingredient absorption. This better contact of materials reduces the tendency of the batch to crumb and gives better reproducibility in mixing.

There is, however, maximum ram pressure beyond which any further increase results in no signification increase in output since the voids in the mixer have been filled and maximum contact has been achieved between the rotors and the materials and between the materials themselves.

Discharge door

Sliding and drop door type. Drop door is better than sliding door. Due to sticking of compound, sliding door causes delays which result in less output.

Temperature control of the mixer

Control of the operating temperature of the mixer is of vital importance to avoid damage to the rubber and to hold the mixing cycle to its optimum time. This is especially true when hard and viscous stocks are being processed. In these cases, lack of control of the temperature of the internal parts of the mixer leads to lack of control over the work that is put into the mix.

The main effect in temperature control is to hold the temperature of the level that will give the lowest temperature combined with a sufficiently high co-efficient of friction between the rubber and the metal of the mixer, imparting high turbulence and shear to the mixer.

To achieve this, it is essential that flow through the parts of the mixer is non-laminar so that full effect of the temperature control medium is felt. Systems now being used include water tempering by mixing steam and cold water.

Thermocouples (two) are generally fitted at the both end frames of the mixer body and average temperature is recorded for the compound.

Handling of batch

Mills, rollers die, extruder and roller die are generally used to process the batch after dumping from the mixer. Batch is taken out in the form of sheet which is cooled either by forced air or by cold water and passes through antitack solution to avoid stock stuck problem after lay down which is done either in wig-wag form or sheet form.

Recent Development in Mixing

The internal mixer has changed only little since *Fernly Banbury* first designed the machine. But it has moved ahead and in recent years moved very rapidly. The simple double rotor system of the *Banbury* has developed into rotor designs that gives higher shear with less power consumption.

The new rotor designs coupled with better understanding of the mixing process have resulted in much shorter mixing times, but with these shorter times some problems have been experienced. In the days when a 10 minute mixing time was a norm, a minute here and there made little difference to the mix. Now, when the mixing times are only a matter of two or three minutes, in many cases a few second variation can spell the difference between a batch being on or off the specification.

One approach to this solution is to dump the batches on power basis. Now-a-day power integrator is being used by many industries. The main purpose of power integrator control is to improve batch to batch uniformity. The first and most obvious improvement is seen by the first batch effect where the mixer has had a long period to cool.

The power integrator is designed so that only the work input to the batch is measured. A no load zero circuit is provided to eliminate the power required to run the mixer empty. The work input is measured as kwhr.

Internal mixers do not normally produce material in the right physical form either for storage or for subsequent handling. In many instances degree of black dispersion obtained is inadequate. Thus internal mixers are followed by additional processing equipment either one or more mills, a roller die, extruder and roller die.

Two mills, especially if are fitted with stock blender, assist in improving the dispersion of carbon black. They also enable to separate out any off specification batch.

The recent trend is to put extruder and roller dies. The thinking is that if the mixing cycle in the *Banbury* can be shortened, giving higher productivity, although at a lower quality level, the extruder underneath can both handle rate and bring the quality up to the level originally desired.

CHAPTER 2

Masticating and Mixing

INTRODUCTION

The raw polymers cannot be used as such. To make them useable a number of operations are carried out on them. The main operations are: (i) mastication; (ii) masterbatching; (iii) remilling; and (iv) mixing.

Wide variety of materials are incorporated into the polymer to obtain desired processing and vulcanisate properties. When rubber contains all the ingredient needed it is known as a 'compounds'. When some ingredient are withheld deliberately, the practically completed compound is termed as 'masterbatch'.

The masterbatch is later on converted into the compound by adding withheld ingredients which are generally the vulcanising or curing ingredients and the complete compound is called 'final mix'.

MASTICATION

Mastication is the breaking down of polymers to make them soft and to reduce their nerves to a point where other mixing ingredients may be added without difficulty. Natural rubber and co-polymers (synthetic rubbers) respond somewhat differently during this processing step but the general procedure and the equipment used is similar with both materials.

Theory of Mastication

Breakdown of rubber at the more elevated temperature is due to oxidative scission. In the lower temperature range and under the stress set up during mastication, mechanical rapture of primary bonds occurs, the free radicals thus formed become stabilised by addition of oxygen. When mastication is carried out in the absence of oxygen i.e., in the internal mixer, little or no breakdown occurs. Under this condition free radicals resulting from mechanical shear recombine. In internal mixer peptisers are used to increases the efficiency of mastication of rubbers, i.e., to increase the rate of molecular breakdown particularly in NR (dosage less then 0.500 phr). Peptisers are added at the start of mastication. The low temperature breakdown of rubber is attributed to interaction of rubber radicals with oxygen terminated fragments of lower molecular weight. Other radical acceptors can be expected to promote breakdown of rubber in an inert atmosphere.

Radical acceptors which promote breakdown of rubber are not to be confused with chemical plasticisers (peptisers). The formers are only effective under the conditions of cold mastication and do not cause breakdown above 100°C in absence of oxygen.

As the temperature reached in a mass of rubber during mastication is increased the rubber becomes progressively softer and the rate of mechanical rapture is correspondingly reduced. Hence, the breakdown efficiency in air or with radical acceptors under nitrogen falls with increasing temperature (Fig. 2.1).

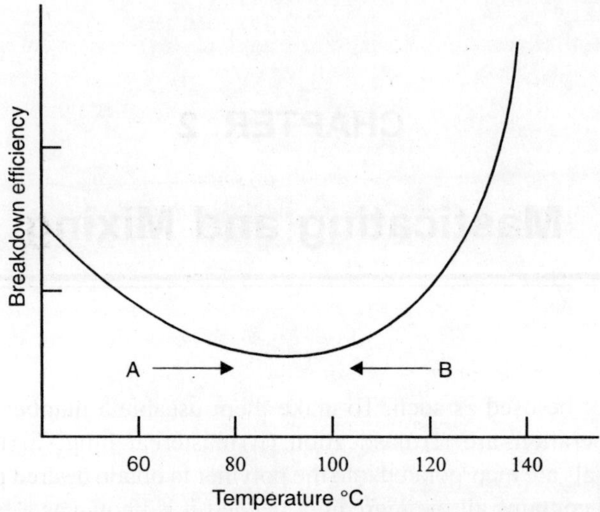

Fig. 2.1. Breakdown of NR after 30 min mastication at different temperature.

A second type of breakdown due to conventional oxidative scission become apparent above 80°C and this increases rapidly with temperature.

The overall behaviour of rubber masticated in air is expressed by the continuous curve in Fig. 2.1, in which stabilisation by oxygen of radicals formed by mechanical rapture is responsible for the left hand part (A) of the curve, and oxidative degradation is responsible for the right hand part (B). Degradation is at a minimum at 105°C.

Mastication process

Breakdown or mastication can be carried out on open mill or in internal mixer. In a modern high factory *Banbury* type mixer is generally used for mastication as open mill mastication is uneconomic due to less production. In internal mixer rubber is loaded with full capacity (full volume of the mixer) and mastication is carried out in short time cycle at higher rpm at an elevated temperature of 150°C. Due to inert atmosphere chemical plasticisers must be added for mastication in internal mixer and are added at zero second.

The difference between the operating speeds of front and back rolls i.e. friction ratio influences the degree of mastication. This also holds true for rotors in internal mixers. The clearances, condition of the rotor surfaces and rotor wings design as well as ram pressure also influences the breakdown.

Masterbatching, Remilling and Final Mixing

The following informations concerning mixing cycles and times can be extended to plant and laboratories mixing:

1. There are definite temperature ranges for mixing rubbers, each specific rubber having an optimum temperature at which the desired dispersion of compounding ingredients in the mix is obtained.
2. Some rubbers require initial breakdown period before the ingredients are added.

3. A specific order (sequence) of incorporation of the compounding ingredients is necessary.
4. The time of mixing for each step in the process is important.
5. A finishing step and control of the final temperature of the mix is necessary to prevent pre-vulacanisation.
6. Banbury mixers require less time in handling and mixing larger batches then corresponding mills.

Although Banbury mixers are capable of handling larger factory batches at a very faster rate then open mills, there are other processing problems involved.

Banbury temperature is higher then mill so scorch problems may be faced which, in turn, reduces plastic properties of the compound so that it can no longer be further processed.

Processing methods

Both in the laboratory and factory the most common method for incorporating the compounding ingredients into rubber involve either use of a mill or Banbury mixer.

Recipe

To illustrate mixing the following NR recipe is chosen:

Ingredients	*Quantity*	
NR	100.00	
ZnO	5.00)	
Stearic acid	1.00)	Master Batch
Carbon	45.00)	
Oil	6.00	
Antioxidant	1.00	
Paraffin wax.	1.00	
Sulphur	2.50)	Curatives
Accelerator	1.00)	
	162.50	

Final Mix = Master batch + Curatives.

Batch weight

For internal mixer batch weight is generally calculated as per following guidelines.

Batch weight = Mixer Gross volume × Compound S.G × Fill Factor

Fill factor is variable and dependent upon viscosity, speed and ram pressure. For a tread type of NR compound fill factor is 0.7 to 0.75.

Open mill mixing

Mixing cycle (Approx.)	*Mill Temp. – 70 ± 5°C*
Load raw rubber on the mill, give cuts from ends makes a smooth band.	12"
Add ZnO, Stearic acid and half carbon, give cuts from both ends. Add rest of the carbon and add slowly throughout width of the mill oil, wax and antioxidant. Give cuts from both ends and mix well. During mixing take out the droppings and add to the mix.	20"
Add sulphur and accelerator throughout the width of mill, give cuts from both ends. After mixing take out the stock in required thickness and keep it for cooling.	25"

On open mill generally total mixing is done at a time. But depending on plant requirement first Master Batch (MB) may be mixed and stored and later on MB can be converted into final mix.

Banbury (Internal) mixing

In Banbury or internal mixer total mixing cannot be done at a time due to high temperature attained during mixing of master batch. So, first MB is mixed and the stock is cooled and stored before converting into final mix.

Master Batch Mixing	*Quantity of ingredients and temperature*
Mixing cycle (approx.) rpm – 40	Temp. 150–160°C
Load masticated NR ZnO, Stearic acid, Carbon black and antioxidant	0"
Add oil and wax	60"
Ram raise	120"
Dump	160"
Close (drop door)	180"

Total mixing cycle = Loading time + Dump time + Drop door opening and closing time.

Say, loading time is 20" and drop door closing time is 5"

So total mixing cycle will be = 20" + 160" + 28" + 5" = 213"

The dumped MB is either processed on basement mill or passed through extruder and taken out in sheet form. The sheet is cooled either by water or air and then stored on pallet. Later on, this stock is remilled, if necessary, and then converted into final mix.

MB can be dumped on power basis (modern technique). For power basis mixing, total mixing cycle may vary by 5 to 10 secs. from batch to batch.

Remilling

For tread type and highly loaded compounds remilling is must to get desired viscosity for final mix for easy and safe operation in further processing stages. For softer compounds remilling is not necessary as viscosity of MB is itself low.

Final Mixing	*Quantity of ingredients and temperature*
Mixing cycle (Approx.) rpm - 30	Temp. 100–115°C
Load MB or MBRM (RM = Remill) and curatives	0"
Ram raise	90"
Dump	130"
Close (Drop door)	150"

Final mix batch is also taken out in sheet form, cooled and stored in pallet.

For final mix dump temperature is the important factor. Temperature should not cross the maximum specified limit. Otherwise compound may turn lumpy and may create scorchy problem in further processing stage.

Modern mixing technique

Now-a-days mixing in internal mixer is either semiautomatic or automatic to obtain higher output and better quality. Fillers, oils are stored in bunkers. Weighing and feeding of fillers and oils are done at auto. Mixing operation is also done at auto. Automatic mixing is also being carried out with the help of computer. Computer not only controls mixing but also spells out off-specified batch.

Efficiency of internal mixer

Efficiency of internal mixer can be increased by shortening the mixing cycle by mainly increasing rpm and also, to some extent by increasing ram pressure. Two-fold increase in production may be achieved by increasing rpm from, say 30 to 60. Modern technique is mixing with warm water to increase the output. Fig. 2.2, shows cooling water temperature influence on the cycle time for natural rubber stock.

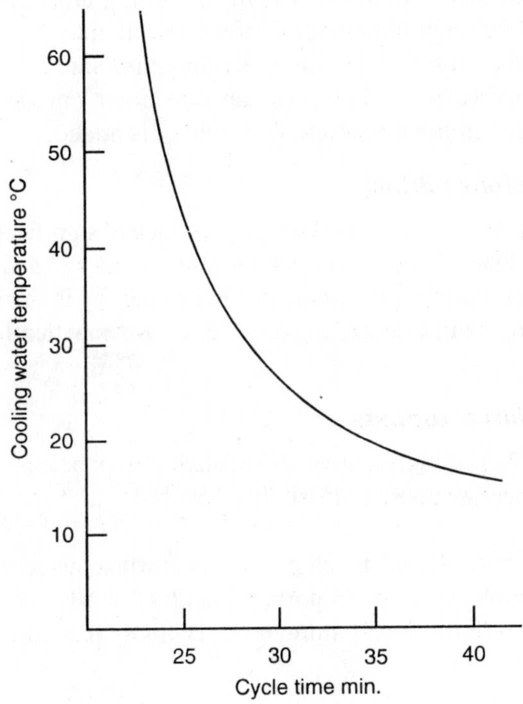

Fig. 2.2. Cooling water temperature influence on the cycle time for natural rubber stock.

Mixing of Synthetic Polymer

SBR (Styrene butadiene rubbers)

In case of NR, it is necessary to plasticise the raw rubber as a separate operation (mastication) before mixing with black and other ingredients. With a SBR compound it is advantageous to make first master batch and then remill the master batch before making final batch, rather then to plasticise the rubber by itself. The masterbatch of certain SBR tread stock must be remilled prior to the final batch. But this is equally necessary for certain NR tread stock.

PBD (Polybutadiene rubbers)

Most PBD rubbers posses inherently high resistance to breakdown, on account of poor mill banding characteristics and rough extrusion appearance. PBD rubber is mostly used as blends with NR and SBR and such blends have been used commercially to obtain adequate processibility and useful balance of physical properties. Cracking and cutting resistance of NR/PBD tread compounds are considerably good compared to 100 NR compound.

EPDM

Ethylene/propylene rubber is usually called 'EPDM' - 'E' for ethylene, 'P' for propylene, 'D' for diene and 'M' for methylene, which are the repeating units.

EPDM polymers can be mixed and processed on all of the machinery commonly found in various rubber fabricating plants. On mill mixing the lower Mooney viscosity - EPDM's in highly loaded stocks are easier to handle. The high viscosity varieties of EPDM are tough and nervy on the mill but their processing improves as they readily accept high loading of oil and fillers.

In Banbury mixing the cycles and dump temperatures are about the same as would be used for SBR. EPDM compounds generally contains more fillers, so 'up side down' mixing is often desirable. In upside down mixing filler is first fed in the mixer and then rubber is added.

IIR-Butyl rubber (Isobutene-Isoprene rubber)

The more chemically inert nature of Butyl rubber is reflected in the lack of significant molecular breakdown during processing. The polyisobutylene portion of Butyl molecule provides high degree of impermeability of Butyl rubber compound. It also contains high loading of filler so mixing is carried out in internal mixer by 'Up-side-down' method. Dump temperature is 165–175°C. No mastication is required prior to masterbatching.

NBR-Acrylonitrile-Butadiene (Nitrile) rubbers

The special function of nitrile rubber is high degree of resistance to attack by oil both at normal and elevated temperatures. Nitrile rubber has poor compatibility with NR, but it can be blended successfully in all proportions with SBR.

The nitrile rubber is less plastic then NR and develops more heat when mixed under similar conditions. Preliminary breakdown of nitrile rubber is more important then that of NR. The more efficient method requires breakdown on a cold tight mill for 5 to 10 minutes. In Banbury it should run as cold as possible.

CR-Chloroprene rubbers

CR vulcanisates are generally recognised to be highly resistance to oxidative ageing and show high level of resistance to flexcracking. Because of the chlorine in the molecule, CR has inherent flame resistance.

Mixing the processing of chloroprene rubbers can be carried out using conventional equipments and methods. In mixing, CR readily bands on a cool mill or masses in an internal mixer. To avoid any scorch problem it is customery in mill mixing to add magnesia first, followed by hard fillers and then soft fillers, followed by oils, waxes, or softeners, with zinc oxide and accelerators, if any, added in the final stage of mixing. With internal mixers, which provide more intensive shearing action, the mixing cycle and heat generation are often minimised by commencing oil addition after only a part of the filler has been incorporated.

RECENT DEVELOPMENT IN MIXING TECHNOLOGY

Early in the history of rubber compounding, it was found that kneading or softening of elastomer was useful to reduce the incorporation time of powders into the rubber. In the early part of rubber industry, this process was achieved through two roll mills. But the introduction of pneumatic tyres associated with raise in popularity of automobile, compound design approach underwent a sea change which demanded more sophisticated and high shear type mixers which can handle very low particle sizes of carbon black and hydrophilic material like silica filler etc.

Every mixer whether it is two roll mill or internal mixer should accomplish a few quality parameters of mixing such as incorporation, dispersion and distribution. All developments in mixing technology are focussed on the improvement in these parameters. In the beginning of rubber mixing, this was achieved through a twin roll mill, a devise for preparing small quantities of compound. This equipment is characterised by two rolls operating at differential speed, counter rotating. Several variations in friction ratios are followed by the manufacturers depending upon the application. But the common ratio followed for rubber mixing mill, breakdown and refiner mills is 1.25:1, and for hold mill and feed mill is 1.08:1. In addition, the slower of the two rolls are usually found on the operators side. The method of mixing is time consuming and labour intensive and the quality is greately influenced by operators. In order to avoid operator variable, blenders are installed to perform cutting action automatically but this is possible only after all the dry ingredients have been incorporated. The temperature rise in the rubber compound on two roll mills is directly related to the sum of speed of the two rolls and friction ratio. Attempts have been made to reduce the heat generation through the reduction of mill speed, friction ratio and adoption of drilled roll in place of bored roll. In the most modern mill the roll speed is maintained at 80-90 FPM and friction ratio is around 1.08:1. Development work in the rubber mixing continued in a systematic manner.

Developments in Premixer Area

Pre-mixer area

In the pre-mixer area, several developments have taken place in material receiving, storage, transportation and feeding into internal mixers. Natural rubber which is received in bale form are cut into pieces and blended properly, compounded with other ingredients and charged to the internal mixer manually for making compounds. But there are a few companies which are using big choppers or granulators of built in capacity upto 20,000 MT per hour for chopping of rubber into small pieces prior to compounding. This process reduces incorporation time and improves the dispersion of carbon black. The practice of compounding carbon black directly from the original bags and dumping them into the internal mixer was changed to automatic carbon black weighing and charging system. Carbon blacks are stored in big silos and the same is transported into the mixer at a predetermined weight either by screw conveyor or air system and charged into mixer. For oils, automatic weighing and charging systems have been introduced in almost all most modern mixers. This process has reduced non-productive mixer cycle time and improved productivity. Several mixers are equipped with split loading of oils which has helped to charge the oil in such a way that slippage of compound due to excess oil is totally avoided. 90° loaders are installed in almost all mixers to reduce the feeding time of second master into the mixer and the same has reduced the incidence of the entanglement of rubber sheets at the mixer throat. The feeding and weighing conveyor of the mixer is automated. One or two check points for weights have been introduced to ensure accuracy of weight of compounded materials and the same is interlocked with the mixer hopper. The mixer hopper control system checks the correctness of the batch and compares it with the specification before it permits for loading into the mixer.

Tempered water system is used for the heating and cooling of the mixer chamber and also to maintain the wall temperature at the specified level. The introduction of tempered water has definitely enhanced the quality of mixed compounds and improved the carbon black dispersion. This practice has helped to reduce batch to batch variation and to achieve greater consistency at extruders especially with cold feed extruders.

Mixing area

Mixing of rubber compound is a very complex process and it is useful to understand the methodology in which the mixing process is completed so that mixed compounds achieve the maximum desired properties. During mixing cycle six basic physical operations takes place. They are: (i) consolidation; (ii) mastication; (iii) incorporation; (iv) dispersion; (v) plasticisation; and (vi) distribution.

It has already been explained in detail about the developmental activities that took place in the premixer area to reduce loading time of ingredients into the mixer. Hoppers are enlarged to accept bales and sheets more readily, the door angle is steepened for more rapid introduction of material into mixing chambers. In order to ease the consolidation process and to drive the material more rapidly into the rotor area, ram design has been modified from time to time. Even though 4 types of configurations are available, the most commonly used is single and double –V- gabled bottom design. Flat design is very old and is seen in old 'Tangential mixers'. Single gable –V- bottom designs is very common and double gable –V- bottom design is used only when the powders and oils are charged over the ram without disturbing the ram. This design has got a disadvantage because it will increase wear and tear of the floating weight and also may cause compound hang up on ram and sides.

Ram pressure is another factor which decides the quality of compound and reduces the incorporation time of ingredients into the rubber matrix. The internal mixers are designed with different ram cylinder diameters. The effective stock pressure for the applied air pressure varies from mixer to mixer and hence it is critical to establish the effective stock pressure for each mixer and regulate ram pressure accordingly. Three probable situations can occur. In the first case, the batch is oversized and hence the ram is finding it difficult to seat and therefore the mixer is struggling to incorporate, disperse and distribute the ingredients. In the second case, the ram is seated properly and the ingredients are incorporated in time and hence good mixing will occur. In the third case, the batch is undersized and the pressure will not be adequate and so good mixing is not possible.

In order to improve the incorporation, dispersion and distribution, several studies have been conducted to understand the process properly and based on the study findings, rotor design, shape and cooling system have been modified. It is understood that rotor design, batch temperature, shear rate and shear stress have remarkable influence on the above parameters.

The geometry of the rotor design is very well understood by the designers. Dispersions of carbon black and other ingredients is governed by the helix angle, lateral overlap and tailing angle. Distributive mixing, which is considered to be a major factor for process consistency is controlled by tip width, tip clearance and wing length.

Other factors like material flow through mixing chamber, transfer across mixing chamber and rotor to rotor, shear rate, entry and relief effect through rotor tip are governed by both helix angle, wing length and the ratio between short to long wing.

After understanding the importance of distributive mixing, several innovations in mixing technology have been adopted. Among the major ones are, the optimisation of helix angle, long wing to short wing ratio etc. Latest developments include the introduction of ST, NST and HDM rotors, elimination of friction ratio between the rotors and tip cooling of rotors.

HDM rotors are major developments in the Mixing Technology. It is also interesting to note that two wing rotor mixers are being replaced with four wing rotor mixers and 20, 30 rpm mixers are getting replaced with 40, 60 and 80 rpm mixers. These developments are aimed at improving productivity because a four wing rotor mixer may increase productivity to 15–30 per cent compared to that of two wing rotor mixer.

The Dust Seal of the internal mixer was always a concern for the compounder because it is a potential area where cured bits are generated. In order to overcome this problem, proved designs of dust seals are available. They are called FYH, SSA and Eco gland self lubricating gland ring. In FYH design, sealing rings are pushed and held towards the rotor ends by hydraulic cylinders and hence the sealing is perfect and there is no chance of entrapment of compound. The frictional heat generated by FYH dust stop is low and heat sensitive stocks can be mixed at high rotor speed without cured particle being formed. All wearing parts including the wearing ring in the rotor end can be removed and replaced without dismantling the mixer. SSA is also a well known Farrel dust stop. In this design the sealing faces are preloaded by spring and additional sealing pressure is generated during mixing by process oil feed which loads the back face of the floating ring which cannot escape because of the compound in the mixing chamber. As soon as the mixer is empty this pressure is relieved. All wearing parts are in halves for onsite replacement. In both cases, it is used to control the feed of process oil to the dust stop so that pump stops when the machine is not mixing which can be signaled by the ram being raised. Another development in the dust seal system is the introduction of Eco gland dust stop self lubricated gland ring for the mixer. This has got the following benefits: (i) eliminate dust stop lube oil; (ii) no waste disposal of used lube oil; and (iii) reduced maintenance cost.

Eco gland self lubricating gland ring is made out of special plastic material and has got high heat and abrasion resistance and it is supplied in split ring form for easy replacement.

Thus we have explained in detail about several developments which took place in the tangential mixer area. In a similar manner, some developments were made in the inter meshing rotor design mixers. The most important of them are the introduction of NR-5 rotor which has helped to improve the fill factor to the tune of 7 per cent and also enhanced the cooling efficiency and batch to batch variation. Another major innovation is the introduction of two cylinder hydraulic system for the control of floating weight.

Mixer Cooling System

Studies have proved beyond doubt that there is an optimum metal temperature for each polymer to avoid slippage and to produce high quality compounds. This temperature is very critical. In a typical mixing cycle for master batch for F270 banbury approx. 55 per cent of applied energy is used for heating the rubber and 10 per cent energy is accounted for electrical and mechanical losses and 35 per cent energy is taken away by cooling water. Proper understanding of heat generation during mixing operation led to the improvement in Banbury design. Drilled configuration for the sides, improved tip cooling for the rotors, better water circulation for discharge door, and well defined and calculated water flow for all the above regions have eliminated the cooling related problems of banbury.

Mixing Process

In conventional system, the compounds are mixed based on time. In this process all the mixing variables such as man, machine and material are reflected on the quality of mixed compound. Gradually, technology moved away from time based mixing to temperature based mixing. In this process the man variable is eliminated to a great extent because the mix has to be kept in the banbury till the specified dump temperature is reached. The introduction of energy based mixing has helped industry to move away from temperature based mixing to PLC controlled or power integrated mixing. This technique is highly useful because it will eliminate the variables such as human and equipment. But the material variable still continues to dominate the quality of mixed compounds. Several experiments are going on to contain material variability. In batch condition controlled mixing, one of the rheological properties of compound is monitored and the mixing cycle, motor rpm, cooling water circulation, ram pressure are adjusted to get desired degree of properties for the compound within the specified mixing cycle. In order to control the batch to batch variation and improved automation, manufacturers are experimenting with continuous mixer and the mixing and venting extruders (MVE). However, these have not been widely accepted by the industry so far.

Post Mixer Area

The material discharged from the mixer is in lump form and the same has to be cooled quickly and also transformed to a shape suitable for handling. This has been achieved with the help of different types of equipment as given below: (i) single or multiple sets of two roll mill; (ii) pork chop extruder and sheeter; (iii) pelletizer extruder; (iv) slab extruder; (v) roller die extruder with sheeter; (vi) roller head extruder; and (vii) twin screw extruder. Two roll mill was very popular from the start of rubber mixing and is still used extensively as a major equipment underneath the mixer for sheeting out operation. But the limitation of the equipment is the human factor. The experience, competence and fatigue of the operator play a vital role in the quality which can be achieved through mill sheet out operation. Improper cutting, blending of compound on the mill and extended residence time can affect viscosity, scorch and cure characteristics of the compound. The mill used underneath the mixer for sheeting out operation is usually designed with low friction ratio and surface speed.

Two roll mills are gradually replaced with extruders and this change over is useful in eliminating human error to some extent. These extruders are of special type. The L/D ratio of these extruder is very short and compression ratio is very low. Different types of extruder heads are in use. Initially, pork chop head was used but it lost its identity in the recent years due to the dryness that is characteristics of some radial tyre compounds. This head is replaced with rubber pelletizer but unfortunately surface coating is required to avoid sticking of pellets to each other. The surface coating has got some effect on the physical properties of compounds. The pelletizer was replaced with roller dies. The roller die consists of two roll vertical calender supported on a stand at the end of extruder. The extruder is outfitted with special head which delivers rubber tangential to top roll. The head is designed in such a way that the stock will spread from the extruder screw to find out the roll face and deliver a smooth sheet to the batch off system. The roller die extruder is equipped with two or three pushers to ensure optimum screw efficiency while conveying the batch to the rolls. The roller die is being replaced with roller head extruder. The roller head extruder differs from the roller die in the manner in which the rolls are positioned to the die. With roller head, the rolls are supported on the head end of the extruder. This eliminates the need for tracks and moving the entire batch of apparatus when the rolls are opened to inspect the head for residual compound. The roller head, since it is hydraulically operated, it can sense the pressure behind the rolls. The major advantage of this is when very dry compounds are mixed, which can exhibit difficulty in knitting together, the rolls will sense the drop in pressure and slow down, creating sufficient back pressure in the extruder to knit the compound into sheet transfer through the batch of system. The latest introduction is twin screw extruder. It consists of two counter rotating conical screws having large diameters at the feeding zone and designed for rapid through put, with low pressure build up, no heat gain; this machine can move and shape discharged stock quickly and automatically.

CHAPTER 3

Extruding of Rubber Products

INTRODUCTION

Extruders are machines which force rubber through a nozzle or die to give a profiled strip of material. They fall into two types: those in which the pressure is produced by a ram, and those in which the pressure is produced by a screw. The latter is the type of machine most generally used in the industry and is known as an extruder, forcing machine, or tuber, whereas the ram extruder is a more specialised machine for short runs.

SCREW EXTRUDERS

A screw extruder comprises a feed hopper, a feed screw operating within a barrel, a head, and a die or a die plate (Fig. 3.1).

The screw is rotated from an electric motor through a reduction gear, and pushes compound through the barrel into the head where it builds up a pressure, this being relieved by allowing the compound to pass through an orifice or die to form the desired shape.

The purpose of the feed hopper is to receive the compound and pass it down into the flights of the screw. The compound may be supplied hot, in the form of an intermittent or continuous feed strip, as 'dollies', or as the discharge of an internal mixer, or it may be supplied cold in the form of a strip or pellets.

The feed hopper may be undercut to assist the feed, or may have a driven roll adjacent and parallel to the screw to give a 'roller feed'. In the case of the internal mixer discharge, a power-operated ram will be needed to push the rubber compound into the flights of the screw. The screw-type extruder is meant to be a continuous operating machine, so most systems are satisfactory with the exception of hand-feeding from dollies or strips.

The screw should preferably have a lower volume in the flights at the outgoing than at the in going end. This can be achieved by (i) a reduction in pitch of the screw; (ii) a reduction in the depth of the base of the screw; (iii) a reduction in overall diameter of the screw and barrel; and (iv) an increase in the number of starts in the screw. Methods (ii) and (iv) are the most commonly used, the objectives being to give sufficient compression to eliminate air and also to ensure a constant pressure in the head. It is most important that an extruder screw is full at the discharge end, otherwise changes in swell of the compound will arise from surges of compound arriving in the head, resulting in dimensional changes in the extrudate which often seem to occur at the same frequency as the speed of rotation of the screw. Using hot compound, the length-diameter ratio of the screw is usually 4:1 or 5:1 whereas, with a cold feed extruder, this ratio is increased considerably, to 15:1 or 20:1 depending upon the type of compound to be extruded. The barrel of an extruder is usually of a hardened steel and is controlled so that a constant compound temperature is maintained in the extruder head. The screw rotates within the barrel and has a clearance of approximately 0.4 mm.

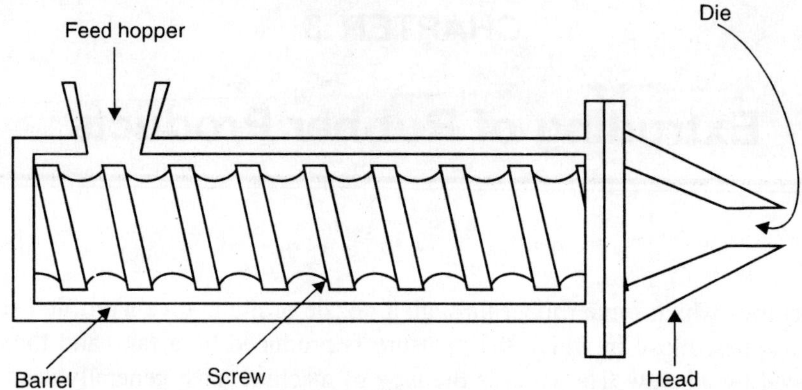

Fig. 3.1. Section of a screw extruder.

If the feed strip to an extruder is gradually increased, a point is reached at which the extruder chokes and will not take all the compound, which begins to back up at the feed hopper. There is an upper limit to the rate of feed for a particular extruder speed, and usually the best conditions are obtained when compound is fed at about 90% of the amount to choke the machine. The choking point increases and decreases as the speed of the screw of the extruder is increased and decreased, respectively, so each compound and screw speed demands its own rate of feed.

The purposes of the head are to equalise the pressure from the screw and barrel and to transport the compound to the die. Again, temperature control is vital. Of all parts of the extruder, the design of the head is one of the most important, and probably receives the least attention. Compound must move smoothly to the die, and ideally at equal pressures and speeds. Any points within the head where compound does not move are known as 'dead spots'. Compound vulcanises in these dead spots, then portions break away from time to time to give bits or 'nibs' of scorched compound in the extrudate.

The last stage is the die, which forms the compound into the desired shape. It should be pointed out, however, that compounds shrink along their length, so must increase in thickness and perhaps width, depending upon (i) the shape of the head and the extrudate; (ii) the pressures in the head; (iii) the head and compound temperatures; and (iv) the compound rheological characteristics.

Extruders can be fitted with various attachments, e.g., a T-head or cross-head for mandrel covering. If a wire gauze is included in the head to remove nibs or other particles of foreign matter, the machines become known as 'strainers'.

A vacuum device can also be fitted in the barrel of an extruder to remove any trace of entrapped air or other volatile matter from the compound. These vacuum extruders are useful in making articles for open steam, hot air, molten salt, and fluidised bath cures. The absence of the final traces of entrapped air and volatile matter eliminates any chance of porosity or blowing arising from this cause.

RAM EXTRUDERS

In the ram extruder, a quantity of warm compound is placed into the cylinder, the die head is attached to the cylinder, and the ram then pushes the compound through the die to form a profiled section.

This type of machine is intermittent in operation, and its operating costs tend to be higher than those of the screw-type extruder. The main virtue, however, is an ability to extrude compound at a lower temperature and/or speed, which can result in marked technical advantages in controlling the conditions for extruding difficult compounds. It is easier to clean and is applicable to short runs. Furthermore, it finds particular application when compounds need to be strained through gauze for high-quality products requiring completely contamination-free material. Other advantages lie in the much reduced heat build-up compared with the screw extruder, and its ability to extrude compounds with poor flow properties.

Vacuum Extruder

Gas (air) or moisture inside the compound can lead to porosity of the extrudate when vulcanisation proceeds without pressure. De-gassing is therefore a pre-requisite for continuous vulcanisation processes which follow the extrusion process.

The extruders with de-gassing ports (vacuum zones) for the de-gassing of rubber compounds for the directly following continuous vulcanisation have been built. Such extruders had already been used in the plastic industry. Their application to the rubber industry requires, however, a special effort since the de-gassing of the rubber compounds which are even highly viscous at high temperatures presents a bigger problem. The vacuum extruder consists of the feed port, a feed zone which exhibits the shallow cut screw typical for cold feeding. A *ring (dam)* is arranged between the feeding and the vacuum zone.

In the *de-gassing zone* which is connected to a vacuum pump, the vacuum has to be between 3 and 5 Torr. This part of the screw is cut either deeper or wider or the space for the vacuum zone has been enlarged by widening the cylinder. The free space in this section of the screw has to be so big that it is only half filled with material during extrusion. This is assured by the ring or dam between the feed and vacuum zones which lets only a small "hose" of compound enters the vacuum zone of the machine and thus also seals the vacuum section. The "hose" is repeated by perpendicularly arranged grooves whereby the surface area is greatly enlarged. After passing the vacuum zone, the extrusion material enters the extrusion section. Even the thin compound layer which enters the vacuum zone connot be completely de-gassed because of the high viscosity of the rubber compounds in contrast to certain plastic compounds. A complete de-gassing is only accomplished inside the machine because the compound rolls off repeatedly from the screw flights of the vacuum zone whereby new surfaces are continuously formed from which air and moisture enclosures can be removed by the vacuum. Due to the relative motion of the compound it rolls off continuously while creating ever new surfaces.

It is therefore of crucial importance that the temperatures of the screw and the barrel in all parts of the machine, but especially in the vacuum zone, are measured accurately and be controlled. This machine thus requires a degree of temperature control and adjustment which is much higher than normally found in extruders. The accurate adjustment of the temperature at the dam is in addition crucial for the viscosity of the material entering the vacuum zone.

Other Extruders

Aside from the types discussed so far, others are also important

1. The *double or triple extruder* where, e.g., the base compound and the tread compound for a protector are found and thus combined in one die to one extrudate.

2. The *mill strainer,* a machine which produces pre-warmed strips, for example for the feeding of calender and extruders.
3. The *strainer,* which carries a screen before the material discharge port with which foreign particles are removed from the rubber compound.
4. The *pelletiser* where the rubber compound is extruded in the form of stands which are chopped by rotating knives.
5. The *Gordon plasticiser* where the front part of the screw is formed so that the rubber is forced to pass several screw flights whereby the rubber is continuously masticated (heavy mastication machine).
6. The *roller model* extruder and the sheets extruder which are used to extrude sheets or slabs after the compound preparation.
7. The *roller head extruder* where an QSM cold feed extruder is feeding a two roll calender directly out of a sheeting die.
8. The *rubber thread extruder* where a paste made from a rubber compound in an organic solvent is extruded through a multiple die into a leaching bath. They barely play a role today.

EXTRUSION TECHNOLOGY

The variability to be expected in the dimensions of the extrudate is at a minimum when the compound has the minimum entrapped stress. The stress can be relieved by means of a small loop immediately following the extruder die and prior to the take-away conveyor, or by means of a system of controlled shrinkage. As the depth of the loop increases, so does the variability of the section, since a large loop introduces stress. The greater the loop, the greater the stress from the compound weight that has to be supported.

Dies should always be designed to operate under conditions of minimum stress and at predetermined running speeds and temperatures, and the extrudates should always be produced under these conditions. Small adjustments in weight per unit length to accommodate minor compound variability can be made by subsequent slight stretching or compression of the section, but here again, the greater the stress in the compound after extruding, the greater the variability in the dimensions of the extrudate.

A different die will probably be needed for each compound, because of differing viscoelastic properties. The design of die is an art, and normally requires a number of attempts to obtain the correct section with any given compound. Sections scan be varied by means of different angles or 'lead-in' at the back of the die, as well as by changing the aperture of the die. Whether the die is a thick or thin one also affects the section obtained.

Whilst attempts are often made to baffle heads to give equal pressures and speeds at all points, this is rarely successful. Die design problems are more complex than this, as different swells are normally obtained at different parts of the die and vary also with the overall die aperture. It is normal practice for the die to be the hottest part of the machine, the temperature being progressively increased to this point from the feed.

Much can be done in compound design to improve the extrudability of a compound. In general terms, as already discussed, the more plastic the compound, and the less its elasticity, the better is the quality of finish of the extrudate. The lower the viscosity of the compound, the great is the throughput to be expected in unit time.

Sheets of compound can be obtained from extruders as well as from calenders, one technique being to extrude a tube, cut it along its length, and open this up to form a flat sheet. Any slight eccentricity in the tube can, however result sheets with humps and hollows. Whilst this system has been and is still used, a type of machine known as a roller die or roller head die has been developed. In this case, the head is used to distribute the compound from the barrel to a pair of calender rolls. The roller die possesses the better features of both a calender and an extruder, namely, the high throughput and freedom from air entrapment of the extruder, coupled with the precision of gauge control and the speed of thickness changes associated with a calender. The range of thickness to be expected from such a machine is the same as that to be expected from a calender or extruder.

The two most important items in the design of roller die are to ensure that it works at the minimum pressures, which must be kept constant as the compound leaves the head, and that the section presented to the cleaner is as near as possible to the ultimate shape required.

VULCANISATION OF EXTRUDATES

The continuous vulcanisation methods for extruded products have been introduced because of economic reasons. While part of the extrudates are processed as intermediates (e.g, tyre treads) before they are vulcanised and other parts are brought to a separate vulcanisation process (for example because of small production volume), the vulcanisation of a considerable part of extruded products occurs immediately after leaving the extrusion die in a continuous fashion. Also the continuous vulcanisation in high pressure steam is important in this connection; it is preferred for the manufacturing of cables. Finally, also the vulcanisation under lead plays a certain role for extruded products.

Vulcanisation in Autoclaves

For the vulcanisation of extrudates in non-continuous processes the heating in air or steam autoclaves under pressure is still of the highest importance. In order to avoid a deformation of the extrudates during vulcanisation, the compounds are filled with sufficient amount of factice or the profiles are placed into support structures or into talc powder. Thin walled hoses are placed onto mandrels in order to avoid deformations and are vulcanised under pressure in vulcanisation autoclaves, of for example, 30 m length and more. It is also possible to heat the extrudates submerged in water within the autoclave. Since these types of vulcanisation are always discontinuous they are only applied when continuous processes are excluded or are uneconomical.

Continuous Vulcanisation in Liquid Baths (LCM Vulcanisation)

This method vulcanises the rubber in a hot liquid. It is called LCM (Liquid Curing Method).

Principle

The principle is very simple: the profiles are brought into a long hot liquid bath immediately after leaving the extrusion die via a conveyor belt. They are kept submerged in the liquid and transported through the bath by a steel conveyor belt. The profiles arrive at the end of the bath fully vulcanised.

The main advantages of the liquid curing method compared to the vulcanisation in vulcanisation autoclaves is not only that continuous profile lengths can be manufactured but also that: (i) scrap rate is lower; (ii) profiles have better appearance; and (iii) the vulcanisation times are shorter.

The lowering of the scrap rate is based upon the fact that a considerably lower distortion and depreciation of the profiles occurs in the most frequently used heating media of the LCM method compared to vulcanisation in the autoclave.

A further decrease of a the scrap rate is achieved for this method because the dimensions and appearance of the finished vulcanised can be controlled and supervised immediately after leaving the vulcanisation bath while one can evaluate them for profiles vulcanised in autoclaves only batchwise and one has to scrap the whole batch if faults are detected. Furthermore, the efficiency of LCM equipment is higher; during vulcanisation in autoclaves heat is always lost when they are opened.

Since powdering materials usually do not have to be used for liquid bath vulcanisation and condensed water sports which occur for autoclave vulcanisation do not form, one obtains by this method especially good looking extrudates.

Design

The LCM vulcanisation is carried out in the salt bath contained in a long trough through which a conveyor belt, for example from steel, runs. This trough is located close to the extruder. The extrudate is submerged into the salt bath and transported through it by the belt.

The required length of a fluid bath is not only dependent upon the vulcanisation rate of the profiles and the temperature of the salt but also upon the output rate of the attached extruder, i.e., the maximum speed. This, in turn, is dependent upon the dimensions of the extruded profile. An excessive time beyond the optimal vulcanisation time is dangerous for the profile in view of the high vulcanisation temperatures. Therefore, a careful balance between extrusion speed and residence time in the heated bath has to be established.

The heating media for vulcanisation baths, are, for example, salt mixtures as well as (more seldom) metal mixture, polyalkyl glycols, glycerine, silicone oil, etc. The method is also called "salt bath vulcanisation" because of the most frequently used salt mixtures.

The eutectic, molten salt mixture most frequently used is composed of (by weight percent). 53% potassium nitrate, 40% sodium nitrite, 7% sodium nitrate.

Advantages and disadvantages of LCM

Using these baths yields very clean looking profiles. The disadvantage of the salt mixture is its high density. Since the density of the rubber compound is practically always considerably lower than that of the salt mixture, a more or less strong buoyancy develops which has to be overcome by the steel band pressing the profile into the bath. For very soft compounds of low density and complicated profile cross-section, deformations can result. In such cases, one can work better, for example, with polyalkyl glycol or silicone oil as the heating bath medium because of their lower density or fluid bed vulcanisation or UHF vulcanisation are preferred.

The *heat transfer* of the heating media is generally quite good and since high vulcanisation temperatures can be chosen (normally 210–240°C) one obtains rather short vulcanisation times. Of the bath liquids the so-called salt mixture has especially good heat transfer characteristics. It is considerably higher than for other continuous vulcanisation methods. However, the relatively slow heat transfer, especially for thick walled articles or for hollow profiles with partitions, is problematic because of the low heat conductivity of the rubber compounds. The larger the profile to be heated, the longer has to be the residence time of the profile in the liquid bath so that it will be heated through. However, this causes a corresponding severe over-vulcanisation of the profile on the surface, especially at very high heating temperatures. Finally, with increasing thickness, an ever increasing gradient of the degree of vulcanisation results, thus also an anisotropy of the vulcanisate properties from the outside to the center. Here the limits of the LCM method are met. For the so-called UHF method a better heating effect over the area of the profile is obtained compared to LCM.

Another problem for the liquid bath vulcanisation is avoiding porosity which can arise because of the very low vulcanisation pressure (only the small liquid columns). Porosity arises from air and moisture contained within the rubber compound It is obvious that this depends to a large degree on the kind and magnitude of the fillers and the other ingredients as well as on the processing technology. The higher the hardness of the unvulcanised compound, the lower is normally the porosity.

The major part of porosity is due to the water content of the compound. Using desicants in the compounds like, e.g., CaO, the porosity may be reduced significantly however, the vulcanisation behavior of the compound will be affected.

Elimination of the remaining porosity caused by entrapped air is more difficult by using extruders with vacuum zones in the screw cylinder, the remaining air may be eliminated. Also by using small amounts (5–10 phr) of factice the degassing of compounds during mixing is much more efficient.

Porous profiles with fine pores can also be manufactured continuously in LCM equipment even though their production with the help of UHF equipment is advantageous. The blowing agents contained in the compound decompose at the bath temperatures and blow the profile to the desired height before the onset of vulcanisation. A premature gas loss in the extruder should be avoided. The temperature history and choice of blowing agent have to be coordinated.

Advantages of LCM vulcanisation

1. No interruption of the continuous extrusion process.
2. No transporting semi-finished goods.
3. Only small temperature losses.
4. No expensive support structure of the extrudates.
5. No anti-stick powder, therefore cleaner operation.
6. Low reject rate.
7. Labour saving.
8. Better appearance of the profiles.
9. Possibility to vulcanise peroxide containing compounds.
10. No oxidation of oxidation prone elastomer, e.g., NR.

Disadvantages of LCM vulcanisation

1. Necessity to use expensive extruders with de-gassing zones.
2. Therefore often lower extrusion rate compared to conventional processes.
3. Certain salt loss in the salt bath depending on the kind of profile, the extrusion speed and the temperature.
4. Therefore maintenance of the bath necessary.
5. Need to clean profiles.
6. Higher danger level working with the salt bath.
7. Occasional deformation of the profiles.
8. Formation of nitrosoamines.

Influence of compound composition

Naturally, the compound composition is of strong influence on the salt bath vulcanisation. All elastomer types can be vulcanised without great difficulty pressureless as long as the compound has been designed properly. When determining the maximum bath temperature one has to consider the heat stability as well as other technological considerations.

From experience the following temperature limits can be given (Table 3.1).

Table 3.1. Maximum vulcanisation temperatures of various elastomer types during LCM vulcanisation.

Elastomer type	*Max. vulcanisation temperature*	*Comment*
NR	Up to 210°C	Above this; surface tackiness, reversion
SBR	Up to 240°C	Perhaps even higher
OE-SBR	Up to 240°C	Above this; formation of pores
NBR	Up to 240°C	Perhaps even higher
CR	Up to 240°C	In exceptions higher
EAM	Up to 220°C	Above this great pores because of peroxide decomposition

In the beginning relatively fast accelerator combinations were used as vulcanisation accelerators for the salt bath vulcanisation to accommodate the short vulcanisation time available. Later experimentation showed that this is not necessary or even not practical since the differences in the heating time are relatively minor at the very high vulcanisation temperatures, however, the possible danger of reversion is increased. The less rapidly accelerated compounds have, above all, the advantage of higher scorch safety in the extruder. Sulphenamides in combination with mercapto accelerators, thiurams and dithiocarbamates are employed. For CR, for example, ETU-thioureas are combined with polyethylene-polyamines. For larger cross-section the use of quick accelerated compounds is preferred because of the low heat inductivity of the rubber compounds and thus the danger of severe under-vulcanisation at the center of the profiles. For elastomers which tend to revert, accelerators are preferred, because of the high vulcanisation temperatures, which improve the version resistance.

For the vulcanisation of compounds with peroxides the vulcanisation temperature has to be limited to 200–220°C since a sudden decomposition of the peroxide at higher temperatures can give rise to bubbles.

Among the fillers, furnace, acetylene and oil blacks do not create bubbles and thus do not create difficulties. In contrast, it is practically impossible tc manufacture bubble free vulcanisates with channel blacks. Negative results are also obtained with Kaoline. By using CaO as well as factice the porosity can be minimised in most cases.

The effect of plasticisers of low volatility on porosity is relatively minor during salt bath vulcanisation. Most plasticisers, especially the mineral oil plasticisers, do not cause porosity even at temperatures up to 240°C. Use of emulsion plasticisers in contrast, almost always results in bubble formation since the water content remaining in the compound cannot be eliminated even in the vacuum extruder.

Fluid Bed Vulcanisation

Aside from the salt bath vulcanisation the so-called "fluid bed vulcanisation" is used to a lesser extent for the continuous manufacturing of hoses and profiles. The principle of the fluid bed vulcanisation is very similar to that of the liquid bath vulcanisation. Also the problems faced by the operator of these two methods are similar. The principle is the following:

Principle

When one lets a gas stream flow upwards through a layer of solid, small particles, the latter are lifted. At a sufficient flow velocity the layer expands and every particle is being suspended by the moving gas and exhibits through the "gas cushioning" a considerable fluidity. In this state the system is called a fluid bed because the small suspended particles behave in many respects like a Newtonian liquid: It fills the volume of the container, exhibits a horizontal surface, can be transported through a pipe and the hydrostatic pressure increases proportional to the depth. It is noteworthy that a profile to be vulcanised can be pulled through this apparent Newtonian liquid analogous to the salt bath. Also, the heat transfer characteristics correspond to those of a liquid and are approximately 50 times higher than those of air.

Such fluid beds may be utilised analogous to the liquid baths for the continuous vulcanisation of extrudates. As particles, small glass spheres of 0.13 to 0.25 mm diameter are used, for example. They are called *Ballotini*. The layer has in loose packing about 40% empty space; when gas is flowing the layer expands by about 10% and the empty space increases to about 45%. This expansion is sufficient to reach a liquid like state and to transform the glass beads into a fluid bed. The extradate to be vulcanised can be pulled through it.

Heat transfer

The heat transfer of a fluid bed depends only to a small extent on the material from which the particles are made, it is determined much more by the degree of fluidisation and the specific heat of the gas, e.g., steam. The degree of polymerisation in turn, determines the "density" of the fluid bed which can be changed and can be adjusted to the density of the vulcanisate. The fluid bed vulcanisation is therefore especially useful for the vulcanisation of complicated profiles which are sensitive to deformations since the profile can "float" freely through the fluid bed. The problems of porosity are the same as those for the liquid bath vulcanisations and are corrected by the same means.

Arrangement

Fluid beds are built in either horizontal or vertical positions.

Continuous vulcanisation in hot air after pre-heating in ultra high frequency AC field (UHF vulcanisation)

Hot air vulcanisation

The vulcanisation in hot air is one of the oldest vulcanisation methods. During continuous operation it is only applied without pressure. The extrudate is pulled through hot air tunnels which can be up to 150 m long; the total length is divided into different sections in order to permit good temperature control and to obtain good speed regulation.

In attempting to reduce the length of the air tunnels drastically of the newest continuous vulcanisation, a pre-heating system for the extrudate in the form of an ultra high frequency "oven" was developed which is coupled with a reactively short hot air tunnel. This continuous vulcanisation in hot air after pre-heating in an ultra high frequency AC field is called not quite correctly-"UHF" or "micro wave" vulcanisation. After UHF preheating the hot air tunnels can be shortened to for example, 22 m or shorter. They serve only to maintain the temperature reached in UHF field.

Principle of the UHF method

For UHF vulcaniation an extruded profile is brought (after leaving the extruder die) through a micro wave oven with an ultra high frequency AC field where it is heated because of the dielectric losses.

The *dielectric heating* which permits a rapid and even heating of all electrically non-conducting substances is based upon the fact that the non-conducting substances brought into the electric high frequency field are subject to a polarisation due to the ultra high frequency AC field whereby dielectric losses are created in these material which in turn heat up the substances.

These dielectric losses, depend on the frequency of the dielectric AC field, caused by the different types of polarisation which consume part of the energy of the field in the form of internal friction which manifests itself in the generation of heat.

It has to distinguish between the following types of polarisation:

1. Electron polarisation, created by a shifting of electrons with respect to the positive nucleus.
2. Dipole polarisation, created by the influence of the field of molecules with dipole character.
3. Interface or ion polarisation, created by the accumulation of free ions at the interface between materials of different conductivity and dielectric constant.

For pure NR, SBR, EPDM, IIR etc. where one deals with nonpolar substances, ultra high frequency heating is only possible by electron polarisation which requires an extraordinarily high frequency. A practical and economical use of pre-heating is therefore not feasible in this case. In contrast, for elastomer types with dipole character, i.e., polar substances such as NBR and CR, considerably lower energy levels are necessary and they are therefore much better suited for an even heating in a high frequency field.

In many cases a heating by interface or ion polymerisation with relatively low frequencies compared to electron polarisation is possible because one is dealing in practical use of high frequency heating with rubber compounds, i.e., with heterogenous material systems. Hereby the explanation of the heating mechanism is still being debated. The "Wagner theory" of the dielectric gains importance which declares the electrical heterogeneity as the cause for the energy loss whereby the comparison is made that an electrically disturbed surrounding absorbs energy in a form similar to an optical disturbed system absorbing light energy.

The heatability of compound is dependent heavily upon the fillers aside from the type of polymer. With increasing amounts of carbon black the heating rate is raised. But also the reactivity of the black give rise to a more rapid heating than less active ones. White fillers yield less rapidly heatable compounds than blacks. Among the white fillers, ZnO yields faster heatable compounds than, for example, $MgCO_3$ and this, in turn, is faster than chalk or silicas.

The equipment contains, furhtermore, the controls to adjust the speed of the conveyer belt and to center the wave guide which permits to lead the profile, depending on a size and form, exactly through the middle of the electric field.

Furthermore, equipment is available to control how much energy of the wave is absorbed by the rubber compound: Photocells, at the end of the vulcanisation tunnel turn off the magnetron when no material passes through the channel anymore.

In order to avoid heat losses of the profile by radiation, a hot air blower is provided which at the same time carries away the developing gasses and prevents any condensation.

The relatively short length of the *pre-heating chamber* requires the addition of an infrared or hot air tunnel in order to finish vulcanisation at the required production speeds. This combination has the advantage of a uniform degree of vulcanisation. Naturally, a complete vulcanisation would also be possible with a correspondingly long UHF tunnel, but this would require uneconomical efforts.

In contrast to other vulcanisation methods, the heat is not supplied in the UHF equipment from the outside but the heat is generated here simultaneously on the surface and the interior so that a pronounced gradient of the degree of vulcanisation does not develop. A uniform degree of vulcanisation results and an isotropic property field over the area of the vulcanisate. Therefore this method requires a considerably lower level of energy compared to the other methods (yield of energy 50 to 60%) and provides, in addition, a substantial shortening of the vulcanisation time for thicker articles for which, for example, the LCM vulcanisation time is problematic.

It is necessary also for the UHF equipment, analogous to the LCM method, to use an extruder with a de-gassing zone in order to obtain bubble free vulcanisates.

Compound design

It is necessary to consider the energy to be absorbed, i.e., the developing vulcanisation temperature, during the design of compounds which are to be vulcanised with the help of UHF equipment.

The heating of a rubber compound occurs the faster, as already mentioned, the more polar the rubber is or the higher the proportion of *polar compound ingredient* like, for example, carbon black. Thus, it is not very difficult to design rapidly vulcanising compounds based upon polar elastomers like NBR or CR which have a high polarity by themselves, or from non-polar polymers like SBR, NR, EPDM etc. as long as they are filled with suitable carbon blacks. Problem arise with non-polar elastomers filled with white fillers. In order to solve this problems the following solutions are available.

1. The blending of non-polar with polar elastomers.
2. Addition of polar materials to non-polar compounds.

Blends of non-polar NR or EPDM with polar NBR or CR lead, for example, to heating rates which make it possible at already small amounts of NBR or CR to develop compounds for continuous vulcanisation. Also, an addition of PVC or PVDC has been applied.

When polar plasticisers are added, for example chloroparaffin, in the customer amounts, no sufficient heating rates can be obtained.

A certain increase in the heating rate is obtained by employing factice, especially castor oil factice. Especially high increases are obtained when filler activates are used like diethylene-glycol and triethanol-amine. It is often already sufficient to add 10 weight percent of the filler loading of diethylene-glycol and triethanol-amine in equal amounts in order to obtain a marked increase of the heating rate. Also, process aids like Zn soaps, fatty alcohol esters etc. are beneficial for the continuous UHF vulcanisation. As a rule, no one single step is sufficient to reach satisfactory degrees of heating. Therefore, usually several of the above mentioned procedures are applied.

When silica fillers are used, always a part of the water is bound so intimately that it cannot be removed in the vacuum zone of the extruder. This water can be set free at the high vulcanisation temperatures and can thus lead to porous vulcanisates unless CaO is added. However, when diethylene-glycol or triethyanol-amine is used, the water normally bound to the active filler surface can be split off and be removed in the vacuum zone. One obtains, therefore, often bubble free vulcanisates even without the incorporation of CaO.

It should be, furthermore, considered that in some cases a change in the polarity of the compound is caused by the vulcanisation which results in an increase in the energy absorption after the onset of and during the vulcanisation.

Vulcanisation by High Energy Radiation.

Since the formation of free radicals on the polymer chains can initiate their cross-linking, essentially all those methods are suitable for cross linking which form radicals on the polymer chain. A radical formation on the polymer chain by dehydration and thus a cross-linking of the polymers can also be accomplished by high energy radiation, which one can thus consider to be an indirect vulcanisation “material”. The energy of the radiation has to be, naturally, higher than the bond energy of the most labile carbon-hydrogen bond of the elastomer. This cross-linking principle which is used frequently in the plastics industry, e.g., for the cross-linking of PE and PVC, is also applicable to the cross-linking of elastomers.

Radiation Sources

Cobalt sources (Co^{60}), *Van-de-Graaff generators* and *resonance transformers*, cascaded and linear accelerators or *betatrons* are used. The electron beam can be controlled by a magnetic deflection system so that one obtains a radiation source in the form of either a spot or a wide band.

Compared to a Co^{60} source, the electron accelerator yields an accurately focusable and constant radiation. The Co^{60} source creates always a diffuse radiation, the energy of which decreases with time according to the half time of the material.

Cross-linking

Cross-linking due to radiation is normally carried out within the electron accelerator so that the article to be cross-linked is carried through it by a conveyer belt. For example, when a 4 mm thick plate needs for full cross-linking an absorbed radiation energy of 8 Mrad and the generator supplies one of only 4 Mrad during one pass, then one pass is required for each side of the plate.

During the irradiation with high energy irradiation (>0.5 MeV), radicals are created in the polymer chain which cause C-C cross-linking. It can, therefore, be expected that the elastomers will have good heat stability. Reduced to the simplest formula, the cross-linking reaction can be described by the following scheme:

$$2\text{—CH} + 2e \longrightarrow 2\text{—}\dot{C} + H_2$$

$$2\text{—}\dot{C} \longrightarrow \text{—C—C—}$$

In reality, the cross-linking reaction is more complicated.

The radiation energy and thus the type of electron accelerator to be used are dependent on the required penetration depth, the density of the irradiated material and the chosen irradiation system.

Since a certain minimum amount of energy is required to form polymer radicals on the one hand, but the radiation energy decreases with the penetration depth of the radiation on the other hand, it is understandable that a uniform degree of cross-linking above a certain thickness of the article is difficult to reach even when the irradiation is carried out from both sides. These problems are of a processing technical nature.

The absorbed radiation energy is considerably lower than that required for the radio-active excitation of the metal atoms contained in the compound.

Compound design

No vulcanisation aids are required for the radiation curing of elastomers like, for example, peroxides. It is sufficient, therefore, to add the usual fillers, plasticisers, colourants etc. to the elastomers. Sometimes activators which are also used as co-activators for the peroxide cure are utlised, like, for example, EDMA or TPTA in order to lower the required amount of radiation energy or to achieve at the same level of radiation energy an increased state of cure.

Vulcanisate properties

At a given cross-link-density, thin-walled vulcanisates cross-linked either by radiation or by peroxide are barely distinguishable as far as their technical properties are concerned. For higher thicknesses - especially at a simultaneous high density of the compounds - it is difficult with the application of older weak equipment to obtain by high energy radiation a state of cure which is uniform over the thickness since the radiation, as already mentioned, is slowed down when metering the elastomer, i.e., it loses kinetic energy. Above certain wall thicknesses the inside of the part can be severely under-vulcanised or remain even completely uncured. The latter is the case when the radiation energy has become lower, because of absorption, than the energy of the most labile bonds. For newer equipment with high energy output a considerably more uniform degree of cure can be obtained because of the greater penetration depth even for only one-sided irradiation.

During radiation cross-linking no decomposition products are created by any vulcanisation chemical which could influence the property spectrum or the physiology of the vulcanisates.

Economics

The use of high energy electron accelerators seems to have become more economical, for example, for the continuous vulcanisation of sheets (according to statements by the manufacturers) than rotational vulcanisation.

CONTINUOUS VULCANISATION IN THE STEAM PIPE

The vulcanisation in high pressure steam can be applied only to those extruded products which have a metallic insert like metallic conductors in cables in order to be able to pull the extrudate. This type of vulcanisation is thus nearly exclusively used in the cable industry.

A sufficiently high steam pressure very short vulcanisation times can be obtained because of the high temperatures.

Principle

The principle of vulcanisation in the steam pipe is the following:

The cable, after having been coated with the jacket is brought into a double walled pipe which is directly attached to the cross head in which the steam pressure is normally 5 to 12 bar, sometimes 20 bar or even more. The steam pipe can be connected via a suitable seal to a cooling water pipe; or it may go over into a cooling trough.

The required length of the pipe depends aside form the gauge of the cable and the thickness of its jacket, primarily on the steam pressure as well as on the choice of the vulcanisation acceleratos. One cannot vulcanise arbitrarily rapidly since one of the variables is the heat transfer of the rubber compound just as for the LCM vulcanisation.

The continuous vulcanisation is only applicable up to certain jacket thicknesses of the cables because of the low heat conductivity of the rubber compounds. However jackets of medium gauge can be manufactured in this manner just as cables of small diameter.

The steam pipes are normally arranged horizontally. But suspended and vertical designs are also known.

Vulcanisation accelerators

For the vulcanisation in the steam pipe only vulcanisation accelerators are suitable which have a sufficiently broad plateau and a high scorch safety but can still vulcanise within seconds. Such accelerators are, for example, the sulphenamides which permit a long flow period in the extruder but act rapidly in steam, especially in combination with thiuram or dithiocarbamates.

Vulcanisation Under Lead

In order to avoid a deformation of extruded articles, for example cables or hoses especially large cross-sectional area, during vulcanisation they are sometimes enclosed in lead before vulcanisation. The lead sheath also protects against on influences, for example the hydrolytic effects of the steam. The lead sheaths can be applied discontinuously by hydraulic presses or continuously using extruders.

The extrudates enclosed in lead are normally vulcanised discontinuously, rolled up onto large drums which are placed into autoclaves. The vulcanisation time is, of course, longer than for a simple steam vulcanisation since the lead has to be heated. After vulcanisation the sheathed vulcanisate is cooled as fast as possible, for example by a water spray. After cooling, the lead is peeled off after the sheath has been slit, separated form the vulcanisate and returned to the lead press.

Silane Cross-Linking

Recently, polymers are also silane modified which self-cross-link in a moist atmosphere. Such methods are employed, for example, in EPDM for the cable and construction industry. The addition of silane is said to improve the water resistance, for example, of cables.

DEVELOPMENT OF CELLULAR COMPOUND FOR EXTRUSION

Development of cellular compound for extrusion with excellent shape retention properties (EPDM Blends) for inner seal of passenger car, having a bulb of approx. 25mm dialocated at the extreme corner of the seal as per Fig. 3.2. Wall thickness of the bulb is approx. 1.5mm, with the density in the range of 0.40 to 0.05. Vulcanisation technique for this product is micro-wave and hot air tunnels.

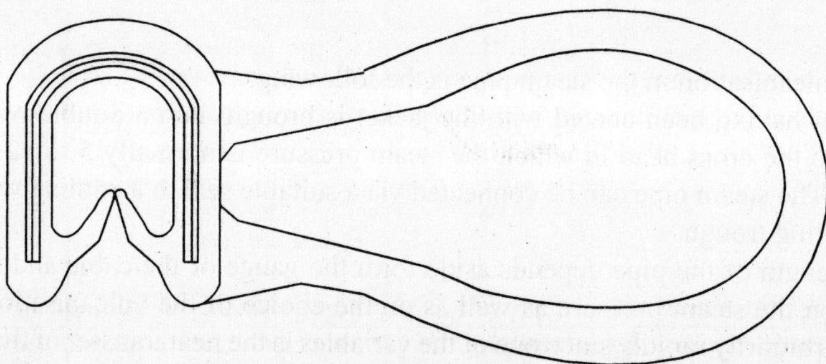

Fig. 3.2. Cellular compound extrusion.

Experimental Details

Initial trials conducted with conventional grade of EPDM (polymer A) is being used for cellular products in the industry having the characterstics as per Table 3.2.

Table 3.2. Conventional grade of EPDM polymer A.

ML 1+ 4 at 125°C	65
ENB content per cent	9.0
E/P ratio	50:50
MWD	Medium

During initial extrusion trials using the formulation A as per Table 3.3 having the rheological and other physical properties as per Table 3.4. The process flow diagram for extrusion line is as per Fig. 3.3. It was observed that when material reaches in the hot air, tunnel sponge tube starts collepsing Fig. 3.3.

Table 3.3. Initial extrusion trials.

Formulation Ingredients	*A PHR*	*Formulation Ingredients*	*B PHR*
Polymer (A)	100	Polymer (A)	100
Polymer (B)	—	Polymer (B)	30.0
Zinc oxide	5.4	Zinc oxide	5.4
Stearic acid	1.2	Stearic acid	1.2
*PEG 4000	2.0	*PEG 4000	2.0
Carbon black N550	50.8	Carbon black N550	50.8
Calcium carbonate	38.0	Calcium carbonate	40.0
Paraffinic oil	62.0	Paraffinic oil	62.0
Processing additive	1.7	Processing additive	1.7
MBT	1.7	MBT	1.7
TMTD	0.6	TMTD	0.6
ZDEC	1.0	ZDEC	1.0
Tedec	0.15	Tedec	0.15
ADC/K	2.2	ADC/K	2.0
OBSH	1.8	OBSH	2.0
Sulphur	1.5	Sulphur	1.5
CaO	4.0	CaO	4.0

* Polyethylene glycol 4000.

Fig. 3.4. Rheological and physical properties.

	Formulation A					*Formulation B*					
Characterstics	01	02	03	04	05	01	02	03	04	05	Remarks
CMB ML (1+4)@ 125°C	31.6	33.2	32.6	31.9	32.9	33.1	32.4	32.4	31.7	32.6	
Ts1	0.65	0.65	0.60	0.62	0.63	0.63	0.63	0.63	0.64	0.61	Tested
By MDR											
Ts2	0.74	0.75	0.71	0.71	0.72	0.72	0.72	0.73	0.74	0.70	@
177°C 0.5 ARC											
T90	3.00	2.98	2.95	2.90	3.08	2.85	2.76	2.88	2.90	2.86	
ML	2.10	2.21	2.18	2.36	2.29	2.26	2.24	2.16	2.15	2.00	
VM*	36.2	37.7	35.1	26.2	37.4	37.5	36.6	37.0	35.2	36.4	Tested by Mooney viscostiy

(Contd ...)

	Formulation A					*Formulation B*					
T5	2.10	2.14	1.92	1.90	1.98	1.98	1.98	1.85	1.87	1.88	New viscometer
T35	3.68	3.42	3.40	3.51	3.38	3.48	3.35	3.40	3.47	3.49	@ 125°C
Tensile strength M.Pa.	3.65	3.83	3.46	3.27	3.56	4.17	4.32	4.27	4.19	4.42	
Elongation at break per cent	490	530	478	488	492	510	519	518	499	511	
Tear strength Kg / Cm	18.2	14.6	15.7	16.5	14.5	16.3	17.5	16.8	18.6	17.6	

*VM (minimum viscosity)

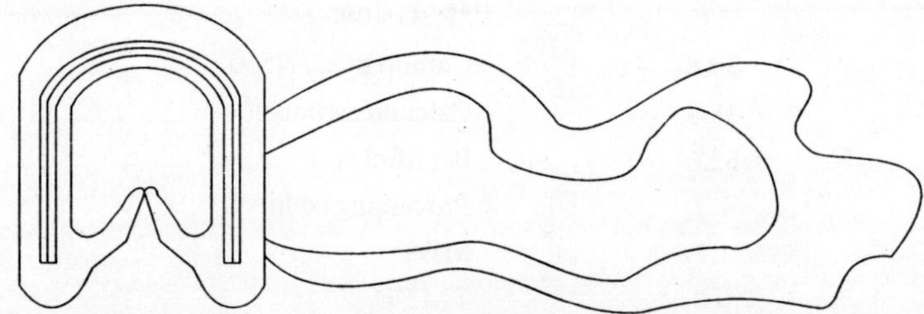

Fig. 3.3. Process flow diagram for extrusion line.

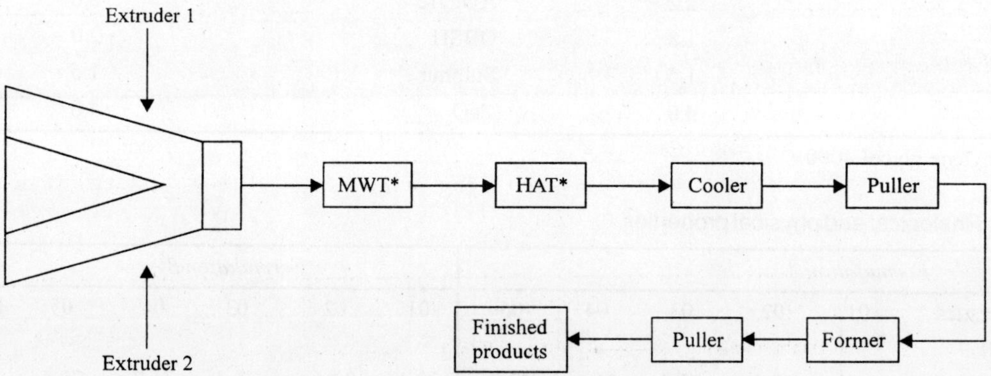

Fig. 3.4. Collepsing of sponge tube.

In the formulation A we have mixed the polymer A (mixing condition and mixing sequence as per Table 3.6 and 3.7 having the E/P ratio 50:50. Not able to retaining the shape of bulb as per requirement may be because of lower ethylene content of polymer, resulting less green strength of the compound, hence we have used the formulation B where we have blended polymer B 30 PHR with polymer, A, having the ethylene and propylene ratio 75:25 and other properties are as per Table 3.5 because higher C_2 value polymer will increase the green strength of the compound, which will help in maintaining the complete shape of inner seal even during the extrusion, blowing and vulcanisation. Table 3.5 is the process flow diagram of mixing. After successful extrusion of the product with formulation B we observed following additional benefits with formulation B in comparison to formulation A except retaining the shape of the product.

1. More out put of compound through extruder even at low extruder rpm(round per minute), results into the energy saving.
2. Better surface finish.
3. Better physical properties Table 3.5.
4. Less load on the extruder during extrusion.
5. Product is showing more snappyness in comparison to formulation A.

Table 3.5. Ingrediates and quantity.

ML (1 + 4) ... 125°C	65
ENB content %	9.0
E/P ratio	50:50
MWD	Medium

Table 3.6. Mixing condition.

Type of mixture	Banbury
Capacity	80 litres
Rotor rpm	32.0
Inlet water temperature°C	21°C
Ram pressure	6.0 Kg/cm^2
Dumping load	164 Amp.
Dumping temperature°C	186°C

Table 3.7. Mixing sequence.

0.00	Polymer (s) + Zno + 2/3 Carbon black + 2/3 $CaCO_3$ + 1/3 P.Oil, REM down
1.30	Stearic acid + PEG 4000 + 1/3 carbon black +1/3 $CaCO_3$ +2/3 P.Oil RA DOWN
3:30	Ram up and Ram down
5:30	Dump

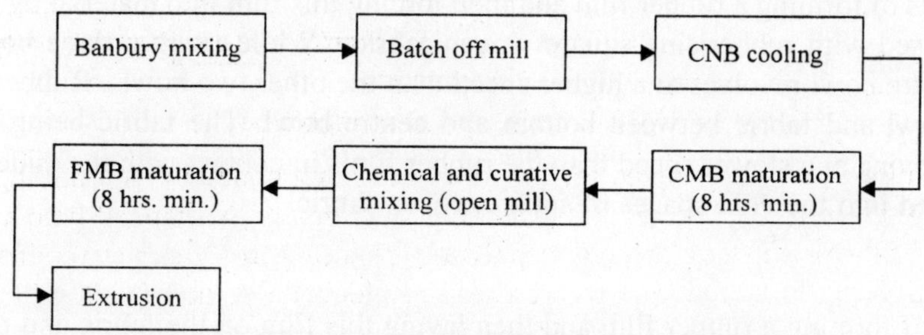

Fig. 3.5. Process flow diagram of compound mixing.

Thus, with above experiment and study it can be concluded that blend of high C2 value polymer with Normal C2 and high ENB content polymer it is possible to achieve complex shape of products in the extrusion process.

CHAPTER 4

Calendering

INTRODUCTION

Calendering is a process used for the continuous manufacture of sheet or film. Granular resin, or thick sheet, is passed between pairs of highly polished heated rolls under high pressure. For the production of thin film, a series of pairs are used with a gradual reduction in roll separation as the stock progresses through the unit. Proper calendering requires precise control of roll temperature, pressure, and speed or rotation. The sheet or film thus produced can be used directly as under-tread, as squeeze application on cut fabric coated plies, as fabric frictioning or topping or it can be batched up in liners for use as insulations later on.

Calender for producing under-tread or insulations may be two bowl or triple bowl machines, but where fabric is to be processed three or four bowl calender is needed. The different forms of calender used are shown in Fig. 4.1.

FABRIC CALENDERING PROCESS

The two main methods of fabric processing are : (i) frictioning; (ii) topping.

Frictioning

Frictioning consists of forming a rubber film and then forcing this film into material by grinding action, and is normally used with rubberising square woven fabrics. While using a three bowl calender for frictioning the centre bowl revolves at a higher speed than the other two bowls. Rubber is fed between top and centre bowl and fabric between bottom and centre bowl. The fabric being in contact with bottom bowl is moving at a slower speed than the rubber films in contact with the middle bowl, and so the rubber is forced into the void spaces of square woven fabric.

Topping

Topping consists of forming a rubber film and then laying this film on the fabric and pressing it at the same time.

A three bowl calender can also be used as a topping calender except that the bottom bowl revolve at the same speed as the centre bowl. Whereas frictioning can duly be carried out on square woven material, topping can be done on either square woven or tyre cord fabric.

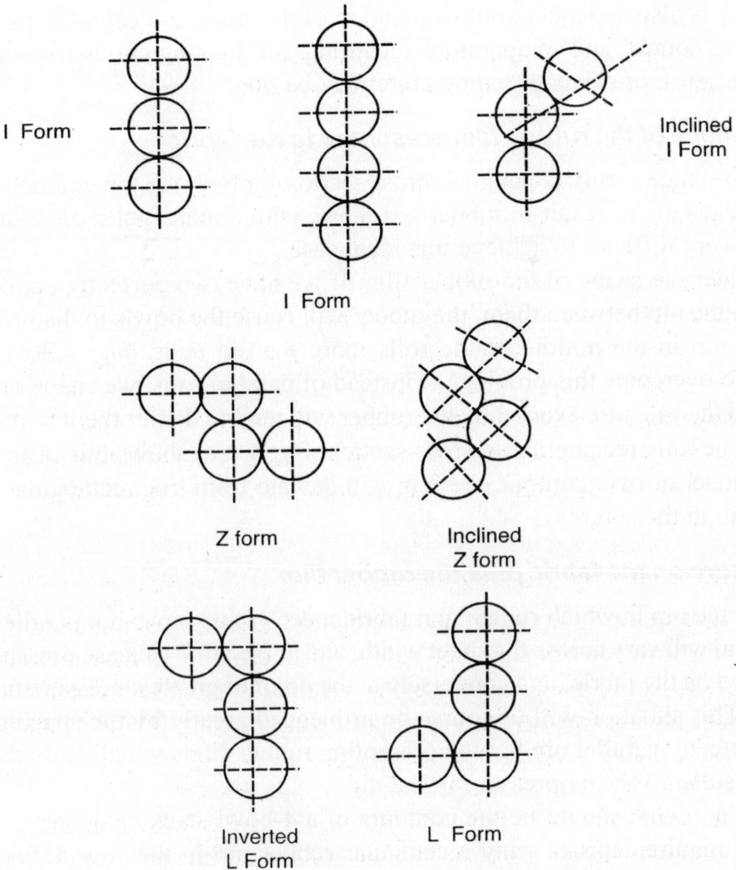

Fig. 4.1. Different forms of calenders used

When fabric is topped it is usually topped on both sides and thus, use of a three bowl calender involves a double pass process. In a three bowl calender, this means that two passes are required, alternatively two three-bowl calender may be used in line where topping on one side of fabric can be done on each calender and this improves efficiency of production. In a four-bowl calender topping can be done on both sides at a time and the bowl speeds are in the ratio 2:3:3:2 for 1 to 4 bowl, respectively. The rubber film go round the inner bowls to the centre nip where they are laid on either side of the fabric as it passes through the nip and pressed on. The operation looks simple, but to maintain uniform calender fabric gauge throughout the width of the fabric and to eliminate defective, e.g., crushed/loose fabric, topping becomes difficult at times leading to enormous economic loss. A dipped fabric has to be coated in such a way that a good textile/rubber composite is formed with even thickness throughout the width. To achieve this the following aspects are worth mentioning

Rubber plasticity and tack

When rubber films are pressed on the fabric, the rubber must have required plasticity to flow so that it can be forced between the cords, and there must be sufficient tack in the two films to adhere when they meet between the cords. If rubber is too much hard with poor tack, loose topping may occur.

Plasticity and tack is also dependent on temperature of the stock and calender roll system must have adequate temperature control and temperature recording and heating/cooling system to be such that quick increase or decrease of calender temperature may be done.

Cross-sectioned shape of the rubber film presented to the fabric

The ideal case is to produce a truly rectangular cross section for both the top and bottom films and press these onto the fabric evenly to result in rubberised fabric with a rectangular cross-section as shown in the Fig. 4.2(a). It is very difficult to achieve this ideal case.

Let us first consider the shape of the rubber film. If we have two perfectly parallel calender bowls and pass rubber into the nip between them, the rubber will cause the bowls to distort, and bend the roll, thus increasing the nip in the middle of the rolls more than at ends. Fig. 4.2(b), shows this in an exaggerated form. To overcome this problem, if instead of parallel rolls, we shape or contour these roll slightly convex, then the pressure exerted by the rubber will tend to distort them to produce parallel nip, and rubber film will be truly rectangular in cross-section. Fig. 4.2(c) shows this in an exaggerated form. If of course we do under or over contour, the film will deviate from true rectangular cross-section, and be either thick or thin in the centre.

Evenness of pressure on the fabric plus the rubber film

Let us now consider the nip in which rubber and fabric meet. If this nip is not parallel then the pressure applied to the material will vary across the sheet width, and at the point of lease pressure the rubber films will be loosely nipped on the fabric, and conversely at the point of greatest pressure the tendency will be to crush the fabric. This situation will of course be influenced greatly by the shape of the rubber film. We could have a perfectly parallel nip but the presenting rubber films which were thicker in one place than another with resultant varying pressure in the nip.

Let us now consider what should be the contours of a 4-bowl stack calender.

If we satisfy our requirements of truly rectangular rubber nip in the way defined earlier, we will require all the bowls to be made convex Fig. 4.2(d). This would mean, however, that bowls B and C would be distorted, and we will not have parallel nip in the centre where rubber and fabric meet.

If we make bowl B and C concave, then pressure exerted by bowl A and D would so distort B and C to have parallel nip in the centre. This type of contour have to be done and is shown in the Fig. 4.2(e).

On an inverted L type calender the pressure exerted by bowl A on bowl B does not affect bowl C as it is in a different plane. We, therefore, have:

Bowl A	Convex	Bowl C	Parallel
Bowl B	Concave	Bowl D	Convex

The arrangement is shown in Fig. 4.2(f). This means that bowl C when distorted will produce a parallel but curved gap between B and C in which the fabric and rubber films meet.

Roll bending

Different viscosity compounds are used for rubber industry and depending upon the viscosity, the force exerted on the roll will be different and hence roll defection and contour will change accordingly. Moreover, due to feeding compound viscosity variation, and variation in the amount of feed in the nip, the roll deflection will change resulting into variation of compound thickness throughout the length of the processed fabric.

This necessitates roll bending device along with roll contour or cambering to meet uniform rubber gauge. For this, large force is hydraulically applied on an extra set of bearings outside the normal bearings to produce roll bending. The force applied tends to bend the roll in a direction opposite to that of roll separating force.

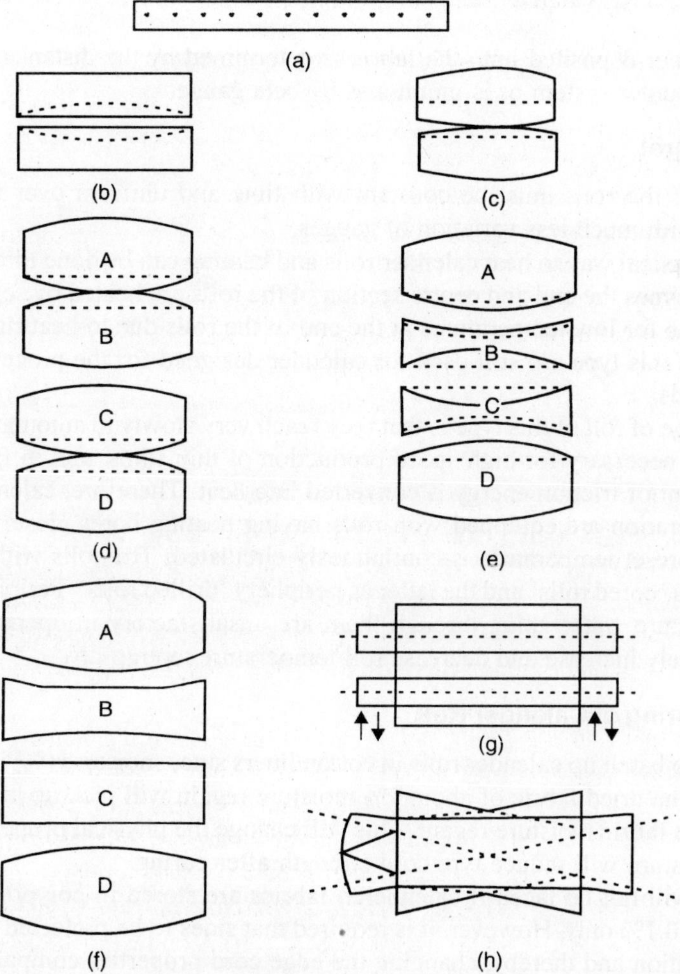

Fig. 4.2. Cross sectioned shape of the rubber film presented to the fabric with rectangular cross-section

Cross axis

This facility is mainly provided for roll No. 1 and 4 of a four bowl calender where these roll axis can be changed thus marking an acute angle at the centre with the original axis. Cross axis or crossing thus does not change the roll separation in the middle, but increase the separation towards the end. By this system edge rubber gauge can be controlled. For example, if edge rubber gauge is less than the body rubber, the axis or roll separation at the edge can be increased. The principle of roll crossing is shown in the Fig. 4.2(h). By suitable combination of number 2, 3, 4 and 5 items in a calender it is possible to achieve uniform gauge and pressure throughout the width.

These combinations of system are must for a calender which, along with efficient calender operation can produce uniform calender gauge throughout the width with excellent quality, with less scrap and correct material control which if not done may run into loss equivalent to buying a new calender in few years time.

Gauge Control

The amount of rubber deposited onto the fabric is determined by the distance between the roll and is taken care of by counter system or is monitored by beta gauge.

Temperature Control

The temperature of the rolls must be constant with time and uniform over the width of the roll to produce material with much less variation of gauges.

Steam is the classical way to heat calender rolls and heating can be done through an axial hole in the bowl centre. Sometimes the end and centre section of the rolls are heated by separate heating system in order to compensate for low temperatures at the end of the rolls due to heat flux through the bearings. Heating systems of this type are still used for calender designed for the processing of thicker sheet at relatively low speeds.

The disadvantage of roll of this type is that they reach very slowly to automatic temperature controls. These controls are necessary for high speed production of thin films, and in fabric calendering, when considerable amount of friction energy is converted into heat. Therefore, calenders primarily designed for high speed operation are equipped with rolls having heating bores close to their surface through which water at a preset temperature is continuously circulated. The rolls with axial hole through the centre are known as 'cored rolls' and the latter as periphery 'drilled rolls'. Periphery drilled rolls without controlled temperature water going through them are unsatisfactory in operation, since hot and cold water used alternately increase and decrease roll temperature too rapidly.

Batching and Storing of Calender Roll

It is not desirable to batch up calender rolls in cotton liners since moisture (%) in cotton can go upto 8% and consequently the dried fabric of about 1% moisture regain will pick up moisture from liners itself with increase in the fabric moisture regain. This will change the physical properties of stored fabric and moreover this moisture will reduce tyre cord strength after curing.

This problem will not be faced if calendered fabrics are stored in polypropylene liners which has moisture regain of 0.1% only. However, it is required that sides to be protected well to reduce the effect of moisture infiltration and thereby, changing the edge cord properties compared with the body cords.

Sheeting operation in calender

Application

Inner liner, insulator or cushion.

Objective

To obtain – Uniform gauge
– Good surface
– Blister free.

Calender—Normally 3 or 4 roll calenders are used.

Gauge Uniformity—3 roll - 4 roll (better)

The calender temperature for different polymers normally used for sheeting are given in Table 4.1.

Table 4.1. Calender temperature for different polymers used for sheeting

Polymer	*Top roll (°C)*	*Middle roll (°C)*	*Bottom roll (°C)*
NR	100-110	85-95	60-70
IR (polyisoprene)	80-90	70-80	55-70
BR (PBD)	55-75	50-70	55-65
SBR	50-70	55-70	55-70
IIR (butyl)	90-110	75-90	75-100

Common problems and counter measures

Shrinkage

Cause - Recovery of elastic memory (not cause evenly) seldom uniform shrinkage and this is transferred to roughness of the surface.

Processing approaches

Reduce plasticity
Higher roll temperature
Higher feed stock temperature

Compounding approaches

Increase plasticiser fillers, add thermoplastic resins (increase tackiness to roll)

Blistering/Air porosity

Processing approaches

Increase the temperature of the feed stock
Minimise the bank
Cross-sectional area of the feed stock.

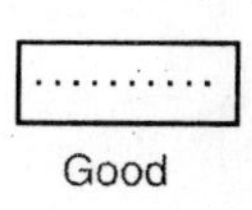
Good

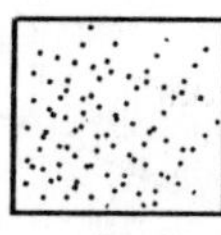
Bad

Calendering

Various approaches of calendering processes are shown in Fig. 4.3(a) to 4.3(g).
Reduce the sheet gauge (maximum 1.0 mm), plyup if higher gauge is needed.
Use larger diameter of calender roll (increased surface)
Air bar, supporting bench of bank and squeeze roll (Fig. 4.3(a)

Compounding approaches

Increase filler
Use large particle size fillers
Incorporate or increase stearic acid
Add waxes.

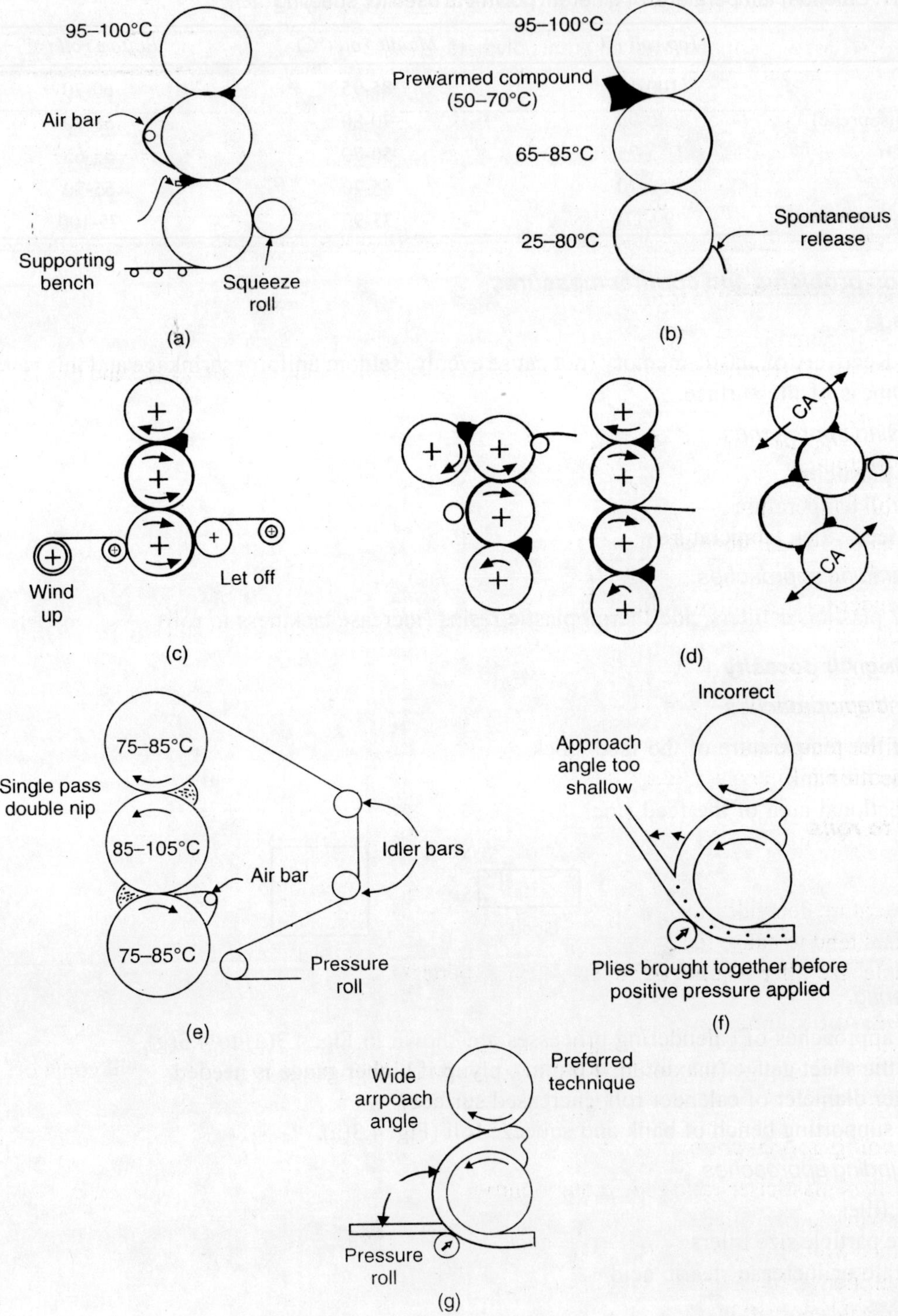

Fig. 4.3. Various approaches of calendering processes.

Relative air permeability of different polymers are given in the Table 4.2.

Table 4.2. Relative air permeability of different polymers.

Polymers	*25°C*	*65°C*	*93°C*
NR	1.0	5.4	9.2
SBR	0.7	4.0	8.8
High styrene (SBR 43%)	0.3	2.2	5.5
IIR	0.07	0.8	2.4

Abnormal surface (Crow's feet)

Cause: Insufficient plasticity
Approaches for Correction - Same as shrinkage.

Calender grain

Cause: Different shrinkage at X-Y direction.

Processing approaches

Not give larger temperature difference between the rolls.
Not give roll ratio.
Limit calendering effect (two rolls)
Slower cooling
Reduce calender speed.

Compound approaches

Increase filler/plasticiser.
Avoid anisotropic fillers, e.g., clay, talc.

Sticking to rolls

Causes

Roll temperature dependency.
White fillers tend to cause it.
Less particle size filler—larger particle size filler (better).

Processing Approaches

Changing the roll temp. (particularly bottom roll temp.) at which a calender sheet will come off the roll by itself Fig. 4.3(b).

Compounding approaches

Increase fillers/plasticiser ratio (stiff compound)
Use larger particle size filler.
Avoid/reduce white fillers.
Reduce tackifier resin (if used)
Use stearic acid, polyethylene wax, waxes.

Doubling up, Ply-up

Where required :
Precision sheeting more than 1.0 mm thick.
Compounds have high impermeability (easy entrapped air).
Components require high impermeability (no pin hole).
In case of using cushion/insulation.
Objective - Ply up two sheet without blister/shrinkage.

Proceeding/Caution

Hot-cold ply up Fig. 4.3(c).
Hot-hot ply up Fig. 4.3(d).
Easier if the two sheets are of same compounds.
Difficult if two different compounds are plied up.
Fig. 4.3(e) suggests solution.

Another caution

Apply wider approach angle of two sheets to be plied up Fig. 4.3(f).
Wind up
Avoid rapid cooling.
Apply minimum tension for wind up.
Use less moisture absorbent liner fabric.
Tightly woven fabrics are preferred.

Common Problems and Possible Solution for Topping of Fabric

Defects	*Causes and solution*
Crush/Loose coat	• Uniform rubber film and parallel nip where fabric and rubber meet as discussed previously. • Sufficient rubber plasticity and tack. • Use of tackifying resins may improve 'oose coat problem. • Avoid crowded/touching cords.
Air entrap	• Use prick rollers. • Use adequate squeeze/pressure in dipping operation.
Over-riding cord	• Less tension of some warp cords at green stage. • Increase tension at dipping/calendering.
Wrinkle	• Improper feeding of fabric during batching up. • Wrinkled liners (replace defective portion of liners).
Split	• Less weft strength and lower elongation (%) weft (use lesser count weft). • Lower fabric width or higher weft tension and lower weft length will cause the problem, as the fabric will not be expanded to desired width, and trying to bring the same to the desired width, split may occur. (Correct fabric width, improve weft tension and increase weft left if possible).
Heavy gum	• Due to improper knife adjustment (Correct knife adjustment. Use good spring or required rubber band for holding knife).
Scorch	• Defective compound and mixing. (Improve mixing, add retarder if necessary). • Higher compound temperature in mills and higher calender roll temperature (Check cooling system of rolls. Do not over load the mills.) Lower calender roll temperature, if possible.

(Contd ...)

Defects	*Causes and solution*
Other Problems	
Elongation	• If higher increase tension for holding and shrinkage, stop zone and if lower reduce tension. • Excessive moisture, if not removed properly, will cause higher elongation (%) particularly in Rayon. (See that drum driers are working satisfactorily. See also moisture % if stored fabric and if too much high requires for redrying.) • Higher temperature of fabric at windup will increase elongation (%), particularly for Nylon fabric. (See that cooling drums are working efficiently or not.)
Gauge variation (length direction)	• Keep uniform and constant feeding in calender nip. (Use required size strip with constant feeding through a conveyor.) • Bring plasticity of feed strip to uniform level. • Use optimum loading to each mill. • Use roller bars on mill. • Avoid manual feeding in roll form. • Keep constant rubber at calender nip. • Avoid calender roll temperature fluctuation. • Increase in bending pressure may also help.
Gauge variation (width wise)	• Use combination of roll chamber, cross-axis and roll bending.
Fabric width	• See spreaders are functioning properly. • Use finger rollers at edge. • Eliminate defective weaving like tight cords, tight bonds, tight edges, crowded edges, bagginess, etc. • Improve weft elongation (%). • Use graduated reeds. • Lower down mooney viscosity of compound.

CHAPTER 5

Coating of Textile Products

INTRODUCTION

Fabrics coated with rubber produced either by impregnation or by surface coating or lamination are called rubberised fabrics or, alternatively, proofed fabrics. However, the name "proofed fabrics" signifies that the coated fabric is proof to the percolation of water, the degree of which may vary. Subsequently, the coated fabrics that are additionally impermeable to air, again varying in degree and with solvents, were developed; these find extensive applications in areas as different as inflatable structures, life jackets and rafts, inflatable boats, flexible fuel tanks, diaphragms and inflatable seals. Again coated fabrics are required in almost every sector of the rubber industry in most of the composite products, such as tyres, belts, hoses, and footwear. In all the applications, the fabric contributes breaking strength, tear resistance and modulus, which are complemented by the elasticity and the flexibility of rubber. New materials such as fibres, fabrics and rubbers and the associated manufacturing processes have been developed and thus have enlarged the capabilities of coated fabrics to the point at which their use in industrial and consumer sectors is growing at a significant and steady rate.

ADHESION OF FABRICS WITH RUBBER

There is one common fundamental rule for all reinforced products, namely, there must be no stress concentrations of any kind. The elastic properties of textile yarn or fabric must be so chosen that there is no infinite stress concentration at the interface between the textile material and the rubber. Since it is impossible to avoid stress concentration at the rubber-textile interface completely, it becomes necessary to provide for adequate adhesion. In fact, it can truly be said that no matter how excellent the physical, chemical and economical aspects of a textile might be, if the material cannot be successfully adhered to rubber, its utility to and in the rubber industry will be very limited. Additionally, the extent of use will be dependent quite heavily on the method of bonding. Adhesion is a subject in the twilight zone between the technologies of textiles, rubber and rubber chemicals and it is impossible to deal with it within any one of these areas without covering some aspects which are logically within the other areas.

Factors Affecting Adhesion

Some of the factors that affect the adhesion of rubber to fabric are (i) mechanical adhesion; (ii) formation of primary chemical bonds; (iii) formation of secondary chemical bonds (e.g., hydrogen bonds); (iv) diffusion or dispersion of adhesive components into the internal molecular structures of the fibre; and (v) fluid process of adhesive application or coating characteristics.

Mechanical adhesion

Textiles made from staple fibres such as cotton, or textiles containing a significant proportion of cotton, do not generally require any adhesive treatment. Synthetic fibres normally produced as continuous filaments confer high strength in relation to bulk, but the resulting textile has a smooth surface finish. The possibilities of mechanical locking are limited, since the only irregularities are due to the interstices in the pattern, as well as the crimp and the twisting of the yarn. Consequently, it is necessary to treat the textile to promote adhesion with rubber.

It has been shown that the area of textile available for bonding rubber is about three times the superficial area for a textile made from continuous filament yarns. Moreover, if in the same textile a spun staple weft is inserted, giving otherwise the same construction, the available surface area is increased by about 10% but the adhesion is increased approximately 10 times, indicating that staple fibres play a dominant role in adhesion.

The complexity of the surface of a woven textile is such that with a vulcanised rubber adhering to it, not all the stresses can fall normal to the interface, nor need they fall across the interface at all. The rigidity of vulcanised rubber added to that of the textile results, through mechanical interlocking, in stress being applied so as to cause failure in the rubber or in the textile as well as, or instead of, failure at the interface. In certain cases the interfacial area over which the stress is distributed is sufficient to with stand considerable shearing force. This happens in the case of spun staple fibres, where, if more than a certain length of the projecting fibre end is embedded in the rubber, the fibre end is broken off instead of being pulled out. In this case the interface is an irregular and tortuous configuration of the fibre end. This is very different from the plane area in which the body of the textile lies. Thus, the increase of efficiency of staple fibres arises not from any change in the physicochemical characteristics but from a redistribution of the mechanical forces. When a filament is partly embedded in the rubber along its length, a small movement suffices to separate the filament completely and the adhesive force ceases to exist. When a staple fibre end is completely embedded in rubber, a small movement may shear the rubber-fibre interface, but this still leaves a force proportional to the length embedded, which must be exerted until extraction of the fibre is complete. Alternatively, the movement will break the fibre in tension if the fibre is sufficiently embedded to provide a total surface force in shear greater than the strength in tension of the fibre.

In dealing with direct adhesion of rubber to the fabric of spun staple yarn, the observed adhesion is a function of the number of protruding fibre ends and their strength. The bond strength arises from the frictional forces that must be overcome to pull out the embedded fibre ends or, in the case of longer fibre ends (where the frictional forces exceed the strength of the individual fibre) to rupture them. The specific interaction between fibre surface and rubber is only of indirect importance in as much as it determines the length of fibre that must be removed before that interfacial shear strength will exceed the strength of an individual fibre. With a fabric made from spun staple yarn, separation occurs with fracture of most of the longer projecting fibre ends and a pulling out of others from the core of the yarn. Where rubber is present in the interstices of the weave, even if it has not penetrated the weave, sufficient fibre ends frequently are embedded in the rubber to permit local fracture to occur.

The most important feature in textile-to-rubber adhesion is the mechanical anchoring of the staple fibre ends into the elastomer matrix, not the penetration of rubber into the fibre structure or fabric interstices. Removal of the fibre ends by employing a fibre woven from continuous filament yarn shows that the rubber penetrates only very slightly between the filament of the yarn of the fibre. There is, however, some contribution to the final bond strength from the penetration of the rubber into the structure or interstices of the fabric.

In coating by the spreading process, the pressure under the doctor blade forces the rubber dough to make intimate contact with the fabric and the rubber coating penetrates the weave but not the yarns. In fact, wetting of one face of the textile takes place, and in spreading coating the wetting properties are determined by the solvent employed, along with the solids content of the dough. After initial contact of the dough and the textile, a few seconds elapse before the heat from steam chest starts the evaporation of the solvent, during which period some progress is made toward equilibrium wetting. Although the dough will show zero equilibrium wetting angle against the fibre, the application of the doctor knife to the moving fabric surface involves appreciable flow rates, thereby increasing the contact angle. The speed of coating determines the angle at which the liquid contacts the fabric surface; for better wetting and penetration, this angle should be as low as possible. An advancing contact angle is always greater than the equilibrium value and with increasing rate of coating, the advancing contact angle tends to 90°, which means that the pressure requirement for penetration tends to infinity. However, limited penetration to the yarns is possible using doughs having low solids content with corresponding high solvent content and low viscosity, coupled with low speed of coating. Ageing, in the spreading, process, before flow into the fabric has ceased, evaporation of solvent starts, thereby increasing the viscosity. Viscosity controls the extent of further penetration under capillary forces, provided conditions are favourable (e.g., low viscosity dough made from solvents having a low rate of evaporation, coupled with low tension on the fabric). As the solvent evaporates, movement or flow of the dough front ceases, and again shrinkage caused by evaporation induces a further increase in contact angle. In coating by the calendering process, the rubber compound is applied with a very high pressure normal to the textile surface for a very short period (milliseconds). The same conditions of wetting and of minimum pressure for penetration used for spreading apply, but the magnitudes of the various parameters are very different.

In calender coating, again the rubber cannot penetrate into the yarns, but only into the weave. However, with a dipping process using either aqueous or solvent-based dilute solutions, penetration into the yarns is possible. Here the conditions limiting the interfacial surface area across which adhesive interaction cannot occur are overcome by maintenance of almost zero contact angle against the fabric surface; the degree of penetration into the yarns is controlled by the tension and time of contact.

Adhesion due to primary bonding

In view of the generally low reactivity of some adhesive systems and the levels of adhesion that should be obtained on a theoretical basis, if true chemical bonding has occurred, it is now considered that covalent chemical bond is only one of several factors contributing to the overall adhesion.

Adhesion due to secondary bonding

The surface of a cellulose fibre, whether of cotton or rayon, is largely hydroxylated and thus provides a dipole surface whose attraction contributes to the adhesion. Natural rubber (NR) has a polarizable molecule; therefore, the attraction between rubber and cellulose consists of the sum of the dipole-induced dipole attraction and the secondary electronic field forces, usually called London dispersion forces, the former being greater in magnitude. All textiles of practical interest have surfaces that contain oriented dipoles, even though, as in nylon, the effective surface concentration of dipoles is reduced by hydrogen bonding between adjacent molecules. All these textiles will, therefore, induce a dipole moment in NR and the specific adhesion between the molecules of rubber on the one hand and textile on the other will depend on dipole-induced dipole interaction.

Among the aspects of secondary chemical bonding, the formation of hydrogen bonds provides a significant contribution to many adhesive systems; for example, in the case of resorcinol-formaldehyde latex (RFL) dipping of rayon and nylon, it is accepted that there is hydrogen bonding between the phenolic groups of RF and the hydroxyl groups of the pyranose rings of rayon and between phenolic and hydroxyl and /or secondary amine groups of nylon. The relative contributions of the various mechanisms to the final adhesion of rayon and nylon are given in Table 5.1.

Table 5.1. Mechanisms contributing to the adhesion of rayon and nylon.

	Percentage contribution	
Mechanism	*Rayon*	*Nylon*
Direct mechanical : penetration of rubber into the structure of the weave and the rayon	20	15
Primary chemical bonds : direct covalent linkages	25	60
Secondary chemical bonds : mainly hydrogen bonding	25	20
Diffusion : microscopic or molecular diffusion into the filament	30	5

The marked difference in the contribution arising from primary chemical bonding in rayon and nylon is noticeable, although the reaction suggested as providing the chemical linkages is much the same for both fibres, namely, condensation between a methoxy group from the resin with a hydroxyl (rayon) or amido (nylon) group on the fibre. This type of reaction may be expected to take place more readily with rayon having more reactive sites because the cellulose backbone contains six such hydroxyl groups per unit of cellulose, which is 10.28 Å long, giving an average distance of 1.7 Å along with the chain between hydroxyl groups, while on the nylon chain there are two amide groups on a unit 17.2 Å long, for an average distance of 8.6 Å between amide groups. The explanation for this discrepancy could be that some steric hindrance effect, or a mismatch of the repeat unit distance between the active methylol groups of the resin and the distance of repeat units of the individual fibres, favours the reaction of nylon rather then rayon.

Adhesion due to diffusion

The diffusion component is ascribed to a much higher contribution with rayon than with nylon. With rayon, the RF resin either actually diffuses into the fibres, giving a true diffusion bond on a molecular level or as a result of microscopic penetration into the porous surface of fibres. With nylon there is no evidence of migration of the RF resin into the fibres. The use of a terpolymer of vinyl pyridine (VP) and styrene–butadiene rubber (SBR) latices provides increased wettability of nylon through the pressure of hydrogen bonding.

Bonding Systems

Resorcinol–formaldehyde Latex (RFL) bonding for different fabrics

With RFL dips, the factors affecting performance are (i) the resorcinol/formaldehyde ration; (ii) the conditions of resin formation; (iii) the pH of the resin solution; (iv) the type of latex used; (v) the ratio of latex to resin; (vi) the catalyst composition; and (vii) the total solids. With rayon, the optimum adhesion is obtained when the molar ratio of resorcinol to formaldehyde is 1:1.5 to 1:2.5; for nylon, the ratio should be between 1:2 and 1:3. Nylon has the higher formaldehyde requirement because the nonparticipating amide groups use up additional formaldehyde to form methylol groups.

The RFL dips are often prepared by rubber processors by mixing the resorcinol and formaldehyde in a definite molar ratio in an alkaline medium. This is stored for approximately 6 hours at room temperature and is then added to NR, SBR, or VP latex or mixtures of these latices. A maturation time of 12–24 hours is given to the dip mix. Since the reaction of resorcinol and formaldehyde is exothermic, it is particularly difficult to maintain constant temperature during the reaction and maturation. Therefore, precondensed and precatalysed resorcinol–formaldehyde novolak-type resins with an optimum and constant degree of condensation are used in combination with NR, SBR, or VP latices. Formaldehyde is next added to cross-link the resin, with elimination of the maturation process. The resin content of the dip is related to the rubber content of the latex, which is usually 20%. The solids content of the RFL dip is tailored to the fibre used (e.g., for rayon, 10–15%; for nylon, 15–20%; for polyester, 20% or more). The dip uptake should be 5–8% for rayon and up to 15% for heavier nylon fabrics. For rayon the latex component comprises a mixture of SBR and VP latices, the ratio varying from 80:20 for the standard tenacity rayons to 20:80 for the higher tenacity. With nylon, 80–100% VP latex is used.

Thermodynamic compatability data show that the solubility parameter of the RF resin and of nylon and rayon are very close, whereas that of polyester is very different. For example, the solubility parameter of RF resin and nylon is 67mJ/m^2, while for polyester it is 43 mJ/m^2. This difference largely explains the difficulty in achieving a satisfactory bond with the simple RFL systems and polyester, which results on a molecular scale, from poor contact between the two phases. With polyester, a two-stage dipping system is typical. In the first stage, the polyester is dipped into an adhesive consisting of water-miscible epoxy added to a blocked isocyanate dispersion, to give a pickup of 0.5%. This blocked isocyanate is then activated at a temperature of 230°C. The polyester is given a second dip with a standard RFL system. A more widely adopted system involves condensation with RF resin of a trimer, formed by reaction of *p*-chlorophenol with resorcinol.

Aramid (aromatic polyamide) is closely related to nylon (aliphatic polyamide) but does not give satisfactory bonding with simple RFL dips. A first dip of an epoxy based on the reaction of epichlorohydrin and glycerol, followed by a second-stage dip in standard RFL dip is advantageous. In view of the very low chemical reactivity of the fibre, it is considered that direct chemical bonding does not contribute significantly. Again good wet adhesion suggests that secondary bonding makes little contribution, the adhesion being mostly due to true diffusion, which is favoured by the high temperature treatment and low molecular weight pretreatment.

With glass fabrics, good adhesion is secured by pretreatment of the fibres with a coupling agent (e.g., γ-aminopropyltriethoxy silane), followed by dipping in RFL. With dips of about 35% solids and with 30% pickup values, good adhesion and product performance result.

Dry bonding systems

The use of dry bonding systems based on resorcinol–formaldehyde donor–silica is the most common method of improving rubber-to-textile adhesion. A combined action of all three components is absolutely necessary to achieve optimum adhesion. As formaldehyde donor, hexamethylenetetramine (HMT) or hexamethoxymethylmelamine (HMMM) is used. This system primarily deals with solid rubber compounds but can also be used in solution in an organic solvent for the pretreatment of fabrics. The process is universally applicable for all types of rubber in combination with all commonly used textile materials, including steel cord. High levels of adhesion with untreated rayon and nylon can be obtained and improved adhesion is attained when dry bonding is used in conjunction with RFL-dipped fabrics: the resultant values are appreciably higher than with either system alone. With polyester, the amine residue from the HMT component can cause significant degradation, by chemical attack, as a result of aminolysis of ester linkages in the fibre. With polyester it is preferable to use HMMM as the methylene donor, since the deleterious effect of this material is negligible.

In the dry bonding system, the RF resin is formed during vulcanisation and migrates to the rubber-textile interface, resulting in an efficient bond between the two components. The silica may function to retard vulcanisation until the formaldehyde donor has reacted with resorcinol, thus allowing time for the migration of the resin and subsequent bond formation. It may also retard the polymerisation of the RF resin, so that cross-linking occurs at the textile interface, resulting optimum bond formation. The normal amounts (in parts per hundred rubber) of resorcinol and HMT are 2.5 and 1.5 phr, respectively, with a minimum thickness of about 0.25 mm of adhesion compound on the textile; otherwise; back migration of the adhesion promoters into the bulk of the standard compound may take place, depleting their concentration at the textile interface below that required for satisfactory adhesion. In coated fabrics with thin proofing, the frequently observed failure to obtain satisfactory adhesion with dry bonding systems is simply due to an insufficient amount of the active ingredient or insufficient thickness of the adhesion compound. Dry bonding may result in discoloration of dyed fabric or of compound. Accordingly, only dyes that are fast to the dry system can be used. However, the characteristic orange-brown discoloration of the compound can be covered up by using up to 25 phr of TiO_2 without impairing adhesion. Apart from ensuring outstanding bond strength, the spreading doughs have a good pot life at normal processing temperatures.

The curing time, depending on the temperature employed, should be adjusted to provide enough time to ensure the release of sufficient formaldehyde. Often the adhesion can be improved by exceeding the optimum cure time or by raising the temperature, especially when nylon fabric is used. The minimum temperature for good bonding is 130°C. Normally the process gives best results when vulcanisation is carried out in the press, where contact between the rubber compound and the fabric is generally close enough. However, it is also possible to cure in hot air, with or without additional air pressure and in steam; good adhesion is obtained without pressure provided the mechanical contact is sufficient, as in the case of spreading or frictioning.

The dry bonding system requires that the composition of the rubber compound be carefully balanced. The sulphur level is not as critical as with other bonding systems, but sulphurless curing systems result in relatively poor adhesion. Scorch safety and cure rate should be balanced, since too much delayed acceleration results in less adhesion. Ultra accelerators are unfavourable because they often do not give sufficient time for the required amount of formaldehyde to be released. Good results are obtained with N-cyclohexyl-2-benzothiazyl sulphenamide (CBS) or 2-(4-morpholinylmercapto) benzothiazole (MOR), used alone or in combination with basic secondary accelerators like N,N'-diphenylguanidine (DPG) or N,N'-diorthotolyl guanidine (DOTG). High doses of stearic acid and benzoic acid enhance adhesion in general. In nitrile rubber (NBR) compound the bond strength is considerably improved by using 2.5 phr of stearic acid. N,N'-Diethylguanidine (DEG) often has a negative influence on the bond strength and is therefore normally unsuitable. Good results have been obtained with accelerator combinations such as mecaptobenzothiazylsulphide (MBTS)/DPG, or DOTG and in case of polychloroprene (CR) compounds with DPTG/tetramethyl thiuram monosulphide (TMTM). In some cases, depending on the formulation of the compound, the type of fabric, and the requirement to be met, both the total amounts of resorcinol and formaldehyde donor and their ratio can be varied to improve adhesion. For example, with SBR-based compounds, increased dosages of resorcinol and formaldehyde of 3.8 and 2.3 phr, respectively, give improved adhesion. Higher dosage of precipitated silica (⩽ 30 phr) is beneficial for higher adhesion; peeling test using woven fabric show a marked improvement, although pull-through test using cord show little advantage. Compounds having high tensile strength give relatively better adhesion values when measured by a peel test rather than a pull-through test, whereas improved flow characteristics, in the early stages of cure, favour the pull-through values, indicating that magnitude of the effect depends on the test method used.

Isocyanate bonding systems

The alternate commercially significant bonding method is the utilisation of isocyanates dissolved in suitable organic solvents, such as ethyl acetate or methylene chloride, in proper proportion to the dilute rubber solution and applying them as initial coatings to the textile. Because the pot life of such solutions is limited, careful control must be exercised. Subsequent coatings are applied without any isocyanate and the storage time before vulcanisation of the coated fabric is kept to a minimum. Contamination with water should be avoided and the solvent used should be free from water. The process has the advantage of allowing just one face of the textile to be treated and coated. The common isocyanates are triphenyl methane triisocyanate and thionophosphoric acid tries (p-isocyanatophenyl) ester, the latter being less staining type. The triisocyanates are especially active and diisocyanates such as diphenylmethane diisocyanate can also be used, though they are somewhat weaker. Isocyanate bonding agents give coated fabrics with better softness of handle, as well as better drape or flexibility than is provided by RFL systems. The superior performance of polyisocyanates in bonding rubber to textiles is due to a bridging function in which there is direct chemical attachment of isocyanate groups with reactive groups on the textile surface and with those (e.g, hydroxyl, carboxyl, hydroperoxide) in the rubber, which are introduced during mastication. Although the reaction of isocyanates and rubber occurs at room temperature, the reaction with a textile appears to need elevated temperatures, and this indicates some form of chemical reaction. Isocyanates have excellent solubility in many organic substances and their small molecular size permits them to diffuse readily. These properties promote adhesion by allowing the isocyanates to penetrate the adherent, where they may undergo the foregoing multiple reactions, all of which enlarge the molecule, entangling it or attaching it to the adherent. Furthermore, the reaction of the di-and polyisocyanates with hydroxyl-bearing polyesters and polyethers, their coreactants in many adhesive applications, produces the polar polyurethanes, which wet, intimately contact and show strong attraction for a variety of textile surfaces.

A variation with isocyanates is the stable "blocked" di-or polyisocyanates in suspension (including aqueous suspension) or solution, which are used with rubber doughs and applied to the textile. Subsequent heating decomposes the blocked isocyanate and regenerates the true isocyanate, which proceeds to bond the adherent members [e.g., methylene-bisphenyl (4-phenyl carbamate), or polyethylene polyphenyl isocyanate (PAPI), dianisidine diisocyanate (DADI)]. Isocyanates give satisfactory bonding for wide variety of rubbers and textiles. CR-based adhesive coating with isocyanate can be used for top coats based not on only CR but also on NBR, chlorosulphonated polyethylene (CSM), or polysulphide. For better solvent resistance purposes with an NBR top coat, an adhesive coat is also based on NBR. For comparatively saturated rubbers like butyl rubber (IIR), both the adhesive coat and top coat are based on IIR. Fabrics treated with adhesive coats containing PAPI or DADI (etc.) can withstand a delay in the application of the top coat for several days with no reduction in the bond strength, thus increasing processing flexibility by permitting the treatment, storage and subsequent processing of large quantities of fabric.

Adhesion and Tear Strength

It has been noted that woven fabrics coated with rubber exhibit direct interactions between the adhesive bonding strength and certain other properties, most particularly the tear strength of the coated fabric. When a fabric is coated as usual in the warp direction, there is no influence of the tensile strength in the warp direction, but there is some loss of strength in the weft direction, a loss that increases with increasing adhesion. The factors that affect the tear strength of uncoated fabrics also apply in coated fabrics; some are magnified and some additional effects come to light once the coating has been applied. The most critical variables are fabric wave, yarn crimp, yarn twist, coating quality, coating location in the yarn system and coating penetration into the fabric interstices. However, most of those variables are interdependent; for example, slight variations in fabric tension during coating upset the balance of warp and weft crimp and the tendency is toward low warp and high weft values; this occurrence then magnifies the extent to which coating penetrates the weft yarn.

The tea strength of a fabric decreases linearly with increase in adhesion as measured by single rip or tongue tear method; roughly, for fabrics in the range of 250–500 g/m^2 weight, doubling the level of adhesion will halve the tear strength compared to original values of the basic fabric. The change from plain weave to twill weave offers some improvement within the same weight range. For high tear strength, it becomes necessary to use either thicker yarns or loosely woven fabrics both of which, within the same breaking strength range, make it difficult to render the fabric airtight. Since, moreover, lightness and surface finish are also impaired, a useful compromise in the construction and weave becomes essential. A closer examination reveals that some of the most important factors associated with coated fabrics are (i) dependence of coating adhesion on the size of the interstices and the consequent influence in tear strength; (ii) influence of twisted yarns in the fabric structure and (iii) effect of various basic fabric weaves.

When continuous filament yarns are used, in the absence of bonding agent, adhesion increases linearly with interstitial area, cutting off an intercept on the adhesion axis for a full "jammed" fabric having no open interstices for penetration.

Increasing yarn twist increases the interstitial area, hence coating penetration; this in turn blocks the "backing up" of yearns in the active region of tear, consequently reducing the tear strength. However, twist is an essential feature of staple fibre yarns. In a twisted structure, the necessary cohesion between fibres is provided by transverse forces, which press the fibres together when tension is applied; the longitudinal forces, which tend to make the fibres slide apart, generate transverse forces, which tend to prevent such separation. By increasing the twist, fibre cohesion and strength are increased, among other effects. For example, the more the fibres are constrained to follow a helical path oblique to the yarn axis, the smaller becomes the proportion of their available strength that is effective in the direction of that axis. Furthermore, increasing shear forces are brought into play, again tending to weaken the structure. Within the lower range of twists, the fibre cohesion factor is dominant; at a certain point, however, the benefits derived from that source are balanced by the losses in stress-transmitting efficiency and any further increases of twist result only in the reduction of strength.

The turning point, or optimum twist at which maximum strength is obtained, depends on a variety of factors including the length, thickness and practical characteristics of the fibres concerned. Extensibility increases with twist and so do abrasion resistance and resistance to dynamic stress. However, the modulus of elasticity decreases with increasing twist. Each of these properties has an optimum for a certain twist. The solution of an optimum twist for a given purpose therefore always represents a compromise between conflicting requirements put on individual properties, including particularly tear strength.

Regarding the effect of weave on the tear strength of coated fabrics, it has been observed that 2 × 2 twill weave increases the tear strength to some extent, but not very significantly, over plain weave. In plain weave, which is most common in proofed fabrics, use of low twist yarns enables a very smooth and flat surface to be produced, the yarn flattening out at the intersections and spreading to give minimum interstices between adjacent yearns, thereby affording a very full cover to the fabric. With twill weave, the longer "float" of the yarn (i.e., the increased length from one intersection to the next) improves the tear strength. At the same time, however, the total number of warp/weft intersections is reduced, thereby making the fabric comparatively less mechanically stable. The easiest way to modify the plain weave is to run two threads together, both in warp and weft, producing 2 × 2 matt weave. In this construction, the tear strength of the basic fabric is improved substantially, with a corresponding significant increase of tear strength in the coated fabric.

The increased tear strength results because under the tearing action, the multiple threads bunch closely together, thereby requiring a higher applied force for propagating the tear. In 2 × 2 matt weave the "float" length and width both increase the tear strength; furthermore, the ratio of number of interstices per units area for matt to that for plain weave, which is 1:9, certainly introduces a penetration blocking effect, thereby increasing the tear strength. Commercially this property is exploited in lighter fabrics such as proofings and tarpaulins.

FIBRE AND FABRIC CHARACTERISTICS

In designing or selecting a fabric for a specific application, three main questions should be addressed: (i) general physical properties to be satisfied (strength, elongation, flexibility, fatigue resistance, tear resistance, heat resistance, etc.), which are mostly dictated by the application; (ii) choice of fibre type to be used in the fabric; and (iii) the actual fabric structure needed to meet the requirements.

While the prime requirements of a textile reinforcing material are strength and bondability, the third and fourth considerations for suitability are dimensional and thermal stability. Certain fabric properties (e.g., strength) are controlled by the characteristics of the basic structure of the fabric such as size and number of yarns; but other properties (viz., modulus and shrinkage) are controlled by the finishing process. The extensibility and elastic recovery of nylon make it a useful material for various types of protective clothing, but these properties are not desirable, for instance, in tyre cord. The elastic recovery of polyester is also quite good. All reinforced rubber articles are finally subjected to a vulcanisation process and it is essential that any reinforcing material be dimensionally stable during cure. Nylon and polyester are both thermoplastic and liable to shrink when heated or if constrained to set up thermal shrinkage forces.

Shrinkage at a curing temperature of say, 160–170°C contributes to dimensional stability by affecting the size and uniformity of the product (e.g., diaphragms) and should be kept to a minimum. The parameters that control shrinkage are the glass transition temperature T_g, the degree of crystallinity and the orientation of the amorphous region. The tendency to shrink can be minimised by "heating setting," which consists of heating a yarn or fabric to a temperature higher than any expected subsequent processing temperature and either holding the material dimensionally stable or allowing a small relaxation.

The initial modulus of a yarn is proportional to the gradient of the load–extension curve at the origin. This figure gives an indication of the resistance to working stretch of the yarn. In some products working under tension stress, growth is observed, which must be kept below the level the rubber can tolerate before it begins to crack. Growth consists of two components–initial extension and creep. Polyester shows some improvement over rayon and nylon, but none of these fabrics is as dimensionally stable as glass or steel. Hence all such low modulus yarns or fabrics are "hot stretched" before being used, to increase the modulus and reduce subsequent growth. Hot stretching is similar to heat setting except that the cord is extended while hot and held free during the subsequent cooling process. Since stretching alone will increased the tendency to shrink, the net result of combined stretching and heat setting operations will be a hot-stretched cord that still retains some heat shrinkage. The process is essentially a compromise in which the modulus is increased to prevent excessive growth, while the thermal shrinkage is reduced to a level that ensures reasonable dimensional stability during cure.

Abrasion resistance is an outstanding characteristic of nylon. Fibre frictional properties are important with respect to resistance to abrasion. Abrasion resistance is improved by increasing the filament diameter but flex resistance tends to be reduced with increasing denier per filament after a certain point. This is because abrasion and flex resistance are related to fibre transverse properties and both can be adversely affected by excessive fibre orientation. Resistance to abrasion is undoubtedly a factor in fatigue life and it may well be that this is the predominant factor in the excellent fatigue properties of nylon.

The proofing industry, which is the oldest user of textiles in the rubber industry, consumes a variety of fabrics made primarily of cotton and nylon, with some polyester, in many different weaves and weights depending on the final use. The quality of the weaving is very important, depending on the specific end use of the coated fabric. In addition to such general requirements as reinforcement and bondability, fabrics for proofing must have an even and good fabric surface free from imperfections (loose threads, missing threads, broken threads, floats, modules, knots, stains, tight or loose selvage, etc.), since all these defects impair such properties as watertightness, airtightness and the finish. The usual standard is a maximum of one fault per 10 meters. Low grade cotton should be avoided because the fabrics made from it may be excessively hairy, causing blemishes on the surface of the proofing as a result of transfer of the hairs. Scouring of the fabrics to remove spinning and weaving lubricant is also required. Again, fastness of colour to washing, rubbing and sunlight, together with copper-and manganese-free characteristics, are important because in their absence, the rubber proofing degrades more rapidly. Furthermore, in the case of inflatable products made from proofed fabrics in particular either cured or uncured, proper heat setting and minimum bow are important to ensure wrinkle-free lay-flat characteristics of proofed fabric and undistorted and dimensionally stable final products.

So far we have discussed the properties of textile material that are important in connection with their use in coated fabrics in general. Obviously the other component in coated fabrics is the coating material, the rubber. Almost all types of rubber are being used, depending on end applications, the latest addition in the list being the polyurethanes (PUs).

COATING PROCESS

The major process for producing coated fabrics are spreading (dealing with rubber doughs) and calendering (dealing with dry mixes). Additionally, dipping is sometimes used and is essentially adopted to give anchor coats. Lighter weight fabrics ($\leq 60 g/m^2$) should be coated by spreading because of the danger of distortion and tearing when the calendering process is used. Proofing contents below 110 g/m^2 are not suitable for calendering. Thus, the choice of coating process is mainly determined by fabric weight and proofing.

Dip Coating

In dip coating, the coating layer or the pickup is quite low, with penetration into the interstices of the fabric as well as within the yarns, and this layer forms the base for further topping materials normally added by the calendering process. The solutions used are generally very thin, and the process obviously coats both faces of the fabric. The impregnation or dip coating is carried out by dipping the fabric into the rubber solution in a V-shaped trough made of stainless steel, followed by passage through padding rolls or polytetrafluoroethylene (PTFE)-coated scraper blades, to remove the excess solution, which runs back into the trough. Finally, the dip–coated fabric is dried, cooled, and wound on batching rolls using plastic liner. All through the process, the tension of the fibre is controlled, because sometimes a stretching process is carried out simultaneously.

Dip coating is advantageously used where more than one layer of fabric is required to be coated, particularly for inflatable products. Dipping prevents wicking action through the edges of yarns of the fabric by inducing saturation by the rubber solution, thereby stopping leakage of air in a built-up structure. Furthermore, since dipping does not stress the fabric, no damage or yarn distortion can take place; rather, the yarns are set more firmly in their original straight-line position, with minimum development of bow through subsequent spreading or calender coating processes, and structures made from such coated fabrics have the least twisting or distortion.

Coating by Calendering

Calenders operate at much higher shear rates (normally s^{-1}) and probably are the only processes that involve simple unidirectional laminar shear flow. A tension is developed, however, between the take-off cooling rolls and the last of the calendering rolls. Consequently, it is likely that the normal stresses within the mass of the molten rubber mix at the point of leaving the last calendering rolls are completely relaxed; again the axial stresses developed by the film during cooling in its travel toward the take-off cooling rolls may introduce considerable monoaxial orientation. The hot plasticised rubber mix starting through the feeding side of the nip accelerators as it approaches the region of minimum clearance, where it moves at maximum velocity. After leaving this region of minimum clearance, it decelerates until it reaches the speed of the roll. Since the material is moving more slowly, it must be thicker than the clearance if a material balance is to be maintained. Again, there is expansion due to elasticity. The increase in thickness, often called the "calender effect," is determined by the molecular weight of the rubber, the calendering temperature, and the speed.

Friction coating by calender

Fabrics to be coated on both sides by calendering are usually frictioned first. This involves unequal low speeds, the fabric entering the nip more slowly than the rubber, resulting in a force that tends to fill the interstices of the fabric with rubber, to secure mechanical bond. This operation is carried out in a three-roll calender with the top and bottom rolls rotating slowly, while the middle roll makes a faster speed. The friction ratio varies from 1:1.5 to 2.0. By keeping the roll temperatures high enough to give certain required characteristics of the compound, a portion of the dough will stick to the middle bowl, whereupon it is returned to the nip. Frictioning cannot be carried out on both sides of a fabric in one operation using a four-bowl calender because the necessary speed difference between fabric and compound cannot be obtained.

Two three-roll calenders in series can serve the purpose of frictioning on both sides, as well as coating on both sides or frictioning on one side followed by coating on the same side. However, a four-roll calender is the most versatile machine for making coated fabric: either one-side coating or both-sides coating can be carried out in a single pass, and the proofed fabric can be doubled, if necessary, to give a sandwich like construction.

Skim coating by calender

The operation of applying a substantial thickness of rubber compound to a fabric on a calender is known as skim coating or topping. Here a sheet of rubber compound is formed in one or two of the nips between the bowls of the calender, and the sheet is passed onto the frictioned or dip-coated or tie-coated (for one-side coating) fabric. The calender bowls are run at even speeds or nearly even speeds and at temperatures some what lower than those used for frictioning.

Calender productivity

For coated fabric production, the calendering speed may vary from 10 to 70 m/min, and it is very much dependent on the consistency of the roll temperature and on the supply position of the hot plasticised rubber mix. If there are few or no compound changes, the maximum output of warmed rubber mix obtained from 1525 × 560 mill is about 300 L/h, while a single mill of same size with a cracker is capable of supplying approximately 500 L/h. When compound changes are frequent, the average hourly rate over a complete shift comes down to about 190 L for a single mill and approximately 400 L for a combined mill–cracker setup. Apart from the actual speed of the calender, which is governed by the considerations above, the productivity of the machines is determined by the amount of time spent actually passing work through the nip (as opposed to the time spent in changing rolls, products, or compounds or between successive passes, as happens during actual operations). Products, changes involving no change of compound usually take a little longer than roll changing or between-pass times of fabric (≈2 min).

Resetting the calender for a different thickness, width, or both is a fairly simple operation and can usually be done without stopping the machine. A compound change can take an appreciable time, especially if the calender and mills must be thoroughly cleaned or run at different temperatures. Ignoring these special cases, however, even a simple compound change can take about 15 minutes if only one mill is available, whereas when two mills are available the job can be done more quickly (≈10 min).

Common calendering problems

For every rubber compound, there is a temperature range in which good quality film and coatings can be produced. In general, for a given nip setting, too low a temperature leads to rough surfaces, and too high temperature leads to blistering or sticking to the rolls. Also, generally, the thinner the film, the higher the temperature needed. As the temperature rises, the stock becomes softer and more plastic, and if this change leads to too soft a stock, the bubbles of entrained air do not burst during passage through the nip and so the slim is blistered. Again, if the stock is too cool, the folding that occurs as the rubber mix enters the feed nip leads to marks on the surface called crowsfeet, caused by imperfect knitting. Most synthetic rubber exihibit less self-adhesion than NR, and so a higher temperature is needed to avoid crowsfeet. The speed of calendering also affects both the gauge and the incidence of these defects. Lower speeds give a longer time under pressure, which tends (i) to reduce elastic recovery and thereby the gauge of the sheet produced; and (ii) to improve adhesion of folded surfaces, so reducing the tendency to form crowsfeet. This explains why synthetic rubbers often must be processed at a lower speed than NR; it also justifies the invariable request to provide a variable speed drive.

Crowsfeet and bank marks are manifestations of differential rates of deformation at the roll–melt interface arising from variations in melt viscosity, which results from partial cooling of the melt or improperly homogenized melt.

Crowsfeet marks originate at the feed because of low shear rates and the lack of further restriction to flow in the subsequent passage through the middle and exit rolls. Bank marks, on the other hand, originate at the nips of the middle rolls. A small bank is always necessary to ensure adequate contact of the film with the rolls and to prevent air entrapment. However, if the bank is too large and the roll speed is sufficiently low, the exposed surface of the bank, as it rotates and reaches the nip, will be at a lower temperature than the material arriving directly at the nip from the preceding rolls. As a result, periodic small depression will be formed along the width of the film. Depressions may also result from slipstick effect created by the periodic change in the coefficient of friction.

The deposit of additives on the surface of the sheet may result in dull patches and surface matting, marring the surface. Furthermore, if the temperature of the rolls is low, a shark skin effect may result when the film leaves the nip, creating surface irregularities, which decrease the gloss.

Another common problem associated with calendering is variation in thickness. Longitudinal variation is due to roll floating, which occurs because of the clearance between the bush and the roll journal. The problem is taken care of either by using close tolerance Timken bearings or by using a hydraulic pullback system in which the roll butt is hydraulically pulled and kept toward the upper semicircle of the bush. Variations across the width are due to roll bending and the simplest solution is to use a roll crown or camber. Since the roll camber is fixed, it cannot compensate for the variable roll deflections due to usage of different rubber compounds and their varying thickness or speed of calendering; this inflexibility alters the separating forces, hence leading to the roll bending. One variable roll bending compensation system consists of roll crossing or skew mounting of rolls; alternatively, hydraulic bending is practiced.

Coating by Spreading

The spreading machine is quite simple in design; it consists of a let-off roll, a spreader roll (to ensure removal of creases from the fabric), a doctor knife (impinging on a small belt or, more commonly, a rubberised roller, which is usually driven), followed by along horizontal steam chest for drying and finally the take-up roller. More exacting metering of coating thickness is accomplished by the knife-over-roller arrangement. The rubber covering of the roll may vary in hardness from 60 to 90 Shore A, depending on the type of fabric and the quality of coating to be done. A chromium-plated steel roll can be used in place of rubber roll if the utmost precision is called for and the fabric is of even thickness. Obviously the roll should be true with the knife surface and it must be free from eccentricity (Fig. 5.1).

Coating knives

Three different types of doctor knife are commonly used: flat or square section, tapered or V-shaped, and bull-nosed, the last being used for heavy coatings, particularly for easily damage fabrics with little penetration into the weave. Tapered or V-shaped knives are used to give a light coating with good penetration. Flat or square section knives commonly have a lead-in chamfer or base on the feed-in side; penetration into the fabric depends on the width of the flat portion, while coating thickness depends on the angle and width of the chamfer, since a compression zone is formed there. Such a knife generally gives good penetration coupled with good coating thickness. The angle between the knife and the roller and the clearance between them control the coating thickness and the degree of penetration of the fabric by the dough. The greater the angle at which the knife meets the moving fabric, the greater the degree of penetration. In modern machines, instead of being fitted with one knife, a knife-supporting beam carries two or three knives of different type, 180° or 120° apart. This arrangement speeds up blade changing considerably and also allows any knife to be set at any angle. Furthermore, instead of mounting knife holders on screw rods and adjusting the gap between knife and roller by hand, modern equipment features two pneumatic cylinders, one at each end of the knife holder; the gap can be adjusted by pressing the electropneumatic control valves and through the forward or backward movement of a tapered wedge placed between the holder and the machine frame. During coating, the knife holder is thus firmly pressed down, and this avoids the sort of minute fluttering of the knife, which takes place with old system because of the clearance between the fixed holding nut and the movable screw rod. Pneumatic quick lifting is provided so that the knife can release lumps, splices, or torn edges. Alignment of the knife with the horizontal axis of the roll is vital; otherwise wedge-shaped coatings are produced.

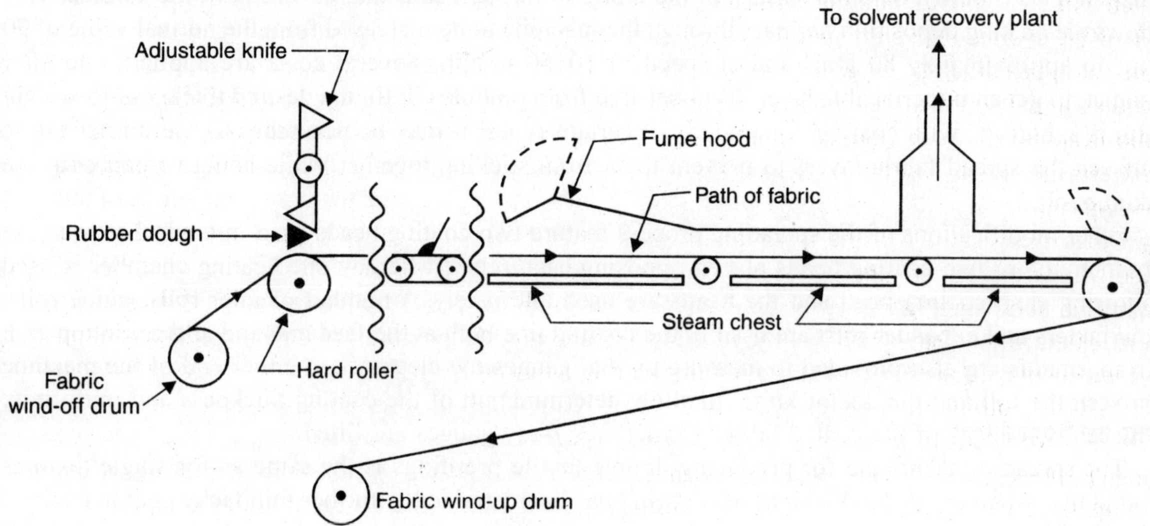

Fig. 5.1. Features of a spreading machine.

Spreading process

The dry and smooth fabric, under even tension, is fed over the driven bearer roller and under the doctor knife. The rubber dough is introduced in front of the knife by means of a ladle or scoop and the machine is started at a low speed. The combined effects of the forward motion of the fabric and the obstacle of the fixed knife give the viscous mass of dough a rotatory motion in the direction of the breadth of the machine. To prevent the dough from spilling over the edges of the fabric, two adjustable iron guard plates are fitted at the end of the knife. Special care is necessary to ensure that the rubber dough does not strike through the fabric, particularly in the case of single-texture proofings. The tendency to strike through may be considerably reduced by using a rather soft rubber roll and a fine-edged doctor knife.

The coated fabric then passes over a steam-heated table, the temperature of which is gradually increased from entry to exit point. The main drive of the machine is connected to the rubber roller at the front, which in turn is connected by a clutch to the back roller at the end of the steam-heated table. Again the drive of the take-up roll at the bottom of the machine is taken from the rubber roll through belt drive and the winding of the coated fabric is controlled by tension straps, adjusted to keep the fabric just taut while passing beneath the doctor blade.

Sometimes a variable-speed dc motor is used to drive the rubber roll and the back roll, while a *dc* series motor or torque motor is used for the take-up roll. The steamheated table is usually 4 m long by

1.5 m wide, but some machines can accommodate fabric in widths up to 4 *mes* and have provision for bending the knife to ensure that the coating thickness is not affected by knife distortion under the pressure of the dough. The speed of the machine is gradually increased, as determined by the rate of solvent evaporation. For example, for a body of dough with ≈ 40–50% solids, using heavier naphtha with a boiling range of 110–150° C on a spreading chest of 4 m, employing a 2 kg/cm^2 steam pressure, a speed of 12 m/min is satisfactory, while with lower boiling naphtha (b.p. 70–110°C), a speed of 15 m/min is practicable.

Solvent evaporation rates are greatly increased in improvised machines using jet type drying ovens, where hot air is blown onto the surface of the fabric or through slots that cover the whole surface. This allows the coating deposition per pass through the machine to be increased from the normal value of 30 g/m^2 to approximately 80 g/m^2 and at speed of 10–50 m/min. Several coats are applied, one after another, to get an impermeable layer of rubber free from pinholes or till the desired thickness-to-weight ratio is achieved. With coating compounds of certain types, it may be necessary to use a liner fabric between the spread fabric layers to prevent them from sticking together while being wound onto the take-up roll.

Other modifications of the spreading process feature two coating heads used in tandem to increase productivity, or two coating heads placed opposing each other, whereby one heating chamber is used following each coating head and the heads are used alternately. A number of idler rolls, guide rolls, dewrinklers and expander rolls are used in the coating line both at the feed end and at the windup end. Arrangements are also provided to measure by dial gauges the clearance at either end of the machine between the roll and the doctor knife, to allow determination of the coating thickness and uniformity between two edges of the coated fabric. Sometimes β-ray gauges are fitted.

The spreading technique for producing double-textile proofings is the same as for single textures except that when the desired weight of proofing has been attained, a further thin tacky coat is applied. The fabric is then transferred to the doubling machine, the essential features of which are two rollers: one covered with rubber and the other with steel, revolving toward each other. The proofed fabric rolls are mounted one on each side of the machine and the two ends are brought up and fed in between the rubber roller and the steel roller, the rubber-covered roller begin fixed to permit regulation of the pressure by means of levers and weights or by using two pneumatic cylinders, one at each end. The doubled fabric is finally wound onto a take-up roller. Frequent arrangements are provided to preheat the proofed fabrics before doubling takes place, thus ensuring coating by the tacky solution.

Spreading dough

The first and most important requirement for good spreading is that the spreading dough have the right consistency. The compounded rubber is cut off thinly into narrow strips or pieces, placed in a box containing the proper solvent and allowed to soak for a few hours. Solvents are selected on the basis of technical and economic reasons and occasionally mixed solvents are used because they exhibit stronger solvating properties and permit the preparation of more concentrated solutions than the single solvents by themselves. The immersed and soaked bulky mass is transferred to Z-blade kneaders consisting of a trough formed by two semicircular tubs, in which the Z-shaped paddles turn at a friction ratio of 1:1.5 to 2.0. The trough is double walled to permit cooling and is kept closed by a movable lid. The Z-shaped paddles disintegrate the swollen mass and are constantly wetted by solvent until total solvation has taken place. The paddles have journals at both sides for bearing support and the bearings are sealed. The drive is from a motor, gearbox and ratio pinions. A tilting device is provided for emptying the trough. In an alternative and modified system, the compounded rubber sheets (≈ 6 mm thick) are covered to granules of varying sizes (3–9 mm) in a special granulating machine and are then fed directly to the Z-type kneader to vertical rotational mixer.

In a rotational mixer, there is shearing between the stationary deflectors and the rotating kneading elements, which serves to fragment the rubber compound and ensure good mixing with the solvent. Here the solvent flow is turbulent and has a cross-over flow pattern, making for rapid solvation of the rubber compound. This mixer design gives lump-free dough in a relatively short mixing time.

Sometimes the prepared dough is transferred to a "wet" mixing mill having evenspeed rolls. The dough is allowed to run until the whole mass has an even consistency. This process is aided by tightening the rolls and bringing the adjustable blade fitted to the front roll down on the same roll, thereby allowing a bank to build up on the front roll. The procedure is repeated until the dough is free from hard lumps and then the smooth dough is ready for spreading. Additionally, the dough is sometimes strained by being forced through a sieve with the help of a hydraulic cylinder, thus eliminating not only hard lumps but also any foreign particles. To ensure consistent behavior and performance, tests for total solids content and viscosity should be carried out before the prepared dough enters the spreading machine. Other properties such as colour consistency should also be checked.

Safety and solvent recovery

From the standpoint of safety and pollution control, it is important that the explosive and partially toxic solvent vapour be recovered. The explosion limit of the naphtha–air mixture is between 30 and 300 g of naphtha per cubic meter of air. It must be assured that the solvent concentration remains under the explosion limit. All motors, switches, lights, etc., must be flameproof and all drives, idler rolls, doctor knives, etc., must be grounded. Static electricity eliminators that work by ionisation methods may be used to prevent the generation of sparks caused by friction of the coated fabric. Stringent fire prevention precautions must be observed in all the areas, including the spreading shop, in which solvents are stored or handled. Solvents should be stored in suitable underground tanks in an area well separated from the spreading shop. Adequate fire fighting equipment must be made available.

Recovery of the solvent is important for reasons of safety, economy and ecology, and two processes–condensation and absorption–are available. In the condensation method the solvent-laden air is pumped to cooled condensers, thereby liquefying the mixture. The recovery rate is less than that achieved by the absorption method and condensation is also comparatively risky because of the possibility of explosion.

In the absorption method the solvent-laden air is sucked from the hoods over the chest of the spreading machine and passed alternately into one of the two containers containing activated carbon. In the passes through the bed of activated carbon, the solvent is completely absorbed, while the air is completely stripped of its solvents. The recovery is approximately 90% and is dependent on the strength of the carbon, the temperature of absorption and the concentration of the solvent in the atmosphere. The percentage recovery, given equal conditions, will always be the same for a given solvent but will vary depending on which particular solvent is being used.

When the carbon has become charged with the solvent to such a stage that it no longer extracts the whole of the solvent, the solvent-laden air is diverted to the second absorption container, while the solvent is being recovered from the first container by charging steam. The steam evaporates the solvent absorbed in the carbon and drives it out to a condenser, to be recovered by subsequent decantation or fractionation. Activated carbon has property of absorbing extremely low concentrations of a solvent from the mixture of solvent vapour and air. For combustible solvents in air, the concentration should be below the explosive limit: thus for gasoline, a concentration of 0.5% should not be exceeded; concentrations as low as 0.0005% are, however, claimed to be within the economic range.

Vulcanisation

The vulcanisation of coated fabrics is carried out on rotational vulcanising machines in a continuous system or by a batch process in hot air vulcanisers. Oxidation may be a problem with hot air curing and precautions should be taken during compounding to protect the rubber from such attack. Before vulcanisation, single-texture proofings are treated with release agent such as french chalk (or starch or silicone oil, depending on the finish desired); this also helps to avoid jamming of the proofed fabric during storage between final coating and vulcanisation. Coated fabrics cured by rotocure give a smooth and polished surface finish on the coating side and also a good feel.

Polyurethane Coating

In the coated fabric industry, the single most important polymer to make an impact has been polyvinyl chloride (PVC). When properly formulated, stabilised and gelled, plasticised PVC provides a durable, tough, flexible and abrasion-resistant coating and thus PVC leather cloth became a major item of coated fabric production. The next polymers to make significant in roads in the coated fabrics field were the polyurethanes. These can be applied to textiles by a wide variety of methods, such as solvent-based doughes, aqueous dispersions, calendered films and hot melt thermoplastics. The rapid growth in the development of PU coatings has been attributed to their excellent film-forming characteristics coupled with their many good properties, which have led to acceptance to a wide variety of end products.

A PU polymer can be considered to consist of two entirely different components: the soft segments of the polyesters or polyethers, and the hard segments of the urethanr or urea groups. Since besides a chemical linkage between molecular chains (e.g., branching and three-dimensional arrangements) there is also physical bonding between the hard segments, the ratio of hard to soft segments is of importance in determining the properties of a given PU. Again the molecular weight of the polyester influences the properties of the PU, a lower molecular weight increasing the glass transition temperature.

In coated fabrics it is always advantageous to have T_g as low as possible because then a softer handle of the fabric can be achieved. In a PU, three general classifications of raw materials are used: polyols, diisocyanates and chain extenders. While each of these has direct bearing on the properties of the PU formed, the polyol and the diisocyanate can be considered to form the polymer backbone and the chain extender is responsible for the final molecular weight and thus for ultimate properties. Polyesters give the best combination of properties for urethanes for coated fabrics and in general, are chosen from their resistance to solvents and outdoor weathering including oxidation.

They also allow versatility in design of molecular structure through varying the chemical composition of the ester. Polyethers impart slightly better low temperature properties and hydrolytic stability. By choosing a polyester of the proper chemical composition, PU has been designed to meet the severe humid ageing test. This has been accomplished in several ways–for example, by lengthening the carbon chain between ester linkages and/or by using glycols containing secondary hydroxyls. In general, FUs based on polyesters give coatings with a more skinlike handle, while those based on polyethers tend to be more rubberlike. Both types are in use.

Polyurethane coating processes

Transfer coatings and direct coating are the most common methods of application of PU to textile. In both methods, the PU is applied from a solution.

Polyurethane transfer coating: Originally designed for the plasticised PVC-coated fabrics, transfer coating is particularly suitable for adaptation to PU coatings, which, for economic reasons, must be applied in relatively thin films (usually < 0.10 mm). The process (Fig. 5.2) involves the application of a PU solution of about 30% solids, normally referred to as the top coat, to a suitably treated release paper, usually embossed. The paper is passed through an explosion -proof oven to evaporate the solvent and leave a film on the release paper. This film becomes the top surface on the coated fabric and provides the colour, texture and wear characteristics.

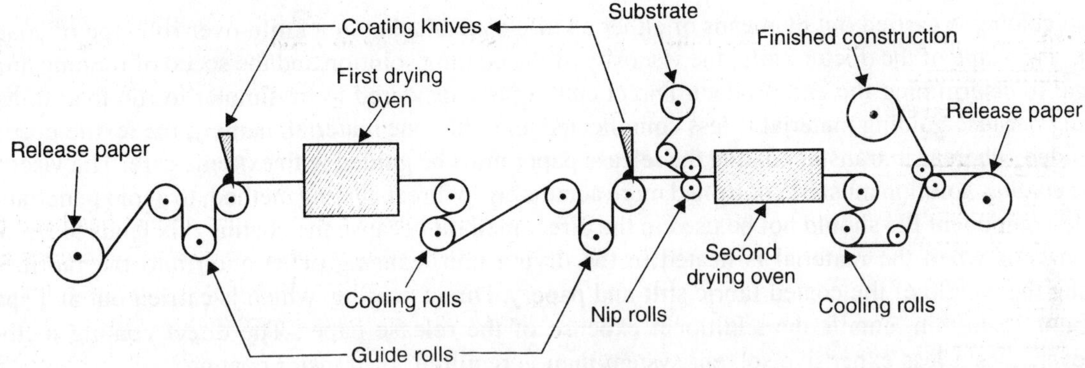

Fig. 5.2. Transfer coating.

A second PU coating, known as the adhesive or tie coat, is applied to the surface of the film and the fabric is pressed into the tacky adhesive layer. The combination of fabric, adhesive, film and paper is passed through another drying oven and cooled, whereupon the paper is stripped off. The resulting fabric is often printed and top coated. The paper can be reused about 10 times. Top coats are normally high modulus, short elongation types of PU, which result in a hard, abrasion-resistant coating. The adhesive or tie coat is responsible for bonding the substrates to the film of top coat and must exhibit a high degree of flexibility combined with maximum bond strength and softness to impart supple handle to the finished product. Generally speaking, for most transfer coated products, light-stable thermoplastic PUs are used for top coats, while light-stable aromatic and cross-linked systems are used for the adhesive coat. In making these choices, it is important to consider the interrelationship of each component (viz., top coat, adhesive coat and fabric) to the finished product, since the flex life, abrasion resistance and other properties are all dependent on their consolidation, not on a single factor.

Basically two types of PU systems are used in coating: (i) two-component systems, in which isocyanates are added to polyol systems shortly before coating, to produce cross-linked nonthermoplastic coatings after the necessary curing operations; and (ii) single-component systems, which do not require a curing operation and produce thermoplastic and weldable coatings. Single-component PU is of the thermoplastic type, having higher molecular weight than the two-component systems to ensure the development of adequate physical properties. Both types are chemically PU prepolymers, having a high degree of chain extension.

PU-coated fabrics can be produced entirely from two-component systems and such systems, offering a choice of polyol and type and amount of isocyanate, can make available a wide selection of drape properties. These coatings also have better resistance to fractional abrasion. In the transfer coating process, however, they must be left on the release paper for a period of time to allow cure to take place. This requirement entails the disadvantage that it is not possible to see the product instantly at the end of the machine. Thus there is no opportunity to correct faults, if any.

The single-component systems do not offer the same ability to select drape properties, but as soon as the solvent is removed the film has fully developed properties. Accordingly, the sheets can be separated from the release paper immediately after making. Again the solutions have virtually unlimited pot life, as opposed to the limited pot life of two-component systems.

In practice, it is common to use a single-component system for the top coat, enabling instance release and a two-component system for adhesive coat, giving a wide choice of drape properties.

Polyurethane direct coating

Direct coating is carried out by means of either a knife over-blanket or a knife-over-roll type of coating head. The shape of the doctor knife, the viscosity of the coating solution and the speed of running are all critical in determining the end product. Direct coating is considered to be simpler to run than transfer coating because guiding material is less complicated and only one material, namely, the textile needs to be guided, whereas in transfer coating the release paper must be guided with extreme care. The viscosity of the coating solution must be controlled more accurately in direct coating method, to avoid penetration. Single-component PU should not be used in the direct method because the coating mix tends to become less viscous when the material is heated in the drying unit, causing penetration into the fabric and making the handle of the coated fabric stiff and papery. Direct coating, which is carried out at a speed of about 15 m/min, entails the additional expense of the release paper. The direct coating method, however, uses a less expensive solvent system than is required for transfer coating.

Another interesting route to PU-coated fabrics involves the use of two-component non-solvent systems. Such a system consists of two liquid prepolymers: one terminated by isocyanate and the other by hydroxyl. The hydroxyl-terminated prepolymer may be pigmented and catalyzed. The two prepolymers are continuously metered by dosing pumps into a mixing dispenser and delivered to the coating head.

Polyurethane dispersion coating

A further development is represented by aqueous PU dispersions, which again are one-component systems. The solid content is about 40–55%. The rigid segments of the PU contain ionogenic groups, so that dispersion is possible without using an emulsifier. The aqueous system derives its importance from the need to pay close attention to environmental controls. One disadvantage is that the energy required to evaporate the water is considerable.

Polyurethane coagulation coating

To meet the need for coated fabrics with a water vapour permeability considerably higher than that of the compact PU-coated fabrics obtained by the usual methods described so far, microporous PU coatings are produced. The permeability of PU-coated fabrics generally lies below 1–1.5 mg/cm^2/h and can be as low as 0.20 when heavy coatings are used. Microporous PU coatings are produced generally by the coagulation method, in which the fabric is coated with a solution of suitable PU products in dimethylformamide (DMF) by dip-impregnation, followed by passing the PU-coated fabric through water as a nonsolvent, thereby coagulating the PU to form a porous structure (Fig.5.3).

Certain specific pieces of calender like machinery have been designed for producing coated fabrics. These are the Bema coater, (Fig. 5.4) and the Zimmer coater, (Fig. 5.5). These machines feed a cold PU thermoplastic compound to the melt rolls, where it is fluxed by the heat and shear of the rolls. The resulting film then presses against the fabric substrate in the laminating rolls. In some cases the fabric must be primed with an adhesive coating. The advantages of this process with respect to calender are direct conversion of granules into a deposited film, quick changes in production and the capability of operating for short runs.

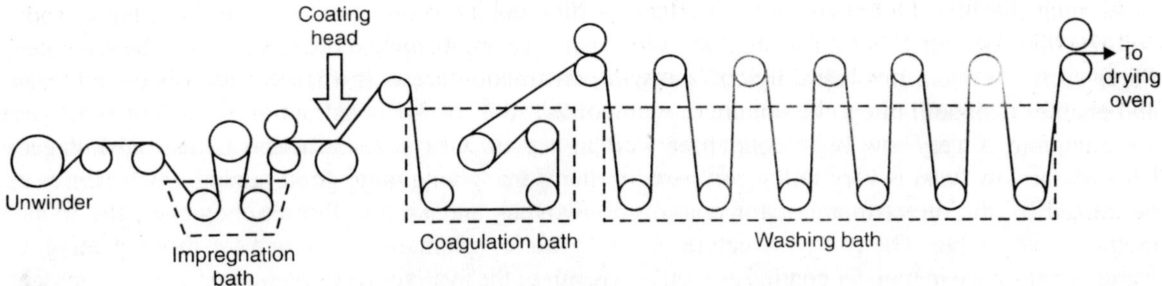

Fig. 5.3. Poromeric coagulation coating.

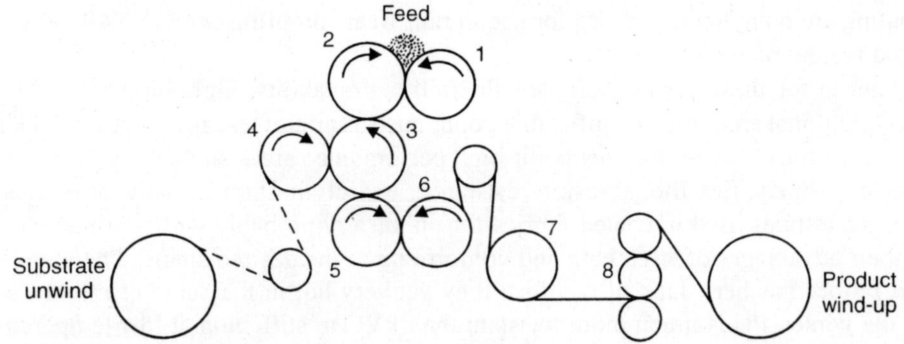

Fig. 5.4. Bema coater: 1,2,3, melt rolls; 4, substrate preheat roll; 5, take-off roll; 6, smooth finish or embossing roll; 7, cooling roll; 8, slitting roll.

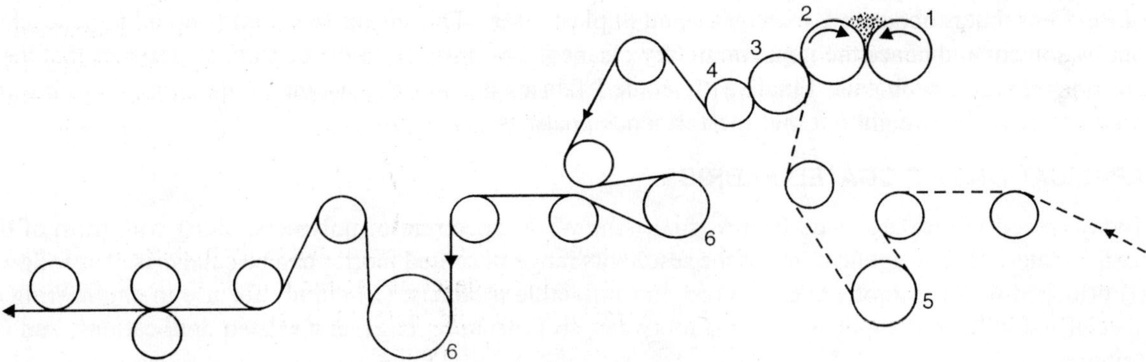

Fig. 5.5. Zimmer coater: 1, movable melt roll (usually with cross-axis); 2, fixed melt roll; 3, take-off roll (usually silicone covered); 4, smooth finish or embossing roll; 5, substrate preheat roll; 6, cooling rolls.

Polyurethane coated fabrics

The breakthrough of PU-coated fabrics is connected with the idea of choosing raised fabrics for the textile support. This construction, featuring an interlayer between the relatively thin PU film and the base fabric, results in the achievement of a soft handle, textile draping qualities and other desirable properties with lightweight and yet tough coated fabrics. The objective of using raised pile fabric is to lay the PU coating on top of the pile, with sufficient penetration to achieve good coating adhesion, but not enough to saturate the pile. Optimum results demand a directionally laid, even pile. Many types of fabric construction and many fabric finishes, ways of raising, cropping and so on, have been developed using fibres such as cotton, rayon and nylon.The twill construction presents a large area so weft for raising and enables a smooth pile to be obtained; twills of 2-1, 3-1, or 4-1 construction using low twist yarns are common. A very low twist condenser weft in a plain weave construction is also advantageous because the low twist is very bulky compared to the warp, giving quite good raising. Such fabrics can be coated by the direct method; for coating nonwovens and knitted fabrics, however, the transfer method is advisable. The woven structure of a pile fabric, originally developed for direct coating, has gained importance in transfer coating as well as a result of the availability of one-component PU systems. This applies especially for upholstery materials, where heavier coating is desired along with possibility of heat sealing. Aqueous PU dispersions are mainly applied by the transfer method. The major applications of direct coating are on nylon substrates for use in rain wear (proofing content » 40–60 g/m^2), skiwear, handbags and recreational tents.

The key needs for these applications are flexibility, drapability, high slip, light stability, and dry cleanability. Additional areas include inflatable goods such as aircraft escape chutes, life jackets, military pouches and tarpaulins. The requirements for high performance areas such as footwear and upholstery are very similar, namely, flex life, abrasion resistance, hydrolytic stability, solvent resistance and light stability. The susbstrates used in coated fabrics for upholstery probably will continue more toward the knits, with their advantages of stretching and conforming to the desired shape. The major drawback of PVC-coated fabrics has been lack of comfort: they get very hot in the summer months and very cold and stiff in the winter. PU is much more resistant than PVC to stiffening at low temperatures and thus a garment made from PU-coated fabric will still feel supple on cold days. Again PU-coated fabric is more resistant to flex cracking at low temperature, an important property for garments, footwear and upholstery. Another superiority of PU-coated fabrics over PVC-coated fabrics is the softness and flexibility of PU films that is obtainable without the aid of plasticisers. This means that their is no oil to be leached out by solvent and hence the item can be dry cleaned. The absence of plasticisers also means that there are no migration problems. Finally, PU-coated fabrics are more resistant to abrasion, enabling the production of lightweight but high performance coatings.

APPLICATIONS OF COATED FABRICS

Today coated fabrics are made from virtually the whole spectrum of polymers, along with most of the textile materials. The applications of the resultants range of coated fabrics can be categorised as follows: (i) principal usage in protective, survival and inflatable products; (ii) mainly for use in engineering or specialised industrial applications; (iii) footwear; (iv) furniture, bags, and related applications; and (v) printing.

Typical applications for protective purposes include covers for trucks, machinery and sports fields; fumigation covers for the protection of food grains and military uses such as ground sheets, revetment fabric, covers for aircraft and missiles, helicopter launching or landing pads and helicopter landing gear. Nylon, the most widely used fabric, is coated with NR, CR, ethylene-propylene-diene terpolymer (EPDM), CSPE, etc. Similar fabrics are also used for inflatable structures and temporary shelters. Another application of coated fabrics is in the production of flexible radomes for the protection of radar and telecommunication equipment. Such fabrics must have excellent resistance to ageing and weathering under wide variations in temperature and they must be transparent to radio waves at the frequencies utilised; finally, on inflation they should maintain the regularity of shape of the radar. Strong lightweight construction using polyester fabric coated with CSPE has been successfully used. A 64 diameter radome installed to track nonsynchronous communication satellites at Andover (Massachusetts) Earth Station has given more than 30 years of service life.

Applications of coated fabrics in survival equipment include crash recovery bags, which are inserted under large vehicles and inflated to life the vehicle back to upright position and aircraft escape chutes, which allow rapid passenger exit in emergencies. Some protective clothing is designed for short-term of emergency used, by many military suits may be used frequently and are often worn for long periods of time. Nuclear, biological and chemical (NBC) protection systems, including personal protective clothing and protective shelters, handle direct blasts but cannot be totally challenged in the vicinity of explosion centers. On the other hand, biological and chemical dangers, which have features in common with radioactivity and some industrial hazards, can be met with appropriate protective equipment. The delayed effects of nuclear blasts, as well as the effects of some biological and chemical agents, are just as deadly as the initial blast and heat.

Individual protective equipment cannot cope with every NBC effect, but very effective clothing has been developed for protection against biological and chemical agents. These individual items of equipment are also useful in a nuclear environment because they offer a better screen to heat flow and prevent radioactive particles from being ingested or coming into contact with the skin. Such protective clothing is made from multiple layers of coated fabric. The outermost layer is high tenacity taffeta fabric coated with aluminised heat-resistance polymer (e.g., CR/IIR) having properties of oil and rain repellency, IR reflectance, nuclear flash reflectance, and repellency to toxic aerosol droplets. The intermediate layer is nonwoven fabric for air filtering, followed by the inner layer coated, with an NR- or CR-based rubber compound containing a high level of activated carbon and finally a very flexible lining with one side coated with IIR. Some specifications for protective suits also require flame resistance and protection against infrared imaging devices. Immersion suits, which use heated water to keep the wearer warm in cold sea conditions, usually are made from a butyl–halobutyl nylon fabric laminate weighting 200–250 g/m^2. Submarine escape suits made from NR- or IIR-coated nylon fabric are of quilted construction, achieved by using molded rubber diaphragms. Attached to the suit is an inflatable stole made of rubberised nylon fabric which by the incorporation of a specially designed flutter valve, controls the rate at which air is discharged from the sole, thereby regulating the rate of ascent of the submariner.

The stole acts as a life jacket when the wearer reaches the surface, and the suit is inflated by a carbon dioxide cylinder. Antigravity suits, for providing protection against variable *g*-values for high altitude pilots, are mostly made from NR- or CR-coated nylon fabric. Inflatable life jackets for naval or merchant marine personnel are made in accordance with the SOLAS regulations, mostly from CR-coated nylon fabric (often both sides are coated, but for single-side coated fabrics, the coating, is on the inside). There are two varieties: an orally inflatable type and an automatically inflating type, which relies on a carbon dioxide cylinder. Life jackets for aircraft use are subject to weight restrictions and PU-coated nylon fabric is favoured.

Inflatable marine life rafts have resulted in significant saving of life in trawler sinkings and gradually their use has been extended to larger vessels, where they supplement the ship's life boats as obligatory survival equipment. The buoyancy tube is made from two-ply find cotton or nylon fabric coated with NR or IIR. The canopy is single ply nylon coated on the outer side, using daylight fluoroscent pigments for visibility. Life rafts for aircraft must comply with minimum weight requirements. Life rafts are rapidly inflated by carbon dioxide, and the impingement of the quickly released gas on the coated fabric quickly lowers the temperature. To disperse the solid carbon dioxide, baffles are built into the buoyancy tube to ensure that the impact is not localised. Rubbers with low T_g such as NR or IIR are favoured to prevent embrittlement. For dinghies, sport boats or larger boats, the buoyancy tube is made from nylon fabric coated with CR or CSPE. Good ageing characteristics coupled with air retention properties and abrasion resistance are the main parameters. The stress developed in the fabric as a result of inflation is low and in consequence, the breaking strength is of less importance than the tear resistance. The tear strength property is important because it is desirable to limit damage in case the fabric is penetrated by, for example, the sharp edge of a rock. In large boats, 2 × 2 matt weave may be used with 1260 dinier yarn. Since there is no significant fatigue problem with the fabric, yarns with little or no twist are used.

A common problem encountered during the manufacture of inflatable boats or any such structures is the tendency of the tube to twist on inflation, causing the structure of the vessel to be distorted. The distortion of inflatable structures can be minimised by using basic fabric with minimum bow and also by using higher modulus yarn (e.g., polyester in place of nylon, since the fabric extends less under that same inflation pressure). The higher modulus of polyester is advantageous in floor fabrics for boats that are driven by an outboard motor. When the boat is driven at a high speed, a slack floor forms a small hump, which sets up a drag. This effect is observed even when a solid floorboard is used above such floor fabric. In seaworthy boats driven by outboard motors, the floor is normally woven from polyester because of its high modulus.

This type of boat, used for river running, would not benefit from the use of stiffer fabrics. Both the buoyancy tube and the floor in such boats are made from coated nylon fabric, to be able to distort readily while running. Lightweight cotton or nylon coated with NR or CR had been employed in the production of inflatable air beds or water beds to be used as leisure items or for high altitude service. Another use is for beds designed for patients suffering from burns or paralysis, to promote comfort and to inhibit the formation of bed sores. A modification of such beds, the ripple action bed, is actually used for recovery from bed sores. Collapsible containers for the storage and transportation of various substances (oils, fuels, chemicals, water) are made in various designs and capacities from coated fabrics. This high performance applications area for coated fabrics is gaining much acceptance in various fields.

Almost any material that is inherently fluid or flowable can be stored or transported in such containers. Since these tanks "collapse" around the contained fluid, almost no air or dust ever contacts the fluid surface. Because they exclude air and water vapour, such tanks are especially popular for containing dielectric fluids. Since unlike rigid tanks, the collapsible tanks do not maintain a vapour space above the liquid level, the risk of explosive vapour formation is reduced to zero, thus protecting against the hazard of explosion. The "collapsing" behaviour prevents "skimming" on such products as inks, latex and adhesives and it also proves beneficial in maintaining deionised and deaerated process water.

Collapsible rubber containers turn open or closed trucks, trailers and vans into dual-purpose tank trucks, because they allow a firm to haul liquids in one direction and general cargo in the other. They are much lighter and, when empty, they are rolled up to only 15% of their filled size. A system of internal restraints controls surging and sloshing during movements of the truck through ups and downs or at sharp bends. Externally a fitted nylon harness surrounds the tank and provides attachment points for a series of ratchet tie-down straps. Because of the complicated dynamics involved, each transport tank is custom-engineered. Design considerations include chemical resistance, liquid density, fill/discharge requirements, volatility and temperature extremes.

Static storage tanks are the logical answer to many fluid storage and dispensing needs. These tanks are pressure loaded and are unloaded by gravity or suction. The flexible and collapsible properties of such tanks make a "breather vent" for filling or drying unnecessary. In fact flow rates in and out can be extremely high without concern for the "buckling" problems of rigid tanks. Rubber containers save on handling and storage coats. They serve as portable warehouses that can be left outside until the user needs the contents. The tanks can be laid out on flat ground with no costly site preparation. Tanks of various capacities are produced, up to 25,000 L or even bigger. Actual applications include tank farms, refining processes, oil spill clean-up automotive and marine refueling operations, hydraulic system reservoirs, aerial fuel transport, oil field operations, auxiliary fuel tankage, power plant dielectric storage, air craft fueling and purging, pollution control and various military support functions.

Combinations of various fabrics and rubbers are used in fabrications. Reinforcing fabrics include various weaves and weights of nylon, polyester, and aramid. Various rubbers such as CR, NBR, IIR, CSPE, PVC, PU, or their hybrids are used. Careful seam preparation, good overlap design, and a seaming technique involving electronic welding, solvent-cement bonding, and hot vulcanisation are critical for successful construction of such tanks. In inflatable structures or collapsible containers, air or oil can be lost through wicking action by the passage of fluid through the fabric threads themselves. With constructions using overlap joints, air or oil can enter the fabric at the internal cut edge, travel through the fabric and perhaps come out at the outside edges. One way to minimise this action is to use tapes of coated fabric or rubber to seal the edges internally. Untwisted yarns used in the manufacture of the woven fabric help in controlling the problems along with adhesive treatment applied by the dip-coating technique.

There are two different types or raincoat fabrics: that are treated with a water-repellent finish and fabrics that are treated on one surface with a continuous layer of rubber. Fabrics of the first type are comfortable to wear because the porosity of the fabric is essentially unchanged, but they provide reactively little protection under severe weather conditions. Also in the normal dry cleaning process, the finish is often removed or rendered ineffective by the residual dry cleaning soap. Fabric of the second type provide protection under all conditions, but this level of performance is achieved at the expense of comfort. The coating that is so effective for keeping the rain out is just as effective for keeping the moisture in, because it functions as an airtight barrier. An essential requirement for comfort is rapid transmission of moisture from the skin. The ability of a fabric to breathe is not identical with air permeability. All textile fibres irrespective of their chemical composition and physical structure are impermeable to air. The air permeability is due entirely to the structure and is independent of the type of fibre used. Water vapour permeability is quite different: most textile fibres are able to absorb a certain amount of moisture from the adjacent air (i.e, the damper the air, the more water vapour the fabric will absorb). When the wearer indulges in strenuous exercise, the concentration of water vapour in the air between the body and the garment will exceed that of the air outside the garment because of the perspiration emitted by the body.

The rate at which water vapour passes through a fabric depends on the nature of the fibre: with hydropholic fibres the rates is very slow, whereas with hydrophilic fibres it is relatively fast–that is, closely woven cotton fabrics have very low air permeability but, because of the hydrophilicity of cotton fibres, they have very higher permeability to water vapour, which, in turn, means that they are comfortable to wear. In the case of garments, the amount of water vapour required to permeate the coated fabric is much greater per unit area of fabric than it is in the case of tents. There is usually a residual condensation inside the garment, which prevents it from reaching the "comfort factor". In recent years coating materials have been developed that have pores so minute that they are claimed to be virtually completely impermeable to wind and water but adequately permeable to water vapour. One such fabric, called

Goretex, is actually a thin film with 3.5×10^9 microscopic pores per square centimeter. Each of these pores is 20,000 times smaller than a drop of water, yet 700 times larger than a water vapour molecule. Such coated fabrics can be divided into two classes: coatings that are essentially gas tight but are relatively permeable to water vapors, and coatings that are microporous in structure. Coatings of the first type take advantage of a molecular mechanism to combine a hydrophilic polymer with a hydrophilic elastomer; specifically, the water molecules can pass through the channel provided by the hydropholic polymer (e.g., a blend of polyvinyl acetate and polyethyl acrylate).

The hydrophilic organic or inorganic pigment may be used in place of the hydrophilic polymer. Coatings of the second type have a porous structure that allows the diffusion of air and water vapour. Liquid water cannot penetrate the pores because of the surface tension. One method of manufacturing these coatings is to apply a thin layer of rubber to the fabric containing a soluble material such as starch. The starch is then leached out to form the porous structure. In another process the rubber is loaded heavily with a porous pigment that provides interconnecting channels at the rubber-pigment interface. In both these processes, the rubber layer has a typical coated appearance with a rather stiff hand and has generally relatively poor abrasion resistance because of the high loading. However, all the materials of these types have water vapour resistances of about 140 mm of still air, which although less than similar materials coated with PVC, PU, or NR, are nevertheless very high.

Silicone/acrylic coatings on fabrics consisting partly of cellulosic fibres offer the possibility of coming closer to this "comfort factor" than is possible with PU coatings. The water vapor permeability depend on coating weight and acrylic resin contents; as the acrylic coating weight is increased, permeability will decrease. The variety of coated fabrics used in engineering and industrial applications is much larger than that used in the protective field. Almost all fibres and rubbers are used and new combinations are constantly being developed.This variety is necessary because the application in engineering design are far more critical and performance requirements are far more exacting than in protective covers. Two main factors determine the choice of a suitable coated fabric in design engineering: (i) the mechanical and physical operation conditions (pressure, flexing, temperature, etc.); and (ii) the media (air, gases, liquids, etc.) with which the coated fabric will be in contact. Ducting made of coated fabrics for conveying flue gases and for use in pollution control equipment is gaining in importance. Such materials must have resistance to gases, chemical, temperature and abrasion. Cotton, nylon, or glass fabrics coated with NR, IIR, EPDM, silicone, etc., are employed. Among the largest users of coated fabric are the gas and automotive industries. The gas industry uses cotton or nylon coated with polysulphide for applications as diaphragms in meters. In the automotive industry, coated fabrics are used for diaphragms for fuel pumps carburetors power control units and air brake controls.

Most of these are based on cotton or nylon coated with NBR. Such coated fabrics are also used for bearing seals; coatings are compounded for resistance to the widest possible range of different gases and oils and the fabric is constructed to provide the required rigidity under operating conditions, rather than the flexibility called for in diaphragms. Both footwear and upholstery applications make extensive used of a wide variety of coated fabrics based primarily on PVC or PU coatings. While knitted construction is very common for upholstery applications, the footwear industry consumes a variety of fabrics such as drapes, dosuti, sheetings, twills, drills, nets and fleeces, as well as nonwoven materials. Since footwear is subjected to complex stresses not only during use but also in the course of manufacture, the properties that matter most are braking strength, tear strength and extensibility. Sometimes extensibility of a fabric is of greater significance than the ultimate strength. Stretchability is a major consideration because the method of shoe manufacture requires the upper material to be subjected to considerable tension in order to achieve the complex geometrical shape of the last.

Blankets used for textile or paper printing constitute a highly specialised area of the coated fabric industry. The blanket for the offset printing of paper is the more demanding in precision. A blanket must possess a range of properties that are both varied and specialised. A laminated carcass of coated fabrics is the strength member and constitutes the bulk of most blankets. Cotton fabrics are still the most extensively used materials. They have, or can be processed to exhibit, the higher tensile strength, low strength and good dimensional stability necessary in all blankets. On the adverse side, cotton fabrics are subject to "packing down" under printing pressures. Their distinctive weave patterns can be impressed on a blanket face and they are also subject to the variations of any natural product. Although cotton remains the preferred fabric, rayon has been successfully combined with cotton as printer's blanket textiles. Higher gauge blankets used in applications for which dimensional stability is particularly important have a layer of glass fibre fabric in their construction, providing exceptional strength and resistance to stretch. The standard blanket carcass consists of 3 plies of high strength warp "Wigan" fabric.

This fabric is specially woven from long fibre cotton and is pretreated to remove residual stretch; its double-warp construction concentrates strength and load-bearing ability in the length direction and around the printing cylinder. The face or printing surface is made from rubber compounds based on CR, NBR, IIR, polysulphide and even PU. The coating solutions are less typical and their characteristics must be controlled carefully. The methods and controls employed in their production, in fact, are akin to those of the printing ink industry. Oil and solvent resistance are obviously of considerable importance, but these properties must be so controlled that sufficient solvent "affinity" is available to ensure consistent and uniform ink retention and transfer. Low compression set, good resilience and immediate paper release are properties of equal importance. The coating must also be easily and safely clearable by any of the various methods used by printers. The success of a blanket represents to a large extent the ability of the manufacturer to reconcile conflicting demands. The blanket manufacturer must accommodate continuous developments occurring in the printing industry as a result of changes in ink technology, faster printing speeds and alternate printing papers.

Not the most recent, but certainly the most significant development is the compressble blanket, providing improved smash resistance, which refers to larger-than-normal compressive forces, usually due to extraneous materials, passing through the printing nip. There are several ways to approach compressibility by introducing a porous layer: (i) the chemical blowing technique; (ii) leaching; (iii) microspheres; and (iv) nonwoven fabric and all these are being used at present. However, the void content varies in the different methods mentioned, namely by 10–20, 70, 40–50, and 70%, for methods 1–4, respectively. Among all these, the use of microspheres gives comparatively consistent results. Another innovation is quick-release blanket, dictated by such developments in the printing industry as faster machine speeds, increased use of multicolour printing, lighter weight substrates and increased use of coated substrates. UV blankets for radiation-curable inks are a specialty and a new addition to this field.

The subject of coated fabrics is quite a vast and wide one, with continuously expanding applications. In general, this breadth of use is requiring newer and specialised fabrics, as well as a wide variety of rubbers: the comparative newcomer, namely, the thermoplastics, along with a wide spectrum of auxiliary chemical and methods of processing and fabrication, must be developed at an accelerated rate to accommodate the needs of the coated fabric industry. Along with the search for newer applications, efforts to attain improved and exacting levels of performance will be directed to the achievement of higher productivity through the use of wider fabrics and faster machines, in ways that facilitate pollution control and result in energy savings.

CHAPTER 6

Moulding and Finishing of Rubber Components

INTRODUCTION

Moulding is the operation of shaping and vulcanising the plastic rubber compound, by means of heat and pressure, in a mould of appropriate form. Fundamentally, all processes of moulding are similar, the ways of introducing the material into the mould distinguishing one technique from another. The basic processes are compression, transfer and injection and these are illustrated diagrammatically in Fig. 6.1.

For the satisfactory large-scale production of components, it is necessary to use carefully designed and well-constructed steel moulds suitably hardened and finished, depending on the surface quality required in the product; small runs can be achieved using moulds of cheaper materials such as aluminium alloys. With compression moulding, to ensure dimensional consistency it is necessary to allow the excess material to move away from the edge of the cavity so that the 'lands' can contact with minimum thickness of rubber (flash) between them. Spew grooves and channels are provided of sufficient dimensions to accommodate this excess and also to allow the escape of air from the mould cavity. In some cases, where the shape is complex, it may be necessary to provide extra venting to allow air to escape from a 'blind' area, where it is likely to be trapped.

Injection and transfer moulds do not require any provision for excess material flowing out of the mould, simply an escape route for the volume of air in the cavities, which the parting faces provide. The mould is closed under pressure and held so while the rubber is forced into the cavities. These two latter methods give parts which carry little or no flash adhering to them at the parting lines of the mould (Fig. 6.2).

BLANK PREPARATION

The type of blank preparation is dependent on the moulding technique employed. Compression moulding requires a blank of accurate weight and shaped so that it will fill the cavity when the mould closes. It can vary from a simple geometrical shape cut from sheeted or extruded rubber to one built up of several pieces.

Transfer moulding requires the compound to be either in pad form or as pellets in sufficient quantity to fill the mould cavities plus an allowance for the transfer pot. Injection moulding machines will accept feed either as strip or pellets.

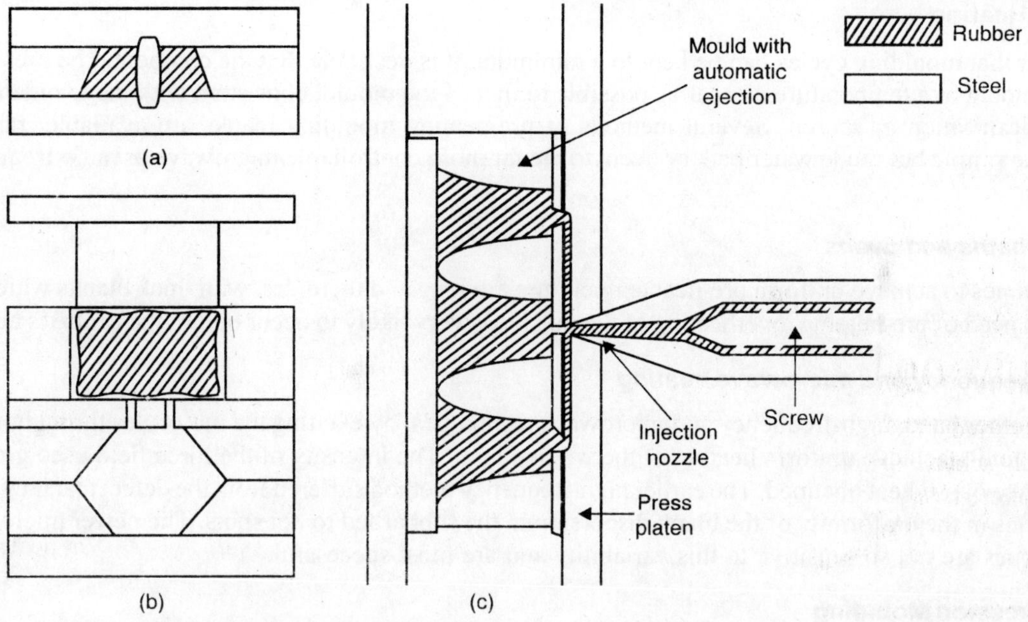

Fig. 6.1. Moulding methods: (a) compression (b) transfer and (c) injection.

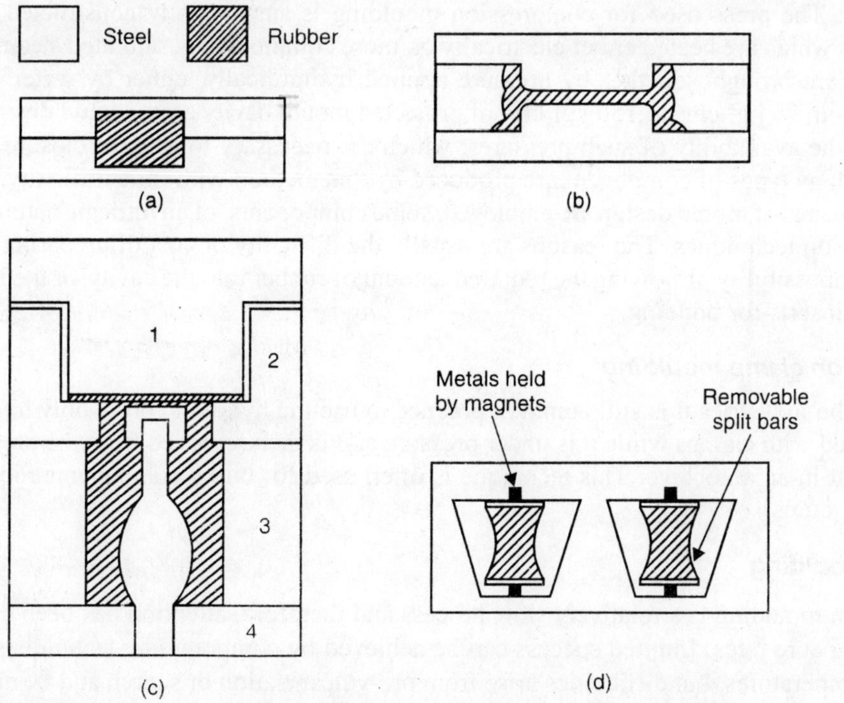

Fig. 6.2. Types of mould design; (a) simple two-piece compression; (b) three-piece compression; (c) four-piece transfer; and (d) split-bar multi-cavity compression.

Blank Heating

In order that moulding cycles can be kept to a minimum, it is desirable that the compound be presented to the mould at a temperature as near as possible to that of the mould, consistent with the avoidance of pre–vulcanisation or scorch. Several methods of pre-heating moulding blanks are available, ranging from the simple but crude water bath or oven, to the far more controllable microwave or radio frequency heater.

Water baths and ovens

Dwell times to achieve uniform pre-heat temperature are long and therefore, with thick blanks which are most in need of pre-heating, overheating of the outside is very likely to occur before the centre is heated.

High-frequency and microwave heating

On the other hand, high-frequency and microwave techniques, by exciting the molecules throughout the mass, rapidly achieve uniform heating of the whole blank. The intensity of the force field used governs the degree of pre-heat obtained. The earlier high-frequency method suffered from the defect that inevitable variations in the uniformity of the blank dispersion in the rubber led to hot spots. The newer microwave techniques are not so sensitive to this variability and are most successful.

Compression Moulding

Compression moulding, the oldest and still most universally used technique, is in many respects and for many products the cheapest process because of its suitability for short runs and because of the low mould costs. The press used for compression moulding is substantially constructed and has two or more platens which are heated either electrically or, more commonly, by saturated steam under pressure. The platens are brought together by pressure applied hydraulically, either by water or oil, to give a loading of from 75 kgf/cm^2 to 150 kgf/cm^2 of projected mould cavity area. Mould designs are generally based upon the availability of such pressures, which are necessary to achieve closure with acceptably thin flash. Many types of component are produced by this method with little difficulty, but, unless very extreme measures of mould design are employed, some components, of an intricate nature, defy moulding by compression techniques. The reasons are usually the difficulty or cost of preparing a suitable blank shape, the impossibility of flowing the required amount of rubber into the cavity, or the inability to retain loose metal inserts for bonding.

Compression clamp moulding

In some of the industries it is still common practice to use the hydraulic press only to close the mould, which is fitted with clamps while it is under pressure and is then removed from the press; vulcanisation is carried out in an autoclave. This technique is often used for bulky components requiring long low-temperature cures.

Transfer Moulding

Compression moulding is a relatively slow process and therefore, attention has been directed to means of increasing cure rates. Limited success can be achieved by compounding techniques and/or increase in curing temperatures, but difficulties arise from pre-vulcanisation or scorch and from the low thermal conductivity of rubber creating uneven cure, the outside being degraded while the centre remains incompletely cured. Two methods of introducing a heated mix to the cavity are common. The first of these, the use of outside pre-heaters, has been discussed above.

Within-mould transfer

The other method is to use a transfer mould where the blank is pre-heated in the one part of the mould as it passes into the cavity Fig. 6.1(b). This method is frequently adopted for rubber-to-metal bonded, large, and complex components. The mould consists of four or five parts. The mould cavity is made in a conventional manner, but, mounted above and connected to it by a series of holes, is another cavity. The prepared rubber is placed in this upper chamber and the integral ram is used to transfer the rubber into the mould cavities under pressure. The energy required to force the rubber into the cavities increases somewhat the temperature of the rubber mass as it flows through the connecting holes. As a result, curing times are significantly reduced; other advantages and disadvantages are stated in Table 6.1.

Table 6.1. Advantages and disadvantages of transfer moulding.

Advantages	*Disadvantages*
Single large blank, giving simplified preparation	High mould costs.
Shorter cure times	Extra weight in mould is cumbersome
Better rubber-to-metal bonds	Longer change over time between cures
Lower level of rejects made	Usually fewer cavities than compression moulding
Easy location of metals	

Transfer press moulding

In commercial use is a wide variety of machines which transfer fixed volumes of compound to clamped moulds, by the operation of a ram in a cylinder fed with pre-heated slugs. These machines are capable of high-speed semi-automatic cycling and are a great advance over the use of loose transfer moulds in conventional presses. The components which are produced are to a large degree flash-free, it only being necessary, in the majority of cases, to remove the injection feed and runner system.

Flashless moulding

Several patented techniques are available which give a completely flash-free component requiring no further handling except for a statistical quality inspection.

The systems used are very costly with respect to moulds and presses, but once in operation these machines are capable of large volume output and the blank is a simple small extruded pad or disc.

Injection Moulding

The main feature of injection moulding, in which it differs from compression moulding, is in the presentation of rubber to the mould at or near moulding temperature. A large part of the cure time necessary to allow the whole of the rubber mass in the cavity to reach the vulcanising temperature by conduction and convection, as in the case of compression moulding, is eliminated.

The types of injection machinery available differ basically in the method of heating and pre-plasticisation of the compounded rubber. The main types of machine are ram or reciprocating screw.

The screw machine generates its own pre-plasticising heat by work on the rubber, which is fed into it in strip form, between the screw and the barrel; this heat is governed by a temperature-controlled fluid medium (water, glycol, or oil) which circulates around the barrel (and in larger machines, also through the centre of the screw). Progression of the rubber up the screw both heats and pre-plasticises it. The rubber, collected in front of the screw, pushes the screw back until a trip switch is operated and the preplasticisation phase is completed. Final heating of the stock takes place during passage through a small orifice (die) into the runner system of the mould. Machines are available which combine both basic concepts. Coupled to any type of injection moulding machine can be single or multi-stage press units to give intermittent or continuous moulding. Compounding of the rubbers for injection moulding requires special attention.

Moulding Shrinkage

Shrinkage is the term used to describe the difference in the dimensions of the mould and the article produced from it, when both are measured at ambient temperature. Shrinkage factors must be determined for individual compounds if strict dimensional accuracy is required and can be obtained by curing a standard test specimen at production temperatures. In general, linear shrinkage figures fall within the range 1.5–3.0%, depending on polymer type and filler loading. Shrinkage takes place because the coefficient of thermal expansion (volume) of steel is 0.3×10^{-4} per C, while that of rubber is from 4×10^{-4} per C to 7×10^{-4} per C, in addition anisotropy may be present in the material and will often influence the dimensions of the moulding. An article moulded from a rubber compound always ends up smaller than the mould in which it was vulcanised. The difference between the dimensions of the finished article and the mould, measured at room temperature and expressed in per cent is called the degree of shrinkage. Since the dimensions of the moulded articles have to be very often within very close tolerances, it is important to know the degree of shrinkage so that it can be considered during the design of the mould. On the one hand the shrinkage of the rubber during cooling can be welcomed. Had the rubber and the mould the same coefficient of expansion, removal of the part from the mould be problematic. On the other hand, this phenomenon complicates the manufacture of articles of correct dimensions.

Basic principle

The degree of shrinkage is primarily determined by the difference of the coefficient of expansion between the vulcanisate and the mould material as well as by the vulcanisation temperature. The rubber part which fills the mould completely during vulcanisation contracts during cooling to room temperature more then the mould because it has a considerably higher coefficient of expansion than the latter. Thus the degree of shrinkage increases with increasing difference between the coefficients of expansion of the rubber compound and the mould material as well as between vulcanisation and room temperature. The difficulties during the calculation of the degree of shrinkage arise mostly from the fact that the coefficient of expansion of vulcanisates changes with the compound composition. In spite of this, degrees of shrinkage can be determined approximately without preparing sample mouldings.

Determination of the degree of shrinkage

Aside from the influence of the vulcanisation temperature the *elastomer content* of the compound is of importance for the degree of expansion and thus the degree of shrinkage. The coefficient of expansion is the higher, the higher the elastomer content of the compound. The coefficient of expansion of the filler are of the same order of magnitude as those of the mould materials because of which essentially only the elastomer content of the compound is considered during the calculation.

Acetone soluble chemicals and additives like accelerators, antioxidants, plasticiser, resins, waxes and sulphur, also factices and reclaim have, in contrast, a coefficient of expansion similar to that of the rubber so that these substances are considered a part of the elastomer during the calculation of the degree of shrinkage.

The *vulcanisation time* and the mastication time during the preparation of the compound are practically without influence on the degree of shrinkage.

Some of the *co-efficients of expansion* important for the calculation of the degree of shrinkage are summarised in Table 6.2.

Table 6.2. Liner expansion co-efficients of some elastomer types, fillers and mould materials.

Mould material	*Linear expansion co-efficient*
Elastomers	
NR	216×10^{-6}
SBR	216×10^{-6}
NBR	196×10^{-6}
CR	200×10^{-6}
IIR	194×10^{-6}
Fillers	*Order of Magnitude: 5 to 10 × 10–6*
Mould materials	
Steel	11×10^{-6}
Light metals	22×10^{-6}

The *calculation of the degree of shrinkage* from the composition of the compound can lead, according to the statements above, only to approximate values the accuracy of which is, however, sufficiently accurate. The following equation can be used for this purpose.

S(degree of shrinkage in %) = $\Delta T \,.\, \Delta A \,.\, K \,.\, \Delta F \,.\, \Delta H$

where

ΔT = the difference between vulcanisation and room temperature

ΔA = the difference between the co-efficient of expansion of the rubber and that of the mould material.

K = vol.% rubber + acetone soluble chemicals

ΔF = difference between the co-efficients of the fillers and that of the mould material

ΔH = difference between the co-efficient of expansion of the acetone soluble chemicals and that of the elastomer.

If one neglects the insignificant factors F and H, one arrives at the following approximate equation:

$S = \Delta T \,.\, \Delta A \,.\, K$

This approximate equation is valid under the assumption that the filler has the one co-efficient of expansion as the mould material and that the acetone soluble components have the same co-efficient of expansion as the rubber and thus do not contribute to the degree of shrinkage.

It Follows from this equation that the degree of shrinkage is the larger for a constant compound composition, the higher one chooses the vulcanisation temperature. One also obtains another degree of shrinkage when one switches from one mould material to another one.

The temperature independent contraction which is a result of the chemical reaction between sulphur and the elastomer is of little consequence for the degrees for shrinkage. Experiments have shown that it accounts for only 0.1% for sulphur dosage which are common for soft vulcanisates but is larger for hard rubber compounds since the contraction is further reduced in filled compounds because of the reduced elastomer content, it is within experimental error and can generally be neglected.

Mould Lubrication

It is usually necessary to provide lubrication at the surface of the mould to facilitate easy stripping of the component and to allow unrestricted flow of the rubber stock. The substances used are termed 'mould release agents' or lubricants, and comprise surface-active material such as detergents, soaps, wetting agents, silicone emulsions, aqueous dispersions of talc, mica and fatty acids, applied by spray or brush. Alternative 'dry' lubricant types are based on PTFE or polyethylene, usually carried in solution and applied from an aerosol. The amount of lubricant required is dependent on several factors: mould complexity, type of lubricant, and polymer type or grade. The main disadvantages of the use of release agents is the build-up of a solid deposit on the mould surface, which results from the constituents of the mould release, stock residues and decomposition products. This deposit must be removed frequently otherwise pitting of the surface will result.

Release agent application should always be minimal, otherwise a number of moulding faults, due to poor knitting together of the stock, will result. In addition, the presence of silicones and similar materials can be extremely detrimental to rubber-to-metal bonds.

Materials for Special Effects

Mould release agents

Problem

Press vulcanised articles tend to adhere and stick more or less tenatiously to the moulds. When they are taken out of the mould, this can lead to damage of the vulcanised article and of the vulcanisation mould. Suitable mould release agents are important for an efficient production of moulded products. Through their use scrap is reduced, the appearance of the products improved, mould cleaning cost reduced and mould life increased.

Requirements for mould release agents vary with the production process, type and shape of the article as well as its post treatment. An ideal release agent that would be optimal for all needs has not been developed yet.

Soap solutions

For many years, simple soap solutions were used - and in some cases they are used even today. Some of the newer synthetic soap substitutes have excellent release properties. These products have the disadvantage that they slowly decompose at vulcanisation temperature causing impurities on the mould surface, that make a clean moulding process impossible. For that reason a frequently cleaning of the mould is necessary that is not only time consuming, expensive and very difficult with complex mould, but frequently leads to premature scrapping of the some times very expensive moulds.

Silicone mould release agents

By the use of suitable silicone mould release agents the disadvantages of the soap solutions can be avoided. Release agents that are based on silicone do not decompose because of their high temperature stability and the vulcanised parts have a high gloss and handle smoothly. The silicones, in contrast to the purely organic release agents, make a subsequent coating, glueing or welding of the surfaces more difficult. Furthermore, the release capacity of silicone release agents is somewhat weaker than that of organic agents.

Combinations of silicones and soaps

An especially favourable compromise can be achieved by combining silicone oil emulsions with good organic release agents. Fluorinated polymers. Recently, fluorinated polymers have been available in spray cans, that are sprayed on the mould and then heat fused. Because of the evaporating solvents (toxicology) and difficulties in repairing damaged coatings, these release agents are not much used.

Powders and release agents to avoid adhesion between unvulcanised compounds

To a large extent talcum, Zn-stearate and other powders are used to avoid adhesion between rubber compound intermediates, for example extruded articles. They have the known disadvantage of dust irritation. One has gone over largely to liquid release agents. A certain success has been obtained with soap solutions, especially with hard, heavily loaded compounds. With medium hard and especially soft compounds the soaps penetrate into the surface and the surface is depleated of the coating. By adding high molecular weight materials that are insoluble, for example cellulose derivatives to such a soap solution, a separating film remains on the surface that prevents sticking together during storage or vulcanisation. Zinc stearate as separation agent (powder agent) has the remarkable property that it prevents undesirable sticking of unvulcanised parts at room temperature, but does not avoid welding during vulcanisation. The release agents mentioned above or solutions of silicone oils, 0.5–1 wt% in benzene, carbon tetrachloride or ethylacetate are often used for example for easy removal of dipped articles from the forms. The vulcanisates obtain a uniform glass and improved appearance.

Single use release agents

This group consists predominantly of non-reactive silicones, soap solutions/suspensions and oils. If we include tyres as moulded parts, the products may even include solids, such as clay, mica, etc. The materials are commercially available or even still 'in-house' produced, as some small and medium size tyre manufacturer still do. The release function of these products can be explained by cohesive failure in the release film when the rubber part is being removed from the mould.

This implies, that a certain amount of release agent is left on the mould surface, which causes most of the failures and problems.

Permanent release agents

During the Seventies, several attempts were made, to coat mould surfaces with a permanent release film, supposed to last the lifetime of the mould.

One material used was PTFE (polytetrafluorethylene). It turned out to be not very practicable. Admittedly, the release effect of PTFE is very good, however, a certain degree of slip needs to be present to aid compound flow in the mould. PTFE coatings, in the opposite, created a rather rough surface, which leads to a very fast wear-of, due to the abrasive nature of most filler containing compounds. Consequently, these systems never became commercially viable on a wider scale.

Semi-permanent release systems

Semi-permanent release agents are designed to achieve multiple releases with one surface treatment. They form a thin, but resistant layer on the mould, that will not be transferred to the surface of the rubber parts.

This becomes possible through the development of cross-linkable polymer systems that can be used in a liquid carrier. The carrier can be based on solvents or water.

After application of a thin layer onto the hot mould surface, a film forms through evaporation of the carrier and by cross-linking at the same time. The adhesion of the film to the metallic surface can by explained by the effect of 'Van der Waals forces'.

Due to the characteristics of the film, the single release effect might be slightly reduced, compared to single use agents (like silicone), but this minor drawback is largely compensated by the numerous advantages of this system: (i) virtually no transfer to the part surface; (ii) prevention of residue built-up on the mould; (iii) provides true multiple releases; and (iv) heat stability.

Semi permanents, in their early stages, were based on solvent carrier that provided excellent film formation and uniformity and could even be applied at lower mould temperatures.

The main problems of these products were the environmental impact and the health and safety in the workplace. Fortunately, most countries have outlawed CFC containing solvents, but highly flammable materials, such as aliphatic hydrocarbons, replaced these; which shifted the hazardous part from one side to the other.

Regulation and increased conscience finally lead to not only the development of water-based materials, but also to their rapidly increasing use.

Today a range of environmentally safe products is available to the industry that is technically equivalent, if not better, than the solvent-based materials.

In this section, we do not wish to elaborate on the development and formulating part of water based semi-permanents, since this is in most cases based on similar building blocks, like surfactants and the polymers itself. The know-how differences are in achieving the right balance between surfactant and types of polymers, which ultimately influences the performance of the product. Usually a catalyst is added to increase the cross-linking speed.

Application of semi-permanents

The most important step for the application of any kind of semi-permanents is the proper cleaning of the mould. This is absolutely essential for a proper adhesion of the coating to the mould surface.

Next important is the application temperature !

With water based semi-permanents, the temperature is somewhat restricted to a minimum of preferably above 90°C in order to evaporate the water as fast as possible.

This is important for two main reasons, to start the cross-linking of the film and to prevent corrosion of the mould surface.

Here, the basic rule applies: The higher the temperature, the shorter the cross-linking time. Since moulding of rubber takes place at or above 140°C–150°C , the cross-linking times for the film are relatively short, 10 minutes or even less, depending on the catalyst being used in the product.

This pre-requirement makes the use of water-based products for low temperature moulding (e.g. PU), difficult if not impossible.

The coating can be applied in different ways, e.g. brushing, wiping or spraying. Spraying is, however, the most commonly used method, which also ensures the most even and uniform film formation.

Semi-permanents should be applied as sparingly as possible; the thinner the film, the more stable and even it will be. After the cross-linking of the first coating, a second layer may be applied. This is mainly to help leveling out irregularities in the mould surfaces, as there are tiny scratches, cavities, etc. It is self understood, that there are limits to what a release agent can do! It does certainly not repair damaged moulds. After these steps, the mould can be put into service. If there is no previous experience with semi-permanents, it is advisable to closely monitor the curing cycles until the release effect diminishes. Now the point of re-application is reached.

This can be done '*in situ*', without cleaning or removing the mould, by simply re-applying one thin film. Semi-permanents prevent material built-up in the mould for extended periods (as compared to non-permanent release agents). These was shown in studies, using methods of surface analysis (e.g. X-ray spectroscopy) to determine the type and amount of built-up after multiple curing cycles. With semi-permanents, it was clearly shown, that whatever volatiles come out of the compound ,it will hardly be transferred to the mould.

The re-application procedure can be repeated to the point where the material residue in the mould reaches a level that prevents a proper film adhesion. The mould has then to be removed for a standard cleaning procedure.

Curing Cycles Achieveable without Cleaning any Mould

It depends on several factors which affect the mould lift between cleaning cycles:

1. Polymer base of the compound: It is well known, that halogenated polymers tend to stick much more to metallic surfaces then standard polymers. Good (bad) examples are FKM, ECO, ACM, CR,.... The numbers of curing cycles to cleaning are considerably lower with these compounds than, for example, for SBR,NR, BR, NBR,....
2. Type of filler: More abrasive fillers, like most mineral filler types, are promoting faster wear on the coating than softer fillers like carbon blacks.
3. The curing system: Accelerator systems or the complexes formed by them can be volatile and might attack the release film. Peroxide systems are especially infamous for such effects.

From experience, moulding cycles can be as many as several hundreds, down to as little as fifty, depending on the above-mentioned factors.

Thus semi-permanent release agents offer a wide range of advantages over non-permanents such as:

1. Reduction of down time, since application after each cycle is not necessary.
2. Prevention of mould fouling.
3. Much extended intervals between mould cleanings.
4. Extreme reduction of flow faults caused by entrapped release agent during the mould filling process (poor knitting).
5. No adverse effect on the surface of the finished part with regards to painting, printing or bonding!

Consequently, this will all result in a reduced scrap rate, which can be counted in money. Since there are differences in polymers, fillers, mould shapes, a range of different products have to be made available with different properties.

A typical range of water-based semi-permanents consists of a minimum of four standard grades, mostly variations in concentration and types of polymers used.

The above-mentioned technical advantages, combined with the environmental aspect, make water-based semi-permanents a must for the responsible moulding operator.

Mould Cleaning

As mentioned above, frequent mould cleaning is essential in long-run production to ensure good part appearance. The cleaning frequency is dependent on the application of the component being manufactured.

The methods used are manual scouring with steel wool and abrasives, wire brushing, grit blasting, vapour blasting and electrolytic oxidation in salt bath. The first two methods are suitable for simple moulds, while the remainder are applicable to any size or configuration of mould. The types of material used for grit blasting are ground vegetable matter, such as nut shells, of 10–30 mesh; the particles after use are cleaned and recycled. Vapour blasting consists of high-pressure application of a detergent solution at high velocity to the mould surface. The abrasive and detergent action effectively removes the deposit.

Salt baths are used in large-scale plant and are extremely efficient but are hazardous in use, requiring the usual safety precautions associated with caustic chemicals.

FINISHING OR RUBBER COMPONENTS

Once a vulcanised product of any process is obtained, a finishing operation is usually necessary to make the part acceptable for consumer use. It is not possible to discuss the finishing of each individual product and therefore, the subject will be discussed under the headings of the techniques employed; namely, flash removal, punching, grinding, painting and surface treatments.

Flash and Spew Removal

Hand trimming

It is still necessary to hand trim many components owing to their awkward or complex configuration or their small quantity production. The usual tools are scissors, both straight and curved and knives of various types suited to the product being trimmed.

Machine trimming (Roller trim)

Various machines are available which trim components, such as discs, lids, or mats, by the action of two rollers or wheels running under power, the rubber being sheared between the two wheels and effectively removed.

Buffing

This process uses a buffing wheel, which may be hard felt, flannel, cotton cloth, or wire brush, according to the type of work. The textile wheels are used to remove flash and also to polish hard rubbers, e.g., ebonites. The wire buff is used to prepare areas of rubber which have to be bonded to other components, to remove rubber from the metal parts of bonded components and in some instances to render them suitable for application of paint to prevent metal corrosion.

Low-temperature tumbling

Low-temperature tumbling is an efficient and widely used means of flash removal in large-scale production. There are three basic types of machine available: the dry ice tumbler the liquid nitrogen tumbler, and the 'wheelabrator'. The dry ice (carbon dioxide) and the liquid nitrogen tumblers are similar in function but differ in the refrigerant used. The action consists of freezing the flash, by means of the refrigerant, until it becomes brittle; steel shot, pebbles, or other media are tumbled in a revolving drum with the frozen parts and break off the flash. The wheelabrator is a modern development in which the flash is removed by shot blast action on frozen parts. Separation of good components from the shot and detritus is usually carried out on vibrating sieves, which may be incorporated in the deflashing machine.

Punching

Independent of quantity, there are some products which require hand punching to clear the flash from an orifice inaccessible to a machine tool. It is now common to find punching being done semi-automatically in reciprocating presses, which only require the operator to locate the sheet of components into the appropriate jig. Punch and die sets are used with this type of machine, which is particularly suitable for thin diaphragms, which require extreme accuracy of punching.

Considerable use is still made of clicker and fly presses with simple cutters, although the latter are now disappearing as more automated equipment is used. The main advantage of the machine punches are the greater throughput possible and the ability to use multi-gang punches.

Grinding

For some products it is necessary to obtain a superior finish to that achieved by extrusion, moulding, or wrap curing. The surface may be marred by the presence of flow marks, cloth impression, sprue lines, or injection pipes, for example. It is then necessary to produce the parts oversize and grind to the correct dimensions. Such products are typewriter platens, printing rollers and tubing from which, for example, parted off washers are to be made. It is necessary to use a lathe grinder, often of the centreless type, in order that the product can be machined successfully. Rubber deforms under the action of the grinding wheel and careful selection of grinding wheels and speeds is necessary for fine finish work.

Shot Blasting

For some applications, it is desirable that sheets of rubber have a clean rough surface. This can be achieved by shot blasting the surface of the rubber with fine aluminium oxide particles of controlled size.

Painting and Lacquering

It is sometimes necessary to treat the exposed unprotected metal surfaces of rubber-to-melt bonderd components. There are many types of protective finishes in use for this purpose, subject to consumers' requirements. Whereas it is not now general practice to apply paints or lacquers to rubber products, there are exceptions, notably such products as footwear and toys, which depend for their market appeal on a high glossy finish. The lacquer, based on rubber, shellac, chlorosulphonated polyethylene, or polyurethane, is applied either prior to the final cure stage or as a final operation.

Polyurethane lacquer

Recently, lacquers made from polyesters and polyisocyanates have frequently been used. With stabilised isocyanates a storage stable solution is formed, that is applied to rubber compounds. Only at the vulcanisation temperature free isocyanate groups are formed that react with the polyester to give an elastic, tightly adhering coating. Solutions with limited potlife made from non-stabilised polyisocyanates and polyesters can be applied to the rubber compound or to the vulcanisates as air drying lacquers.

Influence of composition on lacquer coatings

Rubber compound and lacquer have to be tuned to each other. The rubber compound must not contain paraffine, mineral oil or larger amounts of plasticisers, that can concentrate at the surface and thereby decrease adhesion.

The lacquer must not have components that interfere with the vulcanisation. If the lacquer is applied to the vulcanisate and the vulcanisation is taking place in a mould, mould coating, silicone oils, soaps or similar material cannot be used. On the other hand a powder coating of zinc stearate interferes little with the drying or appearance of the lacquer coating.

One has to be cautious with the use of antioxidants. Discolouring compounds dissolve in the solvents of the lacquer and diffuse in the film and discolour on light exposure. Suitable lacquers do not only improve the looks of the rubber product, but protect it also from weathering. Lacquers can be used with compounds for solid rubber goods as well as latex products.

Drying Oils

Suitable varnishes are for example drying oils, that were blown or sulphur treated. Sometimes they contain alkyl- and other synthetic resins. As base, polymerised vinyl compounds can be used, like polyacrylates, polyvinylacetate as well as plasticised ureaformaldehyde resins.

Depending on its property, the lacquer is applied to the vulcanisate or to the rubber compound and the coating is formed by air drying or during vulcanisation. The latter method has the advantage that a dry (non-sticky) film is formed that allows immediate packaging of the rubber product.

Chemical Surface Treatment

The majority of rubber compounds show a high co-efficient of friction and it is sometimes necessary to lower this for certain applications. This is achieved by the halogenation of the surface of the vulcanisate and usually follows upon the final inspection of the part. Both chlorination and bromination are in general use and the process is carried out by immersing the moulded article in a solution of chlorine or bromine in water for a few minutes. The parts are subsequently rinsed thoroughly to remove all traces of the halogen. Other methods of achieving a low co-efficient of friction are the use of a thin coating of polyurethane or polytetrafluorethylene, applied from solvent or aqueous dispersion or as a film.

COMPUTER-AIDED MOULD DESIGN

Modern rubber moulding relies quite heavily on injection moulding, and the majority of dies in use are multicavity types. There are also applications for multicavity transfer moulds. In both cases, while the design is within the scope of an experienced mould maker, the process rarely produces an ideal mould immediately. Moulds must be altered three, four, or even more times before they can be used satisfactorily, and much expensive machine time is necessary to establish the ideal moulding conditions.

It is now possible to reduce considerably the costs involved both in the design of moulds and in the establishment of ideal moulding conditions by using modern computer-aided design (CAD) developed specifically for injection and transfer moulding tools.

Simple Moulds

The simplest multicavity moulds to design are completely symmetrical and, additionally, have only one cavity for each rubber (Fig. 6.3). Provided the runners are of equal length, this configuration ensures that the rubber flows into each cavity at the same rate, and the only design decisions involve nozzle, runner, and gate sizes. Obviously, the aim is to keep runners as small as possible to avoid wasting rubber and also to make the gate small so that the final trimming of the part is easy. However, if the runners are too small, so much pressure is needed to fill the mould that one risks scorch damage and exceeding the press cavity. If the gates are too small, there will be similar pressure and scorch problems.

Complications arise if the part to be moulded needs more than one gate. The precise position of the gates is a matter for experience and, frequently, experimentation.

The cost increases each time a mould is altered and, once the mould itself has been completed satisfactorily, the moulding conditions must be established. For an injection moulding machine, the relevant variables are: (i) barrel temperature; (ii) dwell time of the rubber in the barrel; (iii) injection time; (iv) mould temperature; and (v) dwell time in the mould.

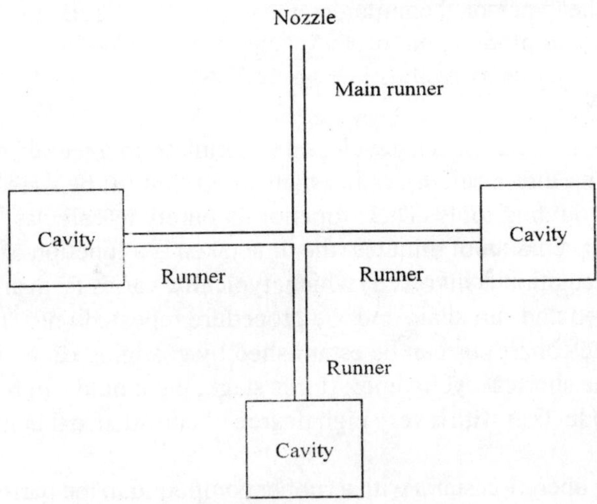

Fig. 6.3. Simple multicavity mould.

A competent operator can make a reasonable guess at all of these variables and alterations during production can be made to bring them closer to the optimum. However, with so many variables and with the understandable reluctance of production staff to interfere with something that appears to be running adequately, optimum conditions are rarely, if ever, achieved.

More Intricate Moulds

More intricate moulds—for instance, those with main runners feeding cavities from side runners—complicate mould design considerably. Moulds in which some runners feed different numbers of cavities to other runners add even further complications.

The main impediment here, compared to a totally symmetrical design, is in balancing the mould. In a balanced mould, all the cavities have the same pressure when they are completely filled. It will readily be seen that if this is not the case, there is a risk of increased flash in the higher pressure cavities and, as the pressure difference increases, there may be gross flash on some parts and short fill on other parts.

There is added difficulty in cavities with more than one gate or in cavities where the rubber flows in more than one path of weld lines. Incorrect gate positioning can cause the rubber to flow in a longer path than necessary, which can lead to premature scorch and knitting problems.

Most designers, having had experience with complicated tools that defy almost every attempt to perfect them, realise that the root of the problem is that every alteration must be tried on an actual moulding machine before it can be accepted. This can be a long, frustrating, and very costly process.

Rubber compounds are complex materials and each one differs from numerous others. From the point of view of moulding, the important properties are: (i) viscosity; (ii) cure rate; (iii) density; (iv) specific heat; and (v) heat conductivity.

Viscosity varies with temperature and also depends on the cure rate because, of course, viscosity is markedly affected by premature cure (or scorch).

It has always been theoretically possible to calculate the behaviour of rubber compounds in an injection moulding tool but, until the introduction of modern computers, so much time would have been needed for just one mould (almost a lifetime's work) that this was not done. Nowadays, with the availability of relatively cheap personal computers, the necessary calculations can be made in a matter of minutes.

Computer Programmes

Modern computer programmes have been developed to simulate an injection moulding tool and to make the design of multicavity moulds a science rather than an art. It must be stressed that these programmes do not design injection moulding tools. Their function is purely to calculate how an existing tool will behave. This they can do in a matter of minutes (the time taken is a function of the complexity of the tool and thus of how much calculation is involved) which, typically, varies from around 5 to 15 minutes. The design can then be changed and run again and the procedure repeated until it is satisfactory. Following this, the optimum machine conditions can be established by changing each variable until a satisfactory moulding is possible in the shortest cycle time. At this stage, the mould can be manufactured and, when it is available, put into production with a very high degree of confidence that it will run perfectly the first time.

An injection moulding operation starts with a rubber compound in the barrel of an injection moulding machine. Here, it is subjected to a particular temperature (perhaps 80°C) for a given length of time, which first determines the viscosity in the barrel and second begins (no matter how slowly) the commencement of cure.

The rubber is then forced through the nozzle into the main runner which, being part of the mould, is at the mould temperature (perhaps 180°C). However, the nozzle usually constricts the flow, and thus there is a degree of adiabatic heating, with increases the temperature still further.

Clearly, the dimensions of each runner affect the flow and thus the pressure required to move rubber through them at a given rate. Every change in dimensions alters this relationship, and finally the gate with it's narrow orifice adds considerably (and if designed properly, beneficially) to the adiabatic heat buildup, thus reducing the cure time in the cavity.

Experiments have shown that even though the mould temperature is, for example, 180°C, the temperature of the rubber in the cavity can be higher by 100°C or more. Uncontrolled, this temperature rise can lead to scorch before the cavity is filled, but controlled, it can help to reduce the cure time to a minimum.

The method by which a programme such as Fillcalc IV is used has two parts. First of all the rubber must be 'characterised'; that is the following essential physical and rheological properties must be measured: (i) flow properties; (ii) cure properties; (iii) thermal properties; and (iv) density.

The flow properties, particularly the viscosity, are the most important in the foregoing list because they control the pressure required for a given filling rate. Cure properties determine scorch and final cure. Thermal properties are needed to simulate conduction and convection within the polymer. Density allows the weight of the moulding to be calculated. Viscosity is measured at several temperatures using known shear rates, and the second-order model is calculated and used.

The second part of the method of using Fillcalc IV consists of describing the mould, beginning with the mould layout. This procedure is best described with several examples (Fig. 6.4 to 6.7). In Fig. 6.4, the flow path is through the nozzle, down the main runner, and then, symmetrically, to each cavity. In tabular form, this configuration is expressed as follows:

Source	Destination	Symmetry
1	2	3
2	3	1

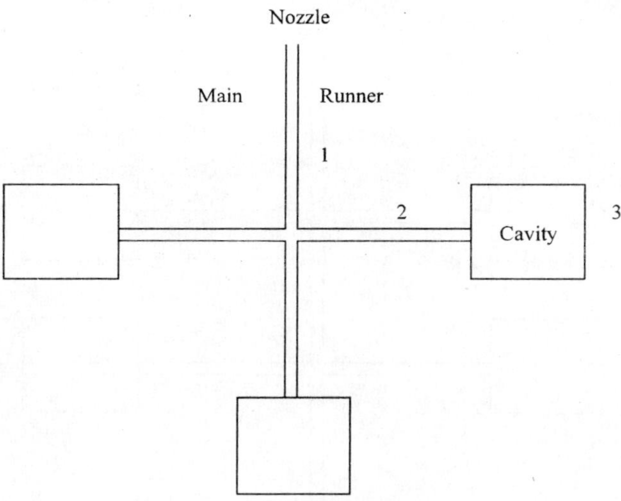

Fig. 6.4. Three-cavity symmetrical mould.

That is, the flow path goes from source 1 (the nozzle and main runner) to 2 (the runner to the cavity). There being three identical cavities, the 'symmetry' is described as 3. The path then goes from the runner to the cavity and, because one runner goes to one cavity, the symmetry is described as 1. Because all the runners are equal in size and all the cavities are identical, from the programme's point of view there is only one flow path. In a situation like this, the programme will not produce data on the symmetry or balance of the mould because there is only one flow path. It can, however, produce a wealth of information about the effects of changing the runner and the gate sizes and the effects of such modifications on pressure, cycle times, and cure. Following this it can calculate the effects of changing the processing parameters, enabling the optimum conditions to be achieved.

There are two flow paths in Fig. 6.5: one through the nozzle, the main runner, and into the first cavity, and second through the nozzle, the main runner, the next runner, and then into a cavity.

Source	Destination	Symmetry
1	2	2
2	3	1
1	4	1
4	5	3
5	6	1

The next complication arises when the cavities have either more than one gate or more than one flow path within the cavity. It is easy to visualise two flow paths if there are two gates, but there can also be more than one flow path if there are constrictions, inserts, or apertures in the finished part (Fig. 6.6 and 6.7).

Source	Destination	Symmetry
1	2	2
2	3	1
2	4	1
1	5	1
5	6	3
6	7	1
6	8	1

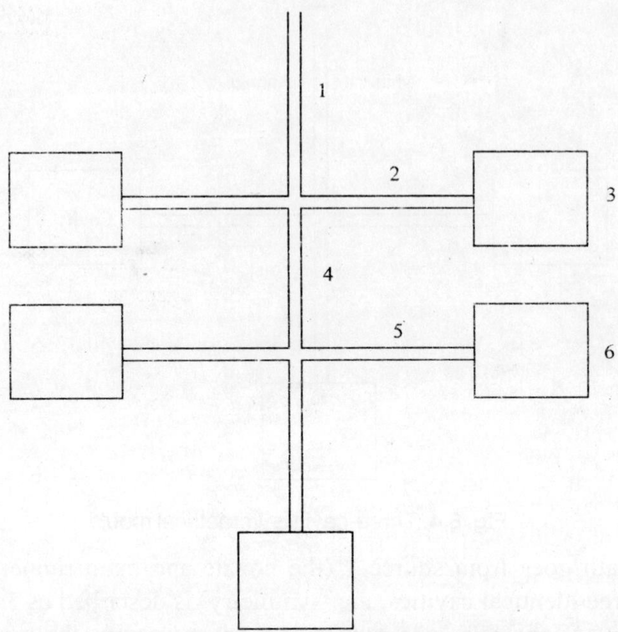

Fig. 6.5. Five-cavity asymmetrical mould.

Main runner 1 connects to runner 2, which is symmetrical to the left and to the right. The symmetry is therefore described as '2'. Runner 2, which includes the gate, goes to the cavity that has, because of the insert, two flow paths, the first one being '3'. As there is only one gate, the symmetry is '1'. Runner 2 also (via the gate) goes to flow path 4 and, again, the symmetry is '1'.

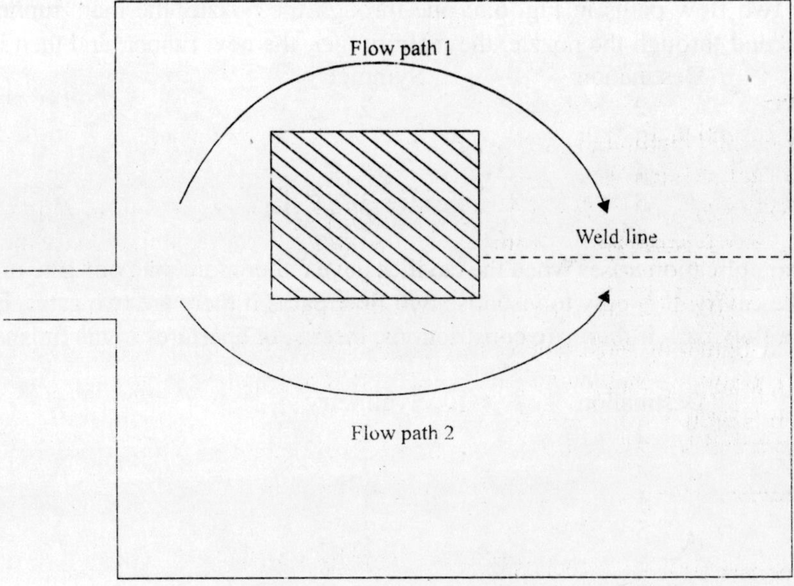

Fig. 6.6. Flow paths within a cavity.

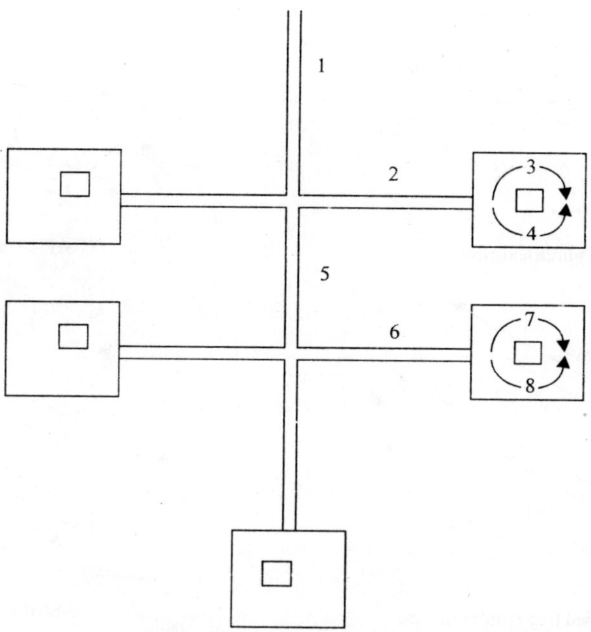

Fig. 6.7. Five-cavity asymmetrical mould with complicated cavities.

Main runner 1 also connects to runner 5 with a symmetry of '1'. Runner 5 in its turn connects to 6, which is identical for the three bottom cavities and therefore has a symmetry of '3'. Finally, 6 goes to 7 and 8 in the same way that 2 went to 3 and 4.

It is easy to see how this system can be used to describe moulds of increasing complexity. The theoretical limit is far beyond anything that might be needed in practice.

The next task is to describe all the physical dimensions of the mould. Fillcalc IV contains a number of standard shapes (Fig. 6.8) that can be combined together to describe any system. Cones and cylinders are used to describe the nozzle and main runner.

Because of the normal method of cutting moulds, half-rounds, trapezoids, or modified trapezoids are ideal for describing other runners.

The mould cavity can sometimes be described using an annulus, a cylinder, a disk, or a slab, but generally this is not sufficient. A more usual method is the 'lay flat' technique, in which the mould is visualised as having been flattened.

One method of dealing with this shape is to use the 'lay flat' technique, splitting, the cylinder in half along its length and flattening each half as in Fig. 6.9. Taking one half, we can see that the shape is similar to Fig. 6.10, assuming that dimension H (Fig. 6.9) is based on half the circumference of a circle, the radius of which is assumed to be in the centre of the rubber.

That is, the first dimension H is based on a radius of 10/2 mm plus 1 mm, giving 6 mm ($H = 6 \times \pi = 18.84$). The next dimension H is clearly based on 10/2 mm plus 2 mm ($7 \times \pi = 22$) and the last one on 10/2 mm plus 3 mm ($8 \times \pi = 25.13$).

This shape can now be modeled using a series of disks and slabs as is shown in Fig. 6.11. Some moulds are quite complicated to describe.

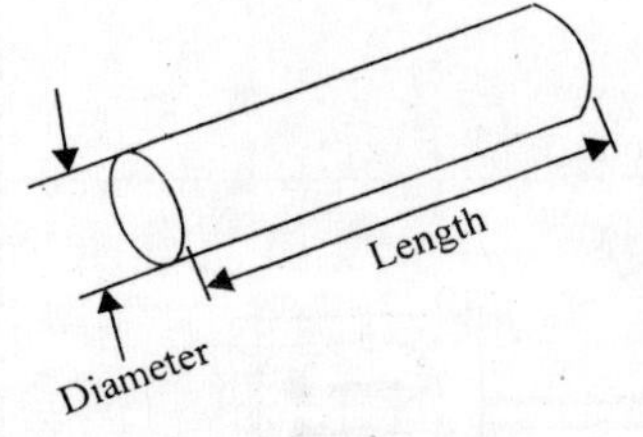

ROD - Cylindrical flow

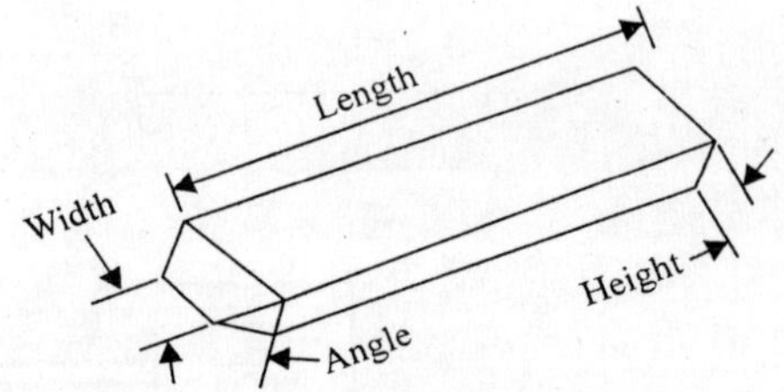

TRAP - Trapezoidal runner

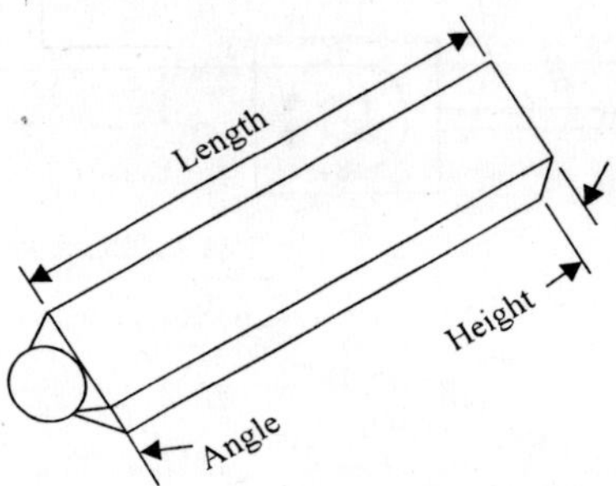

MTRAP - Modified trapezoidal runner

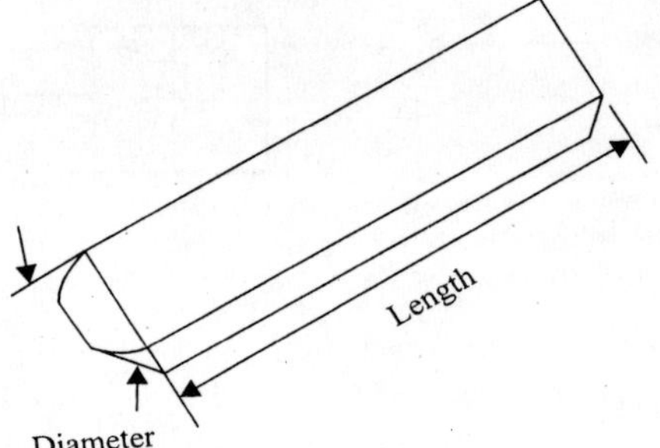

SEMI - Half-round runner

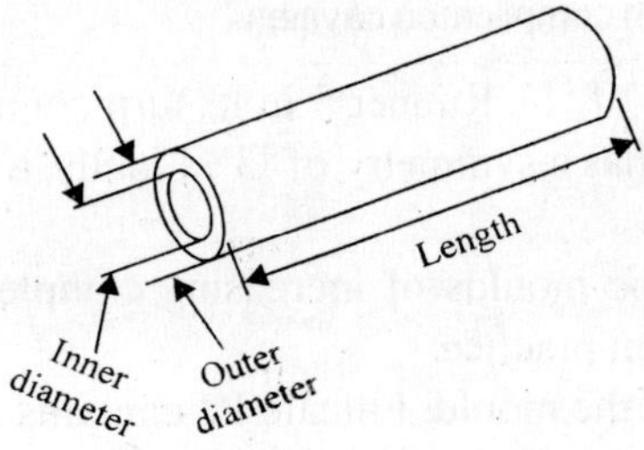

TUBE - Annular flow

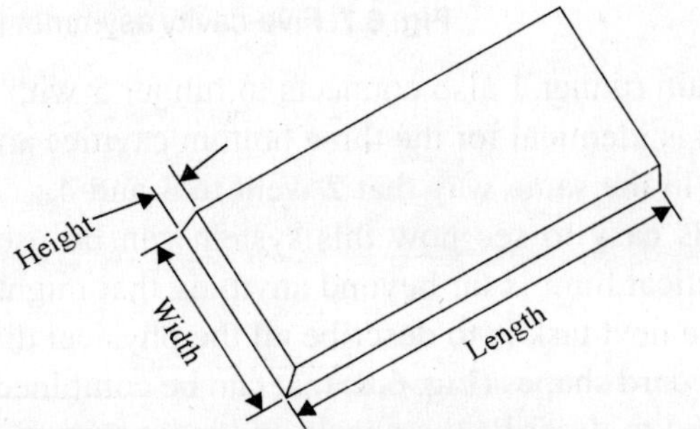

SLAB - Thin wall flow

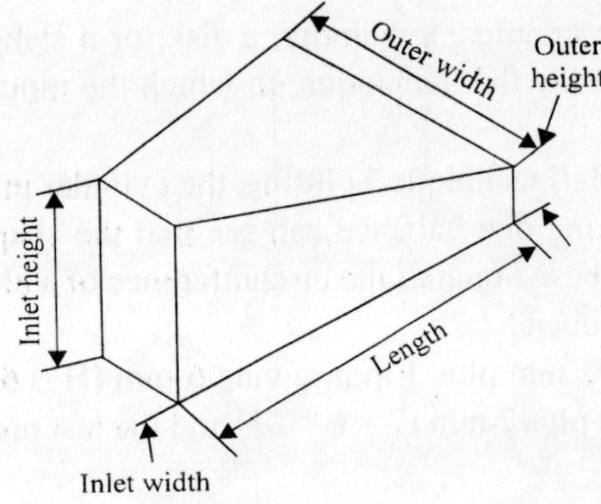

FAN - Transitional flow

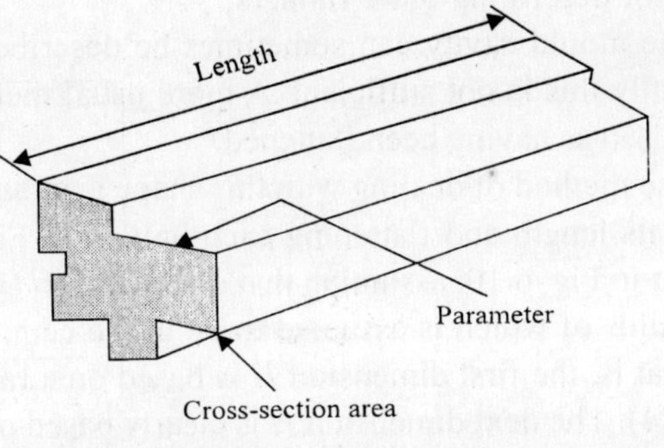

PROFILE - Complex channel flow

Fig. 6.8. Standard fillcalc IV shapes.

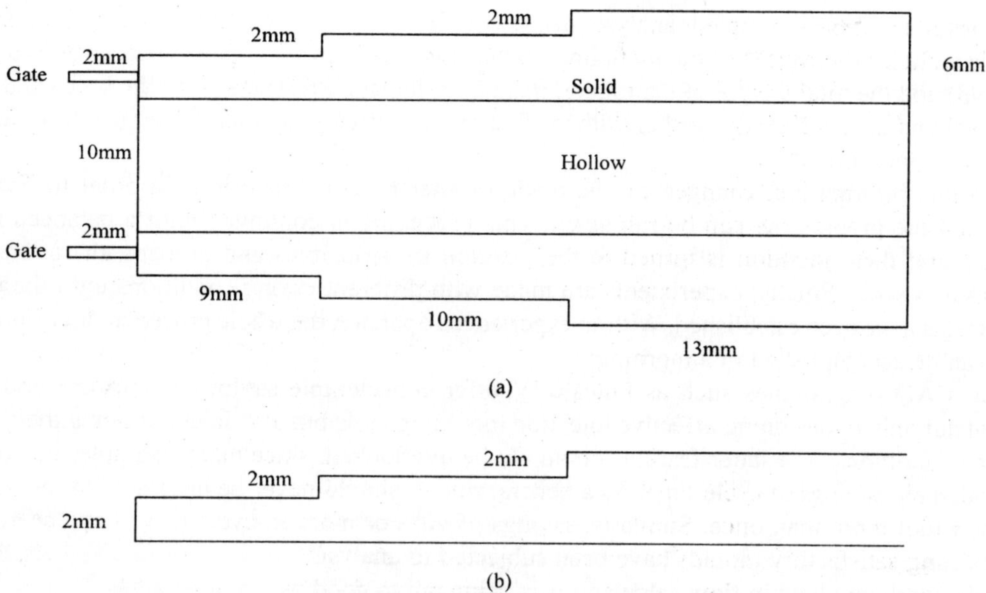

Fig. 6.9. Stepped cylinder: section.

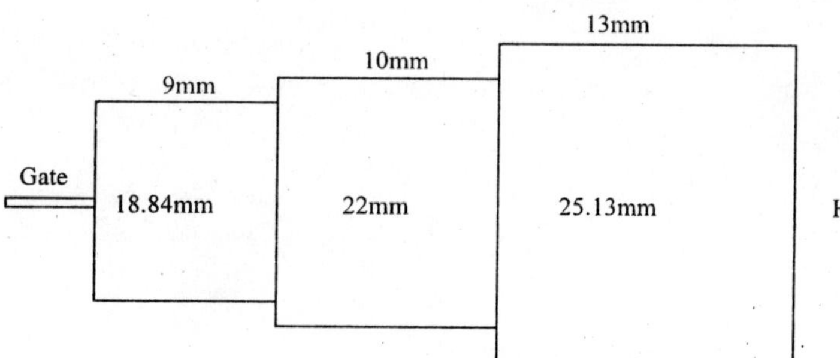

Fig. 6.10. Stepped cylinder: 'laid flat'.

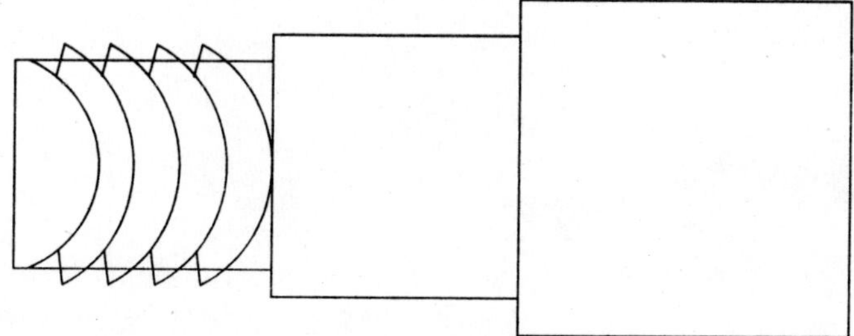

Fig. 6.11. 'Laid-flat' cylinder moulding using a series of disks and slabs.

Once the computer programme has been furnished with the data on the rubber and with the full description of the mould, certain parameters such as barrel temperature, mould temperature, and cycle time can be introduced and the programme can be run.

The results will be a complete analysis of the conditions in every section of the mould. The detail available includes the weight of the moulding (rather useful as a check against gross error in describing the mould) and the total weight of the rubber, including runners. Pressures in all the limbs and sections (individual and accumulated) together with minimum, maximum, and average temperatures are given, as well as reports on scorch.

From this information, changes can be made to alter runner dimensions in order to balance the mould, and the programme can be run again. This procedure is continued until a balanced mould is achieved, and then attention is turned to the position of weld lines and perhaps alterations in gate positions and sizes. Finally, experiments are made with different curing conditions until the optimum requirements have been established. With an experienced operator, the whole procedure for a complicated mould can be accomplished in a morning.

Thus, CAD programmes such as Fillcalc IV offer considerable savings in the time and expense involved not only in designing effective injection moulding tools but also in determining their optimum operating conditions. The latter feature should not be overlooked, since many examples can be quoted of considerable savings in cycle time. As a general rule, it should never be necessary to recut (or build up) a new tool more than once. Similarly, savings of 40% or more in cycle times have been reported when existing satisfactory moulds have been subjected to analysis.

All the work involved in flow calculations is being put to good use in other areas.

SECTION II

Principles of Compounding

CHAPTER 7

Compounding Ingredients and Compound Design

INTRODUCTION

The term compounding describes the selection of additives and their incorporation into a polymer so as to give a homogeneous mixture ready for subsequent processing steps. Except for differences in the nature of the compounding ingredients, especially as required for vulcanisation and reinforcement and described under those headings, the compounding of rubbers is similar to that of most plastics. In the case of natural rubber, SBR, and the other synthetic rubbers produced by emulsion polymerisation, the polymer is first available in the form of a latex. The normal procedure in compounding is to coagulate and dry the latex, and then to masticate and compound the rubber on mills or in other equipment. With SBR, however, two important compounding steps, *oil extending* and *masterbatching*, are commonly carried out before coagulation of the latex.

When designing a mixing formulation for specific end use, a compounder must take account not only of those vulcanisate properties essential to satisfy service requirements but cost of the raw materials involved and the production processes by which these will be transformed into the final product.

The principal task of compounding is therefore concerned with securing an acceptable balance between the demands arising from these three considerations: in brief—final vulcanisate properties, price, and processability. For any particular product, these requirements are not necessarily fixed. Over a period of time the vulcanisate properties may change owing to altered service requirements. Raw material prices fluctuate and, ironically enough, natural rubber itself is one of the most variable in this respect. Mixing, process operations, and vulcanisation cycles tend to shorten as production efficiency increases, and at the same time process temperatures and vulcanisation temperatures tend to rise.

It will be appreciated from these remarks that a compounder finds himself walking frequently on shifting sands in attempting to achieve his task. The time element in formulation development work can sometimes be considerable and preclude, for example, rapid reaction to change in material cost. Certain features of factory processability and service performance cannot be assessed quantitatively by laboratory scale tests alone. Extensive factory trials and sometimes service testing may have to be resorted to before a formulation can be finally proven, and these may be both costly and time consuming.

In this chapter an outline is given of the principles adopted in arriving at a mix formulation to produce particular patterns of properties in both the unvulcanised material and the final vulcanisate.

In consulting the considerable literature dealing with formulations, the newcomer to rubber compounding may often find to his surprise that nominally identical formulations are quoted but appear to possess appreciably different property patterns, the differences being beyond those attributable to interlaboratory testing or to test method variations. It must be admitted at once that this surprise is not always restricted to those unfamiliar with the technology.

The reasons for this, however, are not difficult to understand. A typical formulation may consist of perhaps a dozen individual constituent materials which are brought together by a physical mixing process that will never achieve perfection. Although a written formulation may quote zinc oxide, 5.0 parts, there are many commercial grades of zinc oxide, each having its own particle size distribution and characteristic levels of impurities. Similarly, nominally identical polymers produced on different plants or by different manufacturers will have their own minor variations on the basic characteristics. These considerations apply to a great many compounding ingredients, few of which are restricted to a single source, and, when such differences are summed up for perhaps most or all of the constituents of a formulation, some variations in properties must be anticipated. If to these variations are added the effects produced by differences in the mixing process (e.g., mix cycle times, shear rates, order of additions) and in the total heat history before the final vulcanisate is obtained, then appreciable property shifts are not difficult to understand.

For this reason, any published or quoted formulation should always be cross-checked using the grades of polymers and compounding ingredients which will be employed at the place of manufacture, and the properties determined on materials prepared in this way should only be accepted when they can be reproduced allowing for any reasonable experimental error.

The choice of raw materials is steadily increasing, new polymers and compounding ingredients jostling for a place among those already established and requiring critical evaluation before their worth can be assessed. This review has been drawn up around established materials available at the present time. The picture is, therefore, only transitory, and in the future, with a wider choice, both the range of application and the property cost ratios will shift and a number of the recommendations may no longer apply.

In particular, it should be noted that only the most widely used rubbers are discussed, but in most cases, the general principles will apply.

INGREDIENTS AND FORMULATION OF A MIX

Today, a technical vulcanisate is made up from the following constituents:

1. Base polymer or blend of polymers.
2. Cross-linking agents.
3. Accelerators of the cross-linking reaction.
4. Accelerator modifiers (activators and retarders); zinc oxide and stearic acid form a common activator system for sulphur-vulcanised systems.
5. Antidegradants (antioxidants, antiozonants, inhibitors of metal catalysed oxidation, protective waxes.
6. Reinforcing fillers (black, mineral, and organic).
7. Processing aids (chemical peptisers for polymers, softeners, plasticisers, dispersing aids, tackifiers, factice, and lubricants).
8. Diluents (inert mineral fillers, organic materials, and extending oils).
9. Colouring materials (organic and inorganic).
10. Specific additives, e.g., blowing agents, fungicides, fibrous materials.

In addition to the above, reclaimed rubbers or vulcanised rubber crumb may be included.

Objective in Using a Particular Ingredient

1. To improve the physical properties of the rubber.
2. To improve the processing properties of the rubber.
3. To cross-link the rubber compound.
4. To control the rate of cure.
5. To prolong the service life of the rubber.
6. To extend the service range of the rubber.
7. To reduce the cost of the rubber compound.

Unfortunately, however, any particular additive chosen may have a beneficial affect on one property but a detrimental effect on another.

Table 7.1. Gives a typical rubber formulation and the function of the ingredients.

Table 7.1. A typical compound formulation.

Ingredient [a]	*Ingredient function*	*Amount (phr)*
Natural rubber	Rubber	100.00
Whiting	Diluent white filler	100.00
HAF N330	Carbon black reinforcing filler	50.00
Naphthenic oil	Softener/process aid	20.00
Zinc oxide	Cure activator	5.00
Stearic acid	Cure activator	1.00
Santoflex 13	Antioxidant/antiozonant	1.00
Heliozone	Protective wax for ozone resistance	2.00
MBTS	Accelerator (primary)	1.00
DPG	Accelerator (secondary)	0.50
Sulphur	Vulcanising agent	2.50

[a] For carbon black and accelerators.

Relative Rating of Typical Polymer Vulcanisates

A summary of comparative properties of a range of representative polymer vulcanisates has been drawn up in general terms to show the pattern of properties that may be expected from the types of polymer.

It should be recognised that appreciable shifts in properties are sometimes made possible by special compounding variations–for instance, the heat resistance of NR vulcanisates may be improved considerably by variation of the vulcanising recipe but such exceptional compounding is ignored.

Classification system for elastomeric materials for automotive applications, is a classification system drawn up specifically with the automobile industry in mind. It is based on the premise that the properties of rubber products can be arranged into material designations based on types relating to heat-ageing resistance and classes based on resistance to oil swelling. A two-letter (type and class) designation is followed by three figures, the first indicating hardness, and the second and third tensile strength. Special suffix letters and numbers are used to refer to other characteristics for which a particular mix has been formulated.

Solubility parameter data has been included because of their importance in compounding. It gives an indication of the likely compatibility of a polymer with other polymers or compounding ingredients or with fluids encountered in service.

Since the specific gravities of polymers vary so widely, it is important to remember that any comparison must be made on a volume basis, and this applies also to the fully compounded mixes.

Vulcanisate properties have been rated on a 1:4 scale corresponding to the following property levels; (i) poor; (ii) moderate; (iii) good; (iv) outstanding. It must be emphasised that these general levels are capable of some modification by special compounding.

Cross-linking Systems

For the general purpose polymers (NR, IR, and SBR) and for IIR, BR, NBR, and EPDM, sulphur is the common and most generally used cross-linking agent. Its action is controlled by organic accelerator systems, of which there is a wide choice, and an activating system commonly supplied by zinc oxide in the presence of stearic acid.

The normal sulphur vulcanisation system is capable of many variants which will govern the chemical nature of the sulphur cross-link, i.e., whether it is essentially a mono-, di-, or poly-sulphidic type. The nature of the sulphur cross-link can have a considerable influence on the heat resistance of the vulcanisate.

For CR, cross-linking is induced using combinations of zinc oxide with light calcined magnesia, with or without an additional organic accelerator according to the type of CR. Sulphur is used only occasionally for special effects. CSM can be cross-linked by means of a number of systems, utilising either a metal oxide with organic accelerator or an epoxy resin combined with organic accelerators and retarders. FPM can be cross-linked by means of metallic oxide-amine systems.

Silicones and fluorocarbons and certain other saturated polymers are effectively cross-linked using organic peroxides, and for silicone rubbers a wide choice of peroxides is available.

SELECTION OF INGREDIENTS

For a particular application, the basic polymer, or when desirable and possible the polymer blend, has first to be selected according to the broad pattern of final properties required. This must then be combined with an appropriate cross-linking system, and, whether black or non-black, the reinforcement, if any, has next to be decided, sufficient processing aids being included to ensure satisfactory mixing and processing. It should be remembered that reinforcing fillers of finer particle size are, as a rule, progressively more difficult to incorporate during mixing, yield stiffer stocks, and are also generally more expensive; excessive reinforcement should, therefore, be avoided when not critical. Any permissible diluents or special additives for specific effects are then added to the formulation list.

Protective antidegradants are selected according to the severity and type of service exposure and to their relative effectiveness in the particular base polymer used.

The stages in which the formulation will be mixed in production must also be borne in mind, and the appropriate ingredients grouped accordingly. For instance, protective agents are commonly added early are cross-linking agents in the last.

Compatibility—The Solubility Parameter Concept

The range of possible ingredient is now so large that problems associated with compatibility of polymers with each other or with other compounding ingredients, such as plasticisers, extenders, process aids, or solvents, may often be encountered. In resolving such problems, the concept of the solubility parameter is a useful one. The solubility parameter may be regarded as a measure of intermolecular attraction and is numerically equal to the square root of the cohesive energy density. Polymers and compounding ingredients having closely similar solubility parameter values are likely to be compatible, and this is often of considerable value when dealing with unfamiliar materials.

The concept must be applied with care; no indication of the relative rates of solution will be given since this will be influenced by molecular size, viscosity, and temperature. Where two polymers are concerned, vulcanisation rates may be quite different so co-vulacanised blends are impracticable. Methods of determination of solubility parameters have been reviewed and a calculative method based on summation of molar attraction constants where the molecular formula is known has also been devised. The method however has certain limitations. Typical values for a range of plasticisers, process aids, and solvents are given in Table 7.2.

Table 7.2. Typical values of solubility parameters.

	Solubility parameter ($cal^{1/2}/cm^{3/2}$)		*Solubility parameter ($cal^{1/2}/cm^{3/2}$)*
Process aids		*Solvents*	
Petroleum jelly	7.5	Propane	6.0
Paraffin wax	7.7	N-Hexane	7.3
Low molecular weight polyethylene	7.8	N-Octane	7.6
Stearic acid	8.5	Cyclohexane	8.2
Polyethylene glycol (M = 400)	9.0	Xylene	8.8
Polyethylene glycol (M = 1500)	8.8	Toluene	8.9
Diethylene glycol	10.4	Benzene	9.2
Plasticisers (ester type)		Methyl ethyl ketone	9.2
Dioctyl sebacate	8.7	Ethylene dichloride	9.8
Dibutyl sebacate	8.9	Methyl isobutyl ketone	9.8
Dioctyl phthalate	8.9	Acetone	10.0
Dibutyl phthalate	9.4		
Tritolyl phosphate	9.8		
Trixylenyl phosphate	9.9		

Petroleum oils	*Viscosity gravity constant*	*Aniline point (°C)*	*Solubility parameter ($Cal^{1/2}/cm^{3/2}$)*
Paraffinic	0.78	115	7.2
Naphthenic	0.89	71	8.2
Aromatic	0.96	21	9.2

COMPOUND DESIGN

It can not be overstressed that the key to the successful development of a compound formulation lies in the writing of the "specification" for the compound based on a knowledge of the application for which it is to be used. Experience shows that the majority of part failures are due to an incorrect choice of polymer, because the compounder has overlooked or has not been informed of a small but critical aspect of the service conditions to be met; ozone attack is probably the critical aspect most frequently overlooked.

When presented with a customer's specification, experienced technologists often engage in lengthy questioning to ensure that the document is correct and that a clear understanding of what the part must do has been obtained. When the function is understood, the physical properties and chemical resistance required can be assessed. In this context it is useful to consider the necessary properties given in Table 7.3.

Table 7.3. Properties required of a compound.

Resistance to service temperatures and length of service required
Environmental and chemical resistance
Stress-strain properties, short terms (mechanical properties)
Stress-strain properties, long-term (crepe-stress relaxation)
Dynamic properties (dynamic stress-strain)
Fatigue performance
Friction and wear performance
Electrical properties
Thermal properties
Permeability
Adhesion, staining, and corrosion properties
Processibility
Density and price

Once a clear idea has been formed of the properties required from a compound, and the environment in which it will operate, the first step in the design of a suitable compound is the choice of polymer. Table 7.4 lists the major factors that will govern the choice of the polymer. The items are not given in order of importance, since in any application several of the factors may not be of direct concern. However, the first four generally will be found to be the most important factors. The analysis carried out on the application probably will have yielded one to five key physical properties and these, together with the temperature and environmental considerations, will as a rule very quickly lead to a short list of candidate polymers.

Price and other more secondary considerations (e.g., processibility) can then be used to further reduce the short list to one or two potential polymers for study. Price and other technical considerations may require the inclusion in the study of blends of polymers.

At this stage it is advisable to carry out a literature search on the rubber(s) chosen, with reference to methods of compounding to achieve certain key properties, and on the application; the polymer and the application can be linked to reduce that amount, but to increase the quality, of the references obtained. There is a vast amount of literature published, and access to this material can save a great deal of experimental effort. Computer-based databases with modern communications, and the availability of compact disks to be used with microcomputers mean that information is easily and rapidly accessed. As a minimum, the polymer manufactures' trade literature should be read, since within each rubber type there are many different grades.

The factors that would be considered when choosing a particular grade for study, as listed in Table 31.4, in essence are the same factors that govern the choice of the rubber. Within each type of rubber the whole spectrum of properties attainable can vary, however, and there may be limitations to the choice of cure system and differing rates of cure between the grades. Processibility will be of increasing importance in choosing the grades; for example, if large addition of oil and filler must be made, the choice of a carbon black masterbatch or oil-extended grade will result in reduced further additions at the mixing stage.

Table 7.4. Major factors governing the choice of elastomers.

Elastomer type
Maximum service temperature and service life
Minimum service temperature
Chemical and environmental resistance
Principal properties required
Processibility limitations
Cure system limitations
Price
Elastomer grade
Maximum service temperature and service life
Minimum service temperature
Chemical and environmental resistance
Principal properties required
Processibility
Cure system limitations
Cure rate required
Capacity for extension by added filler and oil (price)

By the time one has chosen the rubber(s) and grade(s) to be studied and obtained as much information as possible from published sources on these materials, a much clearer idea of the difficulties of meeting the compound specification will have emerged. It is quite common to find that the literature contains one or several useful starting formulations; in this case, and with fairly easy specification, it might be decided to mix and test the suggested compounds and then to proceed to develop the final compound by an iterative approach based on the result obtained and your knowledge of compounding. This scenario, in fact, occurs quite frequently, and it is often possible to develop a final compound that meets the specification in fewer than six attempts. This number of trails is less than normally would be required in a statistically designed experimental study, but it must be recognised that the iterative approach may result in the time scale for the trials and successful development being equivalent to, or longer than, the use of the more formal approach.

In many cases, however, the published literature sources will give general useful guidance but the application in which the compound is to be used may be novel, or no useful starting formulations may exist in the open literature. If we are considering the use of blends of polymers, this is even more likely.

Now it can be studied in many variables at once, that is :

1. Variations in the polymer(s) type and ratio if blends are used.
2. Variations in the black type(s) and level.
3. Variations in the softener or plasticisers type(s) and level.
4. Variations in the type and level of cure agent.
5. Variations in the type and level of accelerators(s).
6. Variations in the type and level of antidegradant(s).
7. Variations in the type and level of miscellaneous ingredients.

From the preceding discussion on compounding ingredients, it is clear that this would result in a ridiculously large, unwieldy experimental design if all the commercially available types of each ingredient were included in the study. The number of possible variables therefore must be cut down to a minimum, either by educated guesswork in choosing appropriate types and level for study, or by splitting one experimental programme into two or more separate experimental programmes designed to investigate the key area(s) of uncertainty. For instance, this might entail:

Programme 1: investigation into the effects of the blend ratio of polymers on key parameters using a single filler and a single cure system.

Programme 2: investigation into the interaction of filler/softener (or plasticiser) levels on key parameters using a single polymer, or blend ratio, and a single cure system.

Programme 3: investigation into the effect of the cure system on key parameters by varying the level of cure agent, and accelerator types and levels, but using a single polymer, or polymer blend ratio, and filler level.

Programme 4: an experimental design based on the results of programmes 1–3 but with a reduced number of variables to be investigated.

If a single polymer is used, and the grade is known, programme 1 would not generally be required, but an investigation into the effect of type and level of antidegradant might be needed if the ageing requirement were stringent.

Experimental design is also discussed but before passing on to this subject, and completing the subject of compound design, it is as well to briefly mention the pitfalls in relying on standard laboratory tests for rubber compounds, since even the best experimental designs will be wasted if an inappropriate test is used to be representative of service life. It must recognise that standard laboratory tests can be notoriously bad predictors of service life if used unwisely. For example, abrasion resistance is not a good predictor of service life, and no tyre company ever adopts a compound for use as tyre tread merely on this measure. Positive laboratory test results should be viewed only as a green light to proceed to more complex tests, which more accurately simulate service conditions, or straight service trials.

It goes without saying that factory trials using a new compound should also be run to ensure the absence of processing problems and to establish the process capability.

EXPERIMENTAL DESIGN

It is not generally easy to formulate a compound to meet a specification exactly. Usually, a trial compound is mixed, properties measured, and the formulation amended according to the properties obtained. Often this may involve a cycle of changing the formulation and retesting until usable results are obtained.

Sometimes it may be necessary, or more cost-effective, to design a series of test compounds and extrapolate the results to determine the required formulation. Depending on the number of ingredients whose quantities need to be altered, this may be achieved with or without the use of computer.

Single-Ingredient Variability

With only one ingredient, it is possible to draw the results as a simple line graph of the property value versus ingredient quantity by conducting four or five tests at varying ingredient levels.

Two-Ingredient Variability

With two ingredients, it has been shown that contour graphs may be constructed to give meaningful results, provided the experimental compounds have been designed in accordance with statistical principles. Ideally, the graphs would be constructed by computer for the sake of accuracy. Despite the limitations on the accuracy of results generated manually, however, a computer is not essential.

Multi-Ingredient Variability

With more than two ingredients, the complexity of analysis demands the use of a computer. When the performance of a compound must be resolved by varying a number of ingredient types or quantities, the sheer volume of mixing and testing work involved (not to mention the effort required to try and analyse the results) is prohibitive. This was, originally, the only way in which trends could be determined. The ingredient levels were altered for each ingredient in turn. The results had to be inspected manually and, with such a great amount of data to look at, synergistic effects of ingredient combinations were not obvious and often were lost. Mathematical analysis began to be accepted as a way to inspect these large volumes of data for trends.

MANAGEMENT OF COMPOUND RELATED DATA

For a number of companies, the ability to store a compound formulation and its associated quality tests requires no more then filing cabinet, or even just a good notebook. However, the range of data associated with any one compound formulation can be quite extensive, as illustrated below. This section is presented more in the form of a discussion, to provoke thought and ideas, since it does not deal with the more definite rules that may be applied to compound design. It is intended to aid the design of a practical data management system.

Summary of Compound-Related Data

Formulation

The mix formulation must be stored to ensure that the compound can be repeated in the future. This much is obvious. The formulation, however, does not often remain exactly the same throughout its production life. A variety of factors can produce amendments (e.g., improvement to properties, changes in ingredient prices, improvement in processability). All these changes should be stored somewhere so that if the product exhibits unexpected problems during service, the exact formulation at the time of manufacture can be examined. It may also be a contractual requirement with the customer that such traceability be provided. Formulations are often changed for financial reasons, and as a consequence the properties may begin to drift away from their original values. By saving all versions of the recipe, the fomulator can gain valuable, time-saving help in restoring the properties to their correct required value, should they stray too far. As part of the information stored, the compound reference (which should be unique), the originator's name, the data of design, and the revision number of the formulation are vital items in the traceability path. These data should always be associated with every batch of compound produced.

Weighing instructions

In addition to the compound formulation, the actual weighing instructions need to be produced, since these will give the exact batch weights required for each job. Again, these data should always be associated with every batch of compound produced, if for no other reason than to prove that the batch weights for the constituent ingredients were the same as for the phr in the specification. Provision should be made for the operator to countersign to the effect that weighing has been correctly completed.

Mixing instructions

The mixing instructions must be stored and issued to the factory to ensure that the compound can be mixed exactly the same each time. Also, a fresh sheet of instructions will be far more readable than the dog-eared, black-encrusted forms that often lie around the factory floor for years. Changes will be made to the mixing instructions to obtain optimum processing efficiency. Traceability will be required if the compound fails to meet its specification, to see what changes have been made to the mixing procedure since the compound was first formulated and accepted. Also, for traceability, the batch numbers and quantities will need to be recorded against the specific mixing instructions, formulation, etc., and the mixing instructions should be dated and initialed by the operators as the product is processed.

Provision should also be made for the operator or supervisor to note any peculiarities or irregularities during processing.

Quality control

Each product will have a number of control properties whose value limits must not be exceeded if the product is to be accepted by the customer. Such properties (and these include processing parameters) should be stored together with any value limits, to ensure that the product can be passed as acceptable by quality control. As each batch is made, it should be checked automatically against the quality specification, and the actual results from each test should be initiated by the tester and stored for every batch of material processed, again for traceability purposes. The control specification should define the test parameters as well as the acceptable maxima and minima.

Compound properties

"Compound properties" include any other property values that may have been derived for a compound but do not form part of that compound's intrinsic quality specification. While not necessary for traceability purposes, these values are often quite useful, since they can help the formulator decide whether an existing compound can be used as a formulation for a new job or to form the basis of it.

Government bodies and the larger industrial sectors (e.g., automotive), can demand very high standards of quality assurance. To achieve these standards, not only must the acceptance limits and actual results for quality control properties be stored, but also records of actual production: data (and may be time) of manufacture, which machine(s), which operator(s), processing conditions and machinery performance (e.g, internal mixer temperature and power logging), and finally the property values of both the uncured stock and the vulcanised product.

International standards ISO 9000 and British standard BS 5750 are just two of the definitive requirements to which more and more companies are having to conform in order to continue to attract business.

Ingredients

In addition to compound-related data, some data concerned with ingredients also should be stored for reference. Some companies do not use the ingredient name, or trade name, to identify ingredients, but rather have a form of internal coding system; a master reference to these codings will be required. Cost information can be invaluable when trying to formulate a compound, since while technologists are primarily concerned with producing compounds to meet given specifications, they can quite often be guided by financial considerations imposed from elsewhere. It is often possible to substitute alternative ingredients, and it makes economic sense to choose the cheaper alternative, all other technical factors being equal. Ingredient costs should therefore be stored.

It may be required to identify the storage location of each ingredient, its trade name, and its supplier. It is also useful to have some form of stock control of ingredients, such that warning is given if stocks are low, or there will not be enough stock to complete a batch. This system can be extended to provide automatic warning if any ingredient stock level falls below a set minimum, which naturally will vary according to the type of ingredient. It can also be useful to include a form of cross-referencing to determine which compounds contain a particular ingredient. Circumstances giving rise to this need include a sudden lack of stock, where it becomes necessary to identify, fairly quickly, which compounds will be affected, and a sudden, dramatic, rise in cost, calling for the substitution of alternative materials if possible.

Storage Techniques

With this large amount of data to store and retrieve, it is almost mandatory that the information be stored on computer from the viewpoints of sheer volume of data and ease of recovery. It is almost impossible to catalog this amount of data using the old, established methods (filing cabinets and notebooks). Modern computers can be fast and cheap, and since most computers now either have hard disks fitted as standard or are connected to a network, disk storage is no longer a problem. Even people who used the earlier personal computers with only a few kilobytes of RAM and a tape recorder found that while the searching of data was extremely slow, the results were well worth the effort. It is often assumed that computer data are more secure and accurate than filed paperwork.

There has been a great deal of animosity generated in the past concerning the use of computers in the rubber processing environment. This has been largely due to the historical placement of the older, larger, mainframe-type computer in the firms' book-keeping departments. Much of the information held on those computers, even when they were accessible by technical staff, was of confidential or sensitive nature, and there was marked reluctance by the financial departments to allow technical staff access to the data banks. Also, much of the earlier programming was either carried out or supervised by book-keeping personnel, leading to systems that were not always geared to producing data in the format required by technical departments. This situation has changed for the better in the last few years, especially with the introduction of high performance microcomputers putting real power on the desk of the technologist.

Having made the assumption that a computer is essential to the efficient management of compound-related data, what should we look for in our ideal data management system? The answer to this question depends to a large extent on the requirements of the individual company.

Even so, there are some aspects of compound data management that make a dedicated software package more desirable, even though this places greater reliance on the software supplier than would a standard empty database or spreadsheet.

CHAPTER 8

Compounding to Meet Processing Requirements and Vulcanisate Properties

INTRODUCTION

Polymers in their pure form, as obtained from the manufacturing plants after isolation and purification, are called 'virgin' polymers or 'virgin' resins. Barring a few (such as polystyrene, polyethylene or polypropylene), virgin polymers, as such, may not be amenable for procesing straightaway. Virgin PVC, for example, is a material of horny texture and cannot be moulded without making it soft by the addition of a plasticiser. Natural rubber similarly requires a vulcanising agent to render it mouldable. Most polymers are protected from thermal, oxidative and photo-degradation by incorporating appropriate stabilisers. Many articles are obtained in attractive colours by adding suitable dyes and pigments to the polymeric material prior to its moulding. Lubricants and process aids are added to most of the polymers to reduce friction and improve flow within processing machineries. Fillers are added to impart special properties and reduce the finished product cost.

The process involving incorporation of ingredients such as plasticisers, vulcanising agents, curatives, stabilisers, fillers, colouring agents, flame retardants and lubricants into the virgin resin is known as 'compounding' of the resin.

Before compounding procedures for the various specific vulcanisate properties are described, some general remarks will be made about the requirements for mixing and processing characteristics. This is necessary as a compromise has often to be made to permit a formulation, which has been developed for particular vulcanisate properties, to be mixed and processed in a given manner.

VISCOSITY CONTROL

For a raw polymer type, choice of correct viscosity level is important to ensure acceptable mixing and processing characteristics. Viscosity levels for polymers are commonly expressed in terms of Mooney viscometer readings taken at 100°C. This instrument, however, operates at very low shear rates, and, with present day trends towards faster mixing and processing operations, viscosity data obtained with much higher shear rate instruments may prove ultimately to be more realistic. In the meantime, Mooney data remain the generally accepted criteria, since manufacturers rate their polymers this way. Most types of synthetic polymers are available at various viscosity levels, a nominal Mooney viscosity [(ML (1 + 4) 100°C] of 50 being a common one. For easy mixing and processing, e.g. for formulations with high filler loading or for sponge production where particularly uniform blowing is required, a viscosity of the order of 30 is necessary; at the other end of the scale, a viscosity level of 100 or so should be selected when loading with oil diluents, for economical compounding, or when processing demands a certain level of strength ('green strength') in the unvulcanised material.

Natural rubber produced by conventional means has a typical Mooney viscosity of the order of 90, which is too high for normal direct mixing except by special compounding where oil extension is used. Its viscosity is, therefore, usually reduced to a suitable level by mechanical working or mastication before mixing. For more efficient operation, small additions of chemical peptisers are made. In the case of standard Malaysian rubbers (SMR), viscosity-stabilised grades (known as CV types) with a Mooney viscosity of 55–65 are available and may be suitable for some mixings directly without further mastication.

A number of low molecular weight synthetic polymers can be obtained which are in viscous liquid form. These can be used as part replacement of solid polymer, functioning as general process aids and softeners. After vulcanisation they become part of the cross-linked structure and are, therefore, not extractable by liquids which would extract other softeners or processing aids. Specially depolymerised NR is also available in fluid form and behaves similarly.

Control of 'Nerve'

Some rubbers exhibit a degree of elastic recovery on deformation, even before cross-linking. This 'nerve', as it is usually called, if excessive, must be reduced during mixing processes to a controlled level, so that processing will be reproducible and shaped stocks will exhibit dimensional stability.

Depending on the polymer type, elastic recovery can be reduced by a suitable degree of mastication, by increasing filler loadings, by choice of filler type (e.g. use of high structure furnace blocks in place of normal structure carbon blacks, or by use of additions of materials such as mineral rubber or factice. Another method is to add a proportion of specially produced polymers which have been partially cross-linked during manufacture. These are particularly advantageous in improving shaping operations and ensuring dimensional stability of unvulcanised stocks. Polymers, modified in this way, are produced from NR. SBR, IIR, and NBR: examples are SP (superior processing) grades of NR, and types 1009 and 1018 of SBR.

Adhesion to Mill Rolls

During mill mixing, mill sheeting, or calender sheeting operations, sticking to roll surfaces can sometimes be a problem, and additives to aid release are necessary. Certain soft tacky stocks or ones containing intermediate loadings of mineral fillers are most likely to present this problem. Fatty acids or their derivatives, microcrystalline waxes, or low molecular weight polyethylenes are useful additives to control this, but must be used with particular care in NBR, where, because of limited compatibility, other problems due to bloom or lack of stock coalescence may arise.

Calendering and Extruding

In calender frictioning operations, the stocks used must be soft so that they penetrate the interstices of the fabrics easily and tacky so that ready adhesion to the fabric is ensured. Inevitably, such stocks can give rise to handling problems and call for compromise compounding NR and CR compounded with resins such as coumarone indene or petroleum types produce good frictioning stocks; refined reclaimed rubbers are used frequently in blends for lower-cost stocks.

To obtain good extrusion characteristics, compounding requirements are similar to those described previously for reduced elastic recovery. These minimise undue die swell on shaping and contribute to shape retention until this is made permanent by cross-linking. Internal lubricants assist in reducing die drag, promoting a smooth surface. Waxes, fatty acids, or fatty acid derivatives are also useful additives. For extrusion work generally, it is just as important for easy processing to avoid compounding to too low a stock viscosity as to one that is too high. Too low a viscosity can lead to loss of extrusion efficiency and poor shape retention.

Continuous Vulcanisation

Continuous vulcanisation of extrudates by passage through liquid salt baths or fluidised beds requires special compounding techniques because of the high vulcanisation temperatures and normal atmospheric pressures involved. Volatile softeners or plasticisers and moisture-containing ingredients must be avoided, any moisture traces being eliminated by use of scavengers, such as a quicklime dispersion in mineral oil. Rapid rates of cross-linking are desirable since these ensure a build-up of strength throughout the mass of the stock, causing entrapped air to diffuse away rather than to form expanding pockets which make the material tear internally. Compounding for high-temperature injection moulding operations requires a similar approach with respect to elimination of volatile or unstable compounding ingredients, although moisture scavengers are not normally required.

Tack

Forming operations which involve building layers of stock together require the mix to have tack, the property of auto-hesion or adhering rapidly and strongly to itself when surfaces are brought into contact. This surface condition may be required to last for considerable periods of time, e.g. in tyre retreading stocks or adhesive tapes. Excessive filler loadings are best avoided since they may have a drying effect on the stock surfaces. NR and CR can be compounded readily to give good tack levels. Other general purpose polymers require suitable resins, e.g. coumarone-indene, rosin derivatives, petroleum resins, or phenolic types to give acceptable tack. EPDM is at present difficult to tackify. Waxy compounding ingredients which may form a bloom should be kept at a minimum, and, where necessary, freshening of the surface with a solvent wipe is beneficial. Excessive use of solvent is, however, undesirable since, if traces remain trapped, vapour blows or porosity may result during vulcanisation.

Avoidance of Pre-vulcanisation or 'Scorch'

Vulcanisation systems are normally designed to be as fast as practicable, subject to certain limitations. These are the cost of the vulcanisation system, and the heat unavoidably generated during mixing and processing which causes pre-vulcanisation or scorch problems.

Delayed-action vulcanisation systems must be used to minimise risk of premature vulcanisation, and, for sulphur cured NR or SBR, sulphenamide accelerators are widely used, with an additional chemical retarder such as nitrosodiphenylamine in critical applications. For CR, increased levels of magnesia, additions of sodium acetate, or small amounts of mercaptobenzthiazyl disulphide are effective.

COMPOUNDING FOR VULCANISATE PROPERTIES

The means by which individual vulcanisate properties can be varied are discussed below:

Hardness and Modulus

Hardness is probably the property first specified and, of all vulcanisate characteristics, the one most frequently determined. It gives a measure of the modulus at low strains. Here the hardness values are expressed in international rubber hardness degrees (IRHD), measured according to the ISO method R48.

Hardness and modulus of a vulcanisate are normally increased by the use of particulate fillers, the greatest effect being obtained by the reinforcing types or by the use of cross-linked resin systems. These latter may be phenolic types capable of cross-linking or monomers capable of being grafted *in situ*, e.g. acrylic monomers used with peroxide cross-linking systems. Relatively unreactive resins, such as high-styrene resins are used with NR and SBR for certain hard vulcanisates. PVC may be incorporated into NBR.

Carbon blacks are the most common reinforcing agents used. With non-black fillers, the pattern is less well defined, the varying chemical natures of the materials concerned, in addition to their varying particle shape and surface state, evidently influencing the relative hardening effect.

For a given non-black filler type, decreasing particle size gives a progressively increased hardness.

In addition to the low strain modulus, of which hardness is a measure, the compounder is interested in modulus at higher strains, which is expressed not as true modulus but as the stress required to produce a stated extension, e.g. 100%, 300%, or 500%. Compounding which produces increases in hardness also produces increases in the higher strain moduli values.

Increases in extension modulus in a given polymer-filler system are also sometimes possible by using heat treating cycles during mixing, with or without the use of a chemical promoter, these treatments modify the polymer-filler interaction.

It has been found that, in addition to a reduction in mix viscosity, the vulcanisate properties are modified appreciably, extension modulus and resilience being increased, hardness and compression set reduced, and the rubbery nature of the vulcanisate is thereby enhanced. Because of the magnitude of the property changes, heat treatment cycles must be carefully controlled to achieve reproducible results on a production scale.

Phenolic resin and high-styrene resin additions and the use of *in situ* resin formation with acrylate monomer will increase hardness and modulus. It is interesting to note that all these three systems produce similar increases in hardness at equivalent resin addition. High-styrene resin (or PVC addition to NBR), because of its relatively unreactive nature, has a more adverse effect on resilience and set than phenolic resin cross-linked with HMT.

Changes in accelerator systems produce relatively small differences in hardness and modulus. Sulphenamides tend to produce slightly harder vulcanisates than thiazoles, and boosted accelerator systems, small hardness increases over those produced by straight systems.

Softeners and plasticisers reduce hardness and, at the levels normally used to secure acceptable processing, require additional field or reinforcing agent to counteract their effect. In this respect, where their cost can be accepted, acrylic monomers or phenolic resins can act as effective aids during mixing and processing without recourse to additional softeners. Mineral process oils, which are widely used in general purpose polymers as low-cost softeners produce a hardness drop of approximately one IRHD for each two parts added. Relatively less softening is produced by factice (vulcanised vegetable oil), which not only acts as an efficient process aid but is also capable of enhancing the cross-linking system in some instances.

Elasticity

Vulcanised rubbers show viscoelasticity, and the departures from perfect elasticity are evaluated by measurement of resilience crepe and stress relaxation and set. Compounding which contributes to a more tightly knit cross-linking system occupying the maximum possible volume proportion of the vulcanisate will enhance the elastic properties, as displayed by resilience. If this cross-linking system is stable, with regard to any degradative condition which may be encountered in service, then the associated properties of set, crepe, and stress relaxation will also be minimised. Appropriate antioxidant protection of the polymer will give further improvement.

Polymers such as NR, IR, CR, BR, and certain types of EPDM may all be cross-linked to give high resilience levels, under ideal conditions in excess of 90% with some polymer systems compounded in gumstock form.

Loading with particulate fillers progressively reduces resilience, particularly with the more reinforcing types of carbon black and silica, the effect being more marked at a given loading level as the ultimate particle size is reduced. At the same time crepe, set, and stress relaxation characteristics are worsened.

Softeners and plasticisers at the levels normally used have relatively small effects but, at higher concentrations, generally produce reductions in resilience, high-viscosity petroleum softeners of increased aromaticity being more detrimental than lower-viscosity paraffinic types. Some ester plasticisers in NBR rubbers can increase resilience. In NBR, SBR, and IIR, increases in the proportion of butadiene or isoprene in the copolymer give increased resilience.

Strength

At relatively low hardness levels, say below 50 IRHD, highest tensile strength levels are most easily obtained using high gum strength polymers which crystallise on stretching, such as NR, IR, CR, or isoprene-acrylonitrile co-polymer. Other polymers, such as SBR, NBR, and BR of low gum strength, require fine particle size reinforcing fillers to develop maximum strength, and this reinforcement is accompanied by increased extension modulus and hardness.

In both groups of rubbers the highest strengths are given by using the fine particle size carbon blacks or reinforcing silicas, and to obtain highest levels, it is essential that these are well dispersed. Optimum levels vary according to the polymer filler system but usually fall in the range 30–60 p.p.h.r. Certain resins are also capable of producing increased strength levels in some conditions. Thus high-styrene resins can produce strength increases in SBR, and cross-linkable phenolic resins can also produce strength increases, particularly with NBR.

Resistance to Abrasion and Tear

Abrasion

The resistance of vulcanised rubbers to abrasion is a difficult measurement to carry out since ratings depend on the precise conditions of wear, and a number of features are not understood. As a result, compounding aimed at improving abrasion resistance sometimes produces confusing results.

In general, as a rubber is compounded with progressively smaller particle size carbon blacks, say FEF (N-550), HAF (N-330), or ISAF (N-220), the resistance to abrasion increases, optimum resistance occurring in the region of 50 p.p.h.r. With non-black fillers a similar pattern is seen, maximum resistance being given by reinforcing silica of ultimate particle size around 20 nm.

For vulcanisates based on NR, SBR, and CR, abrasion resistance can generally be improved appreciably by replacing a minor proportion of polymer by polybutadiene. The high abrasion resistance of this polymer is best utilised in this way since, for a number of reasons, it is not normally used as the sole polymer in technical vulcanisates. This practice cannot normally be applied to NBR or IIR because of lack of compatibility and inability to co-vulcanise respectively.

Under mild abrasive conditions NR is superior, whereas under the most severe conditions the BR blend is markedly more resistant. A good level of antioxidant protection, particularly of certain *p*-phenylenediamines, has a favourable influence on abrasion resistance.

Tear

As with abrasion resistance, the determination of tear resistance presents a number of problems which render compounding variations intended to improve this property sometimes difficult to interpret. The nicked crescent test-piece is widely used. This test suffers from the disadvantage that some of the energy applied to the test-piece is dissipated in stretching the arms of the sample rather than tearing it. Methods have been proposed to avoid this difficulty.

Unlike abrasion resistance, reinforcement with carbon blacks of decreasing particle size does not always produce a noticeable effect on tear resistance. With NR and IIR improvement is evident, but with other polymers, such as CR or SBR 1500, the effects are much less marked. In low-hardness stocks the high gum strength polymers, such as NR or CR, give higher tear levels than low gum strength types, such as SBR or NBR. In CR small particle size non-black reinforcement agents such as synthetic aluminium silicate or reinforcing silica give high tear strength figures, silica being capable of producing levels superior to that given by reinforcing carbon blacks.

Coarse fillers of mineral origin, particularly those with relatively large particle fractions, generally detract from tear resistance, but fine clays give some improvement.

Certain resinous processing aids, e.g. coumarone-indene resins, petroleum resins, and phenolic types, are capable of improving tear resistance values when added to a formulation in small amounts of the order of 5%. This is probably due to their ability to enable the polymer to wet out the surface of the reinforcing filler more effectively, thereby improving the polymer filler interaction.

Resistance to Cyclic Stressing, Flex Cracking, Cut Growth and Fatigue

Where rubber products are subjected to cyclic stresses in service through either repeated or intermittent flexing or compression, the initiation and development of cracks is a frequent cause of failure. Furthermore, at high rates of flexing, e.g. in rapidly rotating tyres or drive belts, heat build-up can be an important factor and the compounding involved must allow for this.

Apart from the choice of base polymer, compounding variations can have quite complex effects on relative crack development. The nature and stability of the cross-linking system, the choice of protective agents, and the degree of dispersion of curatives and reinforcing and other fillers, leading to localised strains, can all influence performance on flexing. If compounding reduces resilience, additional heat will be generated on flexing, and, if severe, an increased rate of cracking will result.

The polymer ranking depends on the strain cycle used and that goods resistance to crack initiation is not always accompanied by good crack growth resistance.

Where a risk of physical cutting is involved, then the use of high gum strength polymers such as NR or CR is desirable, reinforced to give a balance between low modulus, high extensibility, and optimum tear resistance. For this purpose, carbon blacks such as HAF-LS (N-326), channel blacks (S-330), reinforcing silicas, or silica-carbon black blends are used.

When the anticipated flexing is of constant amplitude, extension modulus should be kept as low as possible to minimise stresses.

The influence of sulphur cross-link variants may be anticipated from the fact that, for many years, the fatigue resistance obtained using the so-called 'sulphurless' systems, e.g. using TMTD alone, has been regarded as inferior to that of a conventional system. This is presumably because of the relative lack of mobility of elastomer chains joined by monosulphide links as opposed to polysulphidic ones. Results for NR show that, although this is true if the vulcanisate is cycled through zero strain, efficient vulcanisation systems can be superior under non-relaxed cycling, and also under elevated temperature conditions, because of the greater thermal stability of monosulphidic links.

The degree of cure is also important. If vulcanisation cycles are prolonged thermally susceptible polymers such as NR or IR may be affected by degradation, whilst SBR can suffer from high modulus effects due to continued cross-linking with the consequences on cut growth resistance.

Antioxidants vary in their influence on fatigue. At elevated temperatures, crack growth rate increases comparatively rapidly for SBR and CR but less so for NR.

Resistance to Degradative Agencies

Heat

Attention to cross-linking systems and choice of other compounding ingredients can increase appreciably the maximum service temperature of a polymer.

Reference has already been made to the enhancement of heat resistance of NR by using a thermally stable monosulphidic cross-linking system as opposed to the normal polysulphidic type. In the past such a cross-linking system was achieved with TMTD alone, but this method is limited because of pre-vulcanisation risk and objectionable bloom in the final vulcanisate. The new systems avoid these disadvantages and, when combined with a good heat-resistant antioxidant system, are capable of increasing the typical service temperature of NR from around 70°C to 100°C, and even above this for intermittent exposure.

When sulphur cross-linking is not employed, as with CR, other procedures are used. Non-sulphur-modified CR shows greater heat resistance than the sulphur-modified types and is, therefore, preferred. It is used with a cross-linking system based on increased levels of zinc oxide and ETU in the presence of a selected retarder to minimise pre-vulcanisation problems. At the same time, a high level of an octylated diphenylamine antioxidant is added to improve the polymer stability. A typical system incorporating these principles is given in Table 8.1.

Softeners, processing aids, and plasticisers should be chosen so that they have long-term thermal stability at the anticipated service temperature and are of low volatility. High flash point petroleum oils, or monomeric or polymeric ester plasticisers are used frequently, choice being governed by cost and compatibility with the polymer concerned. Somewhat unexpectedly, rapeseed oil has been found to be particularly effective in CR.

Table 8.1. Formulation for heat-resistant mix (non-sulphur-modified CR).

Ingredient	*pphr*	
Octylated diphenylamine	4-0	Antioxidant
Dodecyl mercaptan	1.0	Retarder
Light calcined magnesia	4.0	Vulcanising system
Zinc oxide	10.0	
Ethylene thiourea	1.0	

Flame

By using halogen-containing polymers such as CR, it is possible to produce vulcanisates which, although not non-inflammable, are self-extinguishing when an applied flame is removed. In such formulations tricresyl phosphate and liquid chlorinated paraffin wax should be used as processing aids in place of oils, waxes, or factice, and the carbon black level should be kept at a minimum, the black being used only as a pigment. Reinforcing mineral fillers are used, along with antimony trioxide and the intumescent zinc borate, which assists by forming a hard surface crust on exposure to flame. Recently, certain chloroalkyl phosphates have become available which may prove to be of value. The flame resistance of NBR can be improved by blending with PVC and using tricresyl phosphate plasticisation.

Liquids

The action of liquids on vulcanisates may be regarded as the net effect arising from physical swelling of the polymer, the degradation of the polymer and fillers, and the leaching of, or attack on, plasticising agents.

For optimum resistance to a particular liquid which will have primarily a swelling action on the polymer phase, the polymer should be chosen so that its solubility parameter is as far removed from that of the contact liquid as the required processing and other vulcanisate properties will allow. Chemical attack on the polymer is most likely to arise from oxidising substances, and in this event polymers having a relatively inert backbone structure should be chosen, e.g. IIR, CSM, or EPM.

In addition to the intrinsic swelling and chemical resistance of the polymer, its volume content in the vulcanisate should be kept as low as practicable since this will minimise its contribution to any volume change. Fillers should also be chosen to be as inert as possible under the service conditions, e.g. precipitated barium sulphate (blanc fixe) is useful under acid immersion conditions. Plasticiser stability is also important, and ester plasticisers should be avoided where hydrolysis could occur, e.g. under hot alkaline conditions. Extractable plasticisers or additives are undesirable where shrinkage effects cannot be permitted for instance in certain seals, unless swelling of the polymer by the liquid in which it operates compensates for this. Polymeric plasticisers, low molecular weight polymers which will cross-link on vulcanisation, and oil-resistant factices all have applications in such cases. A high state of cross-linking in the final vulcanisate will also assist in reducing ultimate swelling.

Where resistance to water or aqueous solutions is required, the presence of water-soluble salts in the vulcanisate is detrimental as they facilitate the ingress of water. Thus in SBR manufactured by salt-acid coagulation methods, water swelling is greater than that given by specially finished types (e.g. 1503, 1509, and 1708) and solution-polymerised types, which are for the same reason preferred for electrical applications. CR which is cross-linked using magnesia-zinc oxide systems produces traces of soluble chlorides; so, for maximum water resistance, lead oxide cures are preferred since lead chloride has only limited solubility. Magnesia-cross-linked CSM reacts similarly. Mineral fillers containing traces of soluble salts can also produce these effects. NR and the water-resistant grades of SBR have useful levels of resistance to a wide range of aqueous solutions and the chemicals which do not have any marked oxidising action.

IIR and EPM are resistant to animal and vegetable fats, oxidising chemicals, and some oxygenated organics, e.g. ketones, but not to aliphatic or aromatic hydrocarbons. CR is also resistant to animal and vegetable fats and waxes, as well as aliphatic hydrocarbons. The general oil and solvent resistance of NBR is of a high order. CSM has a high level of chemical resistance, particularly towards oxidising solutions. When very high temperatures are involved, then the best general chemical and solvent resistance is given by the fluorinated polymers.

Light

When considering the influence of light, it should be remembered that the ultra-violet region of the spectrum has a much higher potential energy than the visible or infra-red regions, the level being of an order capable of breaking down organic molecules. In exposed vulcanisates, this manifests itself as discoloration, surface embrittlement, or break-up in which oxygen also plays a part.

To protect rubbers, particularly from intense sunlight, which is rich in ultra-violet radiations, compounding with materials which will either reflect it or absorb and transform it into innocuous wavelengths is necessary. This involves the use of suitable pigments, stabilisers, or possibly ultra-violet absorbing materials. Carbon black is fortunately an outstanding absorber of ultra-violet light and is effective at quite low loadings of a few parts per hundred parts of rubber. Smaller particle size blacks are particularly effective. Among non-black fillers, barium sulphate and calcium carbonate have good ultra-violet reflectance characteristics, whereas zinc oxide and titanium dioxide are relatively poor. The rutile form of titanium dioxide shows greater absorption than the anatase form. Among coloured pigments, phthalocyanine blues and greens, quinacridone reds, iron oxide reds, yellows, and browns, and chromium oxide green have good ultra-violet absorption characteristics. Specific absorbents of ultra-violet light include a wide range of benzophenone derivatives.

For non-black vulcanisates destined for long-term light exposure, CSM and NBR PVC blends are frequently used as base polymers.

Processing aids (such as low aniline point aromatic process oils, aromatic resins, and extenders) which can degrade under exposure conditions should be avoided in compounding non-black formulations. Staining antioxidants are not acceptable, and the use of non-staining phenolic types is essential. Some accelerator residues, e.g. those from certain sulphenamides and from guanidines, can also produce discoloration and may not be acceptable under critical conditions.

Black formulations which will be used in contact with light-coloured paints or lacquers or other coloured surfaces should be compounded similarly if migrating or contact stains are unacceptable.

Ozone

When stretched, vulcanisates are exposed to ozone, even at very low concentrations, and, particularly under dynamic conditions, cracking develops at right angles to the direction of strain, the rate of cracking varying markedly from one polymer to another and being influenced by the compounding employed.

With susceptible formulations, additions of materials, such as petroleum wax blends, of limited compatibility result in the formation of surface blooms which give some physical protection because the flexible protective barrier prevents ozone attack at the polymer surface provided it is not disrupted. Chemical protective agents, notably the substituted *p*-phenylenediamines, give additional protection without suffering from this defect.

The ozone resistance of non-protected polymer vulcanisates increases through the series: SBR, NR, CR, IIR (increases as the degree of unsaturation decreases), CSM, EPDM. In addition, NBR, which is susceptible to ozone attack, can be improved considerably by blending with PVC, suitably stabilised. At NBR-PVC ratios of the order of 60:40, ozone resistance is of the same order as that given by IIR but resilience levels are low at such ratios because of the proportion of non-cross-linked PVC. The ozone resistance of any of the above polymers will be improved by compounding with small additions of selected wax blends or *p*-phenylenediamine antiozonants or both, common levels being in the range 1–5 p.p.h.r. Ozone resistance of EPDM is so high that additional protection is rarely necessary. Levels used with any particular polymer must be chosen with care, because of limited compatibility, to avoid excessive blooms. Heavy wax blooms which crack on flexing are undesirable.

Atmospheric exposure

When vulcanisates are exposed to outdoor weathering, the material is not only subject to the combined effects of light and ozone but also to fluctuating temperatures, rain, perhaps hail and snow, and the eroding action of any airborne dusts. Conditions will be aggravated if the service is a dynamic one. Rain has a leaching action on the degraded surface, and, when mineral fillers are present, these gradually become exposed and may be rubbed off—the phenomenon known as 'chalking'. Compounding which minimises the effects of ultra-violet radiation will reduce this. Geographical location has a marked effect on rates of weathering. Ultra-violet effects will be most noticeable in regions of considerable sunshine, whilst ozone effects will vary since atmospheric ozone concentrations at the earth's surface increase from equatorial regions towards the poles, and with increasing altitudes. Exposure to direct sunlight can give rise to considerable heat build-up at the surface of vulcanisates. The build-up is greatest for black rubbers and least for white, with colours giving intermediate values. In extreme conditions, surface temperatures of black vulcanisates can exceed 100°C, and conditions can become particularly severe inside parked motor vehicles. In such circumstances, protective wax blooms may melt and redissolve in the polymer.

For any severe outdoor weathering application the most satisfactory results will be obtained by combining the most resistant polymer, protective agent system, and filler system appropriate to the conditions.

Low-temperature Resistance

If high gum strength polymers which crystallise easily on stretching, such as NR or CR, are exposed to moderately low temperatures for considerable periods of time, a time-dependent reversible stiffening phenomenon known as crystallisation occurs. For NR this is most rapid at about –26°C, and at –12°C for CR. It is a reversible effect, and the relative rates of crystallisation vary from one CR type to another. A special crystallisation-resistant grade of NR has also been developed. Crystallisation may be minimised by using very tight states of cross-linking and, where possible, by compounding with increased amounts of filler and petroleum plasticisers, including resinous types. In CR, ester plasticisers should be avoided since they tend to increase crystallisation rates the reasons for which are not fully understood. Exposure of all rubbers to low temperatures also produces a reversible stiffening, under static conditions, at a temperature dependent on the glass transition temperature T_g of the polymer. Under dynamic conditions, cracking may occur if the stiffening is severe.

For optimum results with respect to low-temperature flexibility and with regard for other essential properties, formulations should be chosen to combine the polymer having the lowest T_g and, if possible, being free from crystallisation tendencies and having a high degree of cross-linking, with compatible plasticisers, also having the lowest possible freezing point (e.g. adipates, sebacates, paraffinic process oils, or butyl oleate, as appropriate).

Silicone rubbers have extremely good resistance to low temperatures, but, where their cost or mechanical property level cannot be entertained, blends of NR with low-*cis* variants of BR give good results. Compounding with ester plasticisers of the sebacate or adipate type is frequently necessary.

Electrical Properties

Electrical properties of vulcanisates depend not only on the choice of polymer but also on the level and nature of filler and the type of plasticiser. Compounds of NR, those based on specially finished types of SBR with low water extracts and IIR are all capable of giving products characterised by high resistivity.

CR, CSM, and NBR-PVC blends are suitable as insulants in low-voltage work but are more applicable as outer protective coatings over insulating layers where weathering resistance or solvent-resistant characteristics are necessary.

Carbon blacks detract from electrical resistivity, their precise effect depending on ultimate particle size, degree of structure, and level of dispersion in the vulcanisate. The lowest loadings of the least reinforcing types will detract least from insulating characteristics. Of the reinforcing blacks, channel types give highest and high structure furnace give lowest resistance, particularly if compounded into normal grades of SBR or NBR. The super conductive grades of furnace blacks and acetylene blacks are special purpose materials for those applications requiring antistatic or conducting properties in the final product, e.g. for the dissipation of static electricity.

Mineral fillers, such as clays, whitings, calcined clays, and coated silicas, especially those with low water-soluble contents, give appreciably higher electrical resistance and dielectric characteristics than carbon black. Some synthetic aluminium silicates are detrimental and can give non-black anti-static products.

When considering choice of plasticisers or processing aids, petroleum oil softeners will be found to give relatively high volume resistivity of the order of 10^{13} Ω cm, paraffinic oils givin higher resistance than naphthenic or aromatic types. Ester plasticisers vary considerably in their effects, dioctyl phthalate and dioctyl sebacate having resistivities of 10^{11} Ω cm, whereas those of phosphate plasticisers can be as low as 10^{8} Ω cm. Special plasticisers are available for antistatic and conductive rubbers. Possible migration effects of plasticisers into insulation stocks should be borne in mind when compounding outer protective stocks. Compounding for minimum water absorption will assist in maintaining good electrical characteristics, and the use of copper-inhibiting antioxidants, e.g. di-β-naphthyl-*p*-phenylenediamine, in insulation stocks, particularly those based on NR, is desirable. For all insulation applications, the best possible dispersion of ingredients is essential. When considering insulating compositions, the dielectrical strength (the voltage difference per unit thickness at which electrical breakdown occurs) is an important characteristic. Mechanical flaws, filler agglomerates, or other inhomogeneities can have a marked effect on such measurements, and hence the need for a high degree of ingredient dispersion.

In general terms, volume resistivity values above 10^{8} Ω cm are regarded as insulating, between 10^{4} Ω cm and 10^{8} Ω cm as antistatic, and below 10^{4} Ω cm as conducting.

Resistance to Gas Permeation

Permeation of a gas through a vulcanised polymer is influenced by the nature of both the polymer and the gas, and is governed by the diffusivity and the solubility levels of the gas in the polymer concerned.

Solubility of gases

For a particular polymer, gas solubility increases as the molecular weight of the gas increases and is also greater for gases of increased polarity. Solubility is influenced by temperature in accordance with the Clausius–Clapeyron equation. In different polymers, the solubility of a given group of gases follows a similar pattern but tends to decrease as the solubility parameter of the polymer increases.

Diffusivity of gases

The influence of polymer type on diffusivity depends on the presence or absence of polar groups, such as acrylonitrile, and also whether methyl groups are present. Both polar and methyl groupings decrease diffusivity appreciably, and this accounts in part for the low permeability levels both of high-nitrile NBR and of butyl rubbers.

Diffusivity in a given polymer is influenced by the molecular volume and shape of the gas concerned, the state of cross-linking within the normally used range having a relatively smaller influence.

Permeability

In addition to diffusivity, solubility, and temperature considerations, permeation rate varies with the pressure differential initiating the permeation and with the thickness of the polymer barrier involved. By reducing the relative volume content of polymer in the vulcanisate, increase in filler content reduces permeation; fillers having lamellar particles, such as talc, mica, or graphite, produce a proportionally greater reduction in permeability than an equivalent volume of more regular fillers. Other mechanical properties, however, are adversely affected by such fillers, tear resistance being particularly influenced. Wherever thin vulcanisate layers are used to act as permeation barriers, it is also important to ensure that all particulate fillers are well dispersed and free from agglomerates and gritty impurities, otherwise mechanical imperfections in the sheeting can seriously impair gas retention properties.

Contact with Foodstuffs, Drugs

Where rubber products may come into contact with foodstuffs, drugs, or drinks, certain restrictions on choice of polymers or compounding ingredients arise.

ENVIRONMENTAL REQUIREMENTS IN COMPOUNDING

In addition to developing products to satisfy customers, the environmental implications of the technology must be taken into consideration. The environmental impact on compound development must be viewed in two parts: (i) product use and long-term ecological implications; (ii) health and safety, in both product service and product manufacture.

An example of the impact of product usage and the environmental implications is tyre rolling resistance and its effect on vehicle fuel consumption. Reduction in tyre rolling resistance results in a drop in vehicle fuel consumption. This has an immediate impact on the generation of exhaust gases such as carbon monoxide, carbon dioxide, and nitrous oxides.

The crown area of the tyre which includes the tread and belts, accounts for approximately 75% of the radial passenger tyre rolling resistance. Improvements in the hysteretic properties of the tread compound will therefore enable a reduction in tyre rolling resistance and consequent improvements in vehicle fuel economy. The crown area and particularly the tread compound also affects the life cycle of the tyre. Longer-wearing tyres (including retreading) delay the point in time when used tyres must enter the solid waste disposal system.

Critical to a tyre's life cycle performance is the ability to maintain air pressure. Tyre inner liners composed of halobutyl-based compounds exhibit very low air and moisture permeability. Therefore, tyres built with the proper selection of compounds can reduce the rate of premature failure, again delaying entry into the scrap tyre and solid waste streams.

Improved tyre designs have enabled reduction in noise levels. This has become an important environmental consideration. Optimum footprint pressures reduce damage to highway pavements and bridges. All of these improvements in tyre rolling resistance, life cycle duration, noise generation, and tyre footprint pressure have been incorporated into the full range of tyres, from small automobile to heavy truck to large earthmover equipment tyres. Today's radial tyres use 60 to 80% natural rubber as the polymer portion of compounds. Because natural rubber is obtained from trees, it is an ideal renewable resource, and thus as a biotechnology material is preferred to petroleum-based synthetic polymers.

Tyres are one of the most durable technological products manufactured today. They are a resilient, durable composite of fabric, steel, carbon black, natural rubber, and synthetic polymers. The qualities that make tyres or other engineered rubber products a high-value item create a special challenge of disposal. Tyres and other rubber products, such as conveyor belts and hydraulic hoses, are not biodegradable and cannot be recycled like glass, aluminium, or plastic. Four potential applications for such products entering the solid waste stream have been identified :

1. The calorific energy of tyres is higher (35 MJ/kg) than that of coal (24 MJ/kg). With properly designed equipment, tyres can be burned to produce heat in cement kilns.
2. Tyres can be burned in furnaces at power-generating facilities to produce electrical energy.
3. Ground up scrap tyres are beginning to find use in some special asphalt applications.
4. Tyres with the proper installation technology can serve a variety of applications in the construction industry as marine reefs, energy-efficient house construction, highway bank reinforcement, and erosion control.

These four methods of disposal represent the best options for scrap tyre and rubber products disposal. It is anticipated that a variety of new applications for disposal of scrap rubber products will emerge in the future.

In summary, materials scientists must consider the implications of their materials choices, from the quantity of energy to manufacture the product, to the performance during its useful life cycle, and finally to disposal methods. The Environmental Protection Agency (EPA) also provides constraints that the materials scientist must consider in the design of compounds. As most rubber compounds contain approximately 6 to 20 different materials, not only the materials themselves must be clean and harmless, but any by-products that form during product tyre manufacturing must also be harmless to humans and the environment.

The aromatic content of carbon blacks and oils was once considered hazardous. Data were generated that showed that carbon black was stabilised and did not represent a hazard to workers. Resins for cure or tack, antioxidants, antiozonants, and cure accelerators also must be investigated to ensure that the material and any impurities meet changing health and safety standards. Materials safety data sheets and the chemical health and toxicity data must be maintained on all materials.

Nitrosamine-generating chemicals represent an area where suspect materials have been removed from rubber products, even though no governing legislation has yet been drafted. Nitrosamines can be formed when secondary amine accelerators are used to cure rubber. These accelerator changes have a very significant effect on the total rubber industry.

Solvent composition and volatility limits can have significant effects on synthetic rubber production and also tyre manufacturing. Limits of exposure to some trace impurities defined in the Clean Air Act are to be based on the hazard represented, not simply the best available measurement capability.

Thus, the material scientists must continue to adjust to the changes in both the environment and health and safety standards.

CHAPTER 9

Compounding Aspects of Metal to Rubber Bonding

INTRODUCTION

The earliest methods for bonding rubber to metal used a layer of ebonite (hard rubber) applied to the metal surface before vulcanisation. During vulcanisation of the metal/ebonite/rubber composite strong bonds are developed holding the soft rubber firmly in place. This process gives good adhesion but suffers from a number of disadvantages: migration of sulphur from the ebonite to the soft rubber modifies the properties of the latter in particular, ageing performance and the ebonite layer is thermoplastic causing poor performance at elevated temperatures.

The brass plating technique produces excellent bonds by vulcanising soft rubber in direct contact with brass-plated metal. The disadvantages of this process are the high capital cost of plating equipment and the extreme sensitivity of the bond to variations in plating conditions and rubber compound composition.

In more recent years a number of highly efficient chemical bonding agents have been developed and are now widely used in the manufacture of rubber/metal composite articles. These bonding agents are normally solutions and dispersions of specially compounded polymers and resins in organic solvents. The method of use is to apply a thin, continuous film of bonding agent to the cleaned metal surface, by brushing, spraying or dipping. After time has been allowed for the organic solvent to evaporate the rubber is placed in contact with the bonding agent and the assembly is vulcanised.

In many applications excellent performance of bonded units is achieved with a single coat of one bonding agent. However, in applications involving severe dynamic treatment or use in unfavourable environments best performance is achieved with a two coat bonding system, the second bonding agent (cover coat) being applied on top of the dried film of the first (primer). Before discussing the effects of compounding variables, it is necessary to consider how the strength of adhesion can be measured and to reach the alarming conclusion that reliable measurement of the true strength of adhesion is impossible.

MEASUREMENT OF STRENGTH OF ADHESION

Firstly any attempt to measure strength of adhesion requires measurement of the strength of a joint. Application of a breaking force produces non-uniform stresses within the joint and failure occurs when the stress at any point exceeds the strength at that point. In most types of joint the distribution of stress is extremely complex and it is impossible to calculate exactly how the maximum stress is related to the applied breaking force, in other words, the local breaking stress is not known and, therefore the true strength of adhesion (assuming failure at the interface) cannot be known.

As illustration consider the direct tension test (ASTM D429 Method A) used for assessing rubber-metal bonding. The marked difference in stiffness between the rubber and metal causes "necking" of the rubber when the separating force is applied, and this produces stress concentrations in the rubber as indicated in Fig. 9.1.

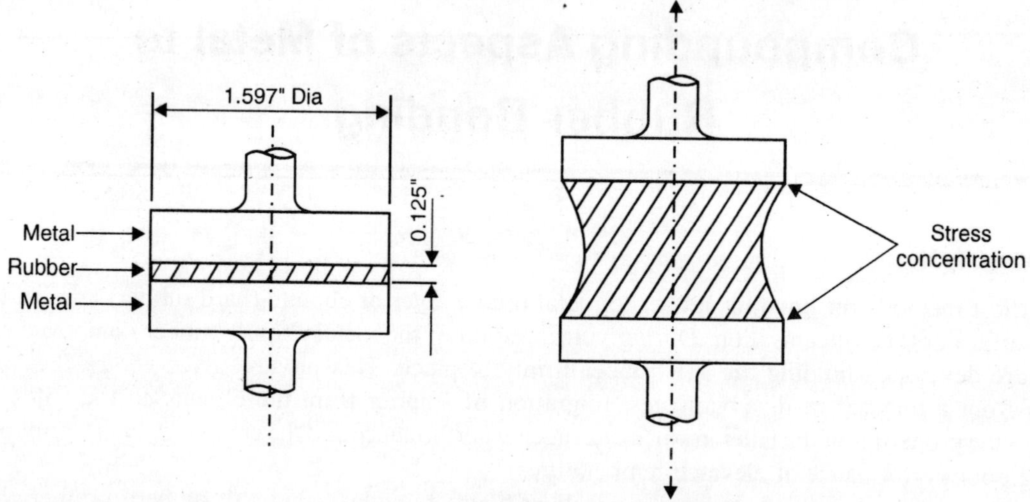

Fig. 9.1. Direct tension test.

The non-uniform stress acting across the rubber-metal junction is at no point as high as that contracted in the rubber.

For the standard test-piece-dimensions the stress concentration increases as the elongation of the rubber increases. That is to say, the concentration of stress at failure depends on the elongation at break of the rubber compound. In this way it can be seen that the concentration of stress in a direct tension test and, hence, the average breaking stress in such a test (breaking load divided by area of metal surface) is dependent on physical characteristics of the rubber compound.

As a second illustration in Fig. 9.2, consider the peeling test (ASTM D429, Method B and BS 903, Part A 21 Method B).

It is generally recognised that the strength of adhesion cannot be calculated from the peeling force, F, because the area over which F is applied is not known, but it is equally generally and erroneously assumed that a comparison of peeling forces constitutes a comparison of strengths of adhesion.

A second reason why joint strength is not necessarily a measure of strength of adhesion is that joints rarely fail cleanly at an adhesive interface: failure usually occurs within either the bonding agent or the rubber.

In the direct tension test, it has been shown that the highest stress in the joint occurs within the rubber: the stress across the rubber metal boundary is non-uniform, but it is at no point as high as the maximum stress in the rubber. The joint fails when the maximum local stress exceeds the maximum local strength, and unless the adhesive strength is unusually low, the high stress concentration will cause failure to start within the bulk of the rubber.

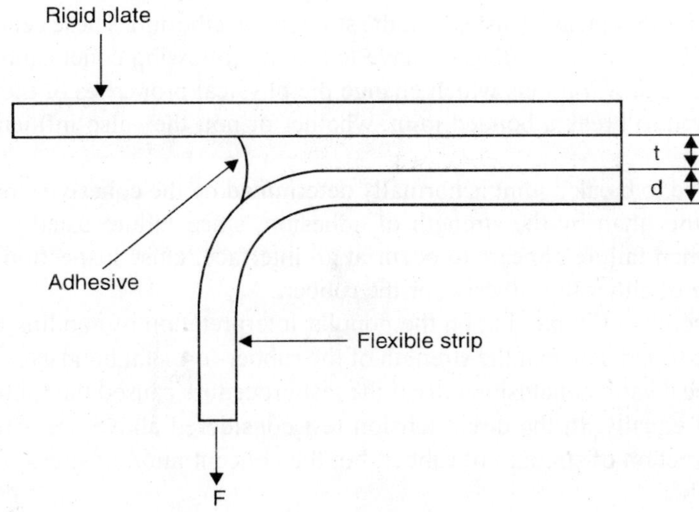

Fig. 9.2. Peeling test.

In the peeling test, also failure is likely to occur within the bulk of the rubber, for an efficient bond always causes a layer of rubber to remain on the metal on completion of peeling. The common occurrence of failure within the bulk of the rubber in standard methods of joint testing led to the development of the cone test. In this test the distribution of stress is such that failure usually occurs close to the bonded interface. The test specimen is shown in Fig. 9.3.

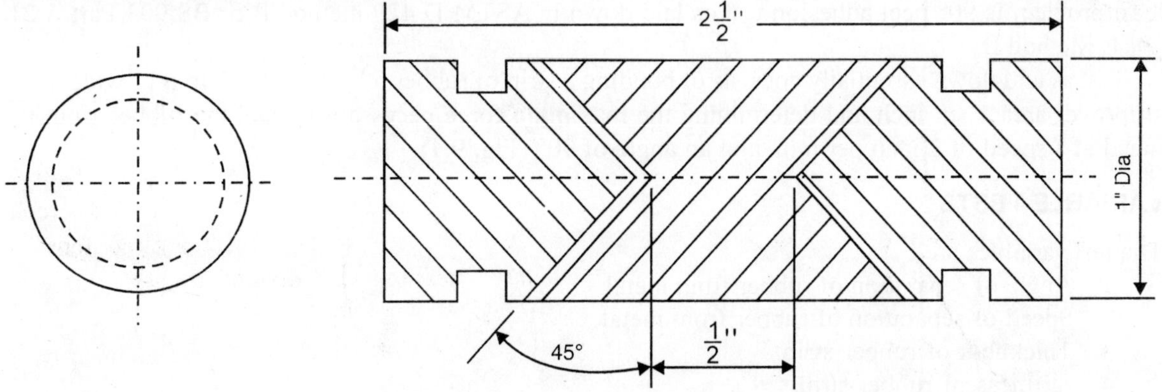

Fig. 9.3. Cone bonding test specimen.

On extension of the specimen bond failure is initiated between the tips of the metal cones, at or very close to the rubber/adhesive interface, and as extension proceeds the separation progresses down the side of the cones. The test is claimed to enable differences to be assessed between the bonding agents which would normally show complete failure in the rubber in 90° peel or direct tension tests.

Even if failure in the cone test, the direct tension test or the peeling test appears to occur at the metal surface, it is most unlikely to have followed exactly the microscopic hills and valleys on the surface, and some failure within the bonding agent is highly probable. So even in such instances, the test is assessing the strength of the adhesive as well as or instead of, the strength of adhesion. These general considerations of methods of testing rubber-to-metal adhesion have led to the following conclusions.

1. Rubber compounding variations which change the physical properties of the rubber will affect the force required to break a bonded joint, whether or not, they also influence the strength of adhesion.
2. The load required to break a joint is normally determined by the cohesive strength of one of the components rather than by the strength of adhesion, since failure usually occurs within the rubber. Even when failure appears to occur at an interface, close inspection is likely to reveal cohesive failure of either the adhesive or the rubber.

It is appropriate to comment in passing on the popular interpretation of bonding tests where failure within the rubber is said to indicate that the strength of the rubber-to-metal bond exceeds the strength of the rubber. This could be a valid conclusion only if the test procedure caused the rubber and the bonded interface to be stressed equally. In the direct tension test considered above, the strength of adhesion might be only a small fraction of strength of rubber, but the concentration of stress would cause failure to occur within the rubber.

Conclusions 1 and 2 above show that it is impossible to use conventional bonding test to study the effect of rubber compound composition on strength of adhesion. However one of such method is necessary to make comparative studies.

First method that is the tension test suffered from its inherent problem of high degree of non-uniform stress concentration and also any spot of weak bond is covered up by surrounding areas of strong bond and hence escapes detection. The third method, that is cone method, though superior to other two methods involves a complicated transfer moulding mould and complicated metal inserts which makes it difficult to get correctly moulded pieces. The most widely used method which has the least problem is 90° peel adhesion test as laid down in ASTM D 429 method B or BS 903 Part A 21, 1961, Method B.

Test as laid down essentially consists of bonding a strip of rubber 5" × 1" × 0.25" to a 1" wide metal strip over area 1 sq. inch and determining the maximum force necessary to peel the rubber from the metal at a speed of 2 inch per min. and an angle of 90° (Fig. 9.4).

VARIABLE TESTS

The test variables are :

1. Angle of separation of rubber from metal.
2. Speed of separation of rubber from metal.
3. Thickness of rubber strip.
4. Stiffness of rubber strip.

The effects of these variables on peeling force are quite independent of adhesion strength.

Angle of Separation

The general trend is for peeling force to increase as the angle of separation increases from 40° to 150°. The pattern of bond failure also changes markedly although in all tests failure occurred within the rubber strip. At separation angles of 30° and 45° the rubber peeled smoothly from the metal leaving a very thin smooth layer of vulcanised rubber on the metal surface. At higher separation angles the phenomenon of "stick slip" tearing became increasingly apparent and the number of peaks occurring on the continuous autographic record of stripping force decreased. These peaks correspond to the changes in separating force caused by the major individual tears.

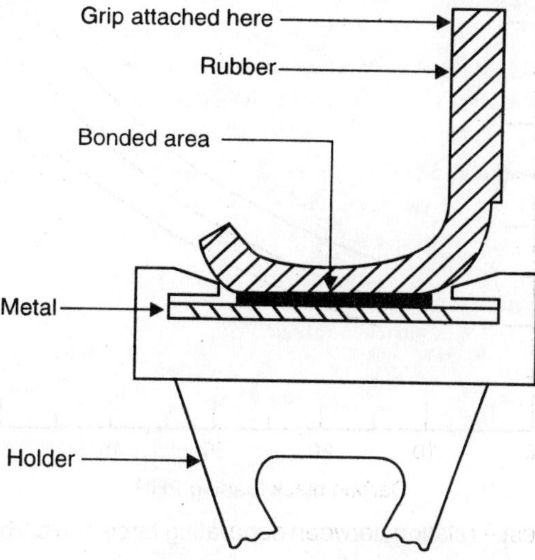

Fig. 9.4. 90 degree peel adhesion test specimen.

Speed of Separation

Peeling force increases steadily with speed of separation. The effect over range of speeds is not so marked as that achieved by increasing the angle of separation. The pattern of failure is not significantly affected by varying the speed of separation in this range.

Effect of Thickness of Rubber Strip

Peeling force increases progressively with the thickness of the rubber specimen. Because of effects of the above variables in ASTM and B.S. standard values for all these variables are specified in order to obtain comparative results.

EFFECT OF COMPOUNDING VARIABLES ON BONDING PERFORMANCE

At the risk of over simplification, it is probably true to say that the basic property of any rubber compound specified for use in the production of a bonded part is vulcanisate hardness. If economics and effects on other physical properties are temporarily ignored, then the number of combinations of different types and amounts of black process oils and curative systems which will produce a stock of given hardness is obviously very large.

In examining the effect of compounding ingredients on bonding performance the following areas have been explored :

1. The effect of varying carbon black loading at a given level of process oil.
2. The effect of varying the process oil content at constant black loading.
3. The ratio of carbon black and oil required to produce optimum joint strength at a given hardness.
4. The effect of varying the type of carbon black, maintaining constant hardness.

Fig. 9.5, shows the relation between cone joint strength and GPF black loading at fixed process oil content.

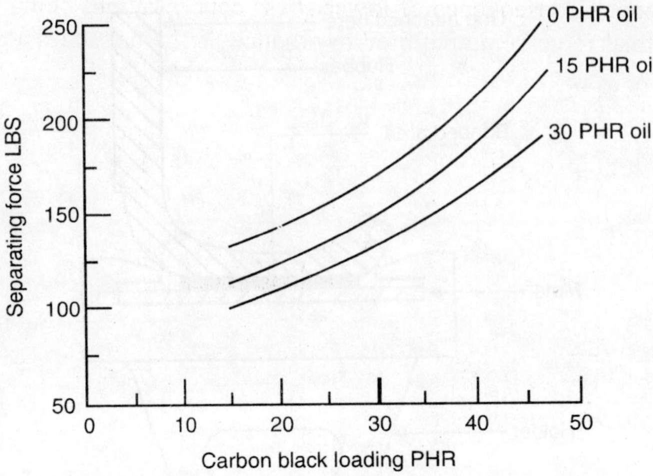

Fig. 9.5. Cone test – relation between separating force and carbon black loading.

Joint strength increases almost linearly with carbon black loading but the presence of process oil reduces the increase in strength resulting from higher black loading.

A plot of cone joint strength vs process oil level (Fig. 9.6) shows a steady decrease in strength with increasing oil level, the effect being most noticeable at higher black loading.

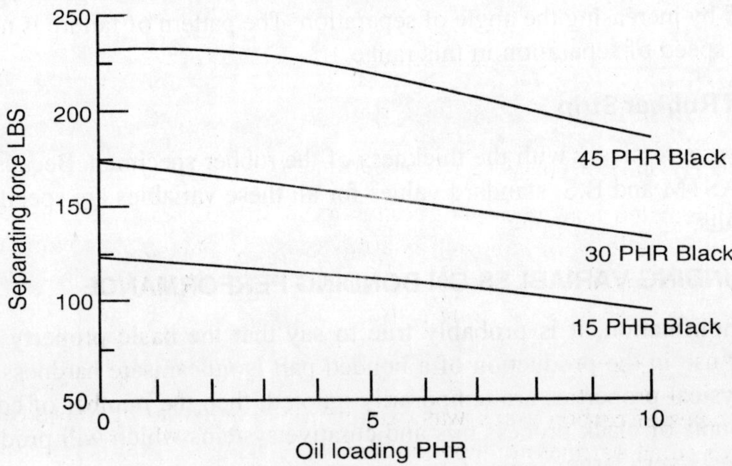

Fig. 9.6. Cone test – relation between separating force and oil loading.

Contours of constant cone joint strength were plotted against black and oil loading in Fig. 9.7. These should be compared with the contours of constant hardness in Fig. 9.8 and constant modulus in Fig. 9.9 It is clear that the changes in black and oil content which produce increases in hardness and modulus also produce increases in cone joint strength. The effect can be more clearly seen in Fig. 9.10 and Fig. 9.11 where cone joint strength is plotted against compound hardness and compound modulus respectively.

From these graphs it is concluded that the strength of a cone joint increases as the stiffness of the rubber compound increases. It is not known whether the variations in black and oil content which caused these changes in stiffness had any effect on the strength of adhesion, but it seems most likely that the observed effect is a consequence of lower stress concentrations being produced in the stiffer compounds, a greater total force being required to produce the failure stress at the tip of the cones.

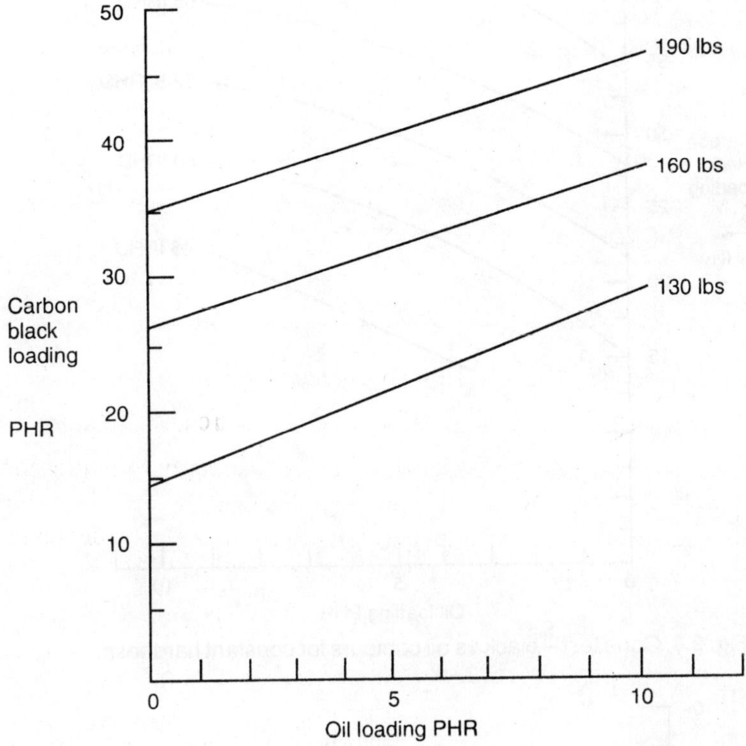

Fig. 9.7. Cone test – black vs. oil contours for constant separating force.

It has already been indicated that increasing compound stiffness in the peeling test will cause the peeling force to increase if the level of adhesion is not changed. The graphs showing the variations in peeling force with the variations in black and oil loading (Figs. 9.12, 9.13 and 9.14) do not show such a simple relationship as was found for the cone tests.

EFFECT OF CARBON BLACK TYPE

The effect of varying types of carbon black was examined by comparing the physical properties and bonding performance of equal hardness compounds prepared using MT, GPF, HAF and ISAF blacks. The compound formulations, physical test data and joint strength results are given in Table 9.1.

The separating forces in both cone and 90° peel tests show a slight downward trend as the particle size of the black decreases from MT through to ISAF.

Joint Strength in Peel and Cone Test

Using the specified bonding system and simple natural rubber compounds it has been shown that joint strength in peel and cone test generally increases with:

1. Increasing compound stiffness (by black and oil adjustment)
2. Increasing compound resilience at constant compound stiffness (by choice of carbon blacks with differing particle size) (Fig. 9.15).

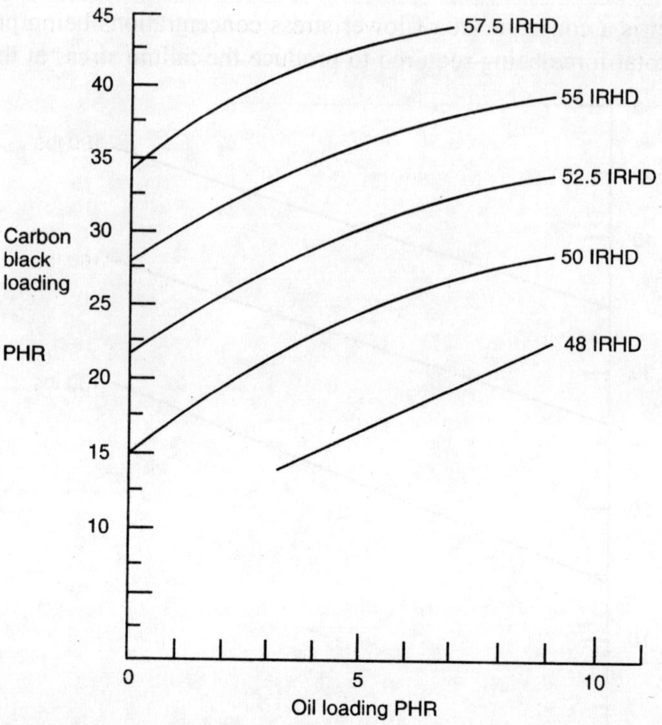

Fig. 9.8. Cone test – black vs oil contours for constant hardness.

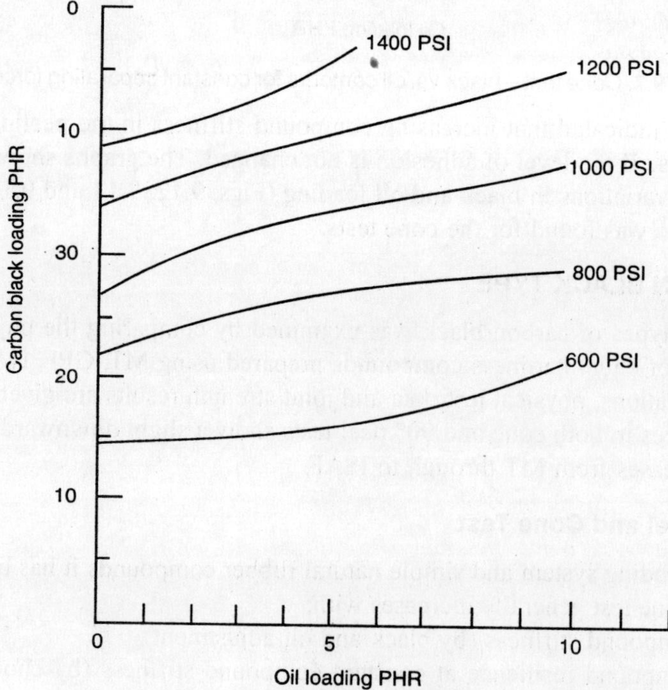

Fig. 9.9. Cone test – black vs oil contours for constant 300 per cent modulus.

Table 9.1. Effect of Carbon black type.

Name of material		*A*	*B*	*C*	*D*
Smoked sheets	— — — —	100.0	100.0	100.0	100.0
Zinc oxide	— — — —	5.0	5.0	5.0	5.0
Stearic acid	— — — —	1.0	1.0	1.0	1.0
Oil	— — — —	5.0	5.0	5.0	5.0
Phenyl beta naphthylamine	— — — —	1.0	1.0	1.0	1.0
Sulphur	— — — —	2.25	2.25	2.25	2.25
Cyclohexyl benzyhiazyl sulphenamide	— —	0.6	0.6	0.6	0.6
MT Black	— — — —	60.0	—	—	—
GPF Black	— — — —	—	30.0	—	—
HAF Black	— — — —	—	—	27.0	—
ISAF Black	— — — —	—	—	—	25.5
Mooney Data					
Minimum viscosity at 121°C	— — — —	33.0	34.5	44.0	49.5
Mooney scorch at 121°C mins	— — — —	46.5	29.5	17.5	11.0
Physical Properties (cure 10' at 153°C)					
Modulus at 100%	lb/in^2	150	100	120	120
Modulus at 300%	lb/in^2	900	830	870	750
Tensile strength	lb/in^2	3020	3600	4080	3500
Elongation at break	%	630	660	700	630
Tear strength	lb/in	350	470	540	570
Hardness	IRHD	57.0	54.5	55.0	54.0
Compression set (22 hrs at 70°C)	%	44.1	40.6	39.7	36.6
Rebound resilience at 50°C	%	80.2	79.8	76.8	74.1
Bonding properties 90° peels test – separating force	lbs/1"w	104	98	87	75
No. of peaks/inch		7.4	9.8	15.6	19.3
Cone test separating force	lbs	178	156	156	148

Since the metals are polar it follows that polar elastomers will be easily bonded than non-polar elastomers and from this the concept of a Bondability index has been developed (Fig. 9.16). The higher the index the more readily will the elastomer give good bonds to metal when using a single coat system. The lower the index the greater necessity for using a two coat system — a primer to give bonding to the metal and a top coat to give good bonding to the elastomer.

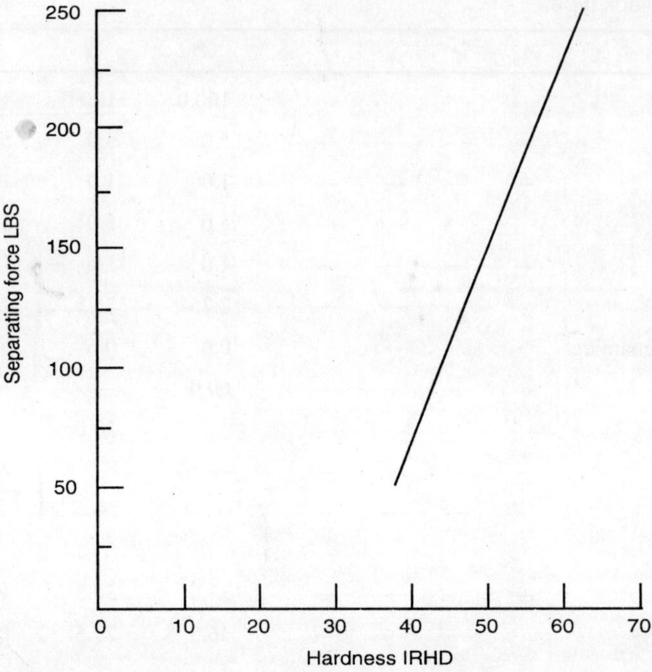

Fig. 9.10. Cone test – relation between separating force and hardness.

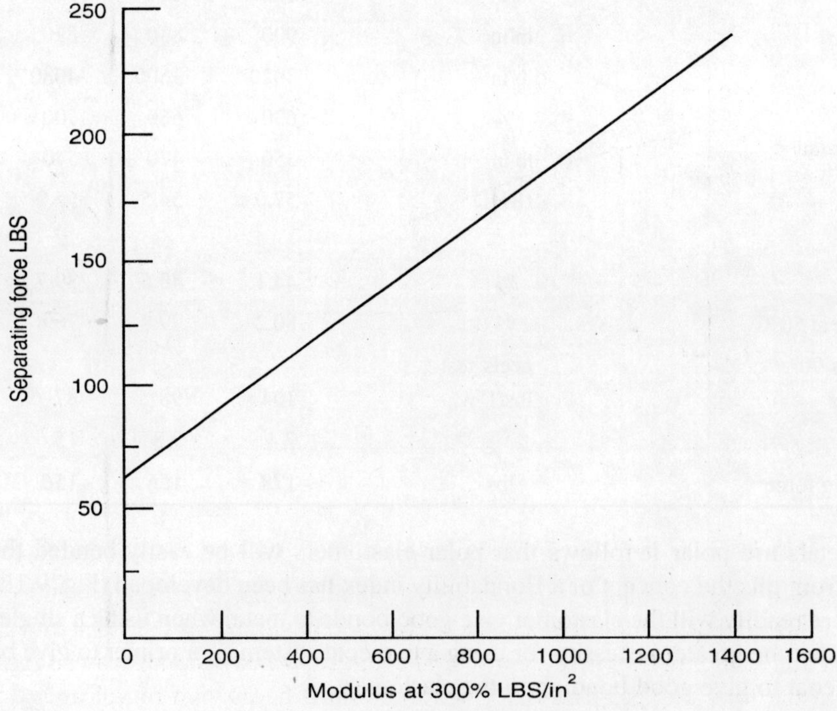

Fig. 9.11. Cone test – relation between separating force and 300 per cent modulus.

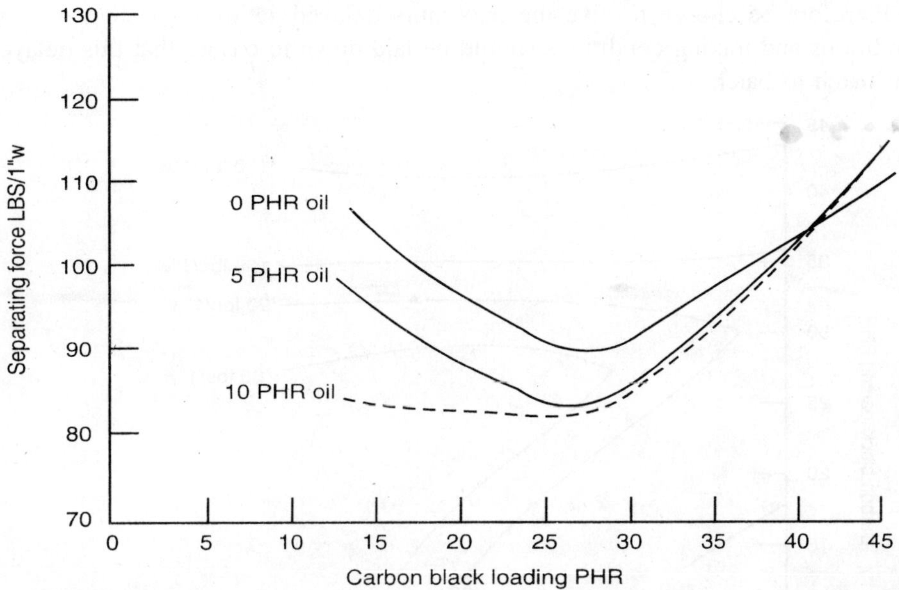

Fig. 9.12. 90° peel test – relation between separating force and carbon black loading.

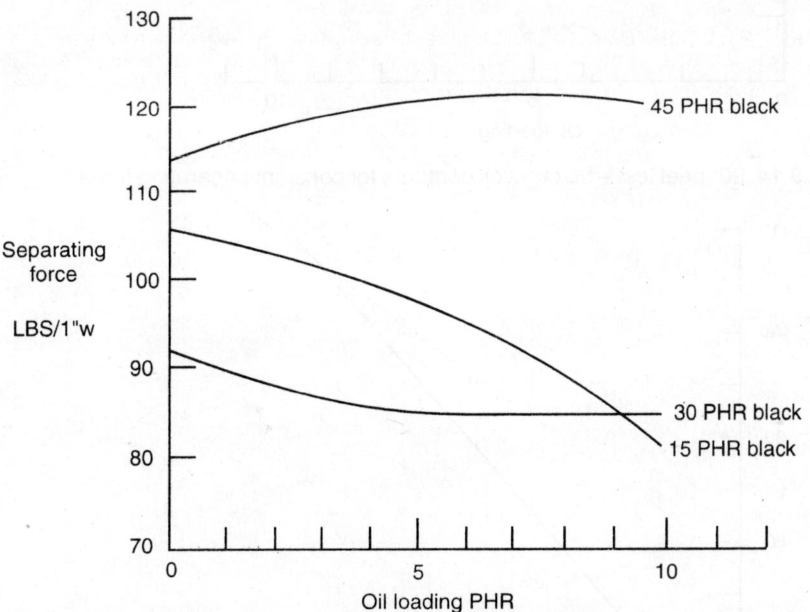

Fig. 9.13. 90° peel test – relation between separating force and oil loading.

In order to obtain strong bonds, it is essential that the elastomer/cement interface should be formed under conditions which ensure that the cement is completely wetted by the stock and that the correct concentration of active ingredients is present so that chemical primary bonds may be formed. From this it follows that at the time the interface is formed there must be no sign of vulcanisation. The curing system must therefore be chosen to give the maximum delayed action consistent with economical moulding conditions and mixing conditions should be laid down to ensure that this delayed action is constant from batch to batch.

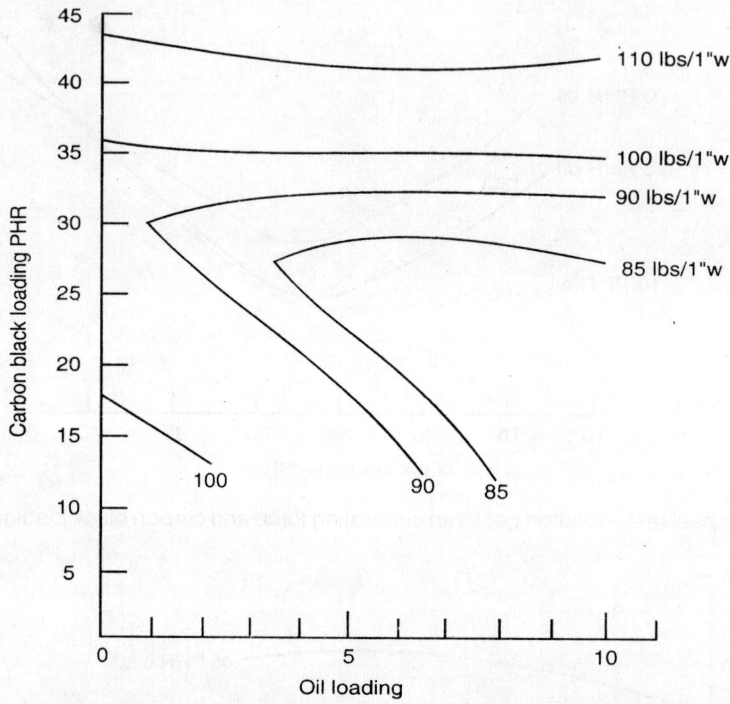

Fig. 9.14. 90° peel test – black vs oil contours for constant separating force.

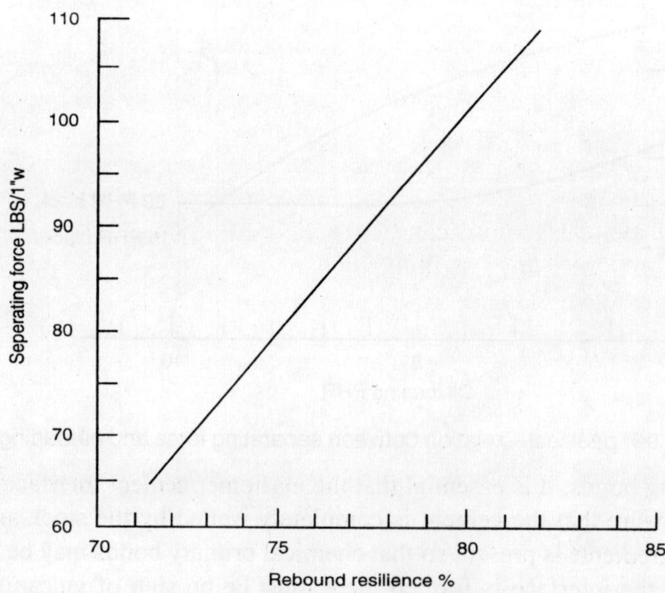

Fig. 9.15. 90° peel test — relation between separating force and rebound resilience.

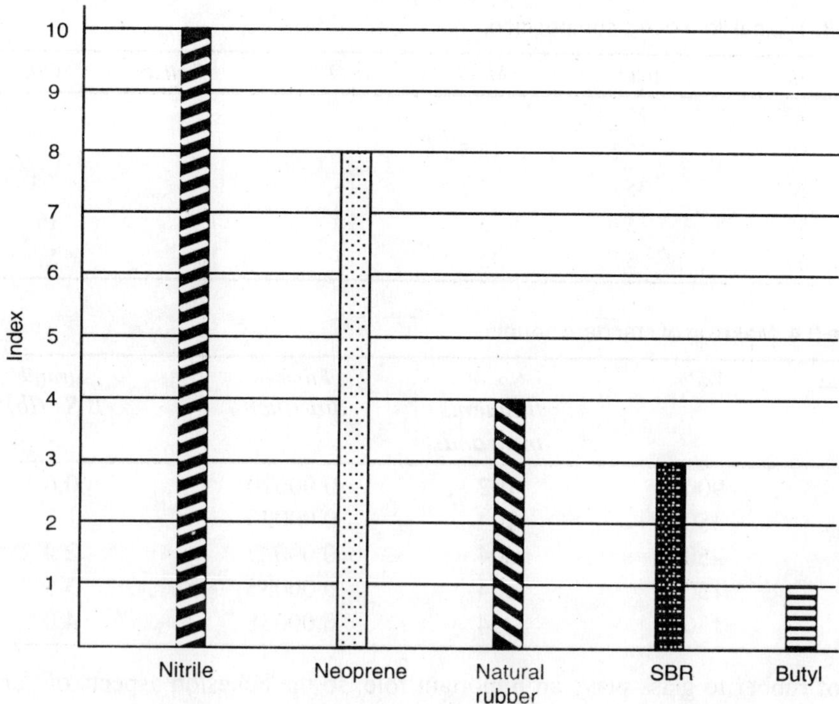

Fig. 9.16. Bondability index of common elastomers.

It is very important that the softening oil should be chosen with considerable care. If the oil is not fully compatible with the elastomer, bonding is severely affected. Aromatic or ester type oils are generally unsatisfactory and best results are obtained with naphthenic types. Highly oil extended grades of synthetic rubbers are not recommended. Much better results will be obtained by using polymer which is not extended or contains small quantities of oil and then adding a naphthenic oil to the desired level.

Any compounding ingredient which blooms to the surface of the elastomer will form a weak layer to which it is impossible to bond. Such ingredients should therefore be avoided or if they are absolutely necessary, the stock will have to be reblended immediately before bonding.

Some special compounding ingredients give severe bonding problems and must therefore be avoided. Copper inhibitors come in this category likewise, stannous chloride should not be used as an activator for resin cured butyl. Fortunately 'Hypalon' can be used in substitution and this gives no bonding problems. Some distinctive properties of glass fibre and their composition are shown in Table 9.2 and 9.3 where as Table 9.4 highlights the standard counts.

Table 9.2. Some distinctive properties of glass fibre.

High heat resistance
High strength to weight ratio
High dimensional stability due to low E.B.
High resistance to most chemicals and solvents
Good weathering resistance
Immunity to bacterial attack
Negligible water absorption and good translucency

Table 9.3. Typical fibre glass composition.

	SiO_2	Al_2O_3	MgO	B_2O_3	*Alkali dx.*	*Other dx.*
I	65	20	15	—	—	—
II	50	32.5	12.5	—	—	5
III	50	35	7.5	—	—	7.5
IV	55	15	3	10	1	16
V	60	5	15	5	10	5

Table 9.4. Make up of standard counts.

Counts	*Yd/lb*	*No. of filaments in strands*	*Filament dia (inch.)*	*Strand B.S. (lb)*
900	90000	102	0.00020	0.6
450	45000	204	0.00020	1.2
225	25000	204	0.00029	2.4
150	15000	204	0.00035	3.5
130	13000	204	0.00038	4.0

Adhesion of rubber to glass plays an important role, so the adhesion aspects of various rubber to glass and glass cord are given in Table 9.5 and Table 9.6. For composition of rubber compounds and their broerties refer Table 9.7 and Table 9.8.

Table 9.5. Adhesion of rubbers to glass cloth treated with 1% aq. silane.

Silane	*Peel strength, ppi*					
Functionality	SBR	NR	NBR	CR	Hypalon	EPDM
No silane	Nil	0.1	Nil	Nil	1.2	Nil
Chloropropyl	0.1	0.6	0.5	0.1	1.2	1.1
Vinyl	0.9	0.7	0.1	0.1	1.8	2.2
Methacryloxypropyl	0.1	0.5	0.1	0.1	2.0	4.6
Aminopropyl	1.1	1.4	0.1	0.1	6.0	5.7
Diamine	1.2	4.4	8.0	1.1	20.0	3.5
Mercaptopropyl	3.0	2.0	1.2	4.0	2.6	8.5
Isothinronium chloride	6.3	3.9	2.0	2.0	2.7	2.9
Cationic styryl	2.0	4.8	5.5	2.2	20.0	20.0

Table 9.6. Adhesion to glass cord 600/5 x 3.

Polymer	Bonding additives Resorcinol—2.4 Cohedur A—2.4	Bond strength (20 mm embeded) in kgf.
NR	Nil	4.6
	Yes	14.5

(Contd ...)

NR/SBR	Nil	2.3
	Yes	10.4
NBR	Nil	7.9
	Yes	18.3
CR	Nil	11.4
	Yes	17.5
EPDM	Nil	1.2
	Yes	9.8
Chlorobutyl		4.8
		11.0

NB:- For the CR mix, Cohedur RK has been used in place of Cohedur A.

Table 9.7. Composition of rubber compounds.

Type of rubber	*A*	*B*	*C*	*D*	*E*	*F*
NR	100.00	50.00	—	—	—	—
SBR	—	50.00	—	—	—	—
NBR	—	—	100.00	—	—	—
CR	—	—	—	100.00	—	—
Chlorobutyl	—	—	—	—	100.00	—
EPDM	—	—	—	—	—	100.00
ZnO	5.0	5.0	5.0	5.0	5.0	5.0
St. Acid	1.5	1.5	2.5			
PAN	1.0	1.0	1.0	1.0	—	—
Ppt Silica	30.0	30.0	30.0	30.0	30.0	30.0
Spindle Oil	3.0	3.0	—	—	—	—
DBP	—	—	10.0	10.0	—	
C.I. Resin	—	4.0	4.0	—	—	
MgO	—	—	—	4.0	3.0	—
Sulphur	2.2	2.0	1.8	0.5		1.50
CBS	0.7	1.0	1.7	—	—	—
DOTG	1.1	—	—	1.0	—	—
TMTM	—	—	—	1.0	—	—
Pet. Jelly	—	—	—	—	5.0	10.0
C.I. Resin	—	—	—	—	—	8.0
TMTD	—	—	—	—	1.0	1.5
MBTS	—	—	—	—	2.0	0.5
Resorcinol	2.4	2.4	2.4	2.4	2.4	2.4
Cohedur A	2.4	2.4	2.4	2.4	2.4	2.4

Table 9.8. Properties of rubber compounds.

Properties	*A*	*B*	*C*	*D*	*E*	*F*
Mooney viscosity ML-4 @100°C	50	70	80	65	95	50
Mooney scorch Time @ 120°C	5	12	42	10	—	748
T.S. (kg/cm^2)	220	200	180	185	101	130
E.B. (%)	650	740	530	790	565	805
MOD @ 300% (kg/cm^2)	51	28	80	46	40	25
Hardness °A	60	60	65	66	51	56

Table 9.9 gives the effect of RPL pickup properties and Table 9.10 briefly highlights the application area of rubber coated fibre glass fabrics.

Table 9.9. Effect of RFL pick-up properties.

Properties	*28% pick-up*	*16% pick-up*
Cord construction	75-5/0 60 TPM	75-5/0 60 TPM
Breaking load (kg)	30	27
Adhesion at 120°C (kg/10mm)	7.3	6.4
Scott-flex (%)	200	100
Mallory tube	130	100

Table 9.10. Application area of rubber coated fibre-glass fabric alongwith advantageous properties.

Application	*Properties*
Covers for high altitude tents used by defence, trench lining (revetment fabrics), shelter cladding, cover for artillery and ground to air missiles, ammunition cover etc.	Resistance to flame, fire, heat cold and weathering.
Storage bins and grain silor	Good strength, antirodent, resistant to bacterial growth, fireproof, good weatherability.
Safety equipments such as fireproof industrial clothing, welder screens, caper aprons, gloves etc.	Fireproof, good strength and good weatherability
Fume ductings	Fireproof, heat resistance, good strengths.

SECTION III

Manufacturing Techniques

CHAPTER 10

Tyre Technology

INTRODUCTION

This chapter covers the interdependence of compound, tyre design, and the system used to make the tyres. Most important to satisfy the customer is the ability to define the purpose and requirements of the tyre, and the required purposes of the individual component parts of the tyre.

With the aid of laboratory testing and simulation tests, knowledge of basic elastomer properties, carbon black reinforcement, and cure can lead to viable compound solutions. Many alternatives may exist. Optimising the compounds to meet the customer tyre requirements and the manufacturing system needs is the purpose of tyre compound development. This very large and complicated subject has been treated primarily as an overview. Focuses on the purpose of the change and the other items that are affected by the change allows change to be managed in controllable amounts.

Complexity of customer requirements and the move away from generic tyres would seem further to complicate the development. However, by understanding "principles" and moving performance prediction from the fleet tyres, to wheel tests, to laboratory testing, to principles, the development process is simplified. Development time and costs are also reduced. Knowledge and understanding of the customer requirements, the total tyre development process, and the manufacturing system become more valuable in this industry change. The use and development of scientific tools to define critical parameters rapidly will become the way of the future. Experiments on tyres will move further to providing knowledge rather than development and release tests for each specific customer.

Customers demands quality products. This will be as true in the future as it is today. Each customer wants the best quality and performance products for the intended use. The diversity of tyres for racing, aircraft, automotive, trucks, farm vehicles, off-road, and miscellaneous industrial and recreation vehicles amounts to various sizes and types. Because of new body and suspension system designs, passenger cars now require specialised tyres. Generally, this means high performance radial tyres and/or all-season types. In many cases, the tyres need to be specifically tuned to operate on a given vehicle model. Passenger car tyres must also meet a multifaceted set of performance requirements to satisfy the customer.

Tyres are also designed to meet a wide spectrum of service and environmental conditions. Aircraft and earthmover tyres are the most advanced global tyres that are specific for service, vehicle, and environmental conditions.

The large number of combinations of service conditions that tyres must be designed for is indicated in Fig. 10.1. Tyre compound design must first address the compound purpose for the vehicle to meet service, tyre component, and composite requirements, as shown in Fig. 10.2. The tyre compound development process is an even broader and more complex subject: a flowchart (Fig. 10.3) helps to visualise the integration of requirements, design, compounding and manufacturing.

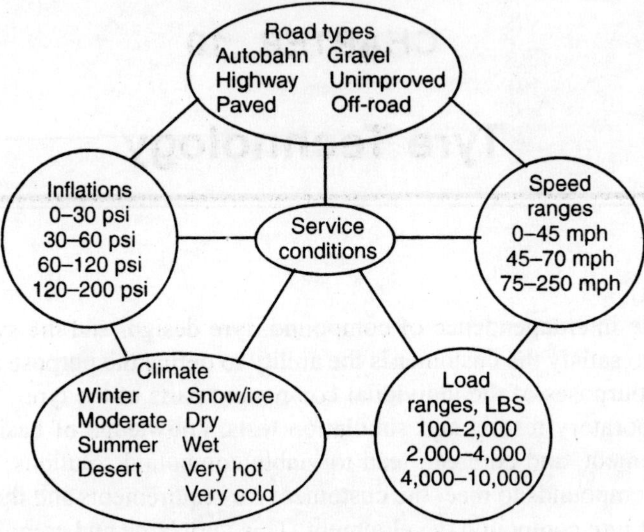

Fig. 10.1. Service conditions.

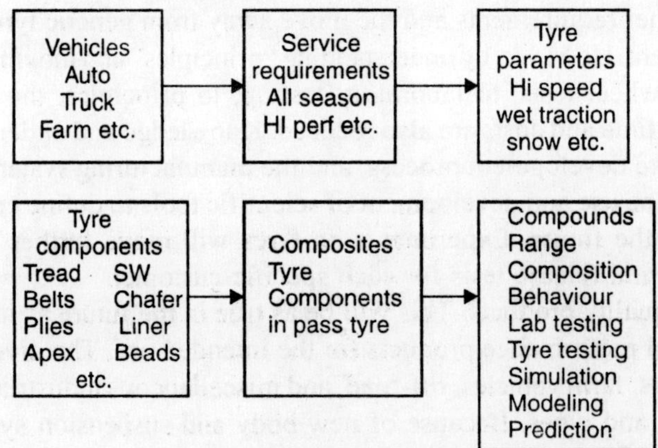

Fig. 10.2. Compound design.

Different methods are used to develop the wide range of compounds required. Some compounds have relatively simple application requirements, and some call for very complex properties that come close to the limits of materials and compounds. The same development process is not required for every compound. Where possible, compounds are standardised for manufacturing process efficiency. Some compounds are developed simply by making small changes in existing compounds. Other compounds require all the steps shown on the flowchart.

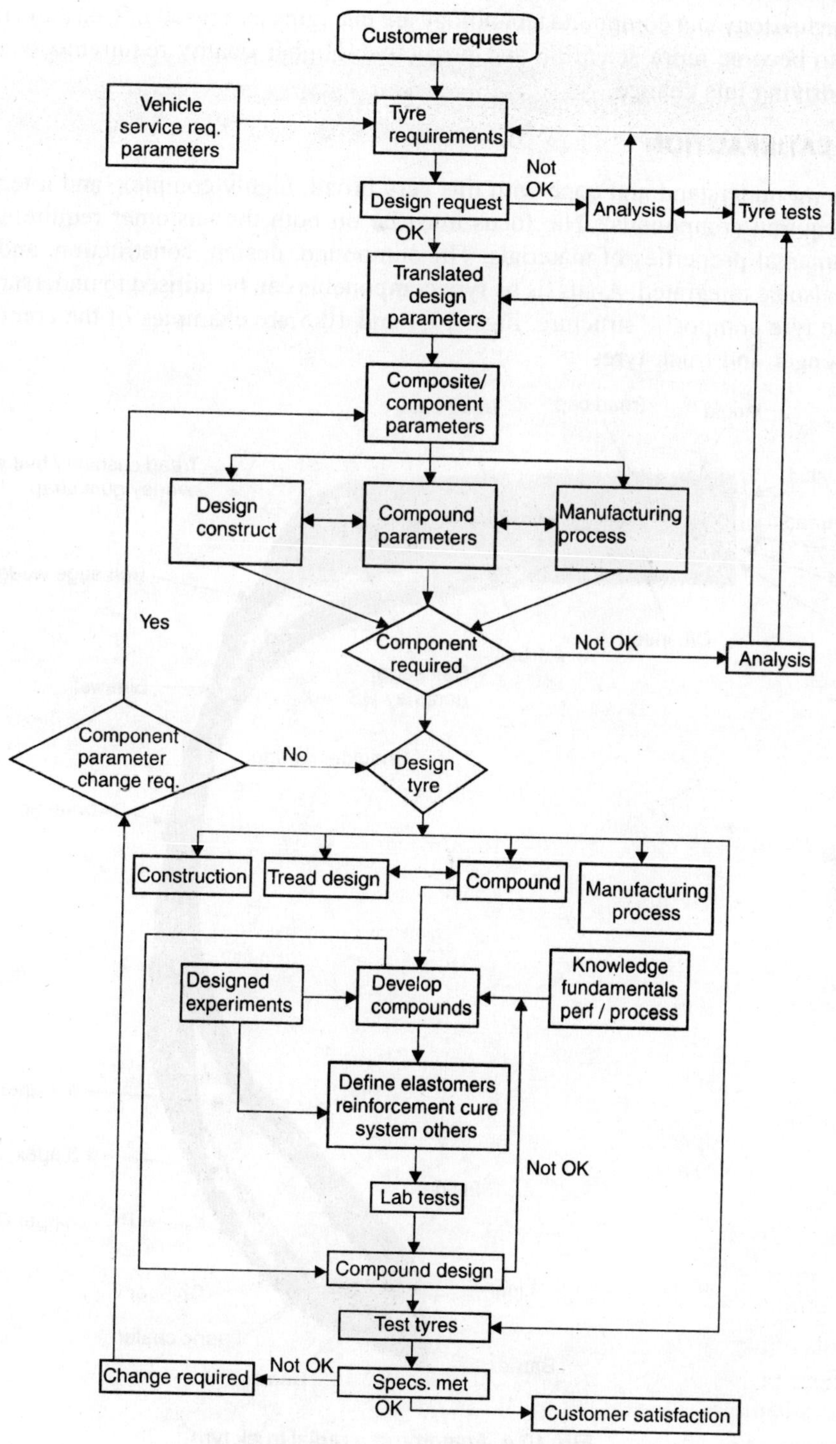

Fig. 10.3. Process flowchart.

Both tyre technology and compound technology are changing as a result of trial and error, experience, and intuition to become more scientific and predictive. Higher quality requirements and competitive pressures are driving this change.

CUSTOMER SATISFACTION

How then can we understand and cope with this very broad, highly complex, and integrated system to develop the required compounds? The focus must be on both the customer requirements (purposes) and the fundamental properties of materials. The compound, design, construction, and manufacturing systems must also be integrated. Analysis by tyre components can be utilised to understand performance in terms of the tyre composite structure. Figs. 10.4 and 10.5 are examples of the component structure of radial passenger and truck tyres.

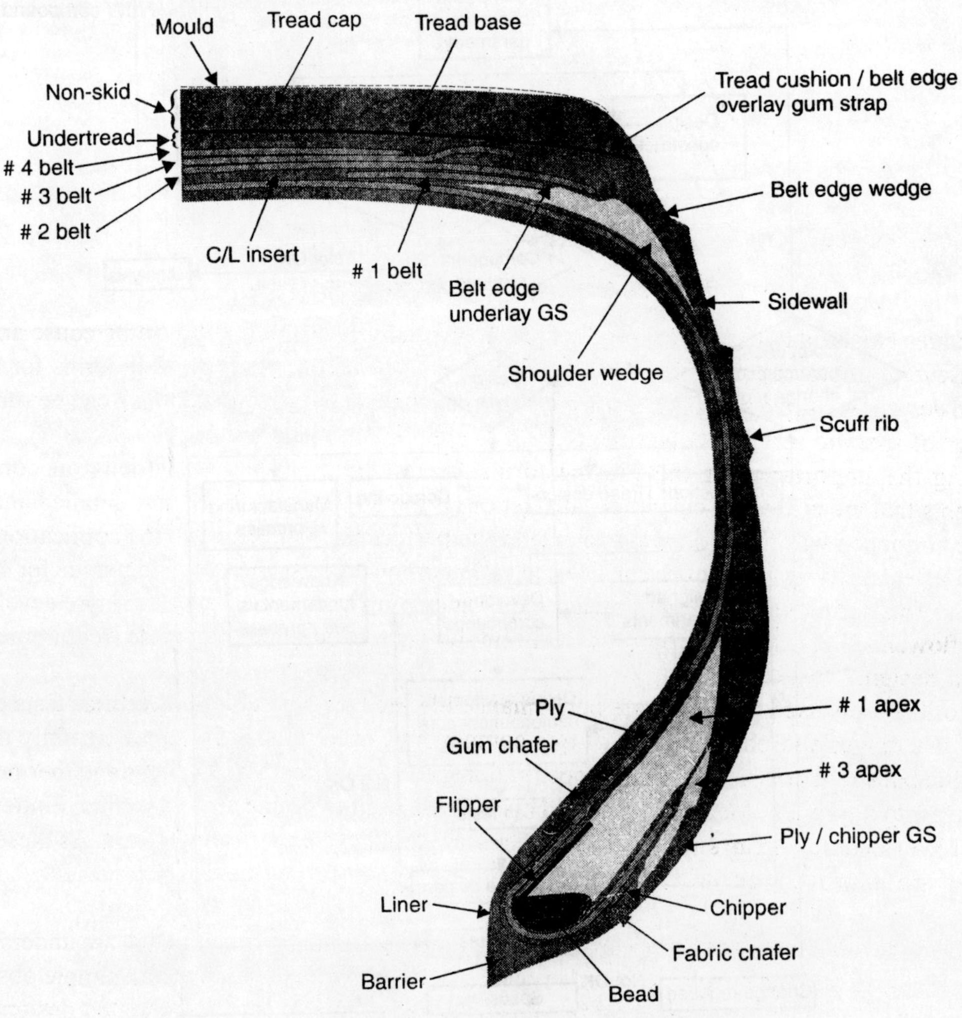

Fig. 10.4. Anatomy of a radial truck tyre.

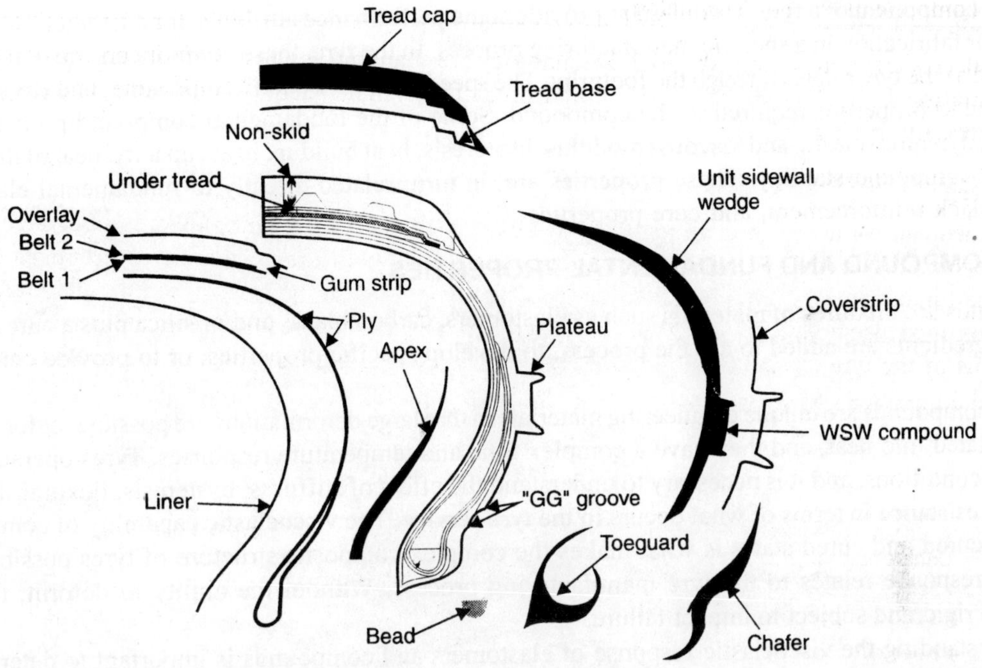

Fig. 10.5. Anatomy of a radial passenger tyre.

Techniques exist today to convert product requirements into design features using cause and effect analysis. Several important steps are involved in defining customer requirements in terms for tyre and compound design. A simple approach starts with defining the tyre requirements, which can be subdivided into ranges of specific service requirements such as speeds and loads.

Ranking the importance of the tyre performance requirements helps to focus on component requirements that are critical to customer satisfaction. If the application is not new, detailed analysis of service performance will focus on ways to extend the performance envelope. If the application is new, cause-and-effect analysis techniques can be used to determine the best alternative to pursue for customer satisfaction. This definition of the performance requirements is a most critical step in the tyre development.

In the flowchart (Fig. 10.3), areas in bold outline show the integration of purpose (requirements) and compound design.

Some of the tools used to investigate performance life are tyre laboratory and vehicle tests designed to exceed use ranges and physical tests of tyre components, used to quantify experimentally the strain cycles, principal strain directions, stress concentrations, and footprints. Strain gauges and thermocouples can be built into a tyre. Photoelastic and moiré fringes are used to define surface strains. Finite element analysis (FEA) techniques can simulate strains, stresses, strain cycles, and temperature. As these various techniques are improved, our ability to model tyre performance and design criteria with computers increases.

The objective of all these analyses is to define customer requirements in terms that are understandable to the tyre designer and the development compounder in an integrated manner. Both simple observation and very complex analysis are used to define customer requirements for the integrated design aspects.

Each component of a tyre is required to provide some performance attributes for customer satisfaction and/or for fabrication in a specific manufacturing process. In the tyre, these components are dynamically deformed as the tyre rotates through the footprint. The speed load-deflection, temperature, and environment will relate to properties required of the compound. Some of the fundamental compound properties are stiffness, dynamic (elastic and viscous) modulus, hysteresis, heat buildup, heat capacity, heat of diffusion, tear, and compound stability. These properties are, in turn, related directly to fundamental elastomer, carbon black reinforcement, and cure properties.

TYRE COMPOUND AND FUNDAMENTAL PROPERTIES

Compounds are mixtures of materials such as elastomers, carbon black, and/or silica plus a cure system. Other ingredients are added to aid the process, to develop specific properties, or to provide compound stability.

Tyre compounds are unique engineering materials in that large deformations are possible; deformations are translated into heat, and they have a complex time and temperature responses. Tyres operate under dynamic conditions, and it is necessary to understand the effect of stiffness, hysteresis, flexural strength, and tear resistance in terms of what occurs in the tyre service. The viscoelastic capability of compounds in the uncured and cured states is what makes the complex composite structure of tyres possible. The uncured response relates to the tyre manufacturing process. Without the ability to deform, the tyre would be rigid and subject to impact failure.

Understanding the viscoelastic response of elastomers and compounds is important to determining how compounds perform in the tyre under various conditions. An important characteristic of viscoelastic response, the Williams–Landel–Ferry (WLF) equation has found significant use in the empirical prediction of some tyre performance characteristics such as rolling resistance and wet traction. The WLF equation deals with the time-temperature super-position that relates properties to the glass transition temperature of the elastomer. Elastomer properties are related to or influenced by test temperature or speed with respect to its glass transition temperature T_g.

Measuring elastomer and compound properties over a range of temperatures from below T_g to the maximum service range has provided additional empirical correlations with some performance characteristics (Fig. 10.6). Comparing the curves of two compounds over a specific temperature range provides some empirical correlation with wet traction and rolling resistance.

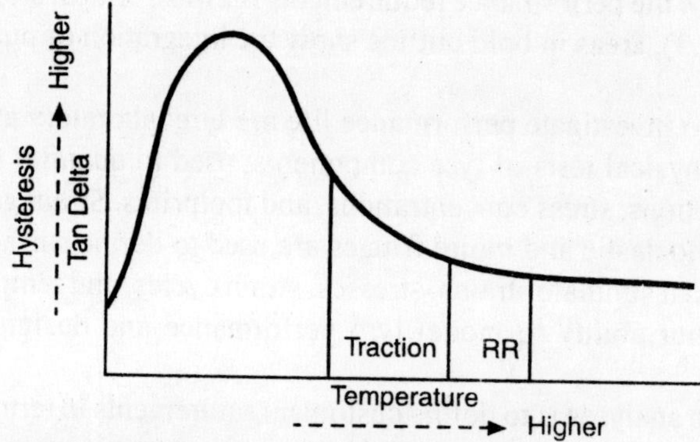

Fig. 10.6. Tan delta temperature sweep: conventional polymer.

Polymer engineering of solution SIBR-type elastomers (styrene isoprene butadiene rubber) has resulted in changing the shapes of this response curve with accompanying changes in wet traction and rolling resistance of the tyre treads (Fig. 10.7).

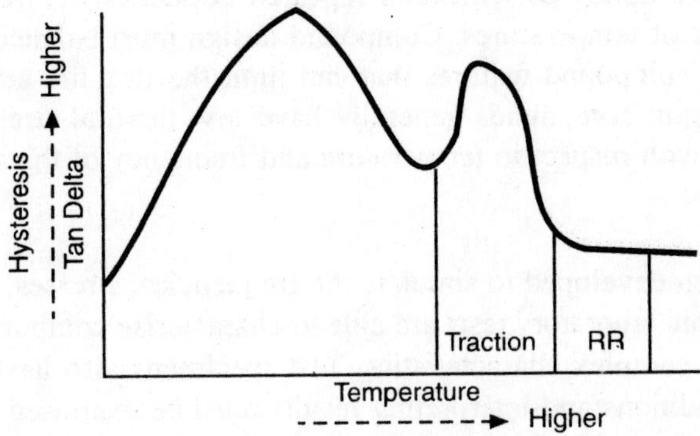

Fig. 10.7. Tan delta temperature sweep: future modified polymers.

Compound stiffness can be characterised by tension modulus, Shore A hardness, and shear or compression modulus. This property is important and relates directly to many tyre performance characteristics. Compound stiffness can significantly change wear and wet and dry traction characteristics.

An understanding of crepe and stress relaxation of compounds is also necessary to relate laboratory properties to tyre performance. The example of springs and dashpots, elastic and viscous components, helps to visualise characteristics of these types (Fig. 10.8).

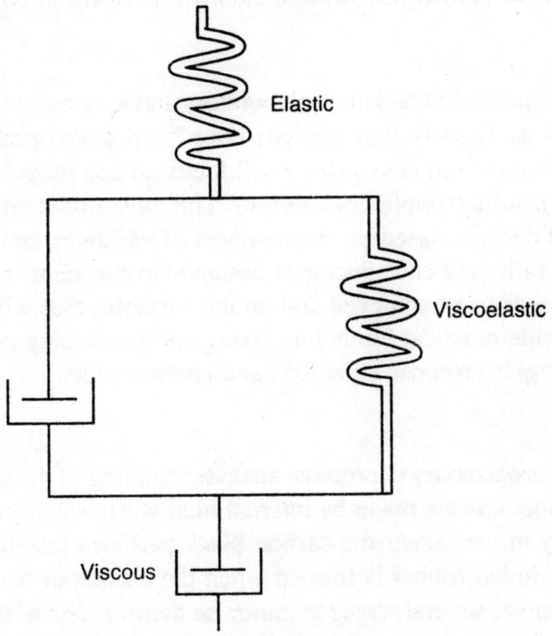

Fig. 10.8. Elastomer model for viscoelectricity: springs and dashpots.

Tear resistance, a compound characteristic related to the ability to stop a cut (tear), is fundamental to the performance of many tyre components. Tear resistance is dependent on the frequency, temperature, and load ranges involved in the tyre service.

Flexural strength is the ability to withstand repeated deflections. Tyre components see repeated strain cycles over a range of temperatures. Compound design must balance the expected life cycle of the tyre application with compound features that can limit the flex life and strength. Very hard high stiffness and low elongation compounds generally have low flexural strength. The flexural strength must be considered also with respect to temperature and frequency of the service conditions.

Laboratory Testing

Laboratory tests have been developed to simulate the frequencies, stresses, and strain cycles observed in tyres. Static and dynamic laboratory tests are able to characterise compounds only in very small and discrete portions of their complex characteristics. Test specimens also have very definite limitations. Care in selecting test conditions and interpreting results must be exercised to assure that the property measured is not simply a function of the sample size and shape, the test conditions, or the instrument. Clearly an understanding of basic viscoelastic and rheological properties is necessary for compound development and compound lab testing.

Tyre component performance requirements dictate the choice of elastomers or blends and the particle size and structure of the reinforcement material. Component properties must satisfy both the performance requirements for that particular part of the tyre and the limits of the manufacturing process. Many laboratory tests have been developed to simulate performance and process characteristics. Testing of rubber compounds in the laboratory has been standardised by American Society for Testing and Materials (ASTM) in Part 37 and by the British Standards Institution (BSI). The purpose of laboratory testing is simply to improve the ability to predict how compounds will perform in tyre service.

Designed Experiments

Compound properties can be quantified both in the laboratory and in tyre tests with designed experiments. Designed experiments such as Taguchi-type designs have been developed to help identify the most important variables. Other designs run in sequence will focus on key properties by developing the next design based on predicting results (simplex procedure). The most quantitative design experiments are the partial and full factorial designs based on the numbers of variables and levels to run. Compounds generally contain from 8 to 14 ingredients. Factorial-designed experiments can be used very effectively by focusing only on what needs to be achieved and on the variables that will have the most significant effects. Ranges that are outside practical limits for mixing and processing generally skew the data into unusual and difficult to interpret response equations and contour plots.

Mixing

A brief discussion of mixing is necessary to promote an understanding of the interaction of the elastomers and carbon black. Tyre compounds are made by internal high shear mixes called Banburys. By mixing under pressure, the Banbury mixers shear the carbon black particles into the elastomer chains with a significant buildup in heat. Bound rubber is formed when the elastomer becomes bonded to the black particles. This mixing can require several stages to minimise thermal degradation of the elastomer and to ensure sufficient dispersal of the materials. Other ingredients are also dispersed and/or melted into the mix. With high molecular weight elastomers, the curatives are generally added in a short, lower temperature stage called the final or productive mix.

COMPOUND DEVELOPMENT

The tyre component function or use in the tyre can be analysed by simple observation or by detailed analytical techniques (physical or finite element analysis simulations) and by experimental wheel tests to define property requirements. The property requirements can be used to start the compound development process. The relationship of fundamental properties in terms of service requirements depends on the viscoelastic response of the compounds. Using selected laboratory tests to simulate these conditions permits screening of compound changes. The compound-designed experiments can also be used to understand compound composition effects. Knowledge of elastomers, blends, carbon black reinforcement, stabilisation, and cure facilitates the compound development process.

Compound development from the beginning should include consideration of the various manufacturing or processing steps that will be used to form the component. These considerations must be viewed as secondary when performance limits are being expanded, however. The choice of alternatives based primarily on existing factory limits may eliminate the better tyre performance compounds. Processing constraints can often be adjusted once the performance criteria have been met or exceeded. Factory processing can be analysed and the important parameter ranges quantified using rheological properties and computer models. These properties and computer simulation models can then be used to screen the compounds developed in the laboratory. Some typical tyre compounds are shown in Table 10.1 to illustrate how the formulations are detailed. Compounds are usually defined based on 100 parts of rubber hydrocarbon.

Table. 10.1. Generic tyre compound examples.

Element or compound	*Passenger tread*	*Heavy-duty sidewall*	*Ply coat*	*Apex*
Natural rubber	–	60	60	75
SBR	75	–	40	25
Polybutadiene	25	40	–	–
Carbon black	75	60	60	80
Zinc oxide	3.0	3.0	3.0	5.0
Stearic acid	2.0	2.0	2.0	2.0
Wax	1.0	2.0	–	–
Process oil	40	5.0	5.0	10
Tackifier	–	2.0	3.0	5.0
Antioxidant	1.0	1.0	1.0	1.5
Antiozonant	1.5	3.0	–	–
Bonding resin	–	–	3.0	–
Sulphur	2.0	3.0	2.5	3.0
CBS[a]	1.6	1.0	1.4	1.4
PVI	0.2	0.2	0.2	0.3
	227.3	182.2	181.1	208.2

[a] *N*-Cyclohexyl-2-benzothiazole sulphenamide.

Raw Materials: Consistent Quality

Both compound performance and the manufacturing process demand that the materials used in tyre compounds be of consistent quality. Specifications for each material are developed to assure consistency. The need to ensure tyre customer satisfaction has made these requirements for constant and specification-correct materials very important.

Elastomers

Elastomers are high molecular weight polymers of repeating chemical structures. Natural rubber and a multitude of synthetic elastomers are characterised by both their microstructure and macrostructure (i.e., bulk properties). Elastomers used in tyres generally contain unsaturation in the form of double bonds between carbon atoms that permit curing or cross-linking. An old method explains viscoelastic response as a series of springs and dashpots. The coil springs (Fig. 10.8) represent the elastic component, and the pots, the viscous or flow type response. When a compound is stretched, it develops more resistance (stress) as more load is applied. The time-dependent portion is controlled by a viscous component.

The viscoelastic characteristics can be further visualised in a bulk state as many high molecular weight polymer chains similar to spaghetti. Many of the chains are entangled. The chains themselves have some elasticity. However, as stress is applied, some of the chains can slide past each other and become untangled. The addition of carbon black with the development of bound rubber and chemical crosslinks produces locks between chains, creating what is termed viscoelastic response. Because of the chemical crosslinks, such a compound is also considered to be thermoset. The sulphur crosslinks in the compound are permanent, but slowly they change from polysulphidic to disulphidic to monosulphidic with service life.

Another type of bonding that can occur is related to crystallinity. Above the melting point of the crystalline structure, the polymer flows more readily. Below the melting point, the material is more rigid, with only some elasticity. The crystallinity is due to the very regular structure of the polymer chains. Natural rubber (NR) has some capability to crystallise, and this is the particular property that makes it preferred for gum bands and elastic thread. NR crystallinity also provides advantages for tyre compounds. Tear resistance, flex fatigue, and hysteresis properties are enhanced by the crystallinity of NR.

Understanding both the micro and macrostructure of elastomers is another key to successful tyre compound development. These data relate directly to the viscoelastic response of the elastomer. The T_g of the elastomer, mentioned earlier, is another key point: T_g and polymer compatibility are particularly important when elastomer blends are made.

Carbon Black and Silica Reinforcement

The reinforcement of elastomers by means of the carbon black and silica is now a science. It has developed from the early channel blacks to the present controlled surface activity, structure, and particle size grades. Approximately 15–20 volume grades exist globally. Several volume grades of silica also exist. Silica reinforcement is similar to carbon black. Low levels of silica in compounds (5–20 phr) can improve the tear resistance with less heat buildup than carbon black loading provides. The choices of black and loading can be approximated from a general knowledge of the effects of particle size, structure, and affinity to the elastomer system. This knowledge provides a good first estimate of compound properties. Stiffness or modulus effects can be achieved with blacks having different levels of particle size and varying structure. Finer particle size blacks improve tear and abrasion resistance, and increase heat buildup. Processing is very difficult and dispersion poor with these fine particle sizes. Increased structure blacks compared at the same particle size increase abrasion resistance, green strength, heat buildup, as well as improving processing and tolerating higher levels of process aids.

The broad spectrum of particle sizes and structures permits a range of compound properties to be developed. The levels and types will depend on the required properties and the elastomer or blends selected for the compound.

Compound Cure Systems

The purpose of the cure system is to cross-link the polymer matrix of the compound mixture. There are many ways to crosslink the unsaturated, double bonds of the elastomers and a few systems to form carbon to carbon crosslinks. Carbon, carbon cross-links can be made with peroxides or resin cure systems. However, these are not used in tyre compounds. The predominant types of cross-link (cure systems) feature both soluble and insoluble sulphur and accelerator systems. In addition, zinc oxide and fatty acid are important cure activators. Retarders are used to improve process safety or to delay cure during processing stages of tyre manufacture.

Sulphur solubility in tyre compounds is limited to approximately 2.0 phr, and higher levels may bloom during the process stages. Insoluble (polymeric) sulphur with an oil coating can be used to increase the sulphur solubility in compounds. This also permits higher processing temperatures, since the conversion rate from insoluble to soluble sulphur is dependent on pH, time, and temperature.

Accelerators are available in eight types: thiazole, sulphenamide, dithiocarbamates, triazine, thiuram, xanthate, guanadines, and thiourea types (the vulcanisation of rubber). Each of these classes has several chemical versions. These chemical modifications affect, in some cases, the solubility of the accelerator in the polymer matrix, as well as the migration rates. More important, the modifications of the basic structure are made to adjust scorch, cure during processing, and cure rates. The accelerators used by the tyre industry are primarily sulphenamide types. Other types are used for special cases where different amounts of poly-, di-, and monosulphide types of crosslink are service required. Some accelerators are used at low levels (0.01–0.04 phr) to further speed up the cure rate.

The presence of nitrosamines in the air became a political and health issue in Germany. The formation of nitrosamines during processing, curing, and tyre storage may cause accelerators of some types and classes to disappear from use. Types CBS (*N*-cyclohexyl-2-benzothiazolesulphenamide) and TBBS (*N-tert*-butyl-2-benzothiazolesulphenamide) appear likely to become larger volume accelerators.

The selection of the sulphur-accelerator system in the compound design is important to the performance. Table 10.2 shows some properties that are affected by sulphur, accelerator levels, crosslink types, and crosslink density; increased level of sulphur will increase the compound stiffness.

The cure system can be approximated based on the selected elastomer or blends, the carbon black reinforcement, and required properties of the compound. The cure system in components such as belts and plies is also important to the adhesive bond formation on steel cord and the adhesion and strength retention in some organic tyre cords.

Accelerators can migrate within the tyre; therefore, the cure systems for individual tyre components must be similar and compatible. Accelerators can be more soluble in some compounds. This is related to the solubility parameters of elastomers, resins, and processing aids.

Significant cure rate differences between components can result in one component curing before another, which results in poor bond strength across the interfaces. Migration of the cure systems can also occur across an interface, resulting in poor bond strength. Such problems can be checked with laboratory simulation tests and solid state NMR spectrometry. Blends of elastomers such as NR and ethylene propylenediene monomer (EPDM) or halobutyl and polybutadiene, which have different cure rates, are candidates for this phenomenon.

Table 10.2. Influence of crosslink type and density[a].

		Increased crosslink type			
Property	*Sulphur*	*Mono*	*Di*	*Poly*	*Cross-link density*
Stiffness	++		+	++	+++
Dynamic modulus	+	–	+	+	++
Hysteresis	–	=	=	–	–
Tear resistance	+	–	+	+	–
Flex resistance	+	–	+	++	=
NR reversion	++	–	+	++	+
Adhesion	+	–	+	++	=
Ageing	–	–	+	–	=
Process scorch	+	=	=	=	=

[a] Symbols: +, property increased; =, little or no effect; –, property is decreased; ++, strong influence.

Laboratory experiments with a rheometer are usually made to fine tune the compound for performance properties, process safety, and assurance that tyre cure conditions will be met. A range of acceptable cure variation is then established for plant tolerance control limits.

Fig. 10.9 shows three types of cure response on the rheometer. Compound A has some process safety followed by a very fast cure rate and very little change with increased cure time. Compound B has more process safety and a slightly lower cure formation. Compound C develops cure very slowly and also shows a continued modulus increase with longer times. Control of the sulphur levels and proper choice of the accelerator system provide the correct type of crosslink (poly-, di-, and monosulphide). The use of retarder can extend the scorch for process safety. These decisions need to be based on the performance criteria and the manufacturing process limits.

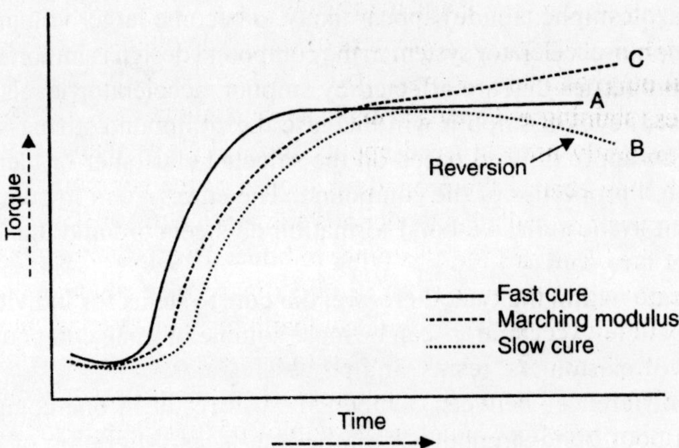

Fig. 10.9. Typical rheometer curves.

Tyre cures

The tyre cure is a simple operation but a very complex chemical process. Thermocouple tests are used widely to set tyre cures and to assure the compounds are cured to meet the designed properties. The tyre cures are largely restricted by thermodynamics and especially because of the poor heat transfer rates of the compounds. The cure reaction rates, which depend on the compound cure system, change with temperature. Activation energy can be approximated from rheometer data obtained at several temperatures. Activation energies ranging from 16 to 26 kcal/mol generally will represent tyre compounds.

A system involving a reference temperature of 300°F is used to calculate equivalent cures (CE) from the amount of cure obtained in each time unit of 0.2 minute and the corresponding temperature at a specific location in the tyre. The calculations are made to match the amount of cure for different temperatures. A summation of the cure increments (CE @ 300°F) is calculated for each location of interest in the tyre. These cure values are then related to laboratory compounds cured for an equivalent time at the reference temperature (300°F). However, compounds cured at different temperatures do not necessarily result in equivalent properties.

Natural rubber compounds tend to revert at high temperatures; that is, they lose stiffness as a result of elastomer scission. Synthetic compounds may become stiffer at higher temperature and longer cure times. Proper selection of the compound cure system can minimise reversion. But most important, the temperature of cure can change the balance of sulphur crosslink types formed, reducing polysulphide to di- and monosulphide forms. Therefore, the compound development must take into account the tyre cure. Thermocouple tests are used widely to set tyre cures and to assure that the compounds are cured to meet the designed properties. The cure process can also be computer modeled to define tyre cures. Cure can be simulated from temperature profiles and computer model using the heat of diffusion, the heat capacities, and the activation energies of the various tyre components. However, complex tread designs can make three-dimensional computer modeling very complex and difficult.

PROCESSING AIDS

Processing or extender oils, used to improve the mixing and processing of the rubber compounds, also have been one of the primary means to reduce compound cost. These oils are classified by composition into polar, naphthenic, and aromatic content. Each type of oil has some content of the others and can vary somewhat dependent on the source and manufacturing process. High polar content oils extend mix times even at low levels (1–10 phr) in elastomers of the NR, polybutadiene, and SBR types. They are more generally used in butyl and nitriles, which offer higher solubility parameters. The naphthenics are generally made to be less staining and are used in non-tread applications. The so-called medium aromatic types are most predominantly used in tread applications. High aromatic oils can also raise the glass transition point, which may change the property response of the compound.

Differences exist between so-called extender oils and lubricating oils. The solubility of the extenders enhances the ability of the black and the elastomer to bond. The lubricating types (oils and fatty acids), generally lubricate the polymer chains so that they can slip over the entanglements. This results in lower mix and process energy requirements. The poorer compatibility with the elastomers (solubility parameters) can also mean a higher tendency for these materials to migrate. Migration to the surface during both the green and cure conditions is called bloom. Therefore, these materials are generally used at low levels. Microcrystalline and microparaffin waxes are examples of lubricating-type extenders. The waxes are also applied to external components, where they form a bloom film that protects the unsaturated elastomer backbone from attack by oxygen and ozone. These waxes are generally blends of different molecular weights to control migration under both hot and cold conditions.

COMPOUND STABILISATION: ANTIOXIDANTS AND ANTIOZONANTS

Additional protection of the elastomer matrix is required. Both antioxidants and antiozonants operate by attaching to the oxygen or ozone (O_3). Oxygen attacks the double bonds of the elastomer, which splits the chain. This also means that with certain conditions of environment, time, and the level used, these stabilising materials can be exhausted, whereupon weathering takes place at an accelerate rate. The antioxidants are usually designed not to migrate to the surface. The antiozonants, generally paraphenylenediamine types, migrate to the surface with time and temperature. The external tyre components usually have the highest levels of protection. Migration may require the use of some barrier components in the tyre to prevent staining of components like white sidewalls. In addition, antiozonants could migrate across interfaces into other components, reducing the amount available at the external surface.

Elastomers with very little unsaturation (e.g., butyl types and EPDM, which have unsaturation only in the side chains) have found some limited application in protecting the external portion of tyres. The use of these materials presents many other complicating factors that must also be developed for satisfactory tyre performance.

The useful life expectancy of the tyre and the range of environmental conditions that must be met determine the system of protection required for the tyre. To satisfy the final customer, alternative solutions must be weighed.

DESIGNING THE COMPOUND MATRIX FOR THE REINFORCED COMPOSITE

Reinforcement with organic or steel cords is very critical to the performance and durability of a tyre. This section deals not with the selection of these reinforcement materials, but with the equally important compound matrix. The compound matrix converts the tyre cord into a composite and then into the complex composite that is a tyre. Composites of various types—plies, belts, beads, chippers, chafers, wraps, and overlays—are utilised in a tyre to provide different performance requirements. The use for each composite is the primary factor in the compound design. Non-bonded cords in a tyre are not likely to meet customer requirements. The bond between the compound and the reinforcement is important in the performance and useful life of a tyre.

Organic Fibres

Organic fibres are dipped in an isocyanate and/or a system based resorcinol-formaldehyde latex (RFL). The dipped fabric is then heat treated. This process permits the dip to penetrate into some layers of the filament bundles, to bind the filament together and to provide a surface that can be bonded to the compound. In the early days of tyres, cotton cords did not require dips for bonding. Rayon, nylon, polyester, polyvinyl alcohol, and polyamide cords all require some type of dip to bond to the compound. The coat compounds may be modified with similar-type RFL resins that can chemically bond to the dipped cords.

The cords are calendered starting with either dipped woven fabric or a creel system. Both systems set the spacing between cords, called the rivet area. Additional rubber is added above and below the cords. The ends per inch (epi) and rivet area are important factors in the compound design. A range of epi fabric can be run with the same stiffness compound. However, if the rivet area is too small, a stiff compound can produce high shear at the cord interface and either debond or have high cut growth splitting between cords in the high flex area of the tyre. Lower stiffness, lower hysteresis compounds are preferred. This is especially true in constructions with higher diameter cords and greater cord spacing. The cord reinforcement in the direction of the cord results in a much more rigid structure than the compounds alone. Therefore, it is important to design transitions of stress between the rigid composite structure and other parts of the tyre. Belts with steel cords require special care to minimise shear stress between the belts. Again, the gages and compound stiffness are important factors. NR and NR/synthetic blends are generally used in coat compounds for low hysteresis, tear, and flex properties.

Coat compounds may be modified with prereacted B stage novolak-type resins that may migrate and bond through sites on the treated dipped cord. Resorcinol and hexamethylene tetramine (HMTA) have also been used in coat compounds to enhance adhesion to the cord.

Laboratory tests of several types (pullout peel, flex) with a range of conditions are used to confirm that the adhesion bond to the dipped cord is satisfactory for the tyre service life. Adjustments are made in the compound until the adhesion to the cord is optimised. Laboratory peel tests on fabrics clearly demonstrate the effect of epi and should be considered with other adhesion data.

Steel Cords and Bead Wire

Both steel cord and bead wire feature alloy coatings that facilitate drawing and bonding to rubber. Specific brass coatings are used almost exclusively on steel cord and only on some bead wire. Most common for bead wire is the bronze coating. The chemistry of developing adhesion to the brass-coated steel cord is very complex.

The brass layer from a compositional zinc and copper content and the gradient composition from the steel cord to the rubber surface are important. This composition is important to the initial adhesion and to the service life of the rubber-brass bond. The theory is that depending on the initial compound reactivity, the brass composition forms intermediate layers of zinc sulphide and copper sulphide. The rates of formation of these layers determine the initial level and the useful life of the bond. Examples have been found under test conditions in which the brass has been completely converted to the sulphide layers.

The rate of formation of the zinc and copper sulphide layers is important then to both initial and service adhesion. If the compound cures up before these intermediate layers can form, only mechanical adhesion to the cord is developed and the cord will pull from the rubber with no attached rubber.

Several alternative compounding techniques are used to develop the bond to brass. Most compounds feature some form of cobalt salt, considered to be a chelating agent, to regulate layer formation. The sulphur level, accelerator system, zinc oxide, and fatty acid grade and types all affect the bond formation rate.

Various forms of hexamethylene tetramine and resorcinol have also been used to develop the bond to brass. A great deal of compounding work has been done to define the bonding system to brass-coated steel cord. Compounds generally feature a high level of NR and, depending on the application, may vary in amount or type of a reinforcing black. The cure and the resin system (if present) are very important factors in the development of bonds to steel cord that can withstand air and moisture, road salt corrosion, and the heat experienced during the service life of the tyre. Lab simulations based on tyre test correlation studies can be used to evaluate compound changes or modifications. Because of the risks involved in changing the steel cord coating compound and the manufacturing system influences, including process and cure, steel cord compound changes are made only after significant volume testing has been completed.

The compound stiffness and tear resistance as well as cord angles, gages, and epi are important factors in belt performance. The ability to identify and minimise stress concentrations in the belt area and under the required range of service conditions is very important to the tyre design.

Functions of the Pneumatic Tyre

The pneumatic tyre performs a variety of functions that are essential to the effective operation of most modes of transportation. These functions include the following :

1. Supporting the vehicle load.
2. Transmitting driving and braking forces to the road surface.
3. Producing lateral forces for cornering and vehicle handling control, to help guide the direction of travel.
4. Providing safety through maneuverability, durability, wet and dry traction, snow traction, and high speed performance.
5. Ensuring that noise and vibration levels from tyre resonances and interactions with the road surface are acceptable.
6. Maintaining dimensional stability by undergoing only inappreciable change of size or shape upon inflation, and providing constant axle height and effective rolling radius when travelling on smooth roads.
7. Providing economy through long tread life (wear resistance) and low rolling resistance (energy consumption).

To perform these functions, the tyre must have enough rigidity to develop substantial forces in all directions, enough flexibility to be able to envelop obstacles without sustaining damage, and a long fatigue life in flexing from a doubly curved shell to a flat surface and back.

Tyre Applications

Tyres can be classified by the type of vehicle on which they are applied. Major classifications include automobile, truck, off-road, farm, aircraft, and race tyres. Other classifications include bicycle, motorcycle, and industrial tyres.

Automobile tyres vary in size, shape, load rating, and speed rating. Automobile tyre types include winter, all-season, summer, and high performance tyres.

Truck tyres are classified as either light, medium, or heavy load truck tyres. Bus tyres are usually included in the medium truck tyre classification. Because of the requirement for economy of operation, longevity, endurance, and retreadability are key demands on truck tyres.

Off-road tyres are used on earth moving, road construction, mining, and some recreational vehicles. Often required to operate on uneven terrains, these vehicles impose special demands on their tyres for floatation on soft grounds and resistance to bruising and cutting by sharp objects.

Farm tyres require floatation on soft grounds, hillside stability, minimum soil compact, and, because of their need for long service life, resistance to damage from weathering.

Aircraft tyres include those for commuter, commercial, and military aircraft. Special demands on aircraft tyres are required by their use at high speeds and high loads relative to their size.

Race tyres are required to provide high speed performance, high reliability during their relatively short service time, and high friction levels in cornering, acceleration, and braking modes of vehicle operation.

Military tyres also have special demands such as an extra margin of safety and a run-flat capability.

This section concentrates on technologies related to the two most common tyre applications, automobile and truck tyres. These technologies are, in general, also applicable to other tyres with modifications to take into account specific loading and service requirements.

Tyre Components

Rubber is an essential element of pneumatic tyres; it offers flexibility, low hysteresis, good friction on most surfaces, high abrasion resistance, and good impermeability by contained air. The basic rubber characteristic of low resistance to tensile forces necessitates the use of an inextensible yet flexible reinforcement, to avoid excessive tyre deformations upon loading. The major components of a typical pneumatic tyre (Fig. 10.10) are discussed in the subsections that follow.

Carcass

The carcass consists of plies made up of many high modulus cords embedded in a rubber matrix. The number of plies is determined by tyre type, tyre size, and the inflation pressure and loads to be sustained by the tyre in service.

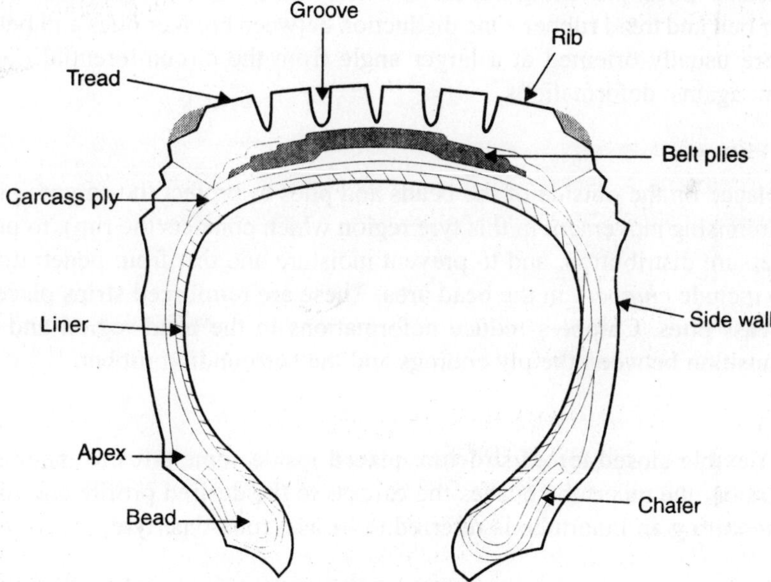

Fig. 10.10. Tyre components.

Beads

The carcass plies are turned around beads located close to the rim flanges. The beads anchor the carcass plies and hold the assembly on the wheel rim, thus preventing the tyre from rocking or slipping. In bias tyres having a large number of carcass plies (e.g., aircraft, earth mover, and truck tyres), two or more beads may be wrapped together, with each bead having a number of plies turned around it.

The rim includes flanges to locate and restrain the inflated tyre, and a central depression to accommodate the beads during tyre mounting and dismounting. Thus the beads must be flexible as well as inextensible.

Tread

The tread is the part of the tyre that contacts the road surface. It protects the carcass and provides the frictional contact required to transmit driving, braking, and cornering forces.

Depending on tyre type, the tread outer surface may be smooth, as in some race tyres, or it may have a pattern. Tread patterns include protruding circumferential elements called *ribs*, separated by depressions called *grooves* or *voids*, individual protruding elements called *lugs*, or a combination of ribs and lugs. Patterned treads also usually include *sipes*, which are slots in the protruding elements aimed at improving traction.

Belts

Belts are fibre-reinforced composite plies forming a hoop under the tread. They restrict deformations of the carcass plies and provide added stiffness to the tread.

Other reinforced hoops may be included in the tread region of a tyre to provide desired enhancement of performances. Such hoops may be *overlay* plies, placed in the tread region to further restrict tyre deformations, or *breaker* plies, providing transition between the first (innermost) belt and the carcass, or between the outer belt and tread rubber. One distinction between breaker plies and belt plies is that the cords in breakers are usually oriented at a larger angle from the circumferential direction and thus provide less restraint against deformations.

Chafers

Chafers are strips placed on the outside of the beads and plies to protect the carcass plies from cutting and wearing (by minimising movement in this tyre region which contacts the rim), to provide improved tyre-rim contact pressure distribution, and to prevent moisture and dirt from penetrating into the tyre.

Some tyres also include *chippers* in the bead area. These are reinforced strips placed near the turn-up ends of the carcass plies. Chippers reduce deformations in the bead region and provide a more gradual stiffness transition between the ply endings and the surrounding rubber.

Innertube

The innertube is a flexible closed-toroid structure placed inside some tyres to maintain contained air pressure. Upon inflation, the innertube pushes the carcass to the desired profile and maintains the tyre on the rim. A tyre requiring an innertube is referred to be as a tube-type tyre.

Inner liner

A tyre not requiring an innertube is referred to as a tubeless tyre. In tubeless tyres an airtight seal is established between the tyre and the rim through compression fitting and sliding the tyre bead on an inclined rim. In addition to this compression fitting, tubeless tyres include inner liners.

The inner liner is a thin layer of low permeability rubber moulded inside the tyre to reduce air loss. For example the permeability of butyl rubber, which is normally used in inner liners of tubeless tyres, is about one-sixth that of natural rubber and one-fifth that of synthetic rubbers used in the sidewall and tread regions of the tyre.

Sidewalls

The portions of a tyre between the tread region and the beads are the sidewalls. The sidewalls support the tread region, protect the carcass plies from damage, and provide the tyre's ride characteristics.

Shoulders

The junction areas of the tread and the sidewalls are called the tyre shoulders.

Crown

The point at the centre of a tread section, halfway between the two shoulders, is the crown.

Tyre Construction Types

Tyres generally have layers of cord-reinforced plies, each containing a series of equally spaced parallel cords. In addition to changing the materials used in the various tyre components, a powerful option available to a tyre engineer is that of changing the cord angles in the plies. Flexibility in this area has led to four main types of tyre construction.

Bias tyres

The body cords in a bias tyre make a rather large angle with the tread centreline, and there are no belt plies in the tread region.

Bias-belted tyres

Bias-belted tyres (Fig. 10.11) have the basic body plies of bias tyres, plus two or more belts between the tread rubber and the carcass. Cords in the belt plies are more nearly circumferential than those in the carcass plies. The belts reduce both circumferential and lateral deformations of the tread area during tyre service, thereby restricting changes in the inflated tyre profile and reducing tread movement in the footprint. By reducing deformations due to inflation, belts enhance tyre durability; by reducing lateral distortion in the footprint during cornering, belts improve tread wear and enhance vehicle handling.

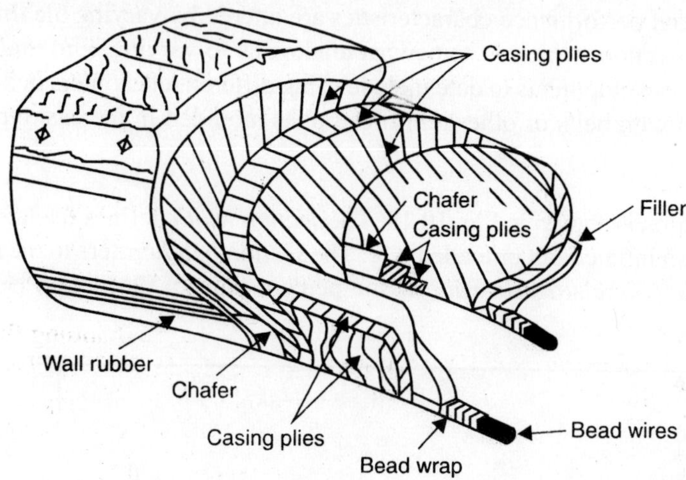

Fig. 10.11. Bias-belted tyre construction.

Radial tyres

Radial tyres (Fig. 10.12) are also belted, but they have carcass cords that lie in the meridianal planes of the tyre (i.e., perpendicular to the circumferential direction). The radial orientation of the carcass cords results in very flexible sidewalls, which act independently of the belts, thus reducing tread movement in the footprint to less than that in bias belted tyres.

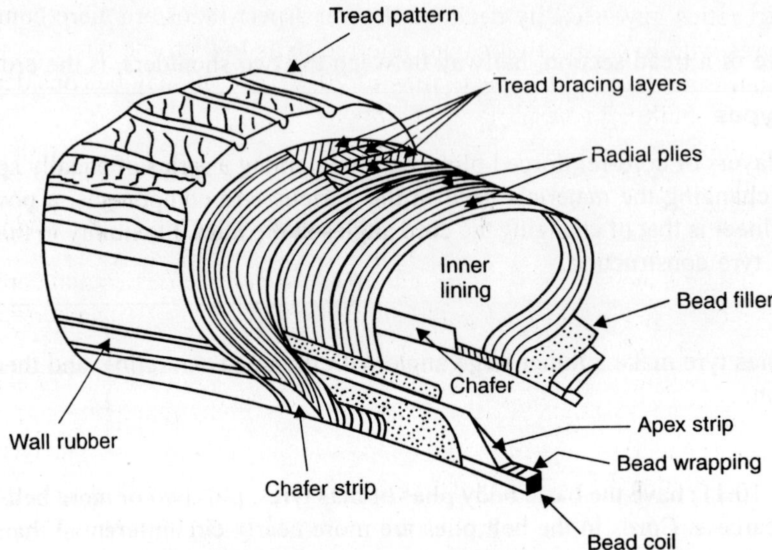

Fig. 10.12. Radial tyre construction.

Cast tyres

Cast tyres contain no fabric, only elastomers to form the air chamber and bead wires for anchoring to the rim. Deformations and performance characteristics are altered by varying the thickness distribution around the tyre's cross-section, which is somewhat analogous to varying cord angles in conventional tyres. Reported cast tyre developments to date include using different elastomers in different tyre regions and hybrid construction with belts or other reinforced plies moulded in the cast tyre.

Tyre profile

A typical tyre cross-section is shown in Fig. 10.13. The *section width* defines the lateral dimension from sidewall to sidewall of an inflated but unloaded tyre. The *section height* refers to the radial distance from the crown to the beads. A tyre's *aspect ratio* is the percent ratio of its section height divided by the section width.

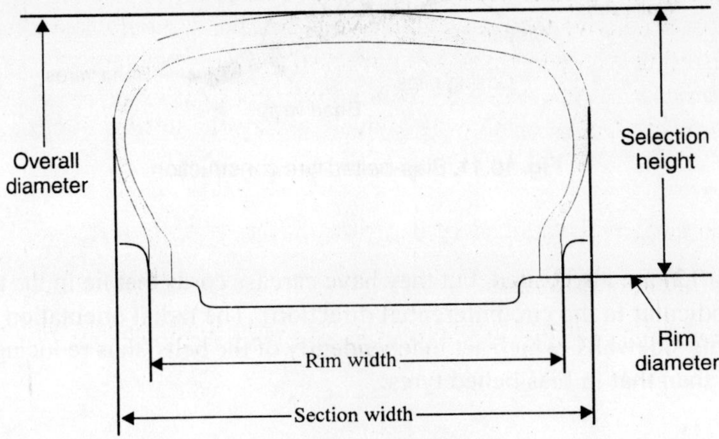

Fig. 10.13. Tyre profile.

In the early days of automobiles, pneumatic tyres had aspect ratios of around 100. With advances in tyre technology, aspect ratios have steadily declined. Lower aspect ratios are more common on high performance tyres. Tyres with low aspect ratios are referred to as low profile tyres.

While radial medium truck tyres with aspect ratios of 70, 75 and 80 may be considered to be low profile tyres, race tyres are built with aspect ratios as low as 30.

Standard tyre designation usually includes size, aspect ratio, type of construction (radial or bias, tube type, or tubeless) and the diameter of the rim on which it is mounted. For example, a 185/70R14 tyre has a nominal section width of 185 mm, an aspect ratio of 70, a radial construction (designated by the letter R), and is to be mounted on a 14-inch diameter rim.

For truck tyres, an 11R22.5 specifies a nominal section width of 11 inches, a radial construction, and a rim diameter of 22.5 inches. A 10.00–20 designation refers to a truck tyre with a section width of 10 inches, a bias construction (no letter R), and a rim diameter of 20 inches.

Additional letters are sometimes included to provide more specific information. For example, a 275/65R17.5HC tyre is intended for use on HC-type rims, while a 7.50R15TR tyre is intended for use on TR rims. An 11–24.5ML tyre is intended for mining and logging applications. Automobile tyre designations are at times preceded by the letter P, referring to passenger cars, and followed by a letter that designated their speed rating.

TYRE DESIGN

Given a set of customer performance requirements and a set of service conditions for a tyre, tyre engineering involves coming up with a tyre design optimally suited to deliver the desired performance.

Performance requirements include operating speeds, length of operation, and the required levels of particular tyre responses. Service conditions include identification of the vehicle on which the tyre would be used; anticipated loads, inflation pressure, and temperatures; the road surface(s) on which the tyre would be used; and the type of tyre maintenance to be expected.

A tyre design specifies its structural details, shape, and tread pattern. Using professional experience and the available data, a tyre engineer develops an initial design, assesses its performance, and then modifies the design for the desired changes in performance; this iterative process continues until the target performance has been achieved.

As tyre technology continues to advance, more and more tyre performances are assessed analytically without having to build a tyre to evaluate every performance feature experimentally. This provides three significant advantages: (i) reduction of the time required for tyre development, (ii) reduction of tyre development costs; and (iii) greater accuracy (by eliminating test errors and variations).

After all available predictive tools and data have been used, a mould is built to produce the chosen tyre design. The resulting tyre is evaluated: if particular performance targets are not met, the design is modified and the process is repeated. The information gained is added to the database and used to upgrade the design technology so as to progressively reduce the needed number of tyre prototypes.

Tyre Structure

Defining a tyre structure involves the steps described in the subsections that follow.

Selection of construction materials

Based on engineering judgement and available data, the cords, beads, and the various rubber compounds are selected.

Tyre cords have evolved by introduction of new materials and new constructions to enhance their properties. Cords may be made of natural fibres, synthetic polymers, glass, or steel. The smallest constituent element of a cord is the filament. A strand, or yarn, is an assembly of filaments. Fig. 10.14 gives examples of steel cord constructions used in truck tyres; the associated designations refer to the size and number of strands in the cord. Table 10.3 shows the breaking strengths, relative fatigue endurances, and relative stiffness of several typical steel cord constructions.

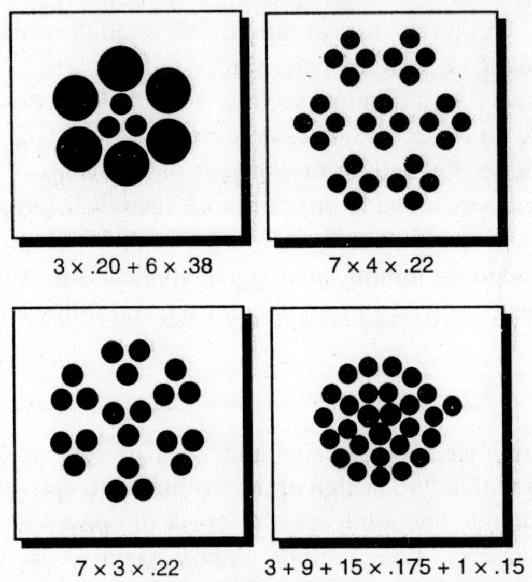

Fig. 10.14. Examples of steel cord constructions.

The bead is composed of strands of wires arranged to provide both high strength and a degree of flexibility essential for mounting and dismounting the tyre. The bead is designed to conform to the wheel contour, to prevent the tyre from rocking or slipping on the rim under severe conditions, and to provide adequate tensile strength to prevent breakage under overload and conventional mounting conditions. Examples of bead designs are shown in Fig. 10.15.

Table 10.3. Typical constructions of steel cords for tyres.

Cord construction	*Breaking strength (N)*	*Fatigue rating*	*Stiffness rating*
3 + 9x.175 + 1	700–900	100	100
3 + 9x.20 + 1	1000–1200	40	80
3 + 9x.22 + 1	1200–1400	30	250
3 + 9 + 15x.175 + 1	1500–1700	100	180
3 + 9 + 15x.22 + 1	2500–2700	15	470
3x.22 + 6x.38	1500–1700	10	900

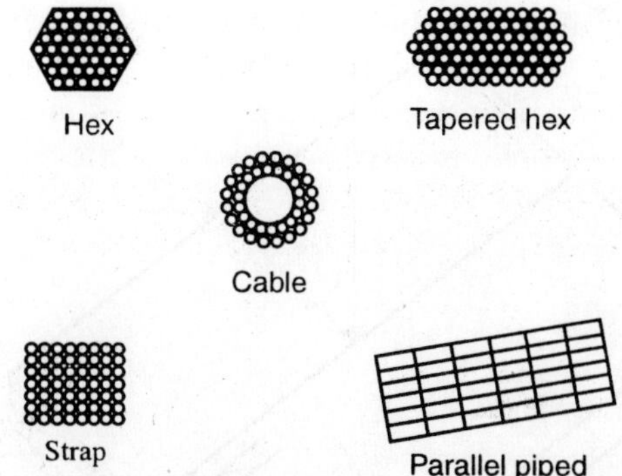

Fig. 10.15. Examples of bead wire configurations.

A typical rubber compound usually includes natural rubber, synthetic rubber, or a combination of rubber types, plus fillers and other systems added to achieve the desired material properties. Rubber compounds for each zone within a tyre cross-section are selected to optimally suit the function performed by the specific tyre component and the service conditions for that tyre zone.

Specification of number of plies

The engineering chooses the number of carcass plies, belts, breakers, chippers, and other cord-reinforced plies to be included in the construction. This choice is influenced by the engineer's experience and by the specific performance requirements for the tyre.

Steel carcass radial tyres often include a single carcass ply; bias truck tyres and earth mover tyres may include a large number of body plies. Radial automobile tyres usually include two belts; radial truck tyres usually include two to four belts.

Designing the cord-reinforced plies

After selecting the cord material and the number of plies, the engineer needs to determine the cord density, as defined by the ends per inch (epi), the cord angle within each ply, and the treatment gage, which defines the coverage of the ply material for adhesion to adjacent rubber materials.

In the simplest material model, rubber is treated as an isotropic material, and the elastic properties of each rubber compound are defined by two elastic constants: the modulus of elasticity E_r and the Poisson's ratio ν_r.

More advanced material models incorporate the incompressibility of rubber. The Mooney-Rivlin and the Ogden models express the strain energy in the rubber as a function of the strain invariants and derive the stress-strain relation in terms of the extension ratios. These models provide better representation of rubber behaviour, especially at large strains.

The cord's multifilament construction yields a complicated nonlinear material response with different properties in tension (extending the strands) and in compression (collapsing the strands). As a first approximation, the cords can be treated as equivalent isotropic materials defined by an elastic modulus E_c and a Poisson's ratio ν_c.

A typical cord-reinforced ply is shown in Fig. 10.16. The parallel cords are oriented along the major principal direction 1; the minor principal directions are designed as 2 and 3.

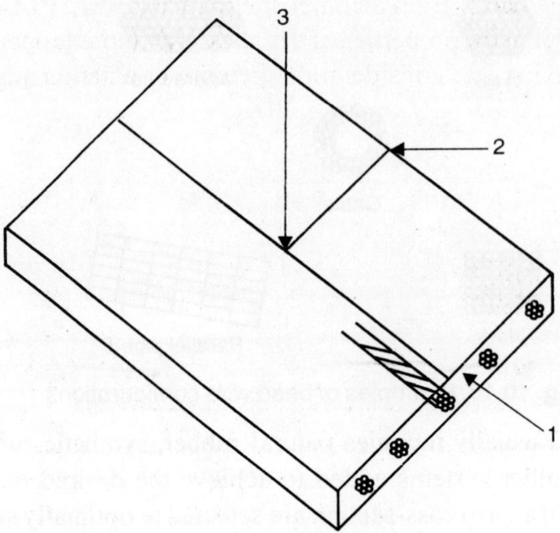

Fig. 10.16. Cord-reinforced ply.

The elastic moduli of the ply along its major and minor inplane axes are given by

$$E_1 = v_c E_c + v_r E_r$$

$$E_2 = E_r \frac{(1 + 2Bv_c)}{1 - Bv_c}$$

where

$$B = \frac{E_c / E_r - 1}{E_c / E_r + 2}$$

and v_c and v_r are the volume fractions of the cords and rubber.

The volume fraction of the cords can be evaluated from their end count and the ply thickness as follows :

$$v_c = a_c \frac{\text{epi}}{t}$$

where a_c is the cross-sectional area of the cord, epi (ends per inch) is the cord end count per unit length, and t is the ply thickness.

The other two in-plane elastic constants of the ply are given by its Poisson's ratio v_{12} and the shear modulus G_{12} :

$$v_{12} = v_c v_c + v_r v_r$$

$$G_{12} = G_r \left(\frac{1 + Cv_c}{1 - Cv_c} \right)$$

where

$$C = \frac{G_c / G_r - 1}{G_c / G_r + 1}$$

When plies are stacked together to form a composite laminate (Fig. 10.17), the elastic properties of the laminate are obtained by adding the properties of the plies, after those properties have been transformed to the laminate axes *X*, *Y*, *Z* taking into consideration the cord orientation angle θ in each ply.

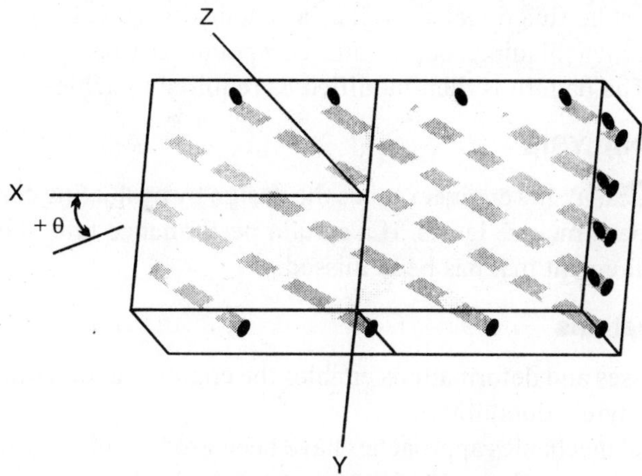

Fig. 10.17. Multi-ply laminate.

In general, there is a need for properties in direction 3, normal to the ply (Fig. 10.16). Detailed theoretical and experimental techniques have been investigated for obtaining these properties. These properties can also be approximated with good accuracy by the following equations :

$$E_3 = E_2 \qquad \nu_{13} = \nu_{12} \qquad \nu_{23} = \nu_r$$

The rationale behind this is that E_2 and E_3 are both in directions normal to the cords, and thus they can be assumed to be equal; both ν_{13} and ν_{12} refer to deformations in directions normal to the cords due to a deformation in the cord direction (and the cord is assumed to be tranversely isotropic); and since directions 2 and 3 are both away from the cords and into the rubber, ν_{23} is close to the rubber property ν_r.

Define dimensions of constituent elements

Finally the dimensions are defined, completing the definition of the structural components. The belts are usually made wide enough to span the tread region (i.e., shoulder to shoulder), with adjacent belts having different widths to provide a step-off for more gradual transition into the surrounding rubber. In the bead region, the turn-up plies are extended away from the high stress, high flex region; the ply endings are varied to maintain a stiffness balance between the tread and sidewalls. The dimensions of each rubber components are chosen to make maximum use of its unique properties.

These dimensions are defined for an initial design. They are modified, as deemed necessary, when the tyre's performances are evaluated.

Tyre Shape

The target tyre size defines certain nominal dimensions for the tyre and an envelope within which the tyre cross section would fit. The tyre engineer needs to define an optimum contour for the tyre within this envelope. The tyre contour is selected to minimise the tyre deformation due to loading and unloading, and to optimise tyre performance.

Tread Design

Tread design involves definition of a tread pattern (shapes and sizes of ribs and/or lugs) and the non-skid (depths of the grooves and/or voids). The tread pattern is designed to provide the desired tyre-pavement interaction. Tread design must be functional as well as aesthetically appealing. Tread element stiffness can be varied through grooving, blading, depth, and compound stiffness. Here too, an initial tread pattern is designed at first. The pattern is then modified as required to achieve the target performance.

TYRE PERFORMANCE ANALYSIS

Having come up with a tyre design, the engineer tunes the design by evaluating certain key performances for comparison with target performance levels. If a certain performance falls short of target the design is modified to meet the requirement that has been missed.

Tyre Stresses and Deformations

Knowledge of the tyre's stresses and deformations enables the engineer to determine the tyre dimensions under load and to assess the tyre's durability.

Various classical structural mechanics approaches have been used historically to compute tyre stresses and deformations. To apply these techniques, the tyre structure is idealised as a network of cords, a membrane, a thin shell of revolution, or a multicurvature shell. However, recent developments in the field of finite element analysis (FEA) and the maturity of that technology, have made it unnecessary to resort to such gross assumptions. A major advantage of FEA is that the different tyre components are recognised and the actual tyre shape is modeled, rather than an equivalent structure with averaged properties.

General-purpose finite element programmes such as ABAQUS, MARC, MSC-NASTRAN, and ANSYS are commercially available; they can be readily used to compute tyre shapes deformed as a result of inflation pressure and other loadings, as well as cord forces and rubber stresses, with better accuracy than that provided by classical methods. Special versions of these programmes can be installed on workstations and personal computers, thus allowing their use by engineers without access to large mainframe computers.

One useful computation involves determining a naturally inflated profile (NIP) for the tyre. This profile is sought by tyre engineers because upon inflation a NIP tyre undergoes very little change in shape, since the deformation of its plies is in a stretching mode rather than bending, and the composite plies are much stiffer in-plane than out-of-plane. By minimising the change in shape, the engineer minimises the strains due to inflation loading. This shape is also sometimes referred to as the neutral contour.

One method for obtaining the NIP profile involves inflating an initial tyre shape via FEA and using the computed inflated profile as the next starting shape. Generally within fewer than five such iterations, a NIP profile is obtained that shows only a slight stretching deformation when the tyre is subjected to the rated inflation pressure. Since inflation loading is axisymmetric, either a simplified two-dimensional analysis involving toroidal finite elements or a three-dimensional model representing a repeating wedge of the tyre need be analysed; this reduces the required time and effort considerably.

Recent work for maximising the durability of tyres goes beyond NIP and uses FEA to compute tyre profiles that minimise the cyclic strains in a rolling tyre. To assess durability, the engineer computes internal stresses in the various tyre components. Rubber stresses can be compared to limits from a selected failure criteria. Cord stresses are compared to limit values to determine the tyre burst pressure (and thus the factor of safety) and the fatigue life of various plies.

Fig. 10.18 illustrates the computed deformations of a typical radial truck tyre. The variation of rubber stress with inflation and deflection loading at a location close to the endings of belts 2 and 3, which experiences very high rubber stresses, is illustrated in Fig. 10.19. Fig. 10.20 shows the computed cord tension in various plies due to inflation pressure, and Fig. 10.21 illustrates the cord loads after a vertical deflection of 35 mm is applied.

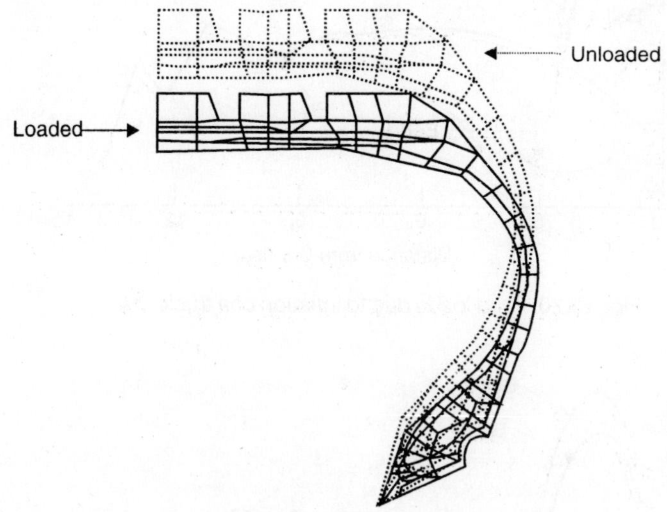

Fig. 10.18. Deflected tyre profile.

Experimental techniques may also be used to determine tyre stresses and deformations. Profilometers and laser devices can be used to measure the deformed shape, which may be compared to the initial tyre shape to obtain the tyre displacements. Rubber strains must be measured with strain gages that are compliant enough to not distort the strain distribution in the vicinity of the gage; liquid metal strain gages are used for this purpose. Cord stresses can be measured by placing transducers on the cords before the tyre is built.

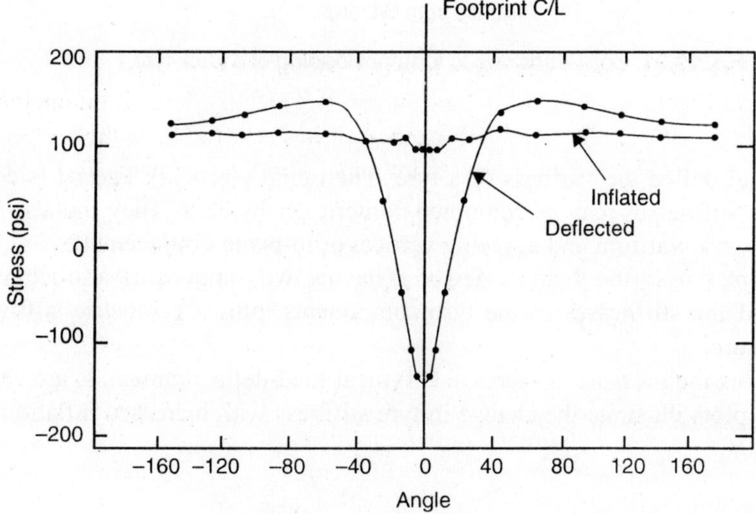

Fig. 10.19. Belt edge stresses in a truck tyre.

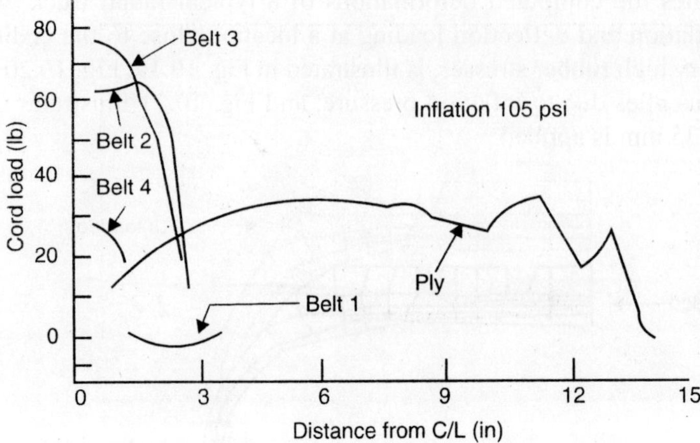

Fig. 10.20. Cord loads due to inflation of a truck tyre.

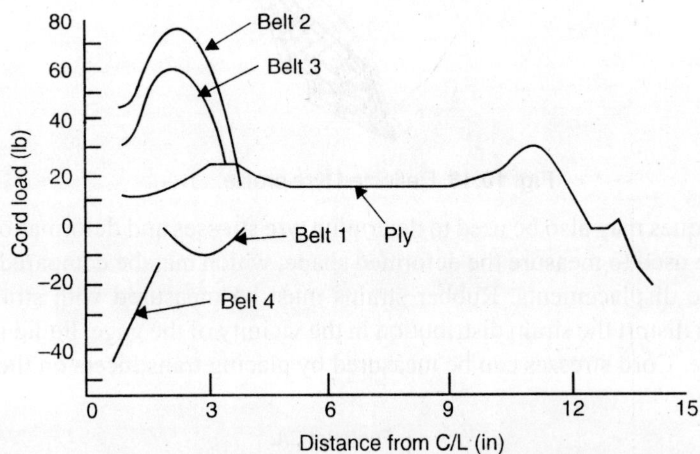

Fig. 10.21. Cord loads due to footprint loading of a truck tyre.

Tyre Stiffness

Deflections per unit load define the stiffness of a tyre. The radial (vertical), laterial (side-ways), and tangential (fore and aft) stiffnesses may be computed numerically by FEA. They can also be measured by loading the tyre against a platform and applying vertical or in-plane displacements.

In a simplified concept, tyre stiffness can be viewed as having two components: a structural component representing the sum of the stiffnesses of the tyre components, plus a geometric stiffness resulting from the inflation pressure.

The latter builds up as the pressure is increased. Typical load-deflection results are shown in Figs. 10.22 and 10.23; these plots illustrate the change in tyre stiffness with increased inflation pressure.

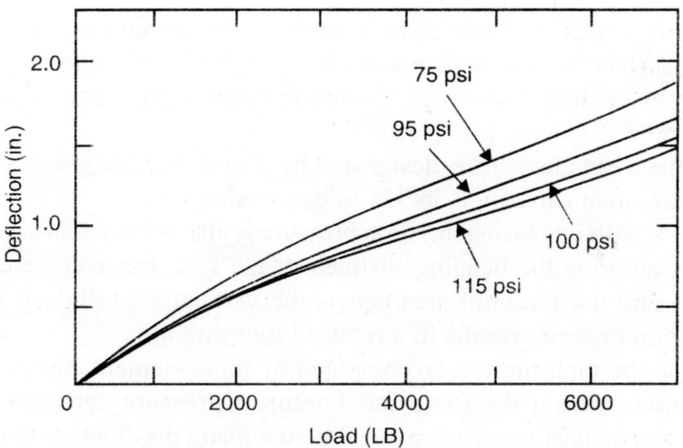

Fig.10. 22. Radial load-deflection response of a truck tyre at different inflation pressures.

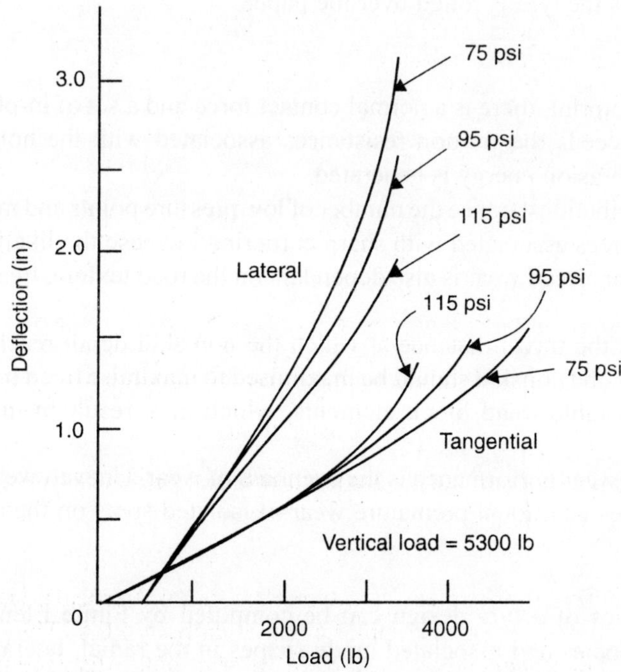

Fig. 10.23. Lateral and tangential load-deflection responses of a truck tyre at different inflation pressures.

Tyre Footprint

As a tyre is loaded against the pavement, the contact area defines its footprint. Engineers normally seek rectangular, slightly convex footprints because such footprint shapes are associated with more even wear of the tread and the generation of desirable tyre-pavement contact forces. On the other hand, butterfly-shaped (i.e., concave) footprints are usually associated with slipping in the footprint and irregular wear patterns. To maintain a stable performance, engineers prefer to minimise footprint shape changes with changing speed.

The footprint area can be computed by finite element analysis, by monitoring the nodes that come into contact with the loading surface as load or displacement is applied to the tyre. Footprints can also be determined experimentally by loading the tyre against carbon paper or other pressure-sensitive paper, or by photographing the footprint through a glass plate.

If the net contact area of the tread elements is designated by N and the total gross contact area is by G, a measure of the footprint groove capacity is its net-to-gross ratio N/G.

The increase in tyre stiffness with increasing inflation pressure is also reflected in a reduced footprint area. In a simplified analysis ignoring the bending stiffness of the tyre, the average contact pressure equals the inflation pressure, and the footprint area equals the vertical load divided by the inflation pressure. Thus a higher inflation pressure results in a reduced footprint area.

The pressure distribution in the footprint can be computed by finite element analysis, by evaluating the contact forces at each node within the footprint. Footprint pressure can also be determined experimentally by loading the tyre against a pressure plate. Pressure plates may include triaxial transducers capable of measuring both normal and in-plane forces, or piezoelectric transducers for measurement of normal forces. Pressure-sensitive paper is also available for measuring the peak contact pressure at each point within the footprint as the tyre is rolled over the paper.

Tread Wear

At each point within the footprint, there is a normal contact force and a set of in-plane forces. When the resultant in-plane force exceeds the friction resistances associated with the normal force at a given point, the point slips and abrasion energy is generated.

More even pressure distributions reduce the number of low pressure points and minimise the likelihood of slippage. High lateral forces associated with sharp cornering increase the likelihood of slippage and thus accelerate the treadwear. Tread wear is also dependent on the road texture, tread material properties, and the temperature.

Tread life is defined as the travel distance at which the non-skid depth reaches a specified value. Thus, all things being equal, the nonskid should be maximised to maximise tread life. However, too deep a non-skid can lead to unstable tread block elements, which may result in uneven wear and poor handling performance.

A key feature of a tread wear performance is the evenness of wear. Uneven wear generates noise and ride disturbances and causes additional premature wear of isolated spots on the tread surface.

Tyre Vibrations

The vibration characteristics of a tyre design can be computed by Finite Element Analysis (FEA), yielding the natural frequencies and associated mode shapes in the radial, lateral, and circumferential directions. The mode shapes and natural frequencies can also be measured by rolling the tyre on drums or flat tracks and measuring the tyre response. To prevent resonance, the tyre's natural frequencies should be different from the vehicle's natural frequencies.

Tyre Noise

Two sources of noise are developed as a tyre rolls: a low frequency structural noise associated with the tyre cavity and distributed over the whole tyre, and a high frequency noise associated with deformations of the tread elements and the pumping and suction of air in the grooves. The structural noise is influenced by the stiffness and damping characteristics of the structural elements of the tyre.

Footprint noise is influenced by the footprint shape and size, the footprint pressure distribution, and the characteristics of the tread. Besides the stiffness and damping properties of the tread material, tyre noise is affected by the shapes and sizes of the tread elements and the grooves separating them. Noise performance generally improves as the non-skid depth is reduced and worsens with higher speeds. Fig. 10.24 illustrates the variation of truck tyre noise with speed; it also illustrates the better performance of rib designs over lug designs (fewer edges of tread elements). In general, radial truck tyres show better noise performance than bias truck tyres.

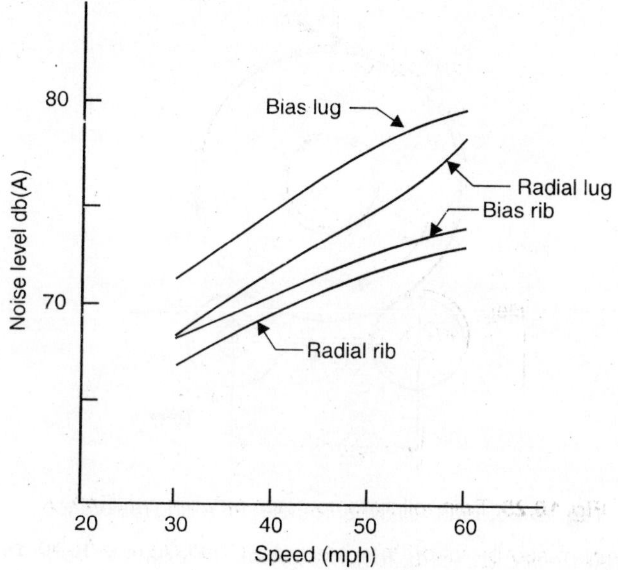

Fig. 10.24. Effect of speed on noise level of radial and bias truck tyres.

Aquaplaning (Hydroplaning)

On a wet surface, contact is maintained when the tread elements penetrate the water film, absorb water in the tread voids, and quickly eject the excess water through the grooves to the outside of the footprint. As speed increases, fast ejection of the water becomes critical. This is accomplished by maximising the volume of the grooves and orienting them to facilitate the flow of water to the outside.

Thus for best aquaplaning performance, the non-skid should be maximised and the net-to-gross footprint area ratio should be minimised. Since aquaplaning is more likely to occur at higher speeds and during sharp cornering, high performance tyres are usually designed with lower net-to-gross ratios. Analytical prediction of the aquaplaning phenomenon is rather complex, calling for further efforts that reflect consideration of both vehicle and road effects.

Rolling Resistance

In a given loading cycle, every tyre component is loaded to a certain level and then unloaded upon removal of the footprint force. Because the materials are hysteretic, the total applied energy in a given loading cycle is not recovered; the lost energy is a function of the material and the strain cycle. Adding the lost energies of the various components yields the energy lost by the tyre (i.e., its rolling resistance). In general, a 5–7% reduction in rolling resistance results in a 1% reduction in vehicle fuel consumption.

Tyre rolling resistance can be computed by finite element analysis, by computing the strain cycles within each element and incorporating the loss properties of the element's material. Summation of the energy loss in all elements yields the rolling resistance of the tyre.

Rolling resistance may also be measured by rolling the tyre on a flat surface or on circular drums. The most common drum radii for this purpose are 67 inches and 2 m. Another test, used, utilises twin drums specified by the Environmental Protection Agency (EPA) for monitoring emissions from vehicles. Fig. 10.25 shows a tyre positioned on such twin drums.

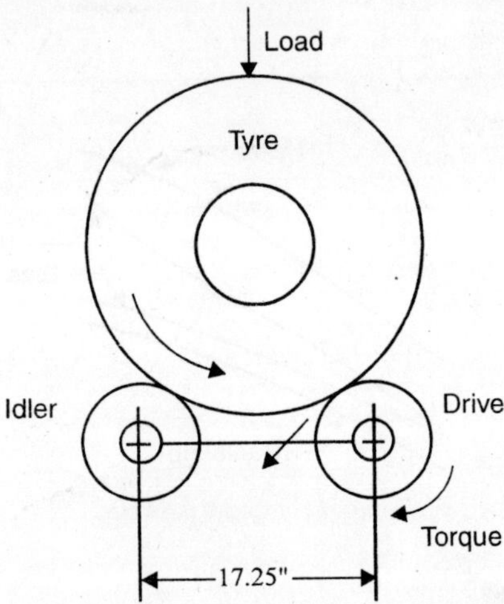

Fig. 10.25. Twin-roll dynamometer for rolling resistance.

Rolling resistance is minimised by using low hysteresis materials to build the tyre and by designing the tyre structure to minimise the strain cycles in the tyre components.

A lower nonskid depth means less rubber subjected to cyclic strains; thus non-skid should be minimised to reduce rolling resistance. A reduced net-to-gross ratio means less tread rubber subjected to proportionally higher strains, and this generally leads to a higher rolling resistance because the deformation energy varies with the square of the strain. Since the rolling resistance energy is converted mainly to heat, it can be used to compute the temperature distribution within the tyre (by FEA). Temperature distribution within the tyre can also be measured, by embedding thermocouples at desired locations. Temperature distribution on the outer surface of the tyre can be measured by infrared thermometers.

TYRE MANUFACTURING

Having implemented all the design modifications indicated by the performance analyses to achieve the target performances, the engineer is ready to manufacture the prototype tyre. The major steps are discussed briefly as under.

Mould Procurement

The engineer may follow one of several approaches: (i) design a finalised mould with all the tyre design details, including the tread pattern; (ii) procure an interim mould for prototype tyre evaluation prior to manufacturing a finalised mould (e.g., less durable moulds utilising materials that are easy to work with may be used at this stage); or (iii) design a mould for smooth treaded tyres: candidate tread patterns may subsequently be carved.

Heat transfer technology (e.g., by FEA) can be used to determine the various mould thicknesses needed to provide uniform heat within the cavity. Since cure temperatures induce stresses in the tyre, which cause it to deform upon ejection from the mould, the desired tyre thicknesses and cross-sectional geometry may be obtained by calculating a mould cavity shape by FEA.

Preparing the Tyre Elements

Brass-plated steel cords and processed fabric cords are received from wire and fabric plants. Brass-plated bead wire is received in spools; bundles of this wire are passed through an extrusion die and given a coating of rubber. The rubber-coated wires are wound into a hoop of specified diameter and sent to the tyre building machine (i.e., assembly machine).

Banbury internal mixers are used to mix the raw materials into rubber compounds. The compounded rubber is then sent to mixing mills; the resulting rubber is pelletised and sent to calender machines or to extruders.

Calender machines include three or more cylinders rotating in opposite directions. These machines take the cords and the compounds and produce a cord-reinforced play. The ply thickness is determined by the set distance between the cylinders. The calendered fabric is then cut in the desired lengths and angles and sent to a tyre building machine.

Treads, sidewalls, and apexes are prepared by extrusion, which involves forcing their rubber through a die having the proper shape. The extruded rubber is then cut and sent to the tyre building machine. Uniform and consistent elements are critical to tyre performance.

Assembling the Green Tyre

The tyre building machine is used for assembling certain tyre components in an initial form, which is different from the final tyre shape; the remaining components are added later. The final change of shape is realised during the moulding stage. The inner liner is applied on a cylindrical tyre building drum. Then the carcass plies are applied, one at a time, followed by setting the beads in place. The chippers and any other bead area plies are also added. Depending on the number of building stages, the belts may be applied at this stage or later. The drum is collapsed and the resulting "green tyre" is removed.

For bias and bias-belted tyres, the green tyre is shaped as a cylinder with belts, if any, in the middle and the beads at both ends. For radial tyres the green tyre is shaped like a barrel with a greater diameter in the middle where the belts are placed; this is closer to the final radius of the tyre.

Tyre Curing

The green tyre is cured in a press in which a cylindrical diaphragms shapes the tyre as the press carrying the segments of the mould closes.

The proper pressure is applied through the diaphragm and is maintained, to keep the components in their desired locations relative to the mould surface. Heat is applied to vulcanise the rubber compounds and achieve the target material properties. The designated cure cycle defines the pressure and temperature variation with time. When the cure cycle is complete, the diaphragm is deflated, the mould opens, and the cured tyre is ejected.

The heat cycle during cure generates strains in the tyre materials. Not all these strains are relieved when the tyre is ejected from the mould. Therefore, depending on the cord material used and the intended application of the tyre, certain tyres are subjected to "postcure inflation" pressure for a while, before leaving the factory.

Tyre Inspection and Finishing

The exterior of the cured tyre is cleaned by removing any flash material (i.e., excess rubber) remaining from the moulding process. Tyre uniformity machines measure the force variations of the tyre. This is usually accomplished by measuring the force required to maintain a constant deflection as the tyre is rotated. Since this impacts potential ride disturbances, the tread may be force-ground to reduce the force variations to preset limits.

Holography and X-ray equipment may be used to inspect the integrity of the various tyre components and to assure the absence of voids or ply separations within the tyre before the tyre is shipped for use.

TYRE EVALUATION

Prototype tyres are subjected to a variety of tests. One set of tests is run to verify the target performance levels and assure safety. As tyre technology develops along many fronts, the hope is that there will be progressively less need for such tests, since tyre testing can be quite expensive and time-consuming.

Other tests are performed to meet regulations and to qualify the tyre for intended use.

Tyre Durability

There are several types of durability test. On-the-road tests are performed by equipping vehicles with test tyres and monitoring the travel distance to failure. Some durability tests are run at actual service conditions. Accelerated durability tests can be conducted by overloading the vehicle and projecting the test distance to what would be expected under rated load conditions. Tyre failure may also be accelerated by testing on roads with more severe surfaces, such as cobblestone.

Durability tests may also be run on smooth drums; here the radius of the drum influences the distance to failure due to the extra severity induced by the drum curvature, which changes the loading on the tyre. Stepped-load testing involves increasing the load after each specified number of test cycles. High speed testing involves increasing the drum speed at specified intervals until the tyre fails, or a preset target speed is met. Another phenomenon monitored in high speed tests is the formation of "standing waves" (i.e., ripples) on the tyre's sidewalls; this is usually a prelude to tyre failure.

Tyre pressure may be either maintained at a constant level during durability tests or left to rise with test distance. Relating failure under severe loading conditions to test distance under normal loads is based either on empirical formulas or on more scientific failure criteria such as the fatigue curves relating the stress-strain level to the number of cycles to failure, taking into consideration the temperature effects. Cleated wheels are used to project tyre durability (belt edge separation, bead area durability) from failure load and test distance. Plunger tests are used to rate resistance to puncture through the tread. Burst tests are performed to evaluate the inflation pressure at which the tyre would burst; for safety reasons, water is normally used to pressurise the tyre.

Uniformity and Vibrations

Tyre uniformity has a direct impact on the vehicle's ride comfort. It also has an impact on the tyre's irregular wear.

As the tyre is rolled against a drum, force variation with speed can be measured in the radial, tangential, and longitudinal (fore-aft) directions. Another uniformity parameter often obtained from the radial force variation data is the "radial first harmonic," which represents the offset of the tread circle from the bead circle. Runout tests measure the deviation of the tread profile from a perfect circle.

Vibration tests involve measurement of the natural frequencies and mode shapes of the tyre. Advanced computer software and electronic hardware allow the engineer to assess ride comfort from a determination of the response of various locations in the vehicle (e.g., steering wheel, floors) to specific tyre inputs.

Ride Comfort Tests

Vehicle ride comfort tests are conducted on a variety of smooth and rough surfaces. Vehicle vibration, bounce, and harshness are evaluated in the vehicle interior. These evaluations are made either subjectively (by human feel) or objectively (by on-board instrumentation). Overall vehicle performance on ride is influenced by the interaction of the vehicle's suspension with the input from the tyre and the road surface. This interaction can mean that a smooth riding tyre on one vehicle can give a harsh ride on another vehicle. This problem must be solved by the use of computer modeling to predict tyre-vehicle interactions.

Tyre Noise

Noise tests are performed either on the road (smooth or rough roads) or in the laboratory. Road noise tests are influenced by the road surface, engine noise, and surrounding environment. Interior vehicle noise is influenced by the vehicle structure. Test specifications by regulating agencies and vehicle manufacturers specify the distance from the road at which the recording microphones should be placed and the height of the microphones. Pass-by noise specifies the steady speed and the gear at which the vehicle is driven on the specified road segment; accelerated pass-by tests specify the initial and final driving condition over the given road segment; and coast-down tests measure the noise as the vehicle passes by with its engine turned off.

Laboratory noise tests are usually performed in semianechoic rooms, on either smooth drums or drums with surfaces duplicating actual road surfaces. Internal noise may be measured by positioning microphones at selected positions inside the vehicle. Either single microphones or two microphones configured to simulate a human head may be placed to record the noise. For both exterior and interior noise, the recorded signals are analysed by the engineer to determine the frequency curves that describe the type and level of noise at different speeds.

Traction

Traction (and braking) characteristics of the tyre are evaluated either subjectively (whereby the driver rates each tyre) or objectively (by instrumented measurement of the forces resisting tyre slippage and the speed at which traction is lost).

Traction tests may be performed on dry roads for dry traction, on wet roads to determine wet skid performance and wet traction at specific water depths, on snow-covered roads for snow traction, or on ice to determine ice traction. Traction tests are performed either on straight road segments or on curved roads. Another traction related test is the "hill-climbing" test.

A tyre performance closely related to traction is aquaplaning, which is also measured either on a straight road segment or on a cured road. For a specific water depth, the progressive loss of contact with increasing speed is measured as an indicator of the tyre's performance.

Handling Tests

Vehicle handling tests may be conducted on either dry or wet roads to evaluate a given set of tyres. Steering response, steering precision, cornering, and tracking are common tests for handling.

As a measure of the handling performance of tyres, the forces and moments generated in the tyre's footprint under different operating conditions are measured on "force and moment" machines. The forces and moments generated at different slip and camber angles, as the tyre rolls, are related to vehicle performance at various driving and cornering conditions.

Forces and moments in wet and icy conditions may be measured by running the tests either inside large drums with proper water or ice films, or on the road with vehicles equipped with special additional wheels and instrumentations.

Tyre stiffness is also measured because of its influence on handling. Handling and force and moment tests are conducted either at constant speed (i.e., steady state) or under transient conditions with applications of braking or acceleration.

Tread Wear

Wear tests may be conducted either on the road, as in the case of durability, or on test drums. The rate of wear is measured by monitoring the change in nonskid depth with distance. The uniformity of wear is measured by stopping the test at intervals and measuring the non-skid depth at different locations across the tread and around the tyre. The footprint shape and the pressure distribution are also evaluated to predict the wear balance.

Rolling Resistance

The rolling resistance of a tyre is determined by measuring the loss of energy as the tyre rotates against a flat surface (most realistic), a drum, or the twin rolls. Although the latter (illustrated in Fig. 10.25) does not correspond very closely to actual vehicle driving conditions, it is often run because of specifications set by the vehicle manufacturers to meet EPA targets for their vehicles, as determined by their corporate average fuel economy (CAFE).

FUTURE DIRECTIONS OF DEVELOPMENT OF TYRE TECHNOLOGY

The pneumatic tyre can be considered today a mature and fully developed primary engineering component of surface transport vehicles and as there is no alternative component as a replacement of tyre even remotely in sight, it will certainly continue in its present form indefinitely into the future. While the traditional requirements of load carrying capacity, cushioning ability, durability and stability etc., will continue to remain in effect, the changes and developments taking place in related areas like vehicle engineering and highway engineering will put heavier demands on tyre to evolve suitably. Therefore it will be pertinent to take a glimpse of such developments and its impact on tyre developments.

The tyre as the intermedium between the road and the vehicle is the critical connection not only in physical terms but also relative to the interchange and application of technology. The tyre technologist has the opportunity and responsibility to acquire the knowledge from Road and Vehicle engineers which relates directly to the safety and efficiency of the road transport system and utilise it in development of his product subsequently.

In vehicle technology, people are becoming increasingly aware of developments like Anti-lock braking system, Traction control system, Semi-active and active suspension system, Ride height control, Tyre deflation warning system. These coupled with tremendous development in electronics have effectively made "driving-by-wire" a feasible concept. Even evolution of "smart tyre technology" can be attributed to demands created by such concurrent technology; in such tyres a micro-chip is embedded in the tyre which can communicate to a dedicated outside source (computer) about load history, inflation pressure, temperature etc.

Most of these ongoing developments in vehicle technology are based on electronics and will inevitably increase the cost of vehicle, complexity and cost of vehicle maintenance. Additionally, the normal as well as peak operating speeds are increasing rapidly and coupled with demand of sophistication, all the automobile products including tyre are facing heavy pressures to evolve to meet such challenging demands.

In highway engineering, noticeable developments like Road monitoring devices (e.g., high speed laser texture meters); drainage asphalt porous road surfaces have posed further demands on tyre performance parameters (reduction of noise, enhancement of wet traction, removal of spray hazard are a few examples).

Thus development in Vehicle and Highway engineering, coupled with legislative and market forces are producing a comprehensive demand list for tyre developers, major of which are :

1. Low tyre weight.
2. Low cost.
3. Low rolling resistance.
4. Improved wet-traction.
5. Better handling and stability.
6. Reduced noise generation.
7. Improved comfort/reduced vibration.
8. Reduction in road damage.
9. Better durability/recycling capabilities.
10. Runflat capabilities.
11. Higher performance consistency throughout the product life cycle etc.

Summarising it, the demand is for "Better technology at a lower price".

It is a well known fact that tyre development is a costly and time consuming process. It may on an average take a period of 12–18 months from concept to commercial release. Conventional design processes embracing the concept of "Design-Prototype-Redesign" is no more suitable to address today's requirement of "Better performing product with reduced design cycle time". Therefore tyre developers have included advanced computational techniques like Finite Element Analysis Simulation, Virtual prototyping into their regular stream so that they can predict performance reasonably well and can suitably optimise it. While these methods have yet to reach a level where complete physical testing can be eliminated, they supplement physical testing well and result in considerable reduction in total cycle time. Development process of tyre today ensure close working relationship with the customer and thereby has become more responsive. Softwares like "Mechanical System Simulation" allows modelling of Road/Tyre/Vehicle together and simulate Ride/Handling/Vibration analysis very early in design process and allows a combined evaluation of corrective methods well before the first physical prototype is build.

We will now critically identify the areas where most of activities in near future will be focused to achieve further improvements. We will first consider tyre weight. It is primarily required for lower rolling resistance and increased response of tyre. Wheel assembly to electronic control. Considerable success has been achieved in identifying alternative reinforcing materials for weight reduction. Development of a non-steel car tyre yielding 30% weight reduction using Aramid (which is six times lighter and yet have tensile strength similar to steel) paves the way of what we know as "ULW (Ultra Light Weight) concept". Research will continue to look for further possible reduction in weight and study of influence of low inertia tyre assembly on overall vehicle performance.

Regarding rolling resistance, it is mainly important from fuel saving and fatigue resistance point of view. Facilities of its measurement are very common now. The ultimate target is to bring rolling resistance down to their theoretical minimum. By experience and evaluation, it is found that majority of energy loss is associated with tread and breaker/belt region (each contributing approximately 40–45% of total energy loss). The developments in near future will target improvements in these areas specifically to attain reduction in rolling resistance.

It is a constant concern to tyre designers that wet-traction capabilities progressively deteriorates throughout the service cycle. Minimum tread-depth legislation tries to address it by making it mandatory to remove a tyre after non-skid depth of 1.6 mm is reached. Efforts to improve wet roadhold performance through changes in pattern design, construction and compounding will be major areas of development attention. Also it is feasible for tyre designers to take the handling and stability characteristics of tyre beyond what is required for particular vehicle variants and indeed to such levels that they are of a minimum concern to the vehicle homologation engineer.

Tyre noise, though not directly related to safety has been increasingly getting legislative attention to address noise pollution and ensure passenger and passer-bys comfort. Here while attention to pattern sequencing/design will be main plank for tyre developers to address noise, development of road surface like drainage asphalt may help further.

For vibration characteristics, the study of the fundamental resonance characteristics of tyres and tyre/wheel assemblies as initiated by road texture, tread pattern or indeed by suspension will be important factors to be resolved. Software based simulation and analysis will be a major thrust area in future and joint work between vehicle and tyre developers will become a reality.

For truck tyres, the performance of the product in terms of grip will envisage enhancement derived from combined polymer and filler developments in order to further reduce stopping distance and thus attempting to reach a value close to those for cars. The area of potential road damage brought about by development of low aspect ratio tyres operating at high inflation pressures/loads and in particular the wide single tyre will be a serious concern which will need to be addressed in near future. While wide single tyre yields important advantages like reduced assembly weight, reduced unit rolling resistance, reduced unit noise, lower maintenance and reduced stock management, enhanced vehicle stability etc., it has disadvantage of increase in contact pressure which may lead to increased road structural fatigue and deformation consequently damage thus requiring increase in road construction and maintenance cost.

The development of drainage asphalt porous road surfaces has become commonplace in developed countries. Such surfaces not only give dramatic reductions in tyre to road noise generation, they also virtually eliminate the problem of spray. However, they have not yielded expected high speed skid resistance advantage may be due to potential reduction in tyre/road contact area and tendency of driver to drive faster on quieter surfaces. Porous surfaces allow the exciting possibility of designing the void distributions to optimise absorption of specific frequency levels originating from tyre and vehicle drive train. Also road rut-depth or transversal/longitudinal road mega texture or unevenness have influence both on stopping distance and energy loss and thus have strong influence on vehicle handling and stability under wet conditions.

There are few major areas in which tyre properties significantly affects vehicle performance. Firstly within the contact patch, where the generation of interfacial friction is essential for the transmission of all the control forces directly governs the degree of safety. It is possibly through the complete understanding of footprint mechanics that further developments of tyre will be forthcoming. Footprint is also the area which initiates the majority of vibrations which are eventually transmitted to vehicle systems. Understanding of vibration transmission becomes significant particularly when attempts for tyre weight reduction are gaining big success. Such understanding will not only have impact on wheel as well as suspension design but will pave way for tyre to be considered a fundamental part of suspension. Such approach in future will enable tyre and suspension system to be considered as one unit for optimum performance in terms of vehicle noise and vibration characteristics.

The major emphasis in tyre development will be on ability to work as teams in terms of communication of existing knowledge, the application of new knowledge and the ability not to repeat the errors of the past thereby paving the way of evolution of "Vertical Engineering" where the disciplines of road, tyre and vehicle engineering communicate more effectively with each other.

In future, development work may attempt towards a cybernetic or "thinking" tyre which changes shape, size, inflation pressure and tread patterns to provide optimum performance under so diverse operating conditions. Further integrated tyre and wheel assembly produced as a single unit may become a reality to yield better uniformity and balancing. Many innovative ways to achieve better run-flat capabilities like inverted bead tyre, Twin-tyre multi-chambered tyres will be tried out and adapted selectively. Coloured tyres to suit vehicle colour will become very common.

Finally, we will summarise features of tomorrow's tyre/vehicle technology. About 50% weight reduction and employment of tyre deflation warning system will be achieved both in car and truck segments. Enhanced comfort, enhanced wet and dry traction capabilities will ensure vehicle performance. While computer chips in tyre (particularly for trucks) will become normal feature (adoption of smart tyre technology), for cars run-flat capabilities (alongwith provision for no bead dislodgement at zero inflation pressure) avoiding necessity of spare wheel, jack etc., will be surely and successfully achieved.

CHAPTER 11

Footwear Technology

INTRODUCTION

Rubber has in the past been associated with utilitarian type footwear, but of late, it has become and established commodity in fashion footwear. Traditionally, rubber footwear was manufactured by hand-assembling pre-shaped component parts and then vulcanising in an autoclave. During the post-war period there have been great advances in machinery equipments and automated processes, thereby reducing the high labour content of the hand-assembled product. The rubber footwear industry has also taken advantage of new developments in polymer science and technology to formulate compounds which meet the new process requirements and give products with practically any desired feature.

The rubber footwear industry has to meet a wide diversity of quality, depending on the end use of the product, the method of manufacture, and the price, with the result the compounds vary considerably and there is no typical footwear formulation.

TYPES OF FOOTWEAR

There are three main types of footwear: rubber, leather, and plastic. The first two types are quite complicated to manufacture, since they involve many stages and numerous operations. Rubber footwear can be (i) plimsolls; (ii) built-up shoes; (iii) all-rubber shoes; (iv) direct vulcanisation process (DVP) shoes; (v) direct injection process (DIP) shoes.

Plimsolls

Plimsolls are the simplest type of hand-built shoes. The uppers are made of frictioned canvas uppers with roll (calendered) soles, edge foxings, toe caps, toe guards, etc. Such shoes are made on continuous conveyors or belt-type conveyors.

Components and their treatment

The components of plimsolls are as follows:

Canvas upper	Filler
Eyelet reinforcement	Sole
Toepuff	Eyelets
Counter	Lace

Binding
Tongue
Insole
Toe cap
Foxing
Toe guard

Preparation of uppers

Basically canvas uppers are made of two textiles. The upper canvas is of cotton, a cotton-synthetic blend, or a synthetic material weighing 200–250 g/m^2, but heavier.

Weights (e.g., 350–400 g) are used for specific purposes. The linings are in general of lighter weight between 120 and 200 g/cm^2, and can be of cotton or a non-cotton synthetic.

Friction formula

The rubber friction compound weighs 250–350 g/m^2 and is typically formulated as follows:

Weight in kilograms: Pale latex Crepe No. 1, 45.000; styrene butadiene rubber (SBR) 1502, 15.000; stearic acid, 0.750; zinc oxide, 2.000; 2-mercaptobenzothiazole (MBT), 0.200; mercaptobenzothiazole disulphide (MBTS), 0.400; tetramethylthiuram (TMT), 0.050; wood rosin, 1.400; titanium dioxide, 20.000; Claret YS, 0.003; Ultramarine blue, 0.300; aluminium silicate, 8.000; calcium carbonate, 88.797; rubber oil, 1.500; salicylic acid, 0.100; and sulphur, 1.500.

After these ingredients have been mixed and the textile combined, uppers are clicked out.

Eyelet reinforcement

Eyelet reinforcements are clicked out from the scrap remaining after the uppers have been made.

Toepuff

Canvas clothes of 120 g/m^2 weight are pasted on one side with a rubber compound to a three-bowl calender. The weight of the rubber compound is approximately 100 g/m^2. Toepuffs are clicked out from the pasted canvas.

Counter

A counter compound may be formulated as follows:

Weight in kilograms: Standard Malaysian rubber (SMR) 50, 15.000; high styrene rubber (60% bound styrene), 6.000; crumb, 3.000; zinc oxide, 0.600; stearic acid, 0.200; wood rosin, 0.500; lime, 0.600; MBT, 0.150; MBTS, 0.050; diorthotolyl guanidine (DOTG), 0.200; salicylic acid, 0.050; wood flour, 14.000; calcium carbonate, 9.250; and sulphur, 0.400.

The main requirements of a counter compound are stiffness and flexibility. The foregoing compounded mixture is pasted on one side of a canvas cloth weighing 200–250 g/m^2 on a three-bowl calender, followed by clicking out in accordance with the sizes desired.

Insole

Insoles are calendered rubber compound pasted onto one side of the textile. The rubber compound can be solid (consisting of crushed textiles) or of cut-out sponge insoles from pre-blown sponge sheets. A typical formulation uses the following chemical compounds.

Weight in kilograms: Ribbed smoked sheet (RSS) No. 5, 5.000; rubber rags, 7.000; reclaimed scrap, 22.000; zinc oxide, 0.500; stearic acid, 0.100; MBT, 0.100; MBTS, 0.100; TMT, 0.010; China clay, 6.000; calcium carbonate, 8.890; and sulphur, 0.300.

Filler

Fillers are made of low cost rubber compounds consisting of crushed textile wastes, wood flour, and so on. The compound is calendered.

Operations Required to Build-up Plimsolls

The following operations are performed serially for manufacturing plimsolls:

1. False vamp marking
2. Pasting in eyelet reinforcements
3. Stitching eyelet reinforcement tapes
4. Stitching backs
5. Rubbing back seams
6. Binding
7. Stitching counters
8. False vamping
9. Eyeletting
10. Lacing uppers
11. Buttonholing
12. Stitching buckles
13. Joining hockey tongues
14. Stitching ankle guards
15. Reinforcing back tape stitching
16. Joining hockey uppers
17. Controlling uppers
18. Fitting toepuffs
19. Cementing uppers (counters)
20. Preparation of lasts and uppers
21. Pulling over
22. Toe lasting
23. Waist lasting
24. Heel lasting
25. Latex edge cementing
26. Insole cementing
27. Cutting toe caps or toe guards
28. Putting toe caps on
29. Leveling toe caps
30. Extruding foxing
31. Fitting foxing
32. Leveling foxing
33. Trimming foxing
34. Cementing heels
35. Fitting heels
36. Pressing heels
37. Fitting soles
38. Pressing soles
39. Leveling soles
40. Trimming soles
41. Fitting and leveling toe guards
42. Quality inspection of unvulcanised shoes
43. Unlasting vulcanised shoes
44. Lacing and cleaning

BUILT-UP SHOES

The construction of built-up shoes is more or less the same as for canvas shoes, but the rubber sole is pre-moulded. Uppers are of cotton canvas, synthetic canvas, cotton-synthetic blend to impart special properties (e.g., strength, low weight, scuff resistance). Shoes of this type are generally used for jogging and sports (e.g., basketball). For the bonding of sole, side foxing, toe cap, toe guard, etc., in built-up shoes manufacturers apply natural rubber latex as cement by dipping or brush of natural rubber base cement. A typical moulded roll sole compound formulation would contain the following ingredients.

Weight in kilograms: Pale latex crepe, 8.000; SBR 1502, 10.000; IR 2000, 8.000; zinc carbonate, 0.600; zinc oxide, 0.500; stearic acid, 0.500; hexamine, 0.200; methylethyl glycol (MEG), 0.500; precipitated silica, 9.900; Si 69, 0.100; benzoic acid, 0.150; rubber oil, 1.000; and sulphur, 0.550.

A rubber compound for cement white might be formulated as follows (weight in kilograms): Pale latex crepe, 25.000; zinc oxide, 7.640; stearic acid, 0.200; DHC, .060; wood rosin, 2.500; titanium dioxide, 4.500; and ultramarine blue, 0.100.

The compound is dissolved in solvent oil to 20% solid content.

ALL-RUBBER SHOES

Wellington boots and waterproof casual shoes are of common types of all-rubber shoes. For shoes made of rubber, special care must be taken in designing formulations that will have good flexibility and good cut resistance without posing problems in the building-up operation.

Rubber shoes are normally made on a conveyor belt. In preparing the upper components, stockinette-type materials are coated on one side with rubber friction dough on a spreading machine and cut to size. The cut pieces are pasted with pre-calendered black waterproof rubber compound 1.0–1.5 mm thick to form individual components of Wellington boots or waterproof casual shoes. To make the shoes, the different components are assembled manually.

The built-up shoe is dipped in lacquer prior to vulcanisation. A typical waterproof rubber compound is as follows.

Weight in kilograms: RSS No. 1, 55.000; SBR 1502, 10.000; semi-reinforcing furnace carbon black (SRF) batch (30% black, 70% rubber, 10% oil), 10.000; Renacit VII, 0.100; zinc oxide, 3.000; stearic acid, 0.750; wood rosin, 0.500; MBT, 0.500; MBTS, 0.600; TMT, 0.300; antioxidant DN, 0.500; microcrystalline wax, 0.350; rubber oil, 2.500; activated calcium carbonate, 69.300; and sulphur, 1.600.

Plimsolls, built-up and all-rubber shoes are vulcanised by hot air cure under pressure in an autoclave. Vulcanisation times and temperatures vary depending on the type of shoe. Typically, however, the vulcanisation temperature is 120–130°C, the time 50–90 minutes, and the air pressure, 3–3.5 atm.

DVP SHOES

In DVP shoes, either canvas uppers are pre-vulcanised in an autoclave or rubber friction compounds of the self-vulcanising type are used.

The uppers are generally sock lasted; that is, the upper is stitched onto a sock, which is directly lasted on a two-part metal last (normally known as a broken last). A rubber sole is placed on the lasted sock and calendered edge foxing is built-up around the lasted margin.

The mould has three parts: a top piece, which is heated to 170–180°C, and two side parts around the last, which are heated to 160°C. The moulding operation takes about 2–3 minutes.

When leather shoes are made by the DV process with direct vulcanisation of moulded sole and heel, to achieve good bonding and acceptable results, the use of uniform thickness, full chrome leather having following properties is recommended :

Fat content not more than 6%
Chrome content not less than 3.5 %
SAWS not less than 0.5
Shrinkage (linear) not more than 5%

Rubber compounds should be designed for fast cure, but they should be stable at the processing conditions.

In fact, very often microwave heating systems (e.g., the Radyne) are used to reduce curing time. Here the sole blanks are activated for 8–10 seconds at 80°C before being placed on the lasted bottom just before vulcanisation. In this case, the rubber formulation should not include carbon black, which tends to generate too high a voltage and too much heat, causing the compound to burn. Normal pressing temperatures are 175–180°C for the top plate and 150°C for the side moulds. Vulcanisation time is 4–6 minutes.

DIP SHOES

As the name implies, the DIP shoe is made by direct injection of polyvinyl chloride (PVC) compound onto lasted upper material. In the direct injection process, any conventional upper material can be used (e.g., as leather, PVC, leather cloth, textiles). When textile shoes are being moulded with foxing strip, the upper must fit tightly over the last without creases. To avoid having the seam of the sock and the upper obstruct the flow of PVC, it is essential that the sock stitching be (a) string lasted, (b) slip lasted, or (c) reverse lasted (seam inside).

Before injection, the upper must be pre-heated to around 60°C to make sure that it is perfectly dry, since this is the surface at which the PVC will come into contact with the textile.

In case of a leather upper, it is necessary to roughen the lasted edge, and one or two coats of adhesive must be applied for good adhesion. As for PVC leather cloth uppers, the lasted margins should be wiped with methyl ethyl ketone before injection for good bonding. Injection moulding machines of various types are being used, from single-station devices to 16–21 stations connected to a rotary table for high output.

The lasts are normally heated to around 100–110°C prior to injection of the PVC compound and cooled when injection is complete. Normal injection time is about 1.5–2 seconds, depending on the *L/D* ratio of the screws. Cooling time on the last is 60–70 seconds.

Typical PVC formula for direct injection of sole is as follows.

Weight in kilograms: PVC resin (K 67), 47,000; P 83, 12.000; diocytl phthalate (DOP), 34.000; epoxidised soybean oil, 2.000; organic phosphite, 0.300; barium-cadmium, or zinc stearate, 0.800; zinc stearate, 0.100; Ba/Cd stearate, 0.500; stearic acid, 0.350; titanium dioxide, 8.100; ultramarine blue, 0.050; and blowing agent, 0.800.

Table 11.1 gives minimum specifications for the soling compounds.

Table 11.1. Minimum specifications for a PVC soling compound.

Property	*PVC leathercloth*	*PVC soling*
Specific gravity	1.6–1.20	1.16–1.20
Hardness, IRHD	55–59	58–62
Tensile strength kg/cm^2	90 minimum	80 minimum
Modulus 100%	25–35	15–25
Elongation at break, %	340 minimum	350 minimum
Tear strength, kg/cm^2	30–34	28–32
Volatility in 3 h at 130°C	1.1%	1.5%
Migration in silica after 24 h at 70°C	5% maximum	5% maximum
Thermal stability after 1 h at 180°C	Pale yellow	Pale yellow
Weathering stability	Light opacity	Light opacity
Adhesion 6 mm on strip	1300 g min	—
Adhesion on SATRA tester	—	Toe, 60 lb; heel, 80 lb
Flexing: no cracks or foxing separation after	1,000,000 cycles	1,000,000 cycles

PLASTIC FOOTWEAR

Although each moulding method differs in many respects from the others, they all have one thing in common—the use of heat and external pressure. When no external pressure is used, the process is known as casting. In casting, the plastic (either thermoplastic for thermosetting) is supplied in liquid, granule, or powder form and may be liquefied by heating or by the addition of solvents, plasticisers, catalysts, etc. The fluid plastic is poured into the mould and is gelled into solid form by the application of heat.

Slush Moulding

Basically, slush moulding is a process whereby a liquid is poured into a hollow mould and then poured out, leaving only a skin or lining of the liquid resin on the interior of the mould. The moulds commonly used are electroformed copper moulds or fine aluminium sand-cast moulds. Heat is applied to convert the plastisol into a solid elastic material, and the mould is cooled until the moulded shoe can be stripped from the interior of the mould.

Most of the slush moulding production lines follow one of two basic systems: single pour or double pour. Both systems are readily adaptable for conveyor or automatic lines.

The single-pour system involves a single filling station where the moulds are filled by the vacuum, gravity, or pressure methods. If the gravity or pressure method is used, the filled moulds are either spun or vibrated to eliminate bubbles at the extremities.

The moulds are then heated, usually by infra-red, steam, or immersion in hot water, to adjust and control the wall thickness. Heating at this point in the system is necessary to induce the liquid resin (plastisol) in immediate contact with the walls of the mould to gel. The mould is then inverted and the plastisol allowed to drain out. The remaining plastisol lining is fused in an oven with temperatures up to 190°C (370°F). The fusing cycle depends on the mould size and the wall thickness desired in the finished article. After fusing, the article in the mould is cooled and unlasted.

The two-pour method is a modification of the procedure above. Again the mould is filled almost to the top, usually by pouring or by pumping, and then vibrated to remove bubbles or air tapped in the extremities. The mould is then dumped, leaving a thin film of plastisol on the interior surface. By filling a cold mould in this manner, entrapped air or gelation that obscures fine details is less likely.

After dumping and draining, the mould is passed through heat to fuse or semi-fuse the first skin. The heating may be of the hot air type or infra-red. There is no problem of fusing or skinning over the liquid resin in a full mould, as in the single-pour method. The heating cycle determines the final wall thickness of the article. After passing through the first pour oven, the mould is filled a second time, dumped, and allowed to drain. The heat of the mould retained from the setting or fusing of the first skin (usually of a different colour from the second) will determine the ultimate thickness of the plastisol deposited on the walls of the mould. The mould is then cooled and stripped as in the single-pour system.

Sponge plastisol may be handled similarly. When plastic footwear is made either completely of sponge or with sponge linings, the two-pour system is used for the latter. A sponge vinyl liquid is poured in the second position. This method is ideal for insulated boots.

Slush moulding has many advantages in the production of such articles of Wellingtons, gumboots, and overshoes. The process is relatively simple, and heavy equipment is not necessary. Accurate replicas of built-up rubber footwear can be produced without seams. When the article is removed from the mould, it is finished: except for fitting studs or decorative buckles, the shoes can be packed for shipping.

A typical microcellular sandal formulation is as follows.

Weight in kilograms: PVC resin (K 67–70), 100.000; (DINP), 75.000; (DIBP), 10.000; epoxy plasticiser, 3.000; BaCd Zn stearate, 1.900; stearic acid, 0.100; Cereclor S 45, 10.000; zinc stearate, 0.100; precipitated calcium carbonate, 10.000; and Cellogen AZ 130, 1.000. The specific gravity is 0.65 and the hardness IRHD 55.

Injection-Moulded PVC Shoes

PVC shoe making is a less labour-intensive operation. Moreover, the shoes can be produced with low capital investment by unskilled workers. The most common form of PVC shoe making is by the injection process. Wellington boots, waterproof shoes, casual shoes, and sandals for men, women, and children are produced in large quantities by injection moulding by this process. Production setups range from a simple vertical hand-operated injection machine to quite sophisticated horizontal injection machines used with single-station to 18–24 stations rotary-type carriers, depending on the output required.

HAND ASSEMBLED AND HOT AIR VULCANISED PRODUCT

Compounds for hand-assembled products are normally based on natural rubber because good building tack is essential and excessive shrinkage can cause distortion of shaped component parts. Adhesives for hand-built products are also based on natural rubber. The grades of natural rubber selected depend on the enduse. For good quality thin upper compounds, where flexing is a factor, a rubber with minimum dirt content is necessary, but lower grade rubber can be used for soles and heels. In moulded products synthetic rubbers are used extensively, such as non-staining low-Mooney SBR, oil-extended SBR, IR, and BR. Where heat- and oil-resistance properties are required, compounds are based on nitrile rubbers, blends of nitrile rubbers and PVC, or polychloroprene, the last of these being used mainly in hand-assembled products because of its superior tack. Whole tyre reclaim is used extensively in black solings and in some moulded black upper compounds. Finely ground vulcanised crumb can be used for their cheapness and to help reduce porosity.

Carbon black is the reinforcing ingredient for black industrial boots, but, to meet the demand for non-black compounds, reinforcing siliceous fillers are used. A high-quality non-black soling will have fine particle size silica as a reinforcing agent, and the medium grade solings either aluminium or calcium silicate. Good quality clays or activated calcium carbonate will give reasonable reinforcement with natural rubber, but to obtain equal reinforcement with SBR and particularly with oil-extended SBR, the addition of some siliceous fillers is required. The main diluents are whiting and cheap clays, and for whiteness it is titanium dioxide.

The accelerator system is normally a thiazole with a guanidine, thiuram, or dithiocarbamate as a secondary, depending on the rate of cure required. For white and light-coloured products, an antioxidant must be selected which is non-staining.

Cotton is still the most commonly used fabric in the rubber footwear industry. Blends of cotton and man-made fibres are sometimes used where additional resistance to abrasion is required. The leg and foot lining of an industrial boot is normally a square-woven fabric of 0.30–0.40 kg/m^2 in weight, rubberised by frictioning and topping, and with sufficient elongation to enable it to be lasted onto the shape of the boot tree. On moulded boots, knitted fabrics (either plain knit or ribbed) are also used, giving the boot leg a more flexible feel.

The upper material of canvas-topped footwear is prepared from a square-woven face fabric combined to a twill backing fabric by means of a thick rubber solution or compounded latex. For top-quality sports shoes, the combined weight of the fabrics is approximately 0.50–0.60 kg/m^2. For slippers, which are very much a fashion article, a variety of fabrics are used such as cotton, wool, nylon, suede, and lurex.

Although this method is labour intensive, it still has its advantages in that it allows for greater flexibility in changing product design without the need for acquiring new and sometimes expensive equipment. Some all-rubber industrial and wellington boots, and canvas-topped sports shoes and plimsolls are still made by this method. A pair of all-rubber industrial boots may consist of 40 rubber or rubberised fabric component parts, each prepared and cut to the required shape prior to assembly. These parts consist of the outer rubber leg, outer rubber vamp, fabric leg lining, fabric vamp lining, anklet, insole, fillings, toe and heel reinforcements, back strips and strengthening pieces, foxing strips, top binding, sole, and heel (Fig. 11.1).

The calendered sheet for the production of outsoles is produced on a four-roll calender where the fourth roll is engraved with the necessary sole pattern and profiled to give the variations in thickness. Any required sole design and thickness is obtained by changing the fourth roll.

The heel of an industrial or wellington boot which is hand-assembled, is usually pre-moulded to the required pattern, size, and shape. The insole, heel-piece, and fillings are made from unvulcanised rubberised fabric waste which has been ground on a tight mill and sheeted to the required thickness or topped onto fabric.

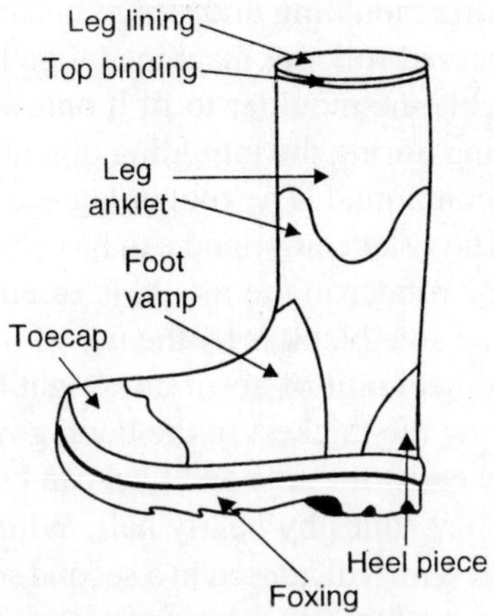

Fig. 11.1. Rubber boot components.

The components, freshened by a solvent wipe on this rubber solution, are assembled, in a specified sequence, tightly on the metal last and then thoroughly rolled to ensure good bonding.

Another method of manufacturing is to omit the outer rubber leg and vamp and to apply a coating of compounded latex by dipping and coagulating to provide the outer surface. After thoroughly washing and then drying off the moisture, the foxing strip, outsole, heel-piece, and top binding are attached. The latex forms a seamless, impermeable layer of rubber around the carcass.

For canvas-topped footwear, the fabric components are die cut and machined together to form the upper, onto which bindings have been sewn and eyelets or gussets and heel-pieces fitted. This is lasted onto the last, and the necessary rubber components attached.

Vulcanisation is carried out in a large cylindrical autoclave heated internally by banks of steam coils fitted in the base and around the sides. The boots are loaded onto carriers and wheeled into the autoclave. In some factories, the carrier peg on which each boot is fixed is attached to a vacuum pump so that, as the boot is vulcanised, the component parts are drawn tightly around the last, thus ensuring good adhesion.

Immediately the autoclave doors are closed, and air, at a pressure of around 3 kgf/cm^2, is admitted to offset any blistering or blowing of the outer rubber parts. The temperature gradually rises to 135°C, where it is then maintained for the remainder of the vulcanising period, which is normally 60–65 minutes in total. Because of the different thicknesses of rubber components and the hot-air method of vulcanisation, the compounds need to have a plateau-type curing system to avoid degradation. Accelerator systems based on thiazoles with TMTD as a booster have proved effective. A good heat-resisting antioxidant is usually added.

COMPRESSION MOULDED INDUSTRIAL RUBBER BOOTS

A complete boot is moulded in one operation using vulcanising presses consisting basically of two lasts, two side moulds, suitably profiled to give the required pattern and thicknesses, and an engraved sole mould. Five components only are required by the moulder, a pre-formed fabric lining shaped to the last, a sole blank, a rubber leg, a rubber vamp, and a filler. A one-piece rubber leg and vamp is sometimes used. The lining fabric is first coated with rubber compound, and then pre-cured to avoid any strike-through of the outer rubber leg during moulding under high pressure. Correctly shaped legs, vamps, and insole socks, die cut from the pre-cured roll, are machined together to form the boot lining. This lining must have sufficient stretch to enable the moulder to fit it onto the last, and yet, when on, it must be a reasonably tight fit to avoid pinching during the moulding operation. Knitted fabric linings may also be used, suitably coated with rubber compound. The rubber leg and vamp compounds are calendered on a profiled roll and die cut to shape. The filler compound can be calendered or sheeted off a mill and die cut to shape. Correct distribution of the rubber in the mould is essential to give good mould definition. An economical method of preparing the sole blank is by the use of the Barwell Precision Performer, which can produce a pre-shaped sole and heel unit of accurate weight and dimensions.

The heel portion of the sole being the thickest in the boot, governs the vulcanisation time. Recently, a method has been incorporated whereby the sole and heel can be given double the vulcanising time set for the completed boot, reducing cure times by nearly half. While one boot is being vulcanised, a sole and heel blank is being moulded and semi-vulcanised in a second sole mould for the next boot. Overcuring the sole must be avoided, otherwise the bond of sole to the upper could be impaired. Two lasts, normally collapsible or split to ease the fitting of the lining, are provided so that, whilst one last is in the press vulcanising a boot, the other is having the lining and rubber parts assembled onto it. The press is fitted with automatic temperature controls and timing devices.

To obtain efficient economical operation, compounds have to be rapid curing yet with delayed action to give good mould flow and perfect definition of pattern. With sole plate and mould temperatures ranging from 150–155°C, a complete boot can be obtained every 4 minutes, the sole and heel having had two 4 minutes cures. Compounds with Mooney viscosity and scorch values at 120°C of 45–50 and 3–4 minutes (for a 10 points rise), respectively, are satisfactory for such conditions. The vulcanised leg compound must be very flexible to avoid any discomfort in wear, have good flex cracking resistance, good atmospheric ageing resistance, and reasonable tear strength. The vulcanised sole and heel compounds must have adequate abrasion resistance and good cut growth resistance. Moreover, the specific gravity of the compound should be no more than 1.2 to 1.25 to avoid excessive weight in the boot.

DIRECT MOULDED PROCESS FOR SHOE BOTTOMING

As the name implies, direct moulded footwear is produced by vulcanising rubber compounds directly to lasted uppers. More shoes are made in this way than by any other method. Moulding presses can be classified into two types; for cellular rubber-soled footwear, internal pressure machines which rely on pressure created by the blowing agent in the compound; and external pressure machines, either pneumatically or hydraulically operated, for moulding solid compounds.

CELLULAR RUBBER-SOLED FOOTWEAR

By far the largest volume of cellular rubber footwear produced is the sponge-soled slipper. There are several different types of internal pressure machines but each consists essentially of (Fig. 11.2) the following structures.

1. A base plate, having the contour of the sole and heel.
2. A ring which sits on the base, split longitudinally, hinged at one end and forming the sole cavity.
3. A last on which the upper is fitted and which acts as the top plate of the mould.

The sole plate and last are fitted with electrical elements, the sole plate being heated to 150–160°C, the last to 125–130°C.

The slipper upper, already prepared and shaped, is fitted onto the inverted metal last and held tightly to it by string or a sock lining which has been sewn to the base of the upper. A semi-vulcanised non-cellular outsole cover is placed centrally in the mould cavity, followed by a shaped blank of the unvulcanised sponge compound and a filling, usually a felt. The last is then lowered until it is clamped tightly over the mould cavity. The total volume of material placed in the mould cavity is usually 60–70% of the volume of the cavity. The gas generated by the blowing agent expands the sponge compound, creating internal pressure sufficient to fill the cavity and form satisfactory bonds between the upper, cellular sole, and outsole cover. Vulcanisation time is from 6 minutes to 8 minutes.

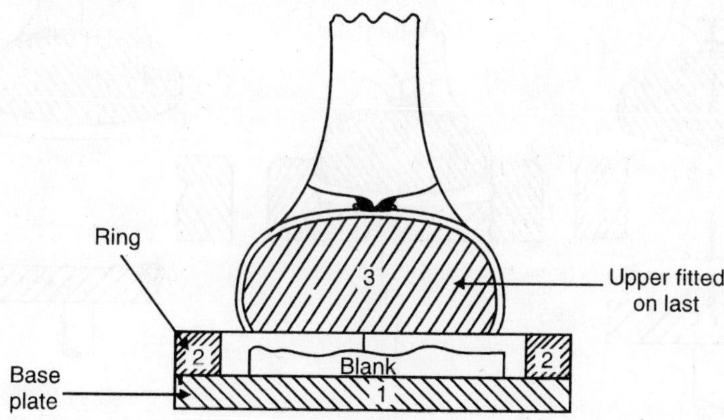

Fig. 11.2. Schematic diagram of internal pressure moulding machine.

Both preparation and presentation of the sponge compound require exacting control. Inorganic blowing agents are used, with a small addition of an organic to give a finer overall cell structure. The polymers are given extended mastication with the addition of chemical plasticisers. The compound must be highly plasticised to allow for ease of expansion by the blowing agent, and vulcanisation must be delayed sufficiently to allow this expansion to take place.

After mixing, several re-mills may be necessary to attain the required plasticity, and the finished batch is then calendered to the required thickness, well dusted to give ease of handling and die cut to the required shapes. The outsole cover compound is also calendered and pre-moulded into sheets 1.0–1.5 mm thick by approximately one metre square from which shapes are cut. This cover compound must have good ageing properties as the cut soles are to be given further vulcanisation in the slipper press.

SOLID RUBBER-SOLED FOOTWEAR

The attachment of solid rubber soling to uppers of either canvas or leather by the direct moulded process has been a revolution in footwear manufacture. Essentially, it is a mass-production process and most efficient when allowed to operate as such. The operation is carried out by the shoe manufacturer in individual presses with electrically heated moulds.

External pressure machines operating at low pressures, between 10 kgf/cm^2 and 20 kgf/cm^2, on the sole area are used for canvas-topped footwear, leather boots, and casuals. For walking shoes, work boots, and heavy footwear, high-pressure machines operating at 55–85 kgf/cm^2 on the sole area are used. The pressure is normally applied by the sole plate acting as a ram.

Machines consist essentially of (Fig. 11.3) of the following structures.

1. A base plate, contoured to the shape of the sole and heel and carrying the required sole pattern.
2. Movable side plates which open and close automatically.
3. The last, which, when fitted with the prepared upper, fits exactly onto the cavity formed by the base and side moulds when closed.

The sole plate, side moulds, and lasts are fitted with electrical elements, and the machines include automatic temperature, pressure, and time controls.

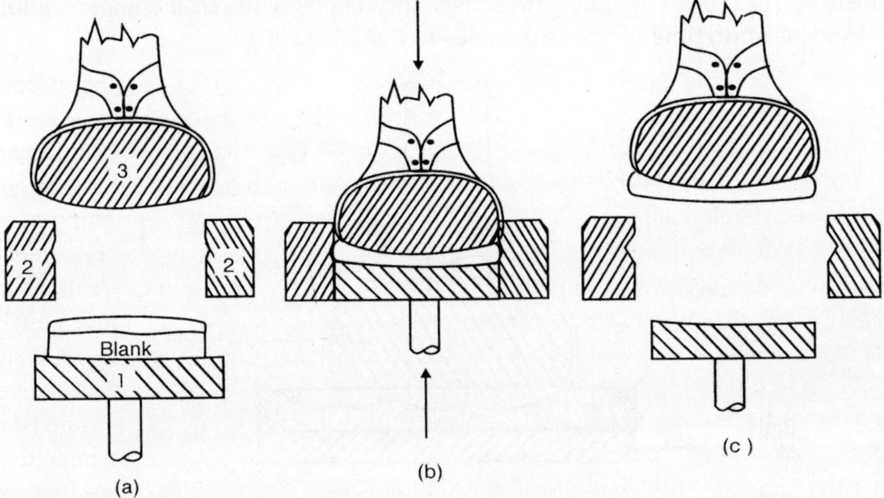

Fig. 11.3. Schematic diagram of external pressure moulding machine (a) mould open, upper on last; (b) mould closed, under curing pressure; and (c) mould open, cure completed.

As in the case of the moulding of cellular-soled footwear, the operation consists of assembling the upper on to the last and loading the mould cavity with the unvulcanised rubber sole and heel blanks. The side moulds are first closed tightly around the upper, and the sole plate is then forced up, acting as a ram, until it reaches a set position to give the required finished gauge of the outsole. The time control is set to the required cure time and when this is reached, the press automatically opens. The footwear is stripped off the last and allowed to cool, and the spew is trimmed off.

With fabric uppers, the pressure applied is sufficient to force the rubber compound into the interstices of the fabric to give a satisfactory bond of rubber to fabric. In the case of leather uppers, the bonding area is roughened to allow better penetration of adhesive into the leather fibres. Two coats of a fast-curing natural rubber cement, sometimes activated with an isocyanate, are applied to the roughened leather of the upper. A wooden or fibre-board heel filler is also included with the rubber blank. The lasted upper must fit correctly within the bite line of the side moulds so that a good edge seal is formed without cutting into the upper. This bite line follows the insole edge. A good bond of sole to upper is essential and on the toe-testing equipment, adhesion tests on finished shoes usually register figures of 32 kgf and over.

Accurate control is essential of the preparation and volume of the rubber blanks which are in cut shapes, strips, or granules, depending on the type of finished product and the sole pattern. The uncured rubber compounds must have fast-curing delayed-action characteristics and a shelf-life of at least 3–4 weeks is also essential. To utilise the equipment efficiently, the cycle times must be as short as possible. Where leather uppers are used, the lasts are normally cool, thus heat transfer into the rubber is from the sole and side moulds only and temperatures at the leather-rubber interface will not exceed 90°C. Temperatures of the sole and side moulds are 155–165°C. With canvas uppers, the lasts are heated to 130°C so that shorter curing cycles are possible. Typical cure times for heavy boots are 8–10 minutes, for lightweight men's and children's shoes 5–6 minutes, and for canvas plimsolls 3–4 minutes. To reduce cure times, pre-heating of the unvulcanised sole and heel blank in an electrically heated oven at around 95°C is commonly practised.

INJECTION MOULDED RUBBER SOLE AND HEEL UNITS

Injection moulded rubber sole and heel units are produced on multi-purpose moulding machines. These machines consists basically of a rubber injector which consists of a reciprocating screw working within a pre-plasticising cylinder, the screw and cylinder being fitted with independent temperature regulators. The moulding stations vary depending upon the machine design, but four, six, and ten-station machines are commonly used, the mould stations being carried on a rotary table. At the beginning of each cycle, the screw rotates and simultaneously moves to the back end of the cylinder, and the uncured compound in strip or pellet form is fed into the injection unit (it is generally found that pellets are more satisfactory for harder stocks). The rotary action of the screw plasticises the compound and the movement backwards of the screw leaves hot plasticised material in front of the screw. The entire injection unit then advances to form a tight seal between the nozzle and the mould sprue. The screw, acting as a ram, moves forward without rotating, forcing the compound through the nozzle into the closed mould. Pressure is maintained for several seconds after the mould is filled, to allow the compound to set up. The injection unit then retracts, the rotary table moves the next station in front of the unit, and the cycle is repeated. At the same time, the station prior to the injection point opens, allowing the moulded units to be removed from the mould cavities.

The number of mould cavities per station depends upon their volume and the maximum shot size of the machine. In addition, the total mould surface area must be such that the upward thrust is not in excess of the closing pressure of the presses themselves, which is in the region of 300–500 Mg, depending upon the machine. Moulds of varying capacity may be used in any one run, since the volume of compound injected into each mould is determined by the back pressure developed in the mould. Effective flash-free mouldings are not produced from this type of machine, but by designing the moulds for tear trimming, minimum spew waste can be achieved.

Each station carries individual temperature controls, so that cure temperatures for mouldings of different sole and heel thicknesses can be achieved. The typical output from a six-station injection moulding machine would be:

Number of stations	6
Pairs per station	2
Mould temperature	177°C
Cycle time	35 seconds
Cure time	2 ¾ min
Output per hour	206 pairs

Whilst the process for bottoming footwear can be adapted as above, the manufacturing of full-length rubber boots presents difficulties in both engineering and rubber technology. Boots of this type are, however, being injection moulded by using PVC, which lends itself readily to this method and has proved to be a very satisfactory material for lightweight wellington boots. Thermoplastic rubbers can be satisfactorily injection moulded and are now used for some casual footwear.

Resin Rubber and Microcellular Soling

Rubber soling can be supplied to footwear manufacturers either in the form of soles cut from pre-moulded sheets or as pre-moulded sole and heel units. High-styrene resin rubber soling is light in weight, attractive in appearance, combines high hardness with good flexibility, and is 4–5 times as durable as leather. Because of these properties, it has practically replaced leather as a soling medium in fashion footwear. Microcellular soling containing minute discreet air cells and having a specific gravity as low as 0.3 is also used extensively. It is made by the addition of an organic nitrogen-generating blowing agent to a soling compound, usually based on NR-SBR blends, with the addition of high-styrene resin to obtain a satisfactory hardness and of siliceous-type fillers for good resistance to abrasion.

The manufacture of microcellular soling requires very careful control in mixing and moulding. Expansion occurs only after release from the press and therefore, the mould is loaded with a blank of volume slightly in excess of that of the mould. As soon as the press is opened, the sheet of vulcanised rubber immediately starts to expand and must not be restrained in any way, otherwise permanent distortion of the sheet occurs. The moulds are usually bevelled along the edges, and the press is made to open as rapidly as possible. The degree of vulcanisation affects the finished density, overcure resulting in near-solid sheets. Accurate control of vulcanising time and temperature is, therefore, essential. After initial expansion has occurred, some degree of shrinkage will continue indefinitely. Stabilisation is, however, achieved by heating the sheet in an autoclave at 120–130°C for approximately 30 minutes. Sheets can be moulded up to 20 mm thick and subsequently slit to any required thickness. Sole shapes are die cut from the finalised sheet.

Adhesive Attachment

The present-day method of attaching rubber soles and heel units is by the use of adhesives, based on CR or NBR. The adhesive is applied to the surfaces of both the upper and the sole, the sole is then accurately placed in position, and pressure applied. Where it is necessary to apply the adhesive some time in advance to meet production requirements, the adhesive-coated sole surface is reactivated by heat from infra-red lamps immediately before contact is made with the uppers.

SPECIAL PURPOSE FOOTWEAR

Safety Footwear

The need for added protection of the foot in the vulnerable areas has led to the manufacture of safety footwear, incorporating a steel toecap in the toe area and a spring steel midsole between the base of the foot and the outsole. The demand for this type of footwear in industry has increased recently. The inserts are fitted during the assembling of the product, which can be hand built, compression moulded, or direct moulded. The steel toecap is shaped to fit over the toe area of the last and is fitted after the boot lining has been lasted. It is essential that the toecap is a tight fit so that no movement occurs during manufacture. The toecap must meet the impact test requirements as laid down in BS 1870. The steel midsole needs to cover as much of the sole area as possible, and BS 1870 specifies that the margin between the protective midsole and the edge of the insole must not exceed 6.35 mm. To ensure that no rusting occurs during service, the steel parts are given a chemical anti-rust treatment and enveloped between two thin layers of rubber compound. Depending on the service requirements of the product, safety footwear is manufactured with steel toecaps only, protective midsoles only, or incorporating both toecaps and midsoles.

Conductive and Antistatic Footwear

Where it is necessary to eliminate the build-up of static charges in the body to avoid sparking, as in factories handling explosives and in hospital theatres, conductive or antistatic footwear is used. Conductive footwear should only be used where there is no danger from electrical equipment. Where potential electrical defects can occur, it is essential to have a lower limit on resistivity to give adequate protection against dangerous electrical shock. This type of footwear is referred to as 'antistatic'.

In the manufacture of conductive and antistatic footwear, the upper parts are of standard materials but the base of the footwear is made of conductive materials. The insole, outsole, and any fillings required are prepared from compounds possessing the required resistivity.

For conductive footwear, BS 3825 specifies an upper resistance limit of $1.5 \times 10^5\ \Omega$, and for antistatic footwear, BS 2506 specifies a lower limit of $5 \times 10^4\ \Omega$ and an upper limit of $5 \times 10^7\ \Omega$.

GENERAL COMPOUNDING OF FOOTWEAR COMPOUNDS

The many different types of footwear which have rubber components common in their manufacture are *hawai* or beach slippers, canvas shoes, rubber-soled leather chappals or shoes, sports shoes, light protective or fashion boots and heavy duty industrial boots, etc. The primary function of any footwear is to protect the foot from external harmful factors such as rough ground surfaces, nails, thorns, stones, dampness and cold during walking or standing. Here we will discuss the manufacture of rubber compounds used in the manufacture of *hawai*/beach slippers, leather uppers, *chappals* and shoes and also manufacture of microcellular soling used for *hawai* slippers and leather chappals and shoes.

The advantages of microcellular expanded soling material are that they are light weight, comfortable to wear and usually have a high flex life. Besides, they are more economical than leather soles.

Types of Expanded Microcellular Soling

Microcellular soling is made in various qualities and can be generally classified into three categories:

1. Soft microcellular soling for *hawai* or beach slippers.
2. Medium hard microcellular for leather top *chappals*.
3. Hard microcellular soling for leather shoes.

Category	*Approx. specific gravity*	*Approx. shore Å hardness*
Hawai/beach slippers	0.2 – 0.4	30 – 45°
Medium hard microsoling	0.5 – 0.6	55 – 65°
Hard microsoling	0.7 – 0.8	70 – 80°

Some characteristics of these soles are discussed below :

Hawai slippers

In western countries these slippers are known as beach sandals and are used on beaches or for indoor use. Because of their low cost, these slippers are within the reach of the poor population in India and hence have become very popular footwear. The hawai slippers consist of three components—expanded coloured sole, white upper topping and solid rubber strap.

Medium hard microcellular soling

These solings are generally used for leather or resin upper chappals. This kind of footwear is in very high demand for fashion conscious urban people who need more elegant and light weight footwear rather than leather sole chappals. This soling has good resistance to wear and tear and has almost replaced leather soles.

Hard microcellular soling

This superior performance soling material is generally used for light wear and heavy duty industrial shoe soles because of its light weight and better wear resistance. It is more comfortable than crepe soles used for leather upper shoes.

Selection of Compounding Ingredients

Elastomers

The hawai and microcellular sheet compounds are normally based on natural rubber and high styrene resin rubber blend or natural rubber, SBR or polybutadiene and high styrene resin blend.

Natural rubber provides good tack to the compound and high tear and resilience to the product. Because natural rubber gives very high tensile strength and elongation, it can take very high loading of fillers with comparatively less drop in physical properties, thus helping to reduce the cost of the rubber compound.

SBR rubber gives better ageing protection and abrasion resistance than natural rubber and is hence blended with natural rubber many times. High styrene resin polymer gives lower density and shrinkage and offers best flex and abrasion resistance. Its use, therefore, in microcellular products is essential. At higher loading it gives high hardness and a leather-like appearance to the product. Besides, high styrene resin rubber such as SBR-1958 gives higher thermoplasticity to the compound.

Polybutadiene rubber is also blended with natural rubber to give good abrasion resistance and resilience to microcellular products.Now-a-days, to manufacture low density microcellular products, a blend of nitrile/PVC resin is used. It also gives excellent abrasion resistance and resistance to oil.

Fillers

The function of a reinforcing filler is to improve the wear and tear properties of microcellular products. Reinforcing precipitated silica fillers and silicates give higher hardness which, to some extent, helps to reduce the use of high styrene resin rubber. Precipitated silica fillers such as Ultrasil VN3 and Hisil-233 give high reinforcing characteristics and are generally used at 20–40 phr level. Silica fillers also help to improve tear resistance, which is the most important property of the microcellular sheet.

The silica filler gives lower specific gravity to the product which, in turn, helps to reduce the volume cost of MC sheets. However, these silicas are very expensive and hence, many a time replaced with aluminium or calcium silicate, which are lower reinforcing than silica fillers. Along with silica and silicate, China clay, whiting or precipitated calcium carbonate are also used. Precipitated calcium carbonate is semi-reinforcing filler and gives better processing characteristics to the MC compounds. China clay and whiting are generally used as extenders to reduce the cost but they make the product heavier due to higher specific gravity and high bulk density.

Plasticisers such as processing oils and petroleum jelly, are used to improve the processing characteristics and reduce the hardness of solings. Cumarone indene Resin and Wood Rosin improve the dispersion of reinforcing silica and silicate and also help to improve expansion and reduce shrinkage of solings. Wood rosin and organic acidic materials also act as activators to DNPT type blowing agents, thus, giving higher expansion to the compound.

Zinc oxide is normally used at 4–5 phr as an activator for vulcanisation. Stearic acid is also an activator and due to its acidic nature helps to decompose blowing agents at a lower temperature and, hence, gives better expansion to the compound. It is generally used at 4–8 phr depending upon the filler loading of the compound.

Antioxidants are mainly used in MC soling to give a longer service life and to improve the flex resistance and storage life of the product. For white and coloured products, Montaclere SP at 1–2 phr is used. For dark coloured and black soling, Flectol Pastilles at 1–1.5 phr or Santoflex DPA at 0.5–0.75 phr are recommended.

The accelerator system for the MC sheet should be such that it should set the compound fast to seal the edges of the mould and should cure slowly to allow decomposition of the blowing agent first and then vulcanise the compound. This type of accelerator system gives higher expansion. The use of Thiotax (MBT) alone is more popular in the *hawai* sole compound at 1.2–1.5 phr level. For higher hardness micro sheets, use of Thiofide (MBTS) at 0.8–1.0 phr and Ureka White F at 1–1.4 phr or Santocure CBS at 0.8–1.0 phr with booster TMT at 0.2–0.3 phr is more common. Sulphur is generally used at 2–2.5 phr level. Higher levels of sulphur give poor tear resistance and higher shrinkage.

A DNPT type blowing agent properly stabilised with a stabiliser should be preferred because it gives good expansion to MC sheets. This material has a very high decomposition temperature—190–200°C. It decomposes at 120–140°C in an acidic medium and, therefore, in the MC compound, a higher dosage of stearic acid at 4–6 parts is used. During the vulcanisation, the blowing agent decomposes to give nitrogen gas and the fishy smelling amine is produced. The nitrogen gas under pressure dissolves into the rubber compound, which on releasing the press blows the compound, giving innumerous micro-cells in the soling. Use of blowing agent depends upon filler loading and hardness required for MC soling product. Generally 4 to 7 phr of blowing agent of DNPT type is used. It should be remembered that a higher dosage of blowing agent not only gives higher expansion but gives higher shrinkage due to higher internal pressure of nitrogen gas in microcell wall of the soling. It is essential to store MC compound for 8–24 hours after mixing of blowing agent in it, to get a uniform cell structure and expansion of sheets.

Organic or inorganic colours and pigments are used to give bright and pleasant colours. Care must be taken in using phthatocyanine blue or green colours, so that they do not contain free copper which poison the rubber, reducing the ageing resistance of the product.

Microcellular crumb is used to reduce the volume cost and it greatly helps to reduce specific gravity and shrinkage of MC soling. However, very high loading of crumb reduces split tear and wear of the product. At high loading of crumb, micro-cell wall strength reduces drastically, which results in the breakage, of micro-cells ultimately resulting in coarse structure in MC soling.

Method of Manufacturing Microcellular Solings

Microcellular soling (MC) sheets are produced from rubber compound containing organic blowing agents of the dinitroso pentamethylene tetramine (DNPT) type. The mould cavity is filled with the compound about 3% excess in volume which helps to seal off the edges of the mould. The moulding is done at high pressure in a hydraulic press normally at 140–150°C. The pre-cure is given in about 75 to 80% of its full cure.

During curing of the MC sheet, the blowing agent decomposes releasing nitrogen gas. The nitrogen gas under the high moulding pressure diffuses into the sheet. On precuring the press is opened. On releasing the external pressure due to high internal nitrogen gas pressure, the compound expands and precured sheet props up from the mould cavity, giving innumerable micro-cells within the body of the sheet. To facilitate easy release of the MC sheet, mould cavity has tapered edges, generally at 45°. The MC sheet on cooling shrinks to some extent. However, to stabilise its size and to optimise physical properties as well as to withstand wear and tear, it is then post-cured in hot air oven.

The quality of microcellular material can be judged from: (i) specific gravity; (ii) hardness; (iii) split tear resistance; (iv) shrinkage control; and (v) flex resistance.

If the basic formulation and curing system of the MC sheet compound is constant, the specific gravity and micro-cell structure of the product depends on:

1. The quantity of blowing agent used.
2. The volume loading of mould cavity.
3. The time and temperature of precure.
4. Moulding pressure applied during vulcanisation of MC sheets..

Quantity of blowing agent

At constant condition of manufacture of MC sheets, higher the dosage of blowing agent, greater the expansion of MC soling and is lower its specific gravity.

Volume loading of mould cavity

If the volume loading of MC compound in the mould is inadequate a coarse sponge irregular cell structure is obtained. The mould is normally loaded 3% excess in volume of the mould. The small excess material forms a thin flash which seals the mould and prevents leakage of gas from the sheet. A large excess compound loading gives thick flash and causes leakage of nitrogen gas giving irregular shape soling sheets.

Time and temperature of pre-cure

The curing of the MC sheet is carried out in press under high moulding pressure. The precure time and temperature are determined by the decomposition behaviour of blowing agent and cure system used in the compound. As cure time increases, modulus of the compound increases which increases the resistance to expansion of sheet. Thus, resulting size of the sheet is lower and consequently the internal pressure of decomposed gas becomes higher.

During storage and post-cure, when internal pressure of gas comes down due to diffusion of gas, the shrinkage of MC sheet becomes more. Hence to control the shrinkage of the product, it is necessary to give low degree of precure as possible. However, too low degree precure gives lower strength to the MC sheet and walls of micro-cells, the weaker micro-cells then burst due to high nitrogen gas pressure and big blow holes are developed in the sheets. Due to poor heat conduction properties of rubber, cure time of MC sheet depends upon the thickness of the sheet in the mould. Deeper mould cavity will need higher precure time. Generally, for medium speed curing system and normal filler loaded MC sheet compound, approximately 0.8 to 1 minute precure time should be given per mm depth of the cavity at 150°C curing temperature. In case of manufacture of MC sheet with uneven thickness or tapered sheets cure time for thickest section is considered. In such cases, thinner section of sheet gets higher degree of cure which obviously gives slightly higher shrinkage. The temperature of 140–150°C is a good temperature.

Moulding pressure applied during vulcanisation

In the case of microcellular compounds, the nitrogen gas liberated by decomposition of blowing agent generates internal pressure in the compound, which tends to open the mould. The external closing pressure on the mould, therefore, should be high enough to keep the mould closed and to keep liberated nitrogen gas dissolved in the microcellular sheet. The minimum specific pressure required will thus depend on the amount of blowing agent in the compound which is approximately 18 kg/cm^2 per phr of DNPT type of blowing agent in the compound.

The specific pressure on the mould area will depend on the ram area of hydraulic pressure × tonnage or gauge pressure.

Specific pressure = Total pressure/Mould area.

But total pressure = Maximum gauge pressure × Ram area.

$$\text{Specific pressure} = \frac{\text{Maximum gauge pressure} \times \text{TT 4} \times (\text{Ram diameter})^2}{\text{Mould area}}$$

Since minimum specific pressure is dictated by the dosage of the blowing agent in the compound, the maximum area of the mould is given by:

$$\text{Max. mould area} = \frac{\text{Total pressure for moulding}}{\text{Min. specific pressure required for the compound}}$$

In case the specific pressure used is lower than minimum or the mould is of the larger area than indicated by above equation, all the liberated nitrogen gas does not dissolve into the compound.

As the part of nitrogen gas remains in the gaseous form in cells, coarser cell structure of MC sheet is obtained due to expansion of cells after precure. This gives poor appearance and lower abrasion resistance to the sheet. Due to internal pressure of the blowing agent in the compound, it exerts a back pressure on the cylinder of the press. This sometimes excessively increases the pressure in the hydraulic system bursting pipeline. Hence, it is advisable to include safety valve in pipeline from the three-way valve to the hydraulic ram.

In a hand press though higher moulding pressure can be applied on the mould, it becomes difficult to open the press against high internal pressure developed in the MC sheet due to the decomposition of blowing agent. Thus due to the slow opening of the press many a times MC sheet gets distorted.

In MC sheet moulding, the maximum moulding, area cannot be used as an opening of the mould expanded sheet. It may hit the guide post of the press and get warped, and edges may be cut.

In the MC sheet manufacture, tapered edges of the mould determine the relative expansion of sheet in the three directions. If angle is lower, it gives higher linear expansion while higher angle gives better thickness to the sheet.

Post-Cure

After the precure, it is essential to give post-cure to microcellular sheet. After post cure the MC sheet is completely cured and becomes permanently stabilised in the expanded size, whereby it achieves maximum physical, wear and tear properties. When the pre-cured sheet, after moulding cools to room temperature, it shrinks due to lowering of internal gas pressure which lower down due to lower temperature. However, the nitrogen gas in each individual cell is still under pressure. The elastic microcell walls are still in a stretched condition due to interior nitrogen pressure. This gas under pressure slowly diffuses through the cell wall to the atmosphere during storage, which causes a fall in internal pressure and stretched cell wall shrink leading to further shrinkage during service life. Post-cure accelerates the rate of diffusion of this gas at high post-cure temperature which brings down internal pressure to atmospheric level and MC sheet stabilises.

Normally post-cure is carried out at temperature of 100–120°C for 1–2 hours. The precured MC sheet from one or two shifts are collected and post-cured in hot air oven having a circulating arrangement. As the heat transfer in hot air is very slow, the sheets take longer time to attain temperature around 100–110°C. After post-cure sheets are cooled and stacked one above another for at least 24 hours, before final use, so that the MC sheet gets stabilised after cooling.

Quality Control of Raw Materials

Natural rubber

Natural rubber should be inspected for low dirt content, lighter colour and should be from fresh stock. Higher dirt or bark content will give poor flex and tear resistance, while dark coloured rubber may give variations in colour shades of the product and old rubber stock will give poor ageing and life to the product.

Synthetic rubber

Synthetic rubber should be checked for any foreign matter or contamination. It should be also tested for moisture content (less than 1%) and constant viscosity. Large variations in viscosity may result in uneven expansion and higher moisture content will give blow holes or blisters in the product.

Crumb

Crumb should be properly sieved to get uniform mesh size and eliminate foreign matter. It should be checked for moisture content and suitably, the same coloured crumb should be used as of the compound to get uniform colour to the cut edges of MC sheets.

Fillers

Fillers should be tested for presence of grit, moisture content (1% max.), pH, specific gravity, oil absorption, sedimentation value and solubility in dilute acids. Large variation in these properties will result in variations in the specific gravity of MC sheet, blistering, poor reinforcement, lower flex and tear resistance.

Processing aids

Liquids (oils) should be checked for colour, odour, viscosity, specific gravity (by hydrometer), flash point and aniline point. Solid processing aids should be checked for consistent colour, odour, melting point/softening point, solubility, acid value and iodine value. Variations in the properties will result in variations in the viscosity of the compound and finally, expansion and hardness of the MC sheet.

Blowing agent

Blowing agent should be tested for grit and gas liberation which controls the expansion of the product. It should be properly stabilised.

Accelerator/antioxidants

Accelerator/Antioxidants should be tested for melting point/range and solubility in solvent to judge the purity of the material.

Sulphur

To get good dispersion of sulphur it should be of fine mesh. It should be tested for acidity, sieve analysis and solubility in carbon bi-sulphide to check its purity.

TROUBLE SHOOTING GUIDE

Table 11.2 given below shows the problems and their possible solutions and Table 11.3 shows typical formulation for *hawai* sole (white upper).

Table 11.2. Trouble shooting guide for footwear.

Problem	*Possible cause*	*Possible solution*
Coarse open cell structure	Too low viscosity	Reduce peptiser and/or plasticiser
		Increase high mooney polymer
	Excess blowing agent	Reduce blowing agent
	Low press cure time	Use faster cure system
	Low pressure on the mould	Increased specific pressure per unit area
Too dense cell structure	Too high viscosity	Increase peptiser and/or plasticiser
		Increase blowing agent
	Cure too fast or overcure.	Use slower curing system
		Reduce booster level
		Use blowing agent with low blow point temperature
Uneven wavy sheet	Undercure	Increase press cure time/temperature
	Difference in the top and bottom platen temperatures or uneven temperature of platen	Faulty heater temperature control system
		Water logging in steam lines
		Insulate steam line
	Uneven pressure	Check for parallel alignment of platens
Irregular shape	Leakage of gas from one side of a mould	Check mould for parallel surfaces
	Due to thick flash from excess of loading	Volume loading of the compound in the mould
	Inadequate compound	
	Low specific pressure per unit area	Increase hydraulic pressure
		Reduce blowing agent.
Blisters and pinholes on surface	Inadequate compound	Volume loading of compound—103%
	Low cure time	Increase cure time/temperature
	Uneven temperature	
Blow holes within the sheet	Under cure	Increase cure time/temperature
	Air entrapment in the blank	Cut blank from calendered sheet
		Or cut sheet out from faster roll of a mill
		Remove trapped air while doubling
		Increase resin level

(Contd ...)

Problem	Possible cause	Possible solution
	Poor dispersion of blowing agent/resin/sulphur	Refine the compound through tight nip
Blow holes in post-cure.	Low cure time	Increase cure time/temperature
	Inadequate decomposition of blowing agent in press cure	Increase activation level and use faster cure system
Cracks at edges	Cure system too fast	Reduce booster/sulphur level
		Reduce cure time/ temperature
High shrinkage	Low content of high styrene resin	Increase high styrene resin content
	Breakdown of nerve incomplete	Increase mastication time
		Add Peptiser
	Poor dispersion of blowing agent	Pass through tight nip
	Excess blowing agent	Reduce blowing agent
	Inadequate reinforcement	Increase filler/crumb loading
	Undercure	Increase curing time
	Less post-cure time	

Table 11.3. Typical formulation for *hawai* sole (white upper).

Ingredients	*Hawai sole* 1	*Hawai sole* 2	*White upper*
RMA	30.00	75.00	80.00
SBR 1958	10.00	25.00	20.00
RB 820	60.00	—	—
Zinc oxide	5.00	5.00	5.00
Stearic acid	5.00	5.00	4.00
Montaclere	1.00	1.00	1.00
Silica	15.00	—	—
Aluminium silicate	30.00	30.00	30.00
Ppt. calcium carbonate	30.00	—	30.00
Whiting		—	30.00
China clay	—	80.00	—
Hawai crumb	—	100.00	—
Petroleum jelly	5.00	5.00	5.00
Process oil	5.00	10.00	5.00
Wood rosin	3.00	4.00	4.00
DEG	2.00	2.00	–
Titanium dioxide	5.00	5.00	20.00
Sulphur	1.40	2.30	2.50
Ureka white F	—	1.20	1.00
Thiofide MBTS	1.40	—	—

(Contd ...)

Ingredients	*Hawai sole*		*White upper*
	1	*2*	
Thiurad TMT	0.20	0.10	—
Blowing agent (DNPT)	8.00	7.00	5.00
Cured at 150°C for mins.	7.00	5.00	6.00
Specific gravity	0.23	0.40	0.55
Hardness shore Å	30	40.00	44
Split tear kg/inch	3.00	3.50	5.00
Shrinkage % (100°C/1 hr)	3.50	1.70	2.50
Ross flex test crack initiation kc	66.00	60.00	75.00
600% cut growth kc	140	140	140

Typical formulation for strap compound, micro compound, ISI specifications for microcellular rubber soles and heels and physical requirements for quality *hawai* sole are shown in Table 11.4 to 11.8.

Table 11.4. Typical strap compounds.

Ingredients	*Quantity*
RMA	100.00
Zinc oxide	6.00
Stearic acid	1.50
Montaclere	1.00
Aluminium silicate	30.00
Ppt. calcium carbonate	30.00
China clay	30.00
Petroleum jelly	6.00
CI resin	2.00
DEG	1.00
Titanium dioxide	5.00
Thiofide MBTS	1.20
Thiurad TMT	0.20
Sulphur	1.80
Santogard PVI-50	0.25
Cured at 150°C (minutes)	4
Specific gravity	1.37
Tensile strength, kg/cm^2	170
Elongation at break, %	430
300% modulus, kg/cm^2	65
Angular tear strength, kg/cm^2	27
Hardness shore Å	54

Table 11.5. Typical formulation of micro compounds.

Ingredients	*Quantity*	
RMA	20.0	20.0
SBR 1958	60.0	10.0
SBR 1502	20.0	—
RB 820	—	70.0
Zinc oxide	4.0	6.0
Stearic acid	5.0	5.0
Montaclere	1.0	1.0
Silica	20.0	20.0
Aluminium silicate	40.0	35.0
China clay	70.0	50.0
Micro crumb	50.0	—
Petroleum jelly	5.0	5.0
Process oil	10.0	5.0
Wood rosin	4.0	4.0
DEG	2.0	2.0
Titanium dioxide	5.0	5.0
Sulphur	2.2	1.5
Thiofide MBTS	—	1.4
Ureka white F	1.2	—
Thiurad TMT	0.1	—
Blowing agent (DNPT)	5.0	5.0
Cured at 150°C (minutes)	6.0	12.0
Specific gravity	0.5	0.4
Hardness shore A°	62.0	50
Split tear strength kg/cm^2	4.0	3.0
% shrinkage	1.5	2.0
Rose flex test		
Crack initiation, kc	35.0	45.0
600% cut growth, kc	70.0	100.0

Table 11.6. ISI specifications for microcellular rubber soles and heels BIS/ISI : 6664–1972.

Physical requirements for microcellular rubber sheets

Tests	*Type 1*	*Type 2*
Relative density	0.60–0.70	0.45–0.55
Hardness IRHD, min.	65°	50°
Change in hardness after ageing at 100 ± 1°C for 24 hours.	+5 to –0	+5 to –0
Compression set at 27 ± 1°C for 24 hours, % max.	12	15
Split tear strength, kg. min.	8.0	6.0
Heat shrinkage at 100°C ± 1°C for 1 hour in both directions, % max.	2	3
Water absorption, % by mass, max.	1.5	2.0

Table 11.7. Physical requirements for standard quality *hawai* sole.

Relative Density	0.3–0.6
Hardness IRHD	45°–50°
Change in hardness after ageing at 100 ± 1°C for 24 hours	+ 5 – – 0
Compression set at 27 ± 1°C for 24 hours, % max.	15
Split tear strength, kg.	4–5
Room temperature shrinkage at 27 ± 1°C for 4 weeks, % max.	1.0–2.0
Flex resistance, number of cycles – Ross Flex testing machine:	
1. Initial crack, min.	40,000–60,000
2. 600% cut growth	60,000–100,000

Table 11.8. Physical requirements for standard hawai strap.

Relative density	1.3–1.4
Tensile strength, kg/cm^2	150–160
Elongation, %	400–500
Hardness IRHD	60 ± 5
Change in initial hardness after accelerated ageing for 24 hours at 100 ± 1°C	+5 – –5
Flex resistance number of cycles on Ross Flex tester :	
1. Initial cracks	60,000–85,000
2. 600% cut growth	100,000–130,000
Tear strength, kg/cm^2	30–45

TESTING OF FINISHED PRODUCT

To ascertain the quality of the product it is essential to test the finished products.

Microcellular products should be tested for following properties :

1. Relative density or specific gravity.
2. Hardness of the sample.
3. Heat shrinkage at 100 ± 1°C for one hour.
4. Compression set.
5. Split tear strength.
6. Ross Flex testing.

The *hawai* strap compound is also tested for stress-strain properties on tensile testing machine.

Specific Gravity or Relative Density

It helps to determine the volume cost of the compound, to keep check on a weighing of compounding ingredients and purity of raw materials. Specific gravity can be found by following formulations. For samples having specific gravity more than one:

$$\text{Relative density or Specific gravity} = \frac{W_1}{W_1 - W_2}$$

For MC sheet or *hawai* sole having specific gravity less than one:

$$\text{Specific gravity} = \frac{W_1}{W_1 + W_3 - W_4}$$

where W_1 = wt. of the sample in air
W_2 = wt. of the sample in water
W_3 = wt. of the sinker in water
W_4 = wt. of the sinker + sample in water

Hardness Test—IRHD (International Rubber Hardness Tester)

This is a widely used hardness test for solid and cellular products other than ebonite to find out the hardness of the product.

Heat Shrinkage

Test piece of 125 × 5 × 5 mm after splitting all sides, is cut. After conditioning at a room temperature of 27 ± 2°C for 24 hours, length of test piece is measured. It is then kept in air circulating oven at 100 ± 1°C for 1 hour. After removing the piece it is cooled for 2 hours at 27 ± 2°C and length is measured again. Shrinkage is calculated as:

$$\text{Heat shrinkage} = \frac{L_o - L_1}{L_o} \times 100$$

Where L_o = length of the test piece in mm before heating
L_1 = length of the test piece after heating

Compression Set

Sample discs of diameter 30 ± 0.2 mm and thickness of 9.5 ± 0.2 mm or alternatively, two discs of same diameter and half of the thickness put one above other. The thickness of the samples after conditioning for 24 hours at 27 ± 2°C, is measured as initial reading. The samples are placed in parallel plate compression apparatus to compression load of 140 ± 1 kg for 24 hours. After releasing the load final thickness is measured after 1 hour.

$$\text{Compression set \%} = \frac{t_o - t_1}{t_o} \times 100$$

where t_o = initial thickness in mm
t_1 = final thickness in mm.

Split Tear Strength

Test piece of 25 × 100 × 7 ± 0.2 mm along and across the direction of sheet is cut. It is split in the centre between the top and bottom surfaces for a distance of 30 mm from one end, thus giving two tongues at the end. The tongues are fixed in a jaws of the tensile testing machine and allowed to separate at a constant rate of 75 mm per minute. The maximum load is recorded and the arithmetic mean of the four tests in kg is split tear strength.

Ross Flex Testing

Flex life of the sample is tested on the Ross Flex testing machine. A sample of 25 × 150 mm is cut and fixed in the machine. Visual inspection is done after every 500 cycles of flexing to check for initial crack. In another test, a 2 mm cut in the sample is made and sample is allowed to flex till it reaches 14 mm in width which gives 600% cut growth resistance in the sample. The number of cycles required to produce initial crack and 600% cut growth resistance is recorded.

SAFETY MEASURES

During the manufacture of the MC sheet, part of nitrogen gas and aldehyde/amines are evaporated on opening the press. This causes irritation of eyes and in extreme cases gives mild skin irritation. It is, therefore, essential that the moulding room should be properly ventilated and should be fitted with exhaust fan to drive away gas fumes. After working hours, operators should wash their hands and faces with plenty of water. In extreme cases of skin or eye irritation, a medical practitioner should be consulted.

Fire hazards due to improper storage of blowing agent have also been reported. It is essential that blowing agent should be properly stabilised with amino compounds. It should be stored away from heat source and at a cooler place.

RECENT DEVELOPMENTS IN FOOTWEAR TECHNOLOGY

Over the years shoe making has become a complex operation. During the last few years tremendous changes have taken place in the area of materials and processes. With the improvement of standard of living, the customer profile has also taken a big change. As in the advanced countries, in India also there is a need for professional shoes in various activities of life. The professional shoes are getting more and more demand in the market due to statutory requirement as well as the production required by actual users. Now a shoe maker has to cater and meet the various needs of shoes. The first thing therefore is to determine the end-use and evolve a shoe accordingly. Fortunately in this country by now we have standard specifications for almost all types of products and therefore it has become easier to plan and execute a product according to the need.

It is also very important to note that before marketing a product one must stimulate various test methods which may happen in actual use. Here again the stimulation programme which can be co-related with the real life use is a important factor to give the customer a guarantee of the product.

Here an attempt has been made to give an outline of shoe production with a stress on recent development. Since material is the most important factor, we shall start with the shoe material.

Material

As far as the material used for the shoe industry is concerned, it is so varied that it is almost impossible to make a proper list and hence unable to cover each and every material in today's discussion. However, from the good old days, right from straw and wood to this day the man-made poromeric materials are in use. The material can be divided into two basic groups—one is upper and the other is bottom.

While upper can be a cotton one or a man-made—woven/knitted with various other synthetic material used as decoration which makes the addition system and vulcanisation system more complex. Even in the area of leather upper, the use of synthetic finishes including the large scale use of P.U. film has rendered the use of leather which is generally a simple material and quite a complex process.

Besides the use of large scale porous/non-porous PVC coated material are used in today's shoe making. It has been noticed, that indiscriminate use of such material without knowing their behaviour, the large scale failure is a common feature.

In the area of cotton material this can be of a rolled sole or a PVC injected sole, both as unit sole or directly injected, each has got its own advantages and disadvantages both in physical properties, customer preference and price.

Process

With the development of technology the shoe making has assumed great varieties of manufacturing process and they are generally known as: (i) built-up process; (ii) stuck-on process; (iii) DVP process; (iv) DIP process; and (v) slush moulding process.

While most of the process is very popularly known one must understand the amount of knowledge and experience required to vulcanise a pair of shoes containing as many as twenty types of compounds and dozens of various distinct materials but one should expect optimum properties of all these materials put under one vulcanisation system. Besides the thickness of various materials are different.

In the area of stuck-on the main important area besides adhesives and preparation of surface is the specific pressure and time. In DVP process generally a very fast curing compound—as low as 45 secs of a thickness of 15 mm is used and this often encounters with either scorching or under vulcanisation and here a good knowledge of accelerated system is very important. In DIP generally there are two materials used—one is rubber and another is thermoplastic like TPR or PVC. While using the thermoplastic material, the process is similar, being thermoplastic in nature whereas the rubber injection process is a difficult one since it creates a lot of frictional heat at the nozzle point before it fills up the cavity and this is a hazard process and generally encounters with short moulding or scorching. Since the PU injection is quite a difficult process than the generally known DIP process, here we deal with the subject separately.

Micro-cellular

It will not be fair if something is not talked about the micro-cellular material which is largely used in our country as soling material. In fact it will not be exaggerated if we say the country walks on *Hawai*, which is nothing but a micro-cellular product. As much as 50% of footwear is constituted of one single shoe—i.e. *Hawai.* The technology is known to this country for many many years and it has become a sort of cottage industry in production of this type of footwear but tough attempt is made by small and large scale industry for this product and market them in common place, the success is not same with everybody. Being micro-cellular in nature there are tricks to achieve the best result.

Recently in India in the production of micro-cellular there are two more materials that has come into large scale use. They are RB-820—a basically polybutadiene of a typical molecular configuration and EVA—ethyl vinyl acetate. In the process of RB-820 a normal rubber machinery can be used and also a similar vulcanisation system as rubber whereas EVA requires a special type of vulcanisation system. From both the materials a similar end product can be obtained by judicial compounding. This material has become quite popular due to physical properties. It is ideal such as low density, high clarity, no shrinkage etc. as compared to rubber compounded micro-cellular product. Whereas rubber compounded micro-cellular has various advantages over RB-820 and EVA such as: (i) low compression set; (ii) low abrasion; and (iii) better split tear etc.

Adhesives

The shoe making is a complex process and basically it is achieved by joining various components together. This by itself is a technology which must be reckoned with. In the manufacture of shoes, the joining is not only for similar materials but many a time various dissimilar materials are to be joined together and more so with the advent of more and more use of synthetic upper materials. The simplest form of adhesive is perhaps a rubber solution but it has got lot of disadvantages because no permanent warming is obtained. Therefore a compounded rubber solution is used with a high content of solvent. On the other hand similar bondage may be obtained by using natural or synthetic latex. The advantages and disadvantages of these two processes are well-known whereas the hibiscus is obtained with a compounded rubber solution and easy to dry due to high solvent composition but on the other hand the latex being low-biscus, the solid content is higher but difficult to dry because of poor evaporation of water.

The main purpose of giving a simple example of this is that one can achieve a very good result depending on what substrait the application of this adhesives is made and sometime it becomes highly technical.

It is also very important factor to understand and know that the success of a particular adhesive system depends on various things such as: (i) preparation of joining surface; (ii) the ventilation system; (iii) the humidity factor; (iv) right temperature; (v) activating factor; (vi) the right adhesion of cross-linking agent; and (vii) determination of pot life etc. The composition of individual adhesive systems are different generally in different company. Most of the adhesives require more than one solvent and by application of more than one solvent one achieves the pull and push system meaning when tack service is still obtained but does not have too much delay in drying.

This is a very complicated area and requires lot of experience for achieving the best result. The latest thing that has come up in the area of adhesive which is generally known as "hot melt" adhesive based on polyamide compound. Also there is a new 'O' rod made of PU which is used with the help of a gun. It is available both for permanent joining and temporary joining. While we talk of the latest development, progress has been made in the area of canvas shoe production by using self-adhesive foxing where there is no need for additional adhesive to join the upper since the foxing itself is coated with a layer of adhesive and simultaneously cross-linked at the time of total vulcanisation of the shoes.

Polyurethane

The latest trend in shoe manufacture is use of polyurethane as soling material. The polyurethane soling materials are generally of two types based on chemical properties—polyester and polyether.

Whereas both are largely used in the manufacture of shoes, the polyester is meant for high performance sports shoes. Unless there is a specific requirement of sportswear, the properties otherwise are similar in both the cases. The mechanism used in this material is quite complicated and unless properly controlled, one can land on great catastrophe.

Therefore serious attention is required with profound backing and knowledge to get the desired properties. Since polyols (polyester/polyether) is a resin which by itself is not used. It has to be combined or mixed with isocyanide. The trick of the game is how best you can mix them together to obtain the best result. It not only depends such as cream time, rise time, tackfree time, pinch time etc. Each is important factor to obtain the best proportion otherwise instead of a proper sole we will get biscuits. The second area of importance is to find out what density is required for the end-use. Obviously it will not be same if the shoe is used just for casual walking as against severe sports performance shoes. The combination of isocyanide and polyol will bring the result as it is mixed in the right proportion. In the modern days the machines have been developed fully automatic and can take care of many bottlenecks normally created in handling. However, the machines can do upto a certain limit. The remaining has to be done by a good technologist who are not too many in our country. They need to be trained to enable them to handle this particular material which is new to this country.

Before we conclude this particular area of production, it will be pertinent to mention some basic features of this particular material which are so important. Here we would also like to mention that this polyurethane is not necessarily always blown material, it can be a solid, semi-rigid material as well.

While in solid or semi-rigid form it develops some outstanding properties such as abrasion and flexing which are very important in shoe. Whereas generally polyurethane is used in the shoes either as a single density or a double density.

Some of the other important physical properties one can derive from this particular material are as follows: (i) high abrasion resistance; (ii) high flex resistance; (iii) lighter weight; (iv) low compression set; (v) better cushioning effect; (vi) high slip resistance; and (vii) attractive colour.

It is understandable that for two colour sole there will be requirement of two injection units on a same rotary machine.

CHAPTER 12

Conveyor Belt Technology

INTRODUCTION

Conveying materials on a conveyor belt remains the most efficient means of moving materials from one place to another. In terms of cost per tonne carried, the conveyor belt is extremely cost effective, saves energy, and is unequaled in the field of transportation. The present-day concern for ecology highlights the environmental friendliness of conveyor belting, which removes from the roads congestion and pollution otherwise caused by trucking.

Belts that are resistant to fire, heat, oil, chemicals, ozone, and even termites, are now available and multigrade belts are requested with heat and fire resistance in addition to adequate wear resistance.

National specifications are available in most countries that have a large manufacturing base. The national standardisation institutions contribute to the debate and specifications are issued by the International Standards Organisation (ISO).

Steel cord reinforced conveyor belting had been attempted in the 1940s, but the belts failed in service with delamination and cord breaks because the cords did not take equal stress during use.

These problems were overcome by developments in Germany, where it was realised that tensioning the cords equally during manufacture allowed uniform transmission of tension and prevented the problems cited above.

Tensioning became the standard procedure for all steel cord manufacture, and steel cord (together with cable belting) was to establish itself on high strength conveyors and long-haul systems.

MATERIALS

Carcass Reinforcement

Belt strength is derived from the carcass material (except for cable belting), and various materials are now used (Table 12.1).

Cotton

Vegetable fibre, 100% cellulose. Short staple lengths are used for conveyor belting (10–20 mm). Cotton offers low strength with high bulk. A natural fibre, cotton is susceptible to microbiological attack (mildew).

Table 12.1. Conveyor belt carcass reinforcement properties.

Property	Units	Cotton	Rayon	Polyester	Nylon	Aramid	Steel
Specific gravity	g/cm^3	1.54	1.52	1.38	1.14	1.44	7.85
Moisture regain	%	8.5	13	0.5	4.5	2.0	Nil
Tenacity	dN/tex	1.5	4.0	8.0	8.5	20.0	3.0
Tensile strength	MPa	230	685	1100	850	2750	2400
Elongation at break	%	8	10	14	16	4	2
Modulus	dN/tex	23	60	90	50	400	250
	MPa	3.5	9	14	5.5	60	200
Moisture regain	The moisture (%) related to the dry weight at standard conditions						
tex	Weight in grams of 10,000 m of yarn; this term has replaced denier in Europe						
Tenacity	Strength per unit weight						
Tensile strength	Strength per unit cross-sectional area						
Modulus	Strength to stretch a given elongation						

Rayon

Usually found as a continuous filament yarn, it can be produced in staple form, as required. Rayon has a higher tenacity than cotton, but its high moisture regain is a disadvantage when processing with rubber as a result of problems of delamination during vulcanisation and in service. An RFL adhesive is required to bond the rubber and is applied as a dip during the heat and set process. Rayon has largely been replaced by nylon and polyester in conveyor belting.

Polyamide (Nylon)

Continuous filament synthetic nylon yarn offers high tenacity and low modulus, as well as excellent resistance to abrasion and fatigue; it imparts good impact resistance, also. During splicing, nylon will stand misuse. It is treated similar to rayon with a heat set and dip (RFL) process tc control shrinkage. Nylon gives the highest adhesions with rubber. Nylon 66 is preferred to nylon 6 because the former has higher melting point (250°C) and can therefore be processed at higher temperatures. Nylon 6, which melts at 225°C, is more unstable and requires careful processing.

Polyester

Polyester is a continuous filament synthetic yarn with higher tenacity than nylon and rayon, higher modulus, and therefore lower extensibility. It is processed similar to nylon. Low moisture regain makes polyester most suitable for light belting in close proximity to water. This synthetic exhibits lower conversion of yarn strength to belt strength compared with nylon.

Aramid

Aramid (aromatic polyamide) is a continuous filament yarn having the highest tenacity of all and the lowest elongation at break for textile yarns. It also has the lowest conversion of yarn strength to belt strength for synthetic yarns, resulting in a high degree of overbuild to attain required belt strength. Aramid has excellent FR properties because it decomposes instead of melting.

Steel

Steel is drawn to the required filament diameter and cabled to produce the cords used in belting. Zinc or brass coating is required to obtain bonding to rubber and to prevent corrosion. Brass coating is reserved for small diameter cords. Steel cords have higher strength, good flexibility, low extensibility, and good fatigue resistance.

Polymers

Natural rubber (NR)

The *Hevea braziliensis* tree, which originated in Brazil, became established in Malaysia, and now is found in Thailand, Indonesia, Nigeria, and elsewhere as the source of natural rubber. Supplied in various forms and qualities, NR is used generally, but it is superior for high tensile strength, wear resistance, and resistance to cutting and gouging.

Styrene butadiene rubber (SBR)

SBR is 71% butadiene, 29% styrene. This synthetic rubber is available in various grades (e.g., standard emulsion polymerised SBR and oil-extended SBR together with solution-polymerised SBR). Standard grades of SBR are replacing NR for general use and give good general properties and heat resistance.

Polyisoprene (IR)

Polyisoprene is the synthetic chemical equivalent to NR, which it replaces in formulations requiring a higher degree of purity.

Polybutadiene (BR)

Polybutadiene, a synthetic rubber used for improved wear resistance, is difficult to mix, hence is normally blended with other polymers.

Butyl (IIR)

A co-polymer of isoprene and isobutylene with superior permeability, chemical, and heat resistance, IIR in standard form is incompatible with other polymers, which makes blends impossible. Chlorinated (CIIR) and brominated (BIIR) rubbers are compatible and are preferred for heat-resistance belting.

Nitrile (NBR)

A co-polymer of butadiene and acrylonitrile with various levels of nitrile (20–50%), NBR offers improving levels of oil resistance.

Chloroprene (CR)

Chloroprene is a long-established synthetic polymer available in many grades to suit processing condition and end use. It has good weather, fire, and oil resistance.

Ethylene propylene diene monomer (EPDM)

EPDM is used when highest grades of heat resistance are required.

Polyvinyl chloride (PVC)

PVC is a thermoplastic co-polymer, available in various grades to suit processing and end use. Example applications include belting grades for plastisols, calender grades for sheeting, and moulding grades.

Polyurethane (PU)

Available in a vulcanised and thermoset grade, polyurethane is used in the light belting industry.

BELT CONSTRUCTIONS

With the exception of light and specialised belting, the usual means of defining a conveyor belt is by belt strength (Fig. 12.1). Belt strengths are expressed in kilonewtons per metre of belt width or newtons per millimeter and working tension is given in pounds per inch of width. For example :

PP 500/3 = polyamide warp and polyamide weft in 3 ply with a whole-belt strength of 500 kN/m.

EP 800/4 = polyester warp and polyamide weft in 4 ply with a whole-belt strength of 800 kN/m.

ST 2250 = steel cord belt with whole-belt strength of 2250 kN/m.

EP 2 ply 220 = polyester warp and polyamide weft in 2 ply or 110 lb/in. fabric working tension with 220 lb/in. working tension of belt.

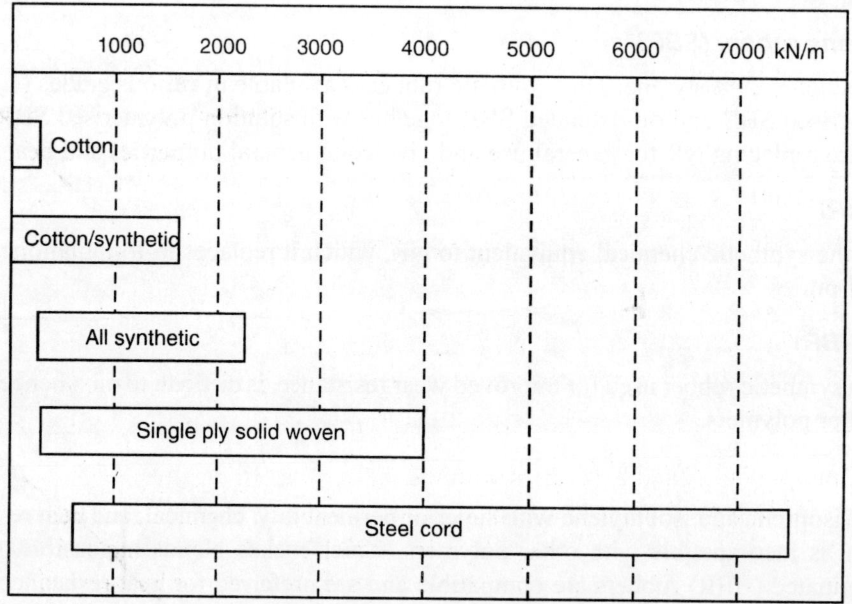

Fig. 12.1. Features of conveyor belt.

Light Belting

Light belting is used in the food and agriculture industries, in supermarkets and department stores, in terminals, in the packaging and paper industries, and in general light conveying. It is made in 1,2, or 3 plies, although the latter is less popular because the trend is toward single-ply, with its advantages of lower bulk and weight. Light belting is available in cotton, cotton-synthetic mixtures, and all synthetic constructions in PVC and PU. PU, suitable for hard wearing applications, is free from surface cracking and requires no plasticiser, PVC is used on softer applications, and there is occasional surface cracking with the loss of plasticiser.

Light belting is used on roller and slider bed conveyors where low extension at working load and low friction are the requirements. Monofilament weft is used on applications called for weft rigidity.

Unlike standard rubber conveyor belts, colours for light belts are generally pastel shades. Surfaces may be moulded or impressed with a multitude of patterns to improve grip. Bonding agents are a necessity with PVC and PU constructions to achieve adhesion to synthetics, especially with monofilament constructions.

Multi-Ply

Multiple layers of plain woven processed fabric are built up to achieve the desired belt strength. The ply strength and composite belt strength conform to standard strengths as specified in national and international standards (Table 12.2.) Increasing the fabric strength is achieved by increasing the number of cords, or by increasing the number of yarn folds within a cord. A limit is reached whereby the volume of yarn required will just not pack into a given space. The weaving pattern for the lower strengths is a plain weave with one warp to one weft, but for higher strengths the Oxford weave is normal, with two warps to one weft to reduce bulking.

Table 12.2. Ply strengths related to belt strengths.

	Belt strengths (k/Nm)				
Ply strength (kN/m)	*2-Ply belt*	*3-Ply belt*	*4-Ply belt*	*5-Ply belt*	*6-Ply belt*
100	200/2	315/3	400/4	500/5	630/6
125	250/2	400/3	500/4	630/5	800/6
160	315/2	500/3	630/3	800/5	1000/6
200	400/2	630/3	800/4	1000/5	1250/6
250	500/2	800/3	1000/4	1250/5	1400/6
315	630/2	1000/3	1250/4	1400/5	2000/6
400	800/2	1250/3	1600/4	2000/5	2500/6

The rubber interply on a plied belt varies with the fabric strength and duty of the application, with thicker interplies for improved impact resistance, especially on 2-ply constructions. Rubber multi-ply belts are used above ground and PVC mutli-ply belts used underground (India). Constructions can be made in 100% cotton up to belt strengths of 250 kN/m, but these belts are rather specialised. Cotton-synthetic mixture belts in strengths up to 1500 kN/m are still popular in applications that required bulk.

All-synthetic plied belting of PP, EP, and some EE (polyester warp and weft) in strengths up to 2500 kN/m is widely used and requires no edge protection. Nylon in the weft imparts good troughability and impact resistance, whereas polyester in the warp is suitable for long-haul conveying, which has limited take-up allowances.

Straight Warp

Alternative designs of weave are possible that depart from the traditional plain weave in an attempt to increase the volume of yarn in the warp without any adverse effects. This construction has straight heavy warp cords with strength weft cords running above and below the warp. The warp and weft are held together with binder cords. The constructions are suitable for rubber impregnation and can be manufactured in single ply up to 500 kN/m with nylon and polyester, and up to 2000 kN/m with aramid. Double-ply constructions extend the strength to 1000 and 3150 kN/m with aramid. Rubber interply is usually 2 mm.

Solid Woven (SW)

Solid woven is a single-ply construction but is not to be confused with straight warp single ply. SW has a weft in two or three layers, with the warp cords interlacing with the weft and holding the whole together. Various yarns may be used: cotton, nylon, and polyester in strengths up to 3000 kN/m and aramid for extending the strengths to 4000 kN/m. The close woven construction will allow PVC to impregnant into the carcass, but not rubber. SW constructions can however be coated with PVC, CR, or NBR and are used extensively underground in Europe because of their outstanding FR properties.

Steel Cord

The basic steel cord construction consists of a number of rectilinear cords in a plane and covered on both sides with rubber. Adhesion to the cords is achieved with a zinc coating and a bonding agent in the rubber adjacent to the cords. All the cords in a particular belt have the same diameter and pitch with alternate twists (Z and S) to ensure that the belt runs straight. Only on steel cord elevator belts in the pitch dissimilar, to allow for the bolting of the buckets.

Cord constructions of 7 × 7, 7 × 19, and 7 × 31 are the norms, where 7 × 7 represents one cord with 7 wires of 7 strands or filaments. Recognised cord diameters range from 2.8 to 14 mm and are related to belt strength. Steel cord belt strengths are derived from the number of cords and the cord strength (diameter) according to the following relation:

$$\text{Belt strength (kN / m)} = \frac{\text{Number of cord} \times \text{Cord strength (kN)}}{\text{Belt width (m)}}$$

Various combinations of cord diameter and number of cords will achieve a given belt strength. The choice is open, but it is influenced by the nature of the splice, minimum edge thickness of belt, and acceptable pitches.

Some countries (e.g., Germany and Australia) declare a range of standard constructions where pitch, number of cords, and cord diameter are specified for a given belt strength. Steel cord belting has a life expectancy of 10 years, and a 20-year life is a common. Joint life is equal to belt life, given normal running conditions.

Steel Woven Fabric

Fleximat, a brass-coated woven steel fabric with an open construction, is available in two types: steel warp and weft for improved impact resistance and steel warp with textile weft. Both Fleximat constructions use synthetic cord to bind warp and weft together. Woven thicknesses range from 3 to 6 mm, in strengths of 350 to 1600 kN/m. The belts have good adhesion, impact, and tear resistance, with satisfactory troughing and splicing; they are suitable for carrying hot materials, where carcass burn-through must not occur.

Breakers

Breakers which are added to most constructions, consist of synthetic plain woven fabrics in an open construction with daylight between cords. They are referred to as leno breakers because in the leno weave the weft passes through the warp in order to lock the loose construction together to improve its stability. Breakers, which are requested by the customer are normally included in the cover thickness, serve the following purposes:

To improve the adhesion of cover to carcass (the breakers improve the modulus of the cover and enhance adhesion when stripping).

To improve impact resistance.

To give some protection against belt tear.

To act as an aid to reinforce and protect the splice joint.

PROPERTIES OF BELT

General

Conveyor belting has serviced a multitude of applications. It is therefore not surprising that belt constructions are many and can be confusing to the end user. Whether categorised as textile/steel cored, rubber/PVC, light/industrial/mining, or single-ply/multi-ply/solid woven. All belting must meet a number of basic requirements which are :

1. Tensile strength to transmit the power required to convey the material over the required distance.
2. Modulus and elastic stretch to accommodate vertical curves, transition distances, and take-up devices.
3. Load support and belt width to carry the type and volume of load required.
4. Longitudinal flexibility to flex around pulleys.
5. Transverse flexibility to trough if required.
6. Adhesion between components to avoid delamination.
7. Tear resistance to withstand damage.
8. Properties to withstand hostile conditions from chemicals, fire, heat, oil, and ozone as required.
9. Fastener holding properties.
10. Cover properties to provide friction for driving and/or slip for slider bed conveyor.
11. Cover properties and gage to withstand the abrasive properties of the material carried.
12. Dimensional stability to run straight.
13. The ability to be spliced and made endless.

Specified

The belt when manufactured is designed to meet a number of properties now specified in standards worldwide. Some properties identified as basic requirements cannot always be determined from laboratory testing. These include cover wear, load support, impact resistance, and the ability to track, which have yet to be specified in a standard.

The recognised, well-established belt properties are as follows :

1. Whole-belt tensile strength and elongation at break.
2. Elongation at reference load (10% of belt strength).
3. Cover properties of tensile strength and elongation at break, both for unaged and aged conditions.
4. Cover properties of abrasion, hardness, tear, and modulus.
5. Adhesion: cover to ply, ply to ply (textile), cord pullout (steel cord), both for unaged and aged conditions.
6. Troughability.
7. Belt tear resistance.
8. Belt modulus of elasticity and elastic stretch.
9. Other special properties (e.g., resistance to chemicals, fire, heat, oil, ozone).
10. Joint efficiency.

Other tests required for steel cord belting but not yet established are centre stand pullout and dynamic cycling. Modulus of elasticity in the rubber industry is given as the tensile strength required to stretch a given elongation. Belt modulus is the linear modulus; it is the belt strength to stretch the belt to the limit of its elastic stretch, which occurs between T_1 (working tension) and T_2 (slack side tension):

$$\text{Belt modulus} = \frac{(T_1 - T_2) \times 100}{\text{Elastic stretch}} \ (\text{kN / m})$$

$$\text{Specific modulus} = \frac{\text{Belt modulus}}{\text{Belt strength}} \ (\text{No units})$$

The test method for elastic modulus is now specified as an international standard, ISO 9856. Both stretch upon mounting the sample, which approximates to initial stretch on site, and belt growth can be determined from the sample test run. Valuable data thus are obtained for application calculations, in determining radii of vertical curves, transition distances, and take-up allowances.

Application

Conveyor belt properties that are related to the application are discussed in the sub-sections that follow.

Troughability

Troughability depends on carcass thickness, troughing angle, belt width, and carcass. Manufacturers indicate width and constructions related to each troughing angle. The extensibility of nylon permits good troughing angles, and steel cord (weftless) can trough up to 60°.

Pulley diameters

Indicated by the degree of flexibility in the belt, pulley diameters depend on belt thickness, construction, and carcass material. Upon rounding a pulley, a plied belt will have extension in the outer ply and compression in the inner ply, relative to the neutral axis of the belt. For solid woven and steel cord belting, the cords are not layered as in plied belting and the stresses induced around the pulley are leveled out; consequently smaller pulley diameter may be used. To maintain the elastic component at the pulley to within safe working limits, pulley diameters are specified by the belt manufacturer.

Take-up allowance

Take-up devices such as gravity and screw take-ups are built into the conveyor system to apply T_2 tension and to accommodate both elastic stretch and belt growth. For steel cord and aramid, the allowance is small, reflecting the low extensibility of the belt, but the allowance for synthetic plied is normally 2.5% and for solid woven it is 3.5%.

Belt growth

After many cycles of use, the carcass fibres on textile belts settle out, and some of this stretch becomes permanent. This belt growth is contained in the take-up, and fasteners are usually fitted to the belt before the final splice is made, to ensure that most of the belt growth is removed.

Transition distance

The distance between the last trough idler set and the pulley is termed the transition because it reflects the action of the belt changing from the trough to the flat. The tensions in the belt at the edge and centre are most critical and are influenced by the belt modulus. High modulus belts such as steel cord and aramid require long transition distances; low modulus belts, such as nylon and solid woven, require shorter distances.

Vertical curves

When a land conveyor changes direction uphill or downhill, the belt must run in a curve. Where the run of conveyor changes from horizontal to incline, or from decline to horizontal, the belt curve is considered to be concave. The edge tension inside the neutral axis will experience compression, while the belt centre, being outside the neutral axis, will be under tension. The radius of curvature is directly related to belt modulus, which means that higher modulus belts such as steel cord and aramid have higher radii than nylon or solid woven.

Impact resistance

The belt is subjected to excessive damage at the loading point, especially in mines and quarries, where the energy released when large lumps fall from great heights can cause belt fracture. Supporting the belt at the loading point is a pre-requisite of such applications, but the belt's resistance to impact damage is influenced by belt thickness, elasticity, and rubber insulation both on the cover and between plies. Low modulus belts such as nylon with adequate insulation between the plies, exhibit the best impact resistance of textile plied belting.

Load support

The ability of the belt, however strong, to support the load is critical. High bulk belts give best load supports, but such belts are difficult to trough. Nipping in the troughed idlers occur with poor load support and can lead to belt damage. Load support, as indicated by the maximum width required to support the load of a given material density, will depend on belt construction, number of plies, and insulation between the plies. For a given belt strength, reducing the ply strength and increasing the numbers of plies is the usual remedy to improve load support.

BELT QUALITY GRADES

General Purpose

For general-purpose belting, most specifications list two or three quality grades giving cover tensile strength, elongation at break, and in some cases abrasion resistance. The polymers used in the cover compounds to meet these requirements are usually NR and SBR. NR is chosen for the higher quality grades, where improved wear resistance and resistance to cutting and gouging are required. Belt wear, as in tyre wear, is a property that cannot be evaluated under test conditions in the laboratory and has failed all attempts to establish regulatory standards. Abrasion carried out on a rotating drum or disk is a measure of volume loss under controlled conditions only: it cannot duplicate site conditions, since material consistency, weight, trip rate of conveyor, and hostile conditions vary from day to day and from site to site. It is also a feature of rubber processing that the more sophisticated the quality grade, the lower the general properties. Cover tensile strength, abrasion, and adhesions are generally lower for oil, fire and heat resistance.

Heat Resistance

When conveyor belts are manufactured, the belt is vulcanised from the uncured to the cured state. The rubber turns from the plastic non-elastic state to the elastic non-plastic state, an irreversible change. The vulcanisation process combines heat (temperature) and time. The rubber compounds have therefore some tolerance to heat ageing, but further heat application in service begins to deteriorate the rubber and belt properties. The vulcanising ingredients will stabilies the rubber over a limited range, but to extend the range to 100–150°C, a specialised selection of accelerator systems is necessary. This regime produces a plateau effect of the properties, such as tensile strength, adhesion, and hardness, and allows the belt to perform at these higher temperatures, or for longer times at lower temperatures. The point is then reached at which the tensile strength and adhesion fall and the hardness increases. Further compounding innovations do little to improve the situation.

For higher temperatures a new classification of polymers is required—butyls. IIR and its derivatives BIIR and CIIR have a much higher tolerance of heat ageing than do NR and SBR. Blending BIIR and CIIR with other rubbers is essential if such physical properties as abrasion, adhesion, and tensile strength are to be maintained. Whereas NR and SBR must be compounded to extend their tolerance of heat, butyl does it naturally. The higher the NR/SBR content in blends of butyl, the higher the tensile strength, adhesion, and tear resistance, but there is a reduction in heat resistance. The polymer technologist must balance these properties. Butyl compounds will stretch the heat resistance tolerance to the 150–180°C range. Beyond these temperatures, EPDM is recommended. This polymer stretches the temperature to 200°C and above. It is difficult to determine the suitability of a particular belt construction for a particular applications with heat-resistant belting. This is not simply a case of measuring the surface temperature of the belt. The conditions to which the belt is subjected (including temperatures of the load), the rate of cooling of the returned belt, and the ratio of fines to lumps must be considered. The trip rate of a belt gives some indication of the rate at which the belt is subjected to high temperatures.

$$\text{Trip rate} = \frac{\text{Belt speed (m / s)} \times 60}{\text{Conveyor length (m)}}$$

The choice of type of belt for a given application entails a consideration of the rate of fall-off adhesion in the belt against the improvement in heat-resistant properties of sophisticated polymers. An arduous application may be suitably serviced with an NR/SBR belt that holds together longer than by a belt of superior heat resistance, whose adhesion deteriorates. Experience of the site and circumstances is as important as temperature measurement within the load carried.

Oil Resistance

A wide range of oil- and fat-impregnated materials are carried by belting. Using NR or SBR results in a higher degree of cover and carcass swelling, with a dramatic reduction in cover hardness, which leads to unfitness for use. Good oil resistance is met with CR belting, but for the highest resistance to oils and fats, NBR is unsurpassed.

Ozone Resistance

Ozone protection is not essential in normal conditions of running, where ozone concentrations are normal. A stationary belt under tension is most vulnerable to ozone attack, however, and ozone protection is now standard practice for belting, especially when it must be stocked for some time. In environments with high ozone concentrations (e.g., Canada), extra protection is mandatory.

Cover Gages

Cover thickness will vary with the material carried. As a rule, a 1 mm top and bottom cover would be the minimum specified, and for sand and gravel applications, a 3 mm top cover is the norm. Special applications such as the mining of the ore requires heavier top cover gages because there is severe cutting and gouging. It is important that the ratio of top to bottom cover does not exceed 3:1; otherwise the uneven shrinkage forces in the cover cause the belt to curl.

Steel cord belting has minimum cover gages of 4 mm, with 6–10 mm top cover recommended for carrying coal.

FIRE RESISTANCE (FR)

FR belts, have grown through necessity to become a major influence on the conveyor belt industry, both above and below ground. Charting the development of FR standards worldwide demonstrates the progress in setting such standards and improving safety.

Steel Cord

Textile PVC belting adequately met the belt strength requirements for the NCB up to 900 kN/m.

BELT SELECTION

Belt Strength Calculations

To determine the necessary belt tensile strength, various tensions on the installation must be considered. The maximum tension, calculated per unit width of the belt, indicate the working tension, and from this the belt strength is derived. The main considerations in the calculations are effective tension and driving tension.

Effective tension

The effective tension—the tension required to move a loaded belt around a conveyor—has three components:

1. The tension necessary to move the empty belt.
2. The tension required to move the load over the conveyor length.
3. The tension required to raise or lower the load for an inclined conveyor. This force is positive for elevating conveyors and negative for descending conveyors.

Driving or slack side tension

The minimum tension that must be induced in the return (empty) run of the belt as it leaves the driving pulley to ensure that the effective tension can be transmitted without slip is called the driving tension.

The maximum tension is calculated from the sum of the effective tension and the slack side tension, with corrections for the various additional factors. The maximum tension thus calculated in kilonewtons, divided by the belt width (m), gives the tension per unit width, and this working tension T_l (kN/m) requires an appropriate safety factor to arrive at the required belt strength.

Safety factors (SF)

The difference between the working tension and the required belt strength is considered to be a factor of safety that varies with the construction and is influenced by the conveyor system and the belt.

Conveyor system

Take-up devices, variable rate of loading, acceleration/deceleration with loaded belt, gradient, and shock loads with steel cord belting, all influence the level of safety factor chosen.

Belt

Joint efficiency, stretch characteristics, fatigue, and impact damage all influence the level of safety factor chosen. Safety factors for textile belts are considered to be 10:1, steel cord belts for normal applications 7:1, and steel woven 8:1, because of their improved joint efficiency and resistance to fatigue. For example, for a working tension of 140 kN/m, we would find:

Textile belt strength (SF 10:1), belt strength 1400 kN/m

Steel woven belt strength (SF 8:1), belt strength 1150 kN/m

Steel cord belt strength (SF 7:1), belt strength 980 kN/m

These strengths would be adjusted to the nearest standard strength available.

Installations

The largest users of steel cord belting are found in the mining of coal: black coal below ground and brown coal (lignite) above ground in open cast mining.

MANUFACTURING

The manufacture of a conveyor belt depends on the basic construction, since each type requires its particular process.

Lightweight Belting

Generally lightweight belting is manufactured by spreading, especially the plied constructions. The fabric passes through a spreading bar on which is deposited the PVC or PU stock. The fabric with the applied coat thickness then passes into a heated chamber. Further the runs are necessary to build up the plies. The final run puts the coating on the surface, which may be left flat using plain rollers or embossed using pattern rollers.

Plied Textile Belting

Cotton and cotton-synthetic fabrics require drying before processing. The dried fabric can be either dipped or spread with rubber solution to achieve the required impregnation, but more usually the fabric is frictioned with rubber by passing between the bowls of a calender, where the adjacent speeds are unequal and give an approximate friction ratio of 1.5:1.

This friction process allows the rubber to be forced into the fabric weave to achieve impregnation, giving good adhesion and tack. For lightweight cottons, no further rubber coating is necessary, but for weights above 28 oz. cotton and its DCN equivalents, a rubber coating is required. This is applied at the calender, but with a 1:1 speed ratio. The required number of plies are built up, covers of the appropriate gages applied, and the green belt consolidated.

Drying is not necessary for synthetic plied belting, and because the adhesion is a chemical bond (RFL), no frictioning is required. The plies are coated with the required gage on a calender, built up, covered, and consolidated as a green belt. For continuous vulcanising, the individual plies and covers do not required a building process.

Solid Woven

It is essential to penetrate the centre of the thick weave with PVC, and various methods are used with plastisols having a high plasticiser content to improve penetration. The penetration is achieved with the aid of rollers to vacuum or high pressure in a tank; alternatively, the PVC plastisol is applied at the weaving stage, but this is not common. The dried or gelled PVC-impregnated carcass is then coated on both sides with PVC to the required thickness, or it may have rubber covers applied as required.

Steel Cord

The steel cords are let off from reels, whereupon they pass through a series of guides and rollers and finally combs, which distribute the cords to the required pitch. The comb pins vary from belt to belt and are designed to match both cord diameter and cord pitch. Tension is applied to the cords, and the cover and bonder for top and bottom, previously assembled, are then applied to the cords and compressed. This pre-press before curing is essential to remove all the air from the green belt, to control cord pitch, and to present a solid belt to the press.

Vulcanisation

PVC belts, being thermoplastic, are not vulcanised but fused. Solid woven belts may be passed through a heating chamber at 160°C and allowed to cool before wind-up. Other process routes press the belt, where stretch, if required, may be applied. The press is raised to 160°C and cooled before opening, applying low specific pressures. To prevent distortion at press ends, the press must have cold sections leading in and out. Rubber covered PVC belts must be pressed.

Press vulcanisation of rubber conveyor belting is similar for steel cord and textile, with both requiring temperature, time, and pressure. Steel cord belting requires higher specific pressures, and curing temperatures are usually 145–155°C. The steel cord belt must be held under tension throughout the cure. Rubber belts moulded in a press must have a cold section upon entry into the press to allow transition from green belt to vulcanised belt. The curing time for each charge length varies with the compound quality, the belt thickness, and the curing temperature. It is essential with moulded edge belting such as cotton, DCN, and steel cord to fit a frame to control the width during curing. Either steel or aluminium metal plates are used, with thicknesses related to belt gauge.

Continuous vulcanisation (CV) allows the individual plies and covers to be fed directly into the unit and eliminates belt building. The belt passes around a large diameter drum and is held against the drum by a closed steel band, which returns via two smaller diameter drums. Heat is applied to the drum, and heated chambers vulcanise the other side. The band is held under tension to consolidate the belt, and low specific pressures are usual with CV units. The cure time is controlled by the belt speed and is related to arc of contact, drum diameter, and temperatures. For given temperatures and drum diameters, cure time is inversely proportional to belt speed.

A new innovation in CV units is the ContiRoll system (registered trademark owned by Siempelkamp GmbH), whereby the standard features of the press are enhanced by continuous steel bands top and bottom, which move with the belt through the press, running on frictionless bearings and returning around large diameter rollers at press ends.

Quality Issues

The challenge from competitors, combined with ever-increasing customer standards and expectations, suggests to manufacturers that reducing costs and improving quality are essential to maintain volume and margins.

Quality control

Traditional methods of quality control with their *adhoc* approach to problem solving and continuous monitoring of production, rely heavily on final inspection, testing, and rejection to prevent defective products from reaching the customer. Such methods have given way to more sophisticated forms of quality improvement.

Quality assurance

Quality assurance emphasises meeting customers' needs and preventing defects and waste by building quality into the product at the earliest stage of design and manufacture. The company must be shown to be capable of supplying customers' needs by demonstrating that internal systems exist and are controlled and the these systems are codified out in a quality manual. The relationship between supplier and customer becomes one of documented evidence, resulting in the establishment of confidence, which in turn leads to supplier acceptance testing and self-certification.

Many large users of conveyor belting had their own quality assurance schemes, granting certification to the manufacturers only after internal standards had been met. Such arrangements have been replaced by accreditation to quality system specifications issued nationally and internationally. These specifications identify the areas of control necessary but leave the interpretation to the company, to permit obtaining a match of product standard, company activity, and customer's demands. The application of statistical process control, with its formal approach to measurement, using basic statistical methods, often with computer-assisted software, allows trends to be seen more readily. Variability is then diagnosed, controlled, and eventually reduced. Single sourcing also has its advantages in reducing variability. This approach has the effect of reducing the degree of overbuild endemic in conveyor belt manufacture.

The quality systems trace the flow through the company from planning to initial order of customer's requirements, to final dispatch of belt. Auditing by major customers on their own quality system is now replaced by in-house auditing or by third-party auditing to ensure continuous monitoring of the quality system. Specific benefits are company-wide involvement, improved productivity and efficiency, reduced costs, the ability to provide a product that consistently meets customers' needs, improved customer confidence, retaining an improving market share, and reduction of testing of raw materials, intermediate testing, and final belt repetitive testing.

Total quality management

Total quality management, which is the aim of quality manufacturers, brings together all the aspects we have considered (viz., inspection, quality control, quality assurance). Its goals are continuous quality improvement involving suppliers, customers, and all operations within the manufacturing plant. Successful firms excel in employee involvement, team work, and performance measurement.

BELT JOINING

Belts are made endless with mechanical fasteners, vulcanised splicing, or cold-cured splicing.

Mechanical Fasteners

Various forms of metal fasteners are available for textile belting and are generally grouped into lacings (hooks) and plate fasteners. Lacings are applied by a jig, whereby the pointed ends are stapled into the belt. Plate fasteners, used for the stronger belts, require holes to be punched to make room for the bolts holding the plates. The fastener may be used up to belt strengths of 1000 kN/m for plied belting and 1400 kN/m for solid woven. Fasteners are also used to form a temporary joint as the belt settles in permitting belt growth to be removed before the permanent joint is made. The joint efficiency is such that the belt must not be run at tensions greater than the working tension.

Belts must have good fastener holding properties; otherwise the fasteners will pull out. The selection of fastener is influenced by its effects on carcass damage, strength retention, carcass thickness, and fastener flexibility to prevent damage to the pulleys and idlers when troughing.

Spliced Joint

A spliced joint is a permanent joint that is vulcanised for rubber belting and fused for PVC; efficiency is higher than for mechanical fasteners, and joint life is longer.

Various splice types are available (stepped ply, 2-ply jump splice, chevron, finger, and diamond splice), although chevron and diamond splice are now less common.

Stepped ply

Stepped-ply splices are used on multi-ply belts, either rubber or PVC: the plies are stepped back at each belt end, but on opposite sides. Thus they can be matched up when joined, resulting in a number of step lengths that is one less than the number of plies. Joints are prepared on the bias with a preferred angle of 26.5°, corresponding to half the belt width. Ideally, vulcanisation is carried out in one heat, but this is not always possible because a press adequate length is not always available. The temptation must be resisted to reduce step lengths in order to accommodate the joint in an available press, for such a joint will surely fail. The area of contact of the splice is related to its strength, with a step length(s) of 150 mm for a multi-ply with 100 kN/m fabric ply strength increasing to 400 mm for 400 kN/m ply strength constructions.

2-ply jump splice

For 2-ply belts, a conventional stepped-ply joint would result in a single step holding the belt together, which is clearly not advisable in the absence of reinforcement with a strong breaker. The steps in a jump splice do not match as in a stepped-ply joint. Rather, they overlap onto face and back ply, resulting in a satisfactory albeit bulky joint. Wedges must be applied during vulcanisation to ensure full, even pressure over the whole splice area.

Finger splice

A finger splice is used for solid woven belting, where stepping is not possible. A zig-zag pattern in the form of triangular fingers is cut from opposite belt ends, which match when fitted; whole-number pitches are necessary. The length and number of pitches are related to belt strength and width. Breakers are usual with finger splices.

Steel cord

The steel cord splice is in effect a finger splice with the steel cords acting as fingers. The cords at each belt end are stripped of rubber after the covers have been removed and interlaced (not welded).

Steel cord splices give a high degree of efficiency. Unlike textile belts, where joints are renewed within belt life, steel cord joints normally last the lifetime of the belt. Other configurations of joints are available for manufacturers specialising in their own joint pattern and mode of preparation, which reduces splicing time and have been shown to be reliable in service.

CHAPTER 13

V-Belt and Fan Belt Manufacturing Technology

INTRODUCTION

V-belts and fan belts are the generic names given to power transmission belts having V-shaped friction power transmitting sides. V-shaped power transmission belts tripled the nominal frictional force compared to flat belts.

Endless, V-belts are manufactured in various types, including the fan belts for automobile use, frictional horse power driven for domestic and light industrial applications, and larger industrial belts.

However, the higher frictional force resulted in increased breakage due to stress coming from sidewall pressure, as each layer of belt material received higher stress during operation. Also, the thickness of the V-shaped belts caused more cracks due to flexing, compared with the thinner flat belts. Thus a higher level processing technique was necessary to avoid such breakage and cracks.

With the emergence of the V-belt, V-shaped belts became popular means of power transmission with motors in a variety of fields. At about the same time, V-belts were used on automobile engines to operate cooling fans and the water pump, which is how the fan belt got its name. The automobile then started to be equipped with a generator, a compressor for the air conditioner, and an oil pump for the power steering. All these accessories required a means of higher power transmission capability, reduced size, and greater operating durability. These requirements triggered improvements in V-belt engineering and processing technologies.

COMPONENTS AND FUNCTION OF POWER TRANSMISSION BELTS

The components and function of power transmission belts are closely related to individual manufacturing processes. When a belt is operated with a dual-axis load, power transmission and reception are conducted on each pulley as shown in Fig. 13.1, and power is transmitted between the span. The function of each material in a power transmission belt is given in Table 13.1.

FUNDAMENTAL MANUFACTURING PROCESS

Fig. 13.2 shows the fundamental belt manufacturing process, which entails four categories, according to the processing technology: (i) material processing; (ii) building; (iii) vulcanisation; and (iv) mechanical processing.

MATERIALS IN V-BELT COMPOSITION

V-belts are made of materials from three groups, namely, rubber, cord, and canvas. Chloroprene rubber (CR) is mainly used as the rubber material, but natural rubber (NR) and styrene butadiene rubber (SBR) have been used for conventional wrapped V-belts for general industrial use. It must be noted that hydrogenated nitrile-butadiene rubber (NEM) was used for the first time for a part of a V-ribbed belt for driving as engine auxiliary device under the hood of a car, at high temperatures and thereafter alkylated chlorosulphonated polyethylene (ACSM) began to be adopted.

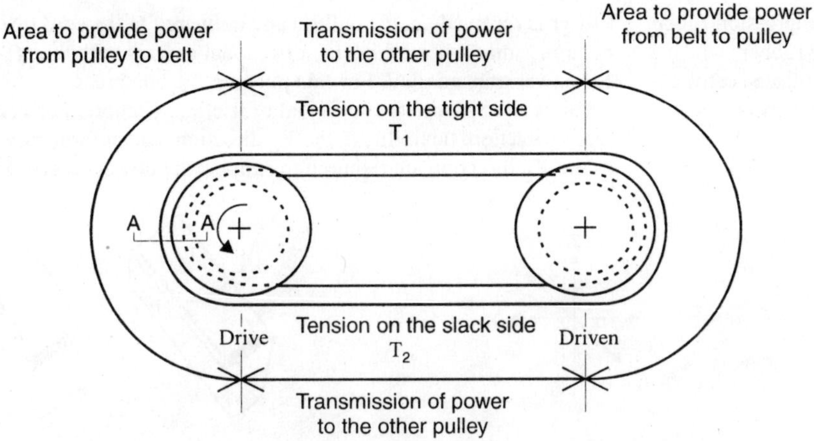

Transmission revolution power, $p = T_1 - T_2$

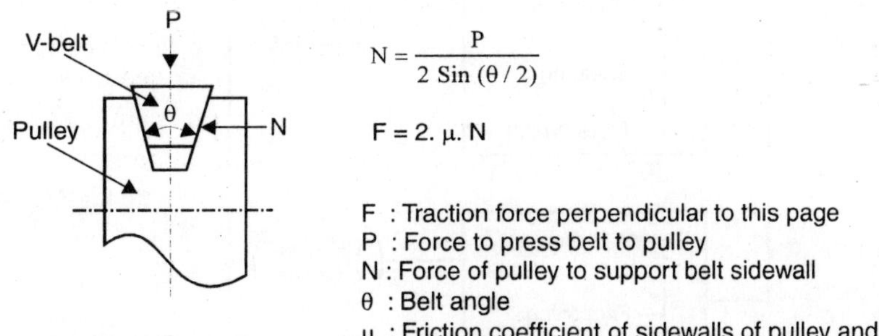

$$N = \frac{P}{2 \sin(\theta/2)}$$

$$F = 2.\mu.N$$

F : Traction force perpendicular to this page
P : Force to press belt to pulley
N : Force of pulley to support belt sidewall
θ : Belt angle
μ : Friction coefficient of sidewalls of pulley and belt

Fig. 13.1. Power transmission and reception relations between two pulleys.

Table 13.1. V-belt functions.

Function	*Material*	*Required material function*
Top conservation	Top canvas Adhesive agent Rubber cement	Protection of cord layer; material should be easy to bend, with minimum cross-sectional deviation.
	Tension rubber	Y_B axis bending rigidity should be low, and X_B axis strain should be low.

(Contd ...)

Function	*Material*	*Required material function*
Tension member conservation	Adhesion rubber	Rigid and firm fixation and holding of cord. Normally, the cord has elasticity more than 100 times greater than other materials, so strong adhesion property is required to prevent cord slippage.
Power transmission	Cord Adhesive agent Rubber cement	To restrain elastic slip (see sketches), the elasticity ratio of the cord should be high in the X-axis, while the bending elasticity ratio in the Y_B axis should be low. Accordingly, structural alignment in the X axis is necessary.
Power reception and delivery	Compression rubber Rib rubber Laminated canvas Rib canvas	Power is received via the pulleys and delivered to the cord by the compression rubber, which is in frictional contact with the other pulleys for power transmission. For this purpose the compression rubber requires an appropriate coefficient of friction. Compression resistance in the X direction, flexibility in the Y_B direction, and deformation resistance in the X_B direction are required to support the power received.

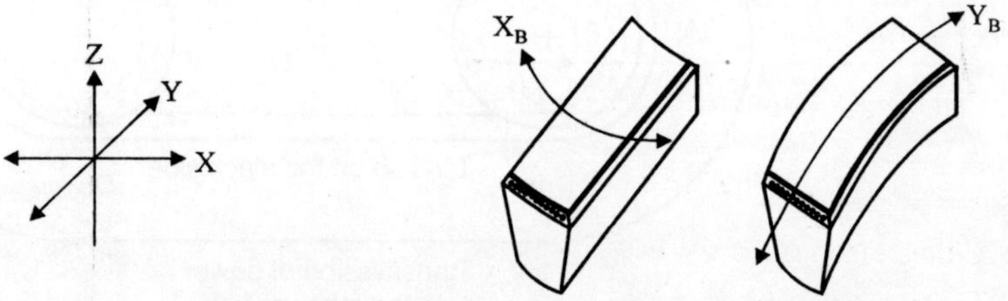

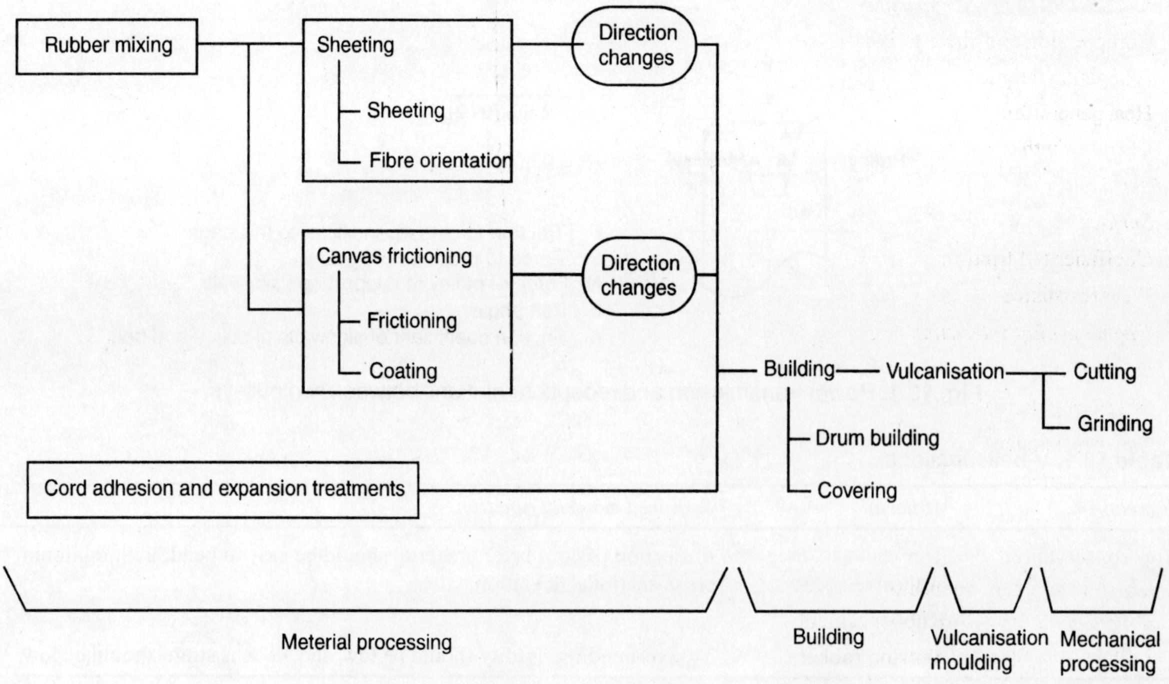

Fig. 13.2. Fundamental manufacturing process for V-belts.

Among rubber materials for V-belts, it is characteristic that short fibre reinforced rubber (SFRR) is used for some of the compression rubber of wrapped and raw-edged V-belts, and for most of the rib rubber of V-ribbed belts. SFRR is used for improving rigidity to lateral pressure and, in the case of raw-edged and V-ribbed belts, for controlling friction characteristics between the belt and pulleys to reduce frictional noise. To increase transverse rigidity and to maintain pliability in the circumferential direction, short fibre is oriented unaxially to a great degree, and the belt is shaped in such a fashion that the orientation direction coincides with the direction of belt width. Table 13.2 shows the required characteristics for V-belt rubber materials.

Table 13.2. Properties required for rubber materials.

	Wrapped V-belt raw-edged V-belt		*V-ribbed belt*		*Canvas coating rubber for belts of all types*
Properties	*Top and compression rubber*	*Adhesion rubber*	*Rib rubber*	*Adhesion rubber*	
Fundamental mechanical properties					
Modulus	1		2		
Tensile strength		2	2	2	2
Ultimate elongation		2	1	2	
Tear resistance	2	1	1	1	2
Loss factor	1		1		
Mechanical fatigue properties					
Compression and shear fatigue	1	2	1	2	
Elongation fatigue			1		
Heat generation	1		1		
Crack growth		2	1	2	
Permanent set	1		1		
Surface characteristics					
Coefficient of friction	1		1		1
Wear resistance		1		1	1
Chemical characteristics					
Heat ageing	1	1	1	1	1
Ozone resistance	2		1		2
Water resistance	2		2		1
Oil resistance	2	2	2	2	2
Mould resistance					1
Others					
Temperature dependence of mechanical properties	1	2	1	2	2
Low temperature resistance	1	2	1	2	2
Fibre deterioration effect		2		2	2

(Contd ...)

Properties	Wrapped V-belt raw-edged V-belt		V-ribbed belt		Canvas coating rubber for belts of all types
	Top and compression rubber	Adhesion rubber	Rib rubber	Adhesion rubber	
Adhesive property to cord		1		1	
Hygiene	2	2	2	2	2
Processibility					
Mixing processibility	2	2	2	2	2
Sheeting processibility	2	2	2	2	2
Solubility					2
Fluidity	2	2	2	2	2
Storage stability (scorching, oxidisation	2	2	2	2	2
Tackiness	1	1	1	2	1
Interface adhesion	1	1	1	1	1
Short-time and flat vulcanisation	2	2	2	2	2

1. Highly required; and 2. required.

Polyester yarn is mainly used for cord (tension member), but *p*-aramid, glass, rayon, cotton, and steel wire cord are also used for limited purposes. V-belt cords are required to have excellent dimensional stability, low stress relaxation, and high fatigue resistance.

Also, cords of high thermal shrinkage force are manufactured by high drawing treatment for the purpose of balancing the mechanical stress relaxation and thermal shrinkage force in using belts. *p*-Aramid fibre is used especially for belts for agricultural machinery, which must withstand high impact loads and for variable speed belts with high load transmission. It is also possible to adopt *p*-aramid fibre for automotive serpentine drive belts. Characteristics required for V-belt cords are shown in Table 13.3.

Table 13.3. Properties required for cord materials.

Properties	Rating
Fundamental mechanical properties	
Modulus	1
Tenacity	1
Elongation at break	2
Catenary	2
Mechanical fatigue properties	
Bending fatigue	2
Stress relaxation	1
Crepe	2
Hysteresis, heat generation	1
Thermal properties	
Thermal shrinkage rate	1

(Contd ...)

Properties	*Rating*
Thermal shrinkage force	1
Thermal expansion	2
Softening temperature, melting point	2
Glass transition temperature	2
Surface characteristics	
Coefficient of friction	2
Wear resistance (among filaments)	2
Chemical properties	
Steam resistance	2
Water resistance	2
Dimensional change due to water absorption	2
Rubber chemical resistance	2
Others	
Adhesion	1
Ease of cutting, fraying	2

1. Highly required; and 2. required.

Plain woven cotton fabric is mainly used as the canvas. Nylon and polyester, and their blends spun with cotton, are also used. Lately, *m*-aramid has been used for fabrics that are required to be abrasion resistant at high temperatures. The fabric is coated with rubber and then brought into the belt building process. Cover, top, bottom, and laminated canvases are used in a 45° bias direction, to maintain pliability in the longitudinal direction of the belt. Wide-angled canvas, where the angle of warp and weft is expanded from 90° to around 120°, is used for the part for which high bending pliability is required. On the other hand, as one of the steps in improving the belt's transverse rigidity, nylon cord fabric, polyester, or the like is used in the same direction as the belt width direction. Properties required for canvases are shown in Table 13.4.

Table 13.4. Properties required for canvas materials.

Properties	*Wrapping, top and bottom canvas*	*Laminated canvas*
Fundamental mechanical properties		
Modulus	2	1
Bending rigidity	2	1
Tensile strength	2	2
Elongation at break	2	2
Tear resistance	2	2
Mechanical fatigue properties		
Bending fatigue	1	1
Compression fatigue	2	1
Elongation fatigue	2	2

(Contd...)

Properties	*Wrapping, top and bottom canvas*	*Laminated canvas*
Thermal properties		
Softening temperature, melting point	1	2
Glass transition temperature	2	1
Surface characteristics		
Wear resistance	2	—
Coefficient of friction	1	—
Chemical properties		
Water resistance	2	—
Chemical resistance	2	2
Steam resistance	2	—
Thermal deterioration resistance	2	2
Others		
Adhesion	2	2
Ease of cutting	2	2
Wrinkle resistance	2	2
Mould resistance	2	—
Antistatic	2	—

1. Highly required; and 2. required.

OUTLINE OF MATERIALS PROCESSING

An outline of material processing is shown in Fig. 13.3. Rubber ingredients are mixed by an internal mixer (Banbury or the like) and then formed into a sheet or the shape of manufacturing materials by calendering or by extrusion. Because SFRR calendered sheet has short fibres oriented in the machine direction, generally it is cut at a right angle, turned 90° and joined. This treatment enhances transverse rigidity and allows longitudinal pliability. Cords are twisted and subjected to an adhesion treatment and a heat treatment at the same time by a cord treating machine. To prevent deterioration of the adhesive agent due to light and to maintain tackiness at belt building, coating with rubber cement is carried out in some cases. Where necessary, the above-described wide angled treatment and adhesion treatment are given to the canvas, which is then coated with rubber cement or frictioning and/or topping by a calender process. Finally, the canvas is cut in the bias direction and is joined.

RUBBER PROCESSING

Main Points in Rubber Processing for V-Belts

As described above, rubber V-belts are manufactured essentially the same as other rubber goods, but the following points are emphasised: (i) processing of SFRR; and (ii) accuracy of sheet thickness.

The mechanical properties of SFRR are greatly influenced by change in length due to breakage and by the level of dispersion and orientation during processing. Therefore, both physical properties and quality stability depend on mixing and the method of sheeting. When building the belt, rubber sheets usually are plied and then vulcanised by an autoclave. Therefore, the variation in thickness of one piece of rubber sheet is multiplied by the number of plied layers, and this is reflected by variations in belt thickness. Thus, high accuracy of sheet thickness is required.

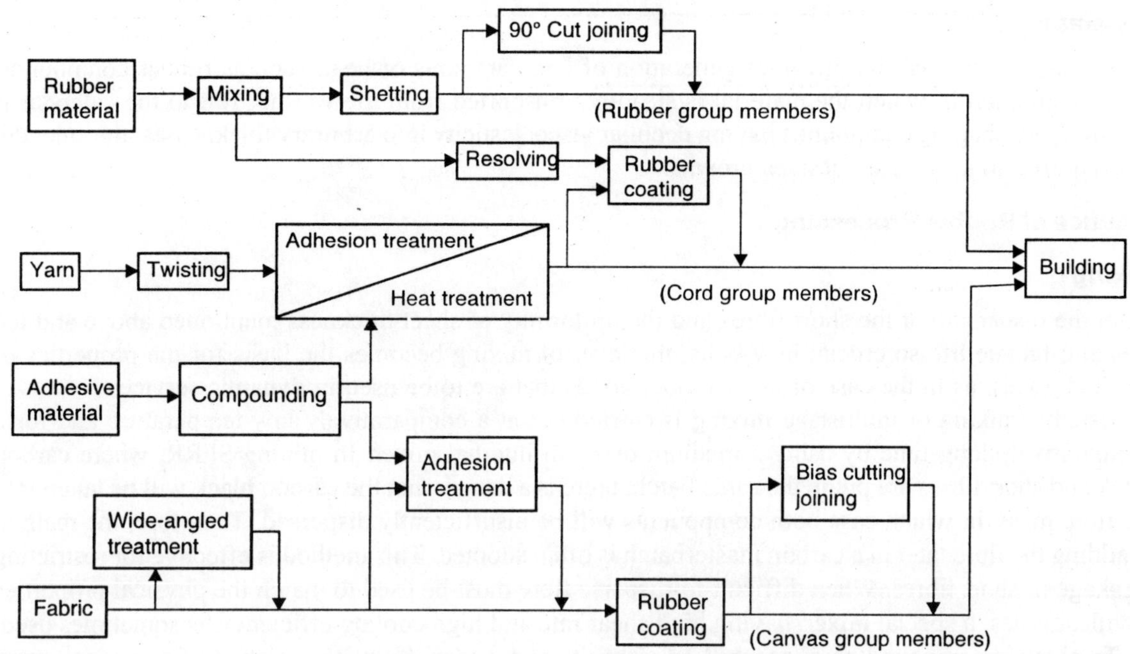

Fig. 13.3. Manufacturing process for materials.

Characteristics and Control Factors of Short Fibre Reinforced Rubber (SFRR)

Among the basic appraisal indices that describe characteristics of SFRR, those that are influenced strongly by processing are length of the short fibres, degree of orientation of the short fibres, and dispersibility of the short fibres.

Accuracy and Control Factors for Sheet Thickness

We shall consider the method of reducing the variation in calender sheet thickness from the material as well as the processing conditions and apparatus viewpoints.

Material

It is, of course, most important in mixing to obtain a uniform compound that is free from within-batch and between-batch variations. Also, provided we admit variations of processing conditions, the compound should be insensitive to temperature and shear rate.

Processing conditions

First of all, it is assumed that processing conditions are established within the area of the temperature in the shear rate range where melt fracture does not occur. If the processing conditions are established outside this area, smoothness is lost and in the case of SFRR, the sheeting produced may have holes. Next, it is important to reduce irregularity of temperature and shear rate (occurring partially or with the lapse of time) to the minimum. (Clearly the irregularity mentioned here is the irregularity not of apparatus condition but of temperature and shear rate of material. This perception is vital in minimising errors when devising improved methods).

Apparatus

In the case of material that involves generation of large amounts of heat, such as rubber compounds, line construction in which the material is steadily transported from the warmer roll to the calender is required. For shaping compounds having peculiar viscoelasticity into arbitrary thicknesses, the calender roll requires an axis cross, besides crowns.

Practice of Rubber Processing

Mixing

From the dispersion of the short fibres and the uniformity of sheet thickness mentioned above and the dynamic fatigue life so crucial in V-belts, the level of mixing becomes the basis for the properties of finished goods, as in the case of other rubber goods that are to be used in dynamic services.

Usually, mixing or multistage mixing is carried out at a comparatively low temperature and for a comparatively long time by using a medium or small internal mixer. In mixing SFRR, where carbon black and short fibres are put in the same batch, there is a danger that the carbon black will be taken into the fibre mass, in which case both components will be insufficiently dispersed. Therefore, the method of adding the fibre later in a carbon masterbatch is often adopted. This method is effective for restricting breakage of short fibres. When difficult-to-disperse fibre must be used to match the physical properties of vulcanisates, a special mixer, having high shear rate and high cooling efficiency, is sometimes used.

To obtain a compound having stabilised viscosity and dispersibility, the method of using integrated power values as control parameters on mixing is generally adopted. In recent years, precise control by the electric power curve has become achievable. Also, studies of the improvement of mixing efficiency by spectroanalysis of the electric power curve have been made. Fig. 13.4 shows an example of such a study. The internal mixer can be regarded functionally as a scaled-up viscometer, and electric power is a direct index of material viscosity. The viscosity is taken as a function of temperature and shear rate, and by controlling power at, and up to, the point of discharge, compounds having constant viscosity can be manufactured.

Sheeting

Rubber material for V-belts is usually sheeted by a calender or extruder. Sheeting of compounds containing no short fibres is the same as for other rubber goods. Sheeting is accomplished by use of a calender, and sheeting or profiling of V-shaped compression rubber is done by an extruder.

Uniaxial orientation in SFRR has been created by using a calender and an extruder. The method of orienting by calender ordinarily employs a roll speed ratio of 1 and can orient fibre in the sheeting direction. The extent of orientation mainly depends on the gap between rolls, and the constant extent of orientation is reached by a gap of approximately 1.0 mm or less. In the case of orientation by extruder, with an expanding mandrel die attached as shown in Fig. 13.5, orientation is in the direction of the sheet width. In this case, orientation to the same extent achieved with the calender method cannot be obtained, but the second method can result in orientation to the extent of $X_i > 80\%$. Sheeting of thick sheet is possible, and this may eliminate the plying up process for belt building.

As shown in Fig. 13.6, the calendered sheet oriented in the machine direction is cut at a right angle, joined in a 90° biased direction, and thus becomes a sheet oriented in the direction of the width. This cutting and joining is the preparation for facilitating the work of matching the direction of the belt width with the direction of orientation of the short fibres. A slight angular deviation causes a large change in physical properties.

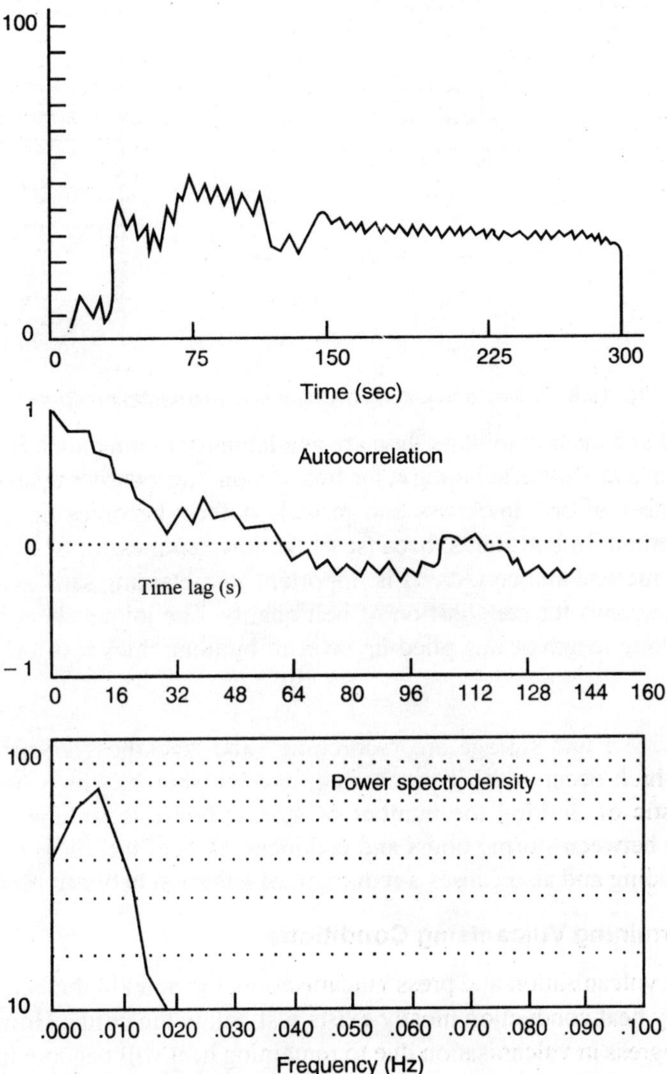

Fig. 13.4. Spectroanalysis of mixing electric power curve.

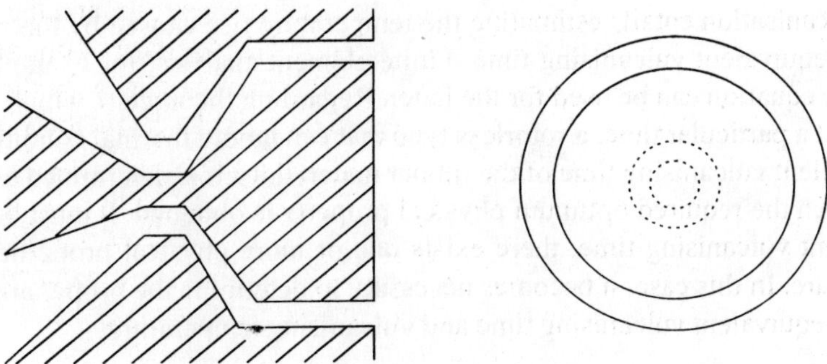

Fig. 13.5. Sketch of expanding mandrel die.

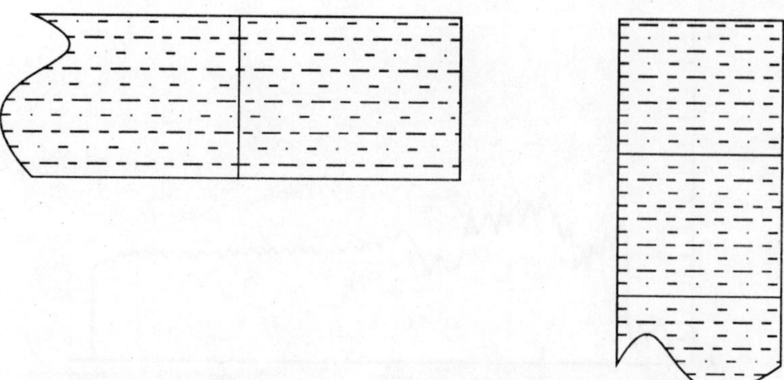

Fig. 13.6. Cutting and joining of uniaxial fibre orientation sheet.

The overlap method and the butt joint method are available for joining material. The overlap method ensures strong bonding and facilitates handling at the time of building, but setting up vulcanising conditions for preventing unevenness of belt thickness and pinhole defects becomes delicate. Strict control on lapped margins is important. In butt joints, because of the low tackiness of SFRR, the establishment of the appropriate joining method and conditions is important for obtaining satisfactory bonding strength. The butt joint is advantageous for stabilisation of belt quality. The joined sheet is supplied to the belt building process in its long length or in a plied-up layer of building thickness, cut to belt length.

Transfer and storage

Problems related to transfer and storage are "scorching" and "tackiness". Scorching is divided into heating time scorch, which occurs during processing, and bin scorch, which occurs during storage.

Another characteristic of limiting the number of storing hours is tackiness. Fig. 13.7 shows an example of the relation between storing hours and tackiness. Loss of tackiness makes it difficult to ply up materials in belt building and also causes a reduction of adhesion between materials.

Preparation for Determining Vulcanising Conditions

In the case of autoclave vulcanisation and press vulcanisation, the point in the belt at which temperature rise becomes slowest by heat conduction mostly exists just below the cords. However, when a cooling process is included, progress in vulcanisation due to remaining heat will become fairly large. Therefore, the proper vulcanising speed for each material is determined on the basis of belt shape and conditions of vulcanisation and cooling. Also, a large difference in vulcanisation rate must not exist between materials touching each other as a result of co-vulcanisation.

Simulation of vulcanisation entails estimating the temperature rise caused by thermal conduction as well as operation of equivalent vulcanising time. Finite element analysis, for example, is used for the former and *Arrhenius* equation can be used for the latter. Regarding rheometers which measure standard vulcanisation speed at a particular time, a rotorless type that can ignore thermal conditions is preferable.

The proper equivalent vulcanising time of the rubber material itself is determined as a function of the conditions under which the required optimum physical property is obtained. It must be noted, however, that besides equivalent vulcanising time, there exists one or more physical properties that depend on vulcanising temperature. In this case, it becomes necessary to determine the proper area within the two-dimensional plane of equivalent vulcanising time and vulcanising temperature.

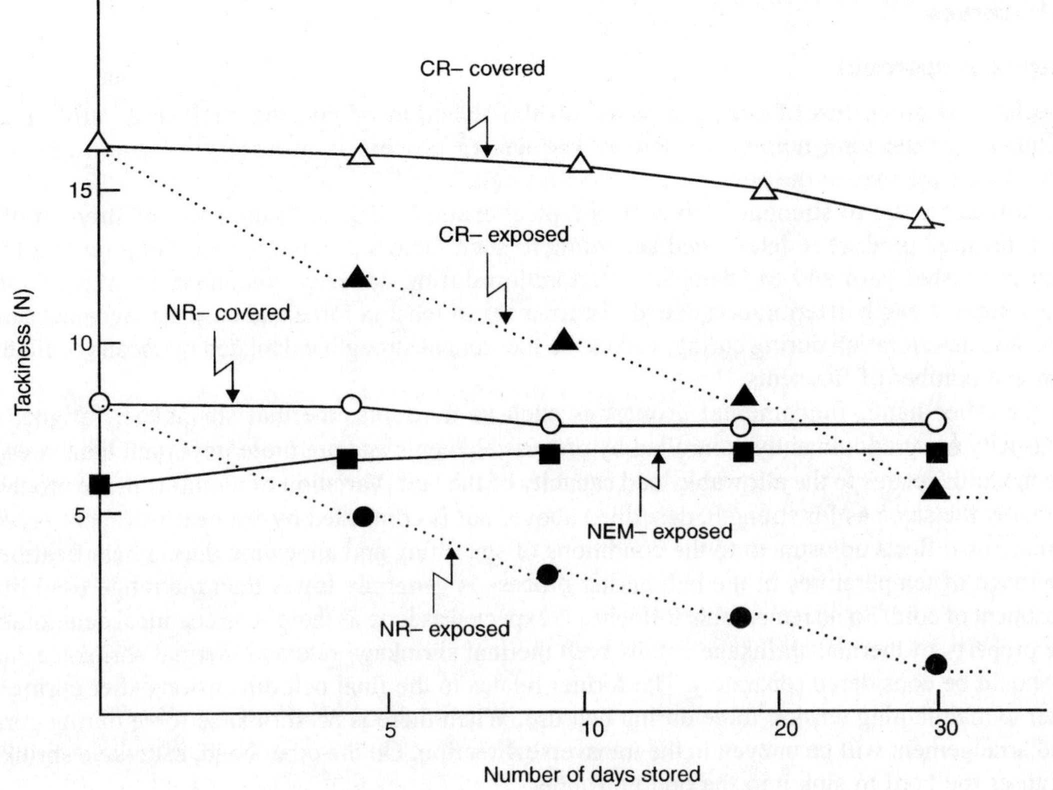

Fig. 13.7. Tackiness of sheet versus storage period.

CORD PROCESSING

Main Points in Cord Processing for V-Belts

Cord processing for V-belts is carried out for the following purposes :

1. To provide mechanical property and adhesion for maintaining belt function.
2. To improve processibility during belt building.

The cord is one of the important components in the basic function of power transmission between pulleys. Belt strength, the force of belt tension, and property changes as the belt is run over time are dependent on the properties of the cord. The friction force raised between the belt and the pulley is also caused by cord tension force, which acts perpendicularly.

These mechanical properties are given by heat treatment, based on the selection of yarn and cord construction, such as the total number of denier and the method of twisting.

Adhesion with the rubber components is achieved by coating with an adhesive and baking. Baking usually is conducted with the same heat treating process used for imparting the mechanical properties.

On the other hand, there is also the function of processibility during building, which makes ensuing processes easier by affording building bindability, smoother demoulding, less fraying during cutting, and so on. Good processibility allows for economical, stable production of high quality goods. It also enhances factors associated with belt performance, such as stability for storage, resistance to heat, moisture, and rubber chemicals during curing, precision of cord position, and finished belt dimensions.

Cord Properties

Mechanical properties

The mechanical properties of cord are based on the selection of raw materials and yarn, and the determination of the total number of denier. The design process is necessary to obtain maximum performance of the yarn in the belt.

The utilisation rate of strength is given as a typical example. The utilisation rate of strength of the cord in a finished product is determined according to such factors as strength deterioration due to the catenary in twisted yarn and to chemical deterioration during adhesive treatment, recovery by heat treatment, total strength deterioration due to disturbances of tension force and cord arrangement during winding, and deterioration during curing, versus the theoretical strength calculated by means of filament strength and number of filaments.

On the other hand, fundamental properties such as modulus, thermal shrinkage, fatigue, and viscoelasticity are predominantly controlled by process elements and are more important than strength.

The modulus relates to the allowable load capacity of the belt. Variation of modulus in the process is conceptually the same as for strength, described above, but is controlled by the heat treatment process. Final modulus reflects adjustment to the conditions of stretching and annealing during heat treatment.

The range of temperatures in the belt curing process is generally lower than the range used in the heat treatment of cord. So no remarkable influence is expected as long as there is no chemical deterioration.

The property of thermal shrinkage entails both thermal shrinkage rate and thermal shrinkage force, which should be considered separately. The former relates to the final belt dimensions after curing and the latter to maintaining tension force during belt use. When there is no shrinkage force during curing, the cord arrangement will be uneven in the transverse direction. On the other hand, excessive shrinkage force causes the cord to sink into the bottom rubber.

Thermal shrinkage force during belt use results from self-generated belt heat and ambient temperature and, for the sake of tension maintenance, offsets tension force deterioration caused by viscoelastic properties (e.g., stress relaxation; crepe). Therefore, a small thermal shrinkage rate and a high shrinkage stress force are desirable, but intrinsic relations exist between the two for each yarn. Thus, it is very important to build suitable property levels, including viscoelasticity, into the heat treatment process. The main control factors in the heat treatment process are tension force, temperature, and time, called 3T.

Fatigue is one of the important properties for belt life, particularly when aramid and glass fibres are used. However, the fatigue mechanism differs from one material to the other; the fatigue mode for aramid fibre occurs as a kinkband on the compression side of bending, whereas breakage in glass fibres is attributable to wear of the filaments. Accordingly, aramid fibre must be treated in such a way that the bending elastic modulus is not increased, and for glass fibre, the filaments must be protected with the adhesion treatment. This process, described next, is important from the standpoint of fatigue.

Adhesion property

The transmission of friction force passes from the pulley to the belt rubber surface and from the belt rubber surface to the cord. V-belt stress concentration in the pulley appears in the underlying rubber area of the cord near the belt sides, and the growth of cracks from this area is one of the factors determining belt life. Therefore, the adhesion between cord and rubber is an important factor in the control of the functions of the belt and belt life. Adhesion affects the dynamic fatigue of the cord itself, as described above, and also affects fraying in the cutting process, which is later described. Moreover, the adhesive itself is expected to act as a protective layer to prevent cord deterioration caused by rubber compounding chemicals.

Bonding strength is determined by chemical bonding force and structural bonding. Chemical bonding is influenced by the selection of adhesion chemicals and occurs during baking. Structural bonding is brought about through the impregnation of the filaments by the chemicals in the adhesive.

Adhesives are generally applied in a two-stage process: a primary treatment with epoxy or isocyanate is followed by a secondary treatment with RFL, which is an adhesion chemical consisting of resorcinol, formalin, and rubber latex. Baking is generally conducted together with heat treatment for mechanical properties, as described above. So a single machine is used, combining adhesion and heat treatments.

Post-processibility

As mentioned earlier, mechanical properties control the cord arrangement and the final dimensions, and adhesion treatment controls fraying during cutting and cord deterioration during curing. A rubber cement coating can be applied to the RFL layer as a means of improving tackiness and preventing RFL deterioration during storage. Particularly in the case of wrapped V-belts, since uncured belts are cut to the specified belt width, a rubber coating is generally applied during the cord winding stage, where strong tackiness is required. RFL-treated cords are greatly affected by ultra-violet rays, as shown in Fig. 13.8. Accordingly, they are packed in a material impervious to ultra-violet rays before being shipped or temporarily stored.

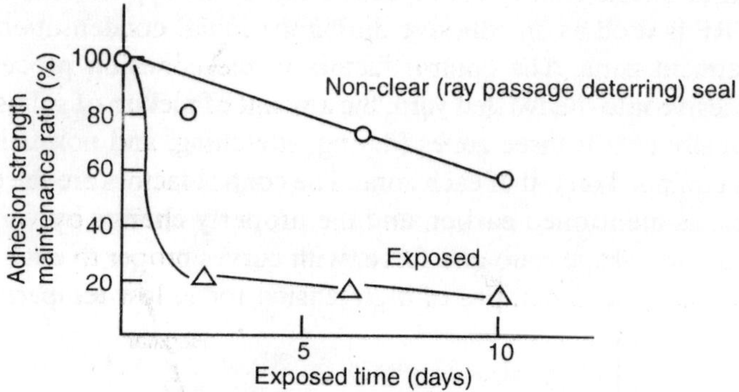

Fig. 13.8. Impact of ultra-violet rays on RFL-treated cord.

Practice of Cord Processing

Selection of yarn

The selection of polyester yarn is determined by the materials' stability toward aminolysis and then its mechanical properties. Aminolysis is the reaction caused by the amine constituent of rubber. Polyester, which has a low concentration of the terminal carboxyl group, is generally advantageous and stable toward aminolysis. Regarding dynamic properties, high speed, spun POY-type yarn offers the advantage of small hysteresis loss, dimensional stability, maintaining tension force, and antifatigue properties. From the point of view of microstructure, POY yarn has low molecular orientation in the amorphous region, is short in the amorphous phase in the fibre axis direction, and has a large distance between crystals in the right angle direction.

Yarn twisting

The twisted construction of yarn plays a role in mechanical bending fatigue, modulus, and elongation. A higher number of turns gives lower modulus, larger elongation, and better resistance to bending fatigue.

The three methods of yarn twisting are single twisting, plied yarn, and Lange's ply. Plied yarn is generally used because of its stability during belt operation and the stability of the cord winding pitch during belt building. Also, good shape stability is given by combining the primary and final twisting coefficients. An ordinary yarn twisting machine is used.

The occurrence of the catenary defect in the twisting process has a strong influence on aramid and glass fibre yarns in that it will decrease the utilisation rate of strength and fatigue resistance, since these yarns cannot be expected to exhibit recovery effects after heat treatment.

Adhesion and heat treatments

Adhesion and heat treatments are carried out by a cord treating machine combined with a heat setter. The cord treating machine normally consists of three zones of repetitive processes of impregnation with adhesive followed by heating.

Adhesion treatment involves pre-treatment by epoxy or isocyanate, then treatment with RFL. Vinyl pyridine (VP) is used as the latex part of RFL for NR, or SBR. VP or CR is used for CR, and other latexes, (NEM, CSM, NBR, etc.) are used for NEM. Other examples include abolishing the pre-treatment process by adding a chlorophenol derivative in RFL and adding amine intake components, such as carboxylated latex in RFL, for the control of polyester aminolysis. Typical RFL compositions are R/F = 1:1.2 and RF/L = 1:6. RF is used as an adhesive during the initial condensation stage, and it is cross-linked in the heat treatment zone. The control factors in the adhesion processes are the extent of impregnation of the adhesive into the twisted yarn, the amount of pickup of solids, and the reaction rate.

Heat treatment basically entails three zones (drying, stretching, and normalising), and a system of adhesion application is equipped on top of each zone. The control factors are the three Ts (tension force, temperature, and time), as mentioned earlier, and the property change by variation of 3T is shown in Fig. 13.9. The thermal shrinkage ratio and force, with curves proper to each yarn, become large as treatment is conducted under the conditions of high tension force, low temperature, and short time.

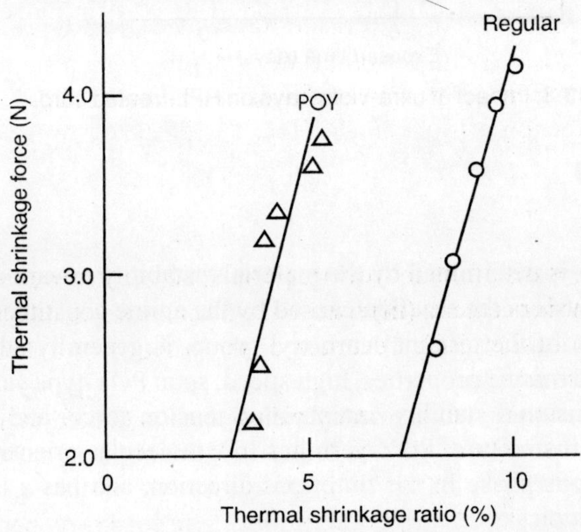

Fig. 13.9. Relation between thermal shrinkage force and thermal shrinkage ratio.

Post-processing

Cementing is continuously carried out in the adhesion heat treatment process. Adhesion rubber dissolved in solvent usually serves as the rubber cement, and a tackifier is added, if needed.

The treated cord is immediately packed in material impervious to ultra-violet rays and shipped or stored dry, under constant temperature conditions. If the storage temperature increases, shrinkage will occur, causing a change in the mechanical properties of the high stretching yarn.

Important factors are tension control during cord winding in the building process for cord arrangement, demouldability after curing, and the finished dimensions of the belt. Moisture entering the cords during curing obstructs the adhesion and causes cord deterioration. Furthermore, it is necessary to demould the cured belt after cooling to a temperature lower than the glass transition temperature region of the cord material.

CANVAS PROCESSING

Biasing and Lapping

Examples of methods of manufacturing bias cloth (warp and weft intersecting angle being 90°) are shown in Fig. 13.10(a). Method (a)-1 consists of trimming the selvage part of the rubber-coated plain woven cloth, cutting in a 45° direction, and joining the trimmed parts in succession. The overlap system is usually adopted for joining, but each company has its own proprietary approach to ensure accuracy of lap margin, cutting (trimming) angle in the direction of thickness, and so on. In method (a)-2 the tubular woven cloth, having no edges, is opened 45° in the direction of the width to produce biased cloth having no joint. The jointless cloth has, however, such defects as fixed canvas width, disordered yarn at the selvage, and weaving speed; in addition, frictioning in which tension is applied to the longitudinal direction is impossible. Therefore, method (a)-2 is used only for coating.

An example of the method of manufacturing wide-angled bias cloth (warp and weft intersect angle ≈ 120°) is shown in Fig. 13.10(b). Method (b)-1 consists of altering the texture by applying shear strain on an ordinary plain woven fabric in the direction of the width with high tension. To prevent the angle from reverting when tension is removed, the degree of angle is fixed at stage (ii) by carrying out, for example, an adhesive treatment. In method (b)-2, bias cloth is made at a 90° intersecting angle obtained at (a)-2 wide-angled by stretching it in the direction of the width by using a stretching tenter.

Cord fabric is subjected to adhesion and heat treatments as in the case of cord and is fixed by rubber topping. Then it is cut in the direction of right angles [Fig. 13.10(c)] in conformity with the building width of the belt.

Rubber Coating

Canvas is coated with rubber by means of the following: (i) an adhesive treatment using (e.g.) RFL; (ii) coating (soaking, coating) with rubber cement; (iii) frictioning by calender; and (iv) topping by calender. For materials that require high adhesion, RFL treatment or the addition of a direct bonding adhesive (HRH) system to coating rubber is adopted.

As stated above, treatment of canvas that is already bias shaped is limited to methods 1 and 2, and such a canvas as-is, maintaining the angle of yarn by tenter. The hot air oven or heat-drum systems are used for dryers, and coating is carried out mainly by means of a knife coater and a roll coater.

Soaking rubber is required to have penetrability between fibres, and coating rubber must have thixotropy and must not scorch during drying.

Frictioning causes elongation in the warp direction as a result of the application of shear stress to the canvas, with resultant decrease of weft density and decrease of elongation of warp. Therefore, to maintain the balance of warp and weft directions in the final biased canvas, a griege fabric (weaving) design in consideration of changes in processing is important.

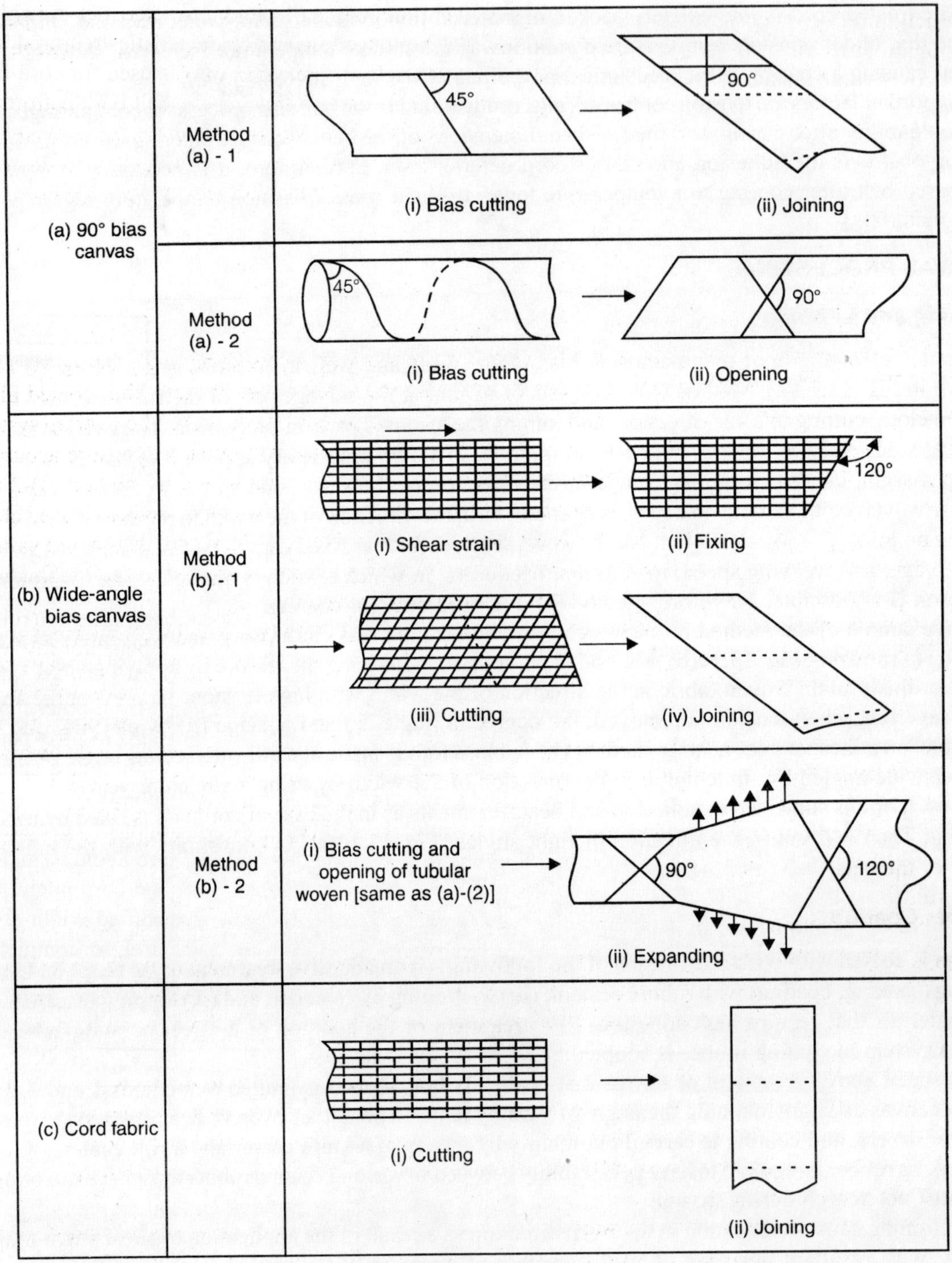

(a) 90° bias canvas	Method (a) - 1	(i) Bias cutting	(ii) Joining
	Method (a) - 2	(i) Bias cutting	(ii) Opening
(b) Wide-angle bias canvas	Method (b) - 1	(i) Shear strain	(ii) Fixing
		(iii) Cutting	(iv) Joining
	Method (b) - 2	(i) Bias cutting and opening of tubular woven [same as (a)-(2)]	(ii) Expanding
(c) Cord fabric		(i) Cutting	(ii) Joining

Fig. 13.10. Canvas biasing and lapping method.

Friction rubber is designed by taking into consideration both compound characteristics suitable for frictioning process, such as low viscosity and low green elongation, and characteristics required in the final cover rubber, such as wear resistance and shear strength. Topping not only is used for cord fabric but also is applied to open side of coated and frictioned canvases to simplify the belt building process.

METHOD FOR PROCESSING WRAPPED V-BELTS

The process flow for wrapped V-belt production is shown in Fig. 13.11.

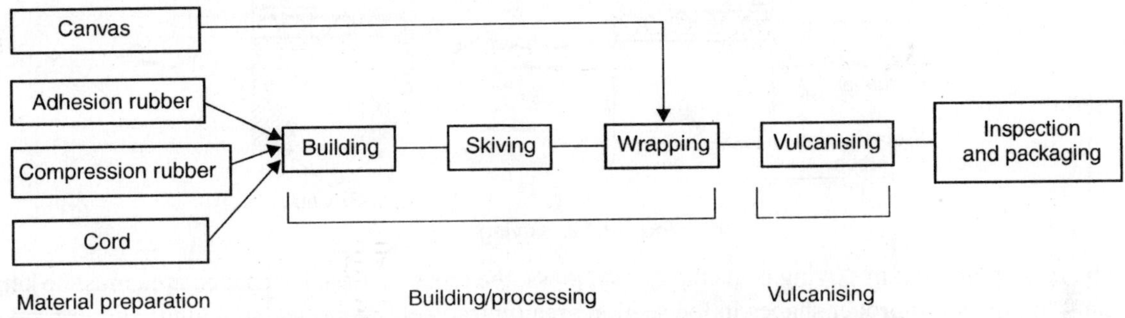

Fig. 13.11. Manufacturing process for wrapped V-belts.

Building

Building is accomplished by plying and/or wrapping rubber sheets (compression rubber, adhesion rubber), cord, and canvas. There are building machines designed exclusively for wrapped V-belts, for raw-edged belts, and for V-ribbed belts, which are described later. Wrapped V-belts are made by placing, plying, and winding the component materials on an expanding mandrel, which is different from other types of belt making. The expanding mandrel can accommodate a certain range of belt lengths in its expanding mechanism. Rubber sheets, cord, and cover canvas are plied on the mandrel, rotating at a specified speed. The applicable belt length of each expanding mandrel number is given in Table 13.5.

Plying, winding, and wrapping the materials evenly and minimising the amount of air entrapped between the layers are the keys to good building work. It is particularly important to avoid wrinkles on the rubber or canvas sheets. It is also important to control the specified tension and cord pitch.

Upon completion of layering (building), the layered material is cut to the specified width at right angles to the mandrel. Then the cut pieces are removed from the deflated mandrel, to complete the building process.

Table 13.5. Mandrel diameter and belt length.

Mandrel	*Diameter (mm)*	*V-belt inside length (mm)*
1	160-195	580-675
2	180-225	645-770
3	210-270	740-910
4	255-330	880-1100
5	310-400	1050-1320
6	380-500	1270-1635
7	480-625	1585-2025
8	605-780	1980-2515

Skiving

Skiving is a process to further cut the square pieces into V-shaped cross-sections. Fig. 13.12 shows the general concept of skiving. The square, unfinished belt is placed inside out around the two pulleys of the skiving machine. Two rotating cutters, with angles set, cut the running belt into a V-shape.

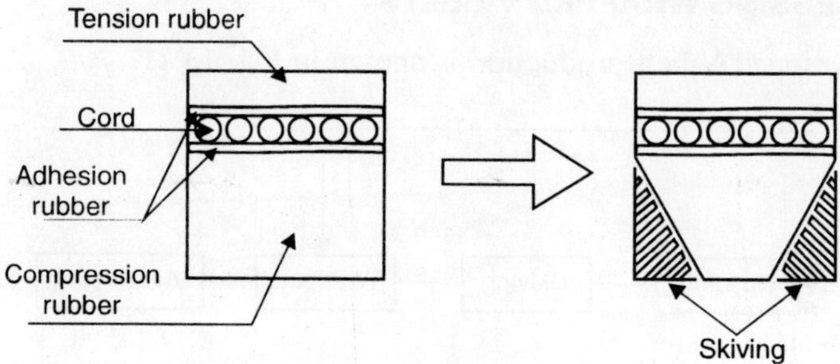

Fig. 13.12. Skiving.

The important point in skiving is in cutting exactness: the cross-sectional V-shaped area must be kept constant to prevent improper shapes in the section area (entrapment in mould, sponging) and improper cord alignment due to rubber flow during vulcanisation.

It is almost impossible to check the sectional shape of all belts, so as a practical alternative, the weight of each belt is measured at the production site. A weight allowance of ± 1.5% seems to be generally acceptable. Skiving angle varies by belt type; so it is necessary to adjust the angle of the cutting blade to accommodate belts of various types. A cutting angle range of 40–60° is popular.

Wrapping

After skiving, the V-shaped belt is wrapped with highly flexible rubber-impregnated canvas. This canvas is usually cotton fabric, biased in the longitudinal direction of the belt. Some high quality belts use 120° biased canvas for greater flexibility.

The most important point at this stage is to wrap the canvas evenly and tightly around the whole circumference. The overlapping width at the material ends is also an important characteristic of the finished belts. To maximise lap joint strength and belt flexibility, an overlap of 3–10 mm is generally used. Fig. 13.13 shows the wrapping processing and a typical wrapping machine is shown in Fig. 13.14.

Vulcanising

Type of vulcanisation

In the vulcanising stage, heat and pressure are applied to the belts after building for performance and shape.

Steam, electricity, oil, and hot air are among the heat sources used. Steam is the most popular and economical for belt manufacturing, as for other rubber products. To develop pressure, oil, air, and steam have been used. Since autoclave-type vulcanisation is predominantly used in belt production, just as steam is the most popular source of heating, steam is also generally preferred as a pressure source. Table 13.6 gives general grouping of vulcanising methods for V-belts and fan belts.

Canvas pressing
Canvas folding
Canvas wrapping
Canvas cutting
Finishing

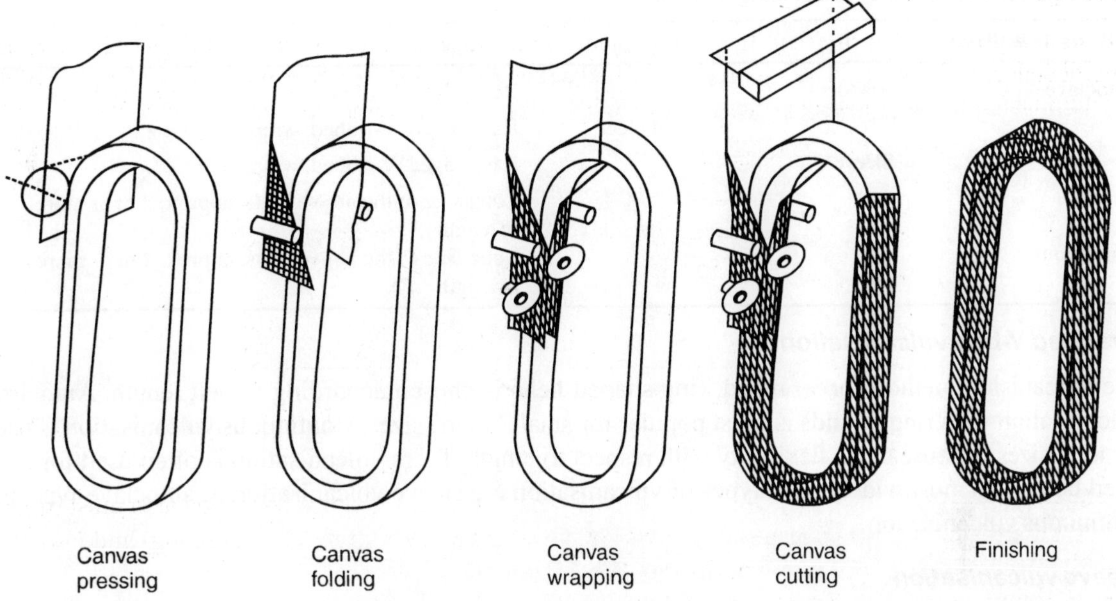

Fig. 13.13. Wrapping process.

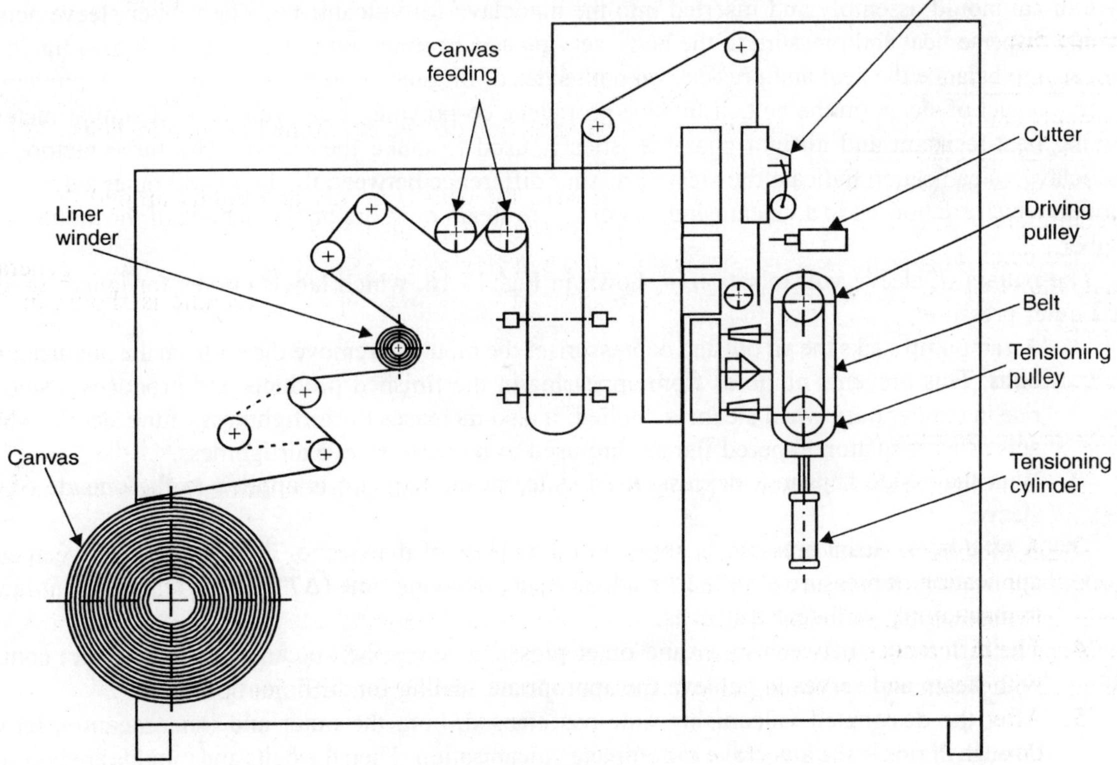

Canvas guide
Canvas feeding
Cutter
Liner winder
Driving pulley
Belt
Tensioning pulley
Canvas
Tensioning cylinder

Fig. 13.14. Wrapping machine.

Table 13.6. Classification of vulcanising method.

Vulcanising method			*Belt type*
Autoclave	Bladder	Outer bladder	Raw-edged, V-ribbed
		Inner bladder	Raw-edged, V-ribbed, wrapped V
	Sleeve		Raw-edged, V-ribbed, wrapped V
Press			Long size of the above belts, approx. 2 m or more in length
Continuous			Long size of the above belts, approx. 2 m or more in length

Wrapped V-belt vulcanisation

The vulcanising method for wrapped, ring-shaped belts is chosen according to belt length. Autoclave vulcanisation with ring moulds is most popular for small (short) sizes. Continuous vulcanisation is used for long sizes because of its flexibility with respect to length. Press vulcanisation is often used for mid-sized belts. The most widely used types of vulcanisation are sleeve vulcanisation of autoclave-type and continuous vulcanisation.

Sleeve vulcanisation

As shown in Fig. 13.15, sleeve vulcanisation begins with 10–50 pairs of built-up, ring-shaped belts and ring moulds, piled, one after another, in succession. They are tightened with flanges at the top and bottom, using the flange tightening bolts provided. Then the rubber sleeve is slipped over the piled-up cylindrical mould assembly and inserted into the autoclave for vulcanising. The rubber sleeve acts to evenly disperse heat and pressure to the belts, serving as a heating and pressing plate. It also functions as a seal to balance the heat and pressure on both sides of the sleeve itself, at the same time, preventing direct contact of steam on the belts. Ethylene propylene co-polymer, butyl rubber, or a similar material having heat-resistant and airtight characteristics is used to make the sleeve. The three methods of autoclave vulcanisation indicate the steam pressure difference between the inner and outer sides of the mould, which are housed in a vulcanising vessel where steam pressure on the outside of the mould is set higher.

The pattern of sleeve vulcanisation is shown in Fig. 13.16, which labels curves for inner pressure and outer pressure:

1. An air pump sucks the air out and depressurises the mould to remove the air from the unvulcanised belts. This prevents pinholes from appearing in the finished products and produces a smooth rise in temperature when steam is applied; it also increases the airtightness of the sleeve, which acts as an insulator. Tapered flanges are used to increase sleeve airtightness.
2. With the inside kept in a depressurised state, steam pressure is applied to the outside of the sleeve.
3. A little later, steam pressure is applied to the inside of the sleeve. The time gap between each application of pressure is called "inside pressure delaying time (ΔT)" and is an important factor in maintaining well-sealed sleeves.
4. The differences between inner and outer pressures saves the vulcanisates from direct contact with steam and serves to achieve the appropriate sealing for sufficient pressure.
5. After the designated vulcanising time has elapsed, both the outer and inner steam is let out through drains in the autoclave to complete vulcanisation. Then the belts and moulds are removed from the autoclave.

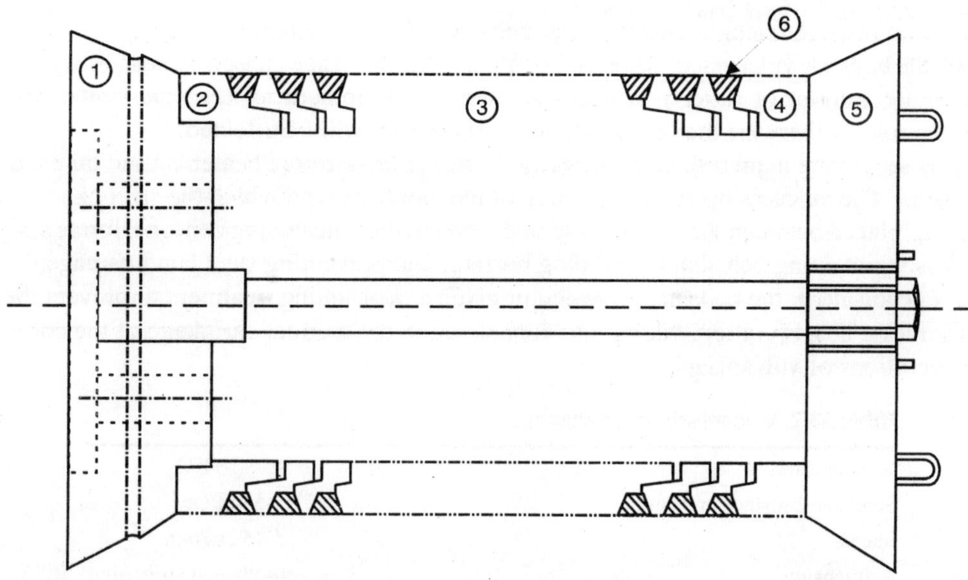

Fig. 13.15. Ring mould. 1. lower flange; 2. lower ring; 3. middle ring; 4. upper ring; 5. upper flange; 6. unvulcanised belt.

The explanation above is for sleeve vulcanisation. Bladder vulcanisation is quite similar, but substitutes a bladder for the sleeve. The sleeve system depends on the sleeve and pressure difference to prevent the belts from exposure to the steam, whereas the bladder system is an improved, safer system, because the steam is contained in the bladder.

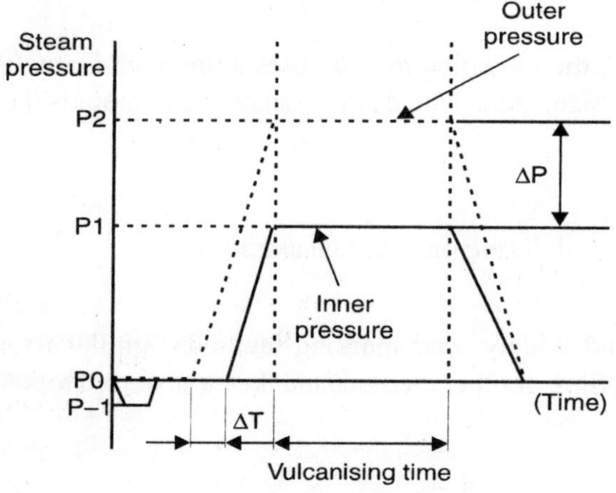

Fig. 13.16. Pressure and time chart for sleeve vulcanisation.

Continuous vulcanisation

Continuous vulcanisation is popularly used for conveyor belts and rubber sheets, and also for long V-belts. When making V-belts using pressurised autoclave vulcanisation, the belt length is determined by the mould size. Continuous vulcanisation on the other hand, is suitable for making comparatively long belts. Since this type of vulcanisation uses two rolls on which the belts are loaded, it clearly offers flexibility with respect to length. With several belts running on the main (V-mould roll) and tension rolls in parallel, the belts are vulcanised. To resist and avoid cord shrinkage caused by heat during vulcanisation, an appropriate amount of rigidity is necessary for the equipment to keep the centre distance fixed between the rolls. Otherwise the length of the finished belts will be affected.

Steam is sent to the main roll, and an electric heater or far-infrared heater is used on the other side of the steel band. The vulcanising zone is the area of the mould around which the steel band is wrapped. A rubber sheet, placed between the belt and the steel band in the vulcanising zone, performs such important functions as maintaining belt shape, avoiding burring, and preventing steel band damage.

After vulcanisation, the belts are annealed or given a cool setting treatment to prevent the cord from stretching during belt operation, taking into consideration the thermal shrinkage of the cord. Table 13.7 lists the conditions of vulcanisation.

Table 13.7. Vulcanisation conditions.

Temperature	150-160°C
Steel band surface pressure	3-5 kgf/cm^2
Speed	5-15 cm/min
Belt tension	0-5% belt stretching
Cold setting	50-60°C or lower

Marking

Methods for marking brand name, type, size, and other informations on the belt surface are briefly described in the sub-sections that follow.

Moulding method

Also called "stencil marking", the moulding method uses a thin metal plate that has been engraved or pressed with the brand name (size, etc.). This device transfers and moulds the letters or marks onto the belt during vulcanisation.

Printing method

Printing is a popular method used during or after vulcanisation.

Transfer label method

This is the most common and widely used marking methods, similar to moulding. The necessary symbols are printed on paper, film, or other medium and then transferred to the belt during vulcanisation by way of heat and pressure.

Stamping method

Using a quick-drying ink, marking is stamped on the vulcanised belt surface with a pad.

Screen method

In this well-known method, ink is rubbed onto the vulcanised belt through a fine-meshed marking plate.

Ink jet method

As in office printers, quick-drying ink is jetted onto the vulcanised belt by electric signal, with many tiny dots forming the letters. This method is expected to gain popularity in the future.

METHODS FOR PROCESSING RAW-EDGED V-BELTS

Raw-edged belts are made by the process shown in Fig. 13.17. The material plied up on the mandrel for raw-edged belts is similar to that of wrapped V-belts. Unlike the process for wrapped V-belts, which are cut before vulcanisation, the materials plied on the mandrel are vulcanised as-is, then cut to width for raw-edged belts. With raw-edged belts the slabs are vulcanised prior to cutting, which results in long belt life and high reliability due to improvement of positional stability of each material.

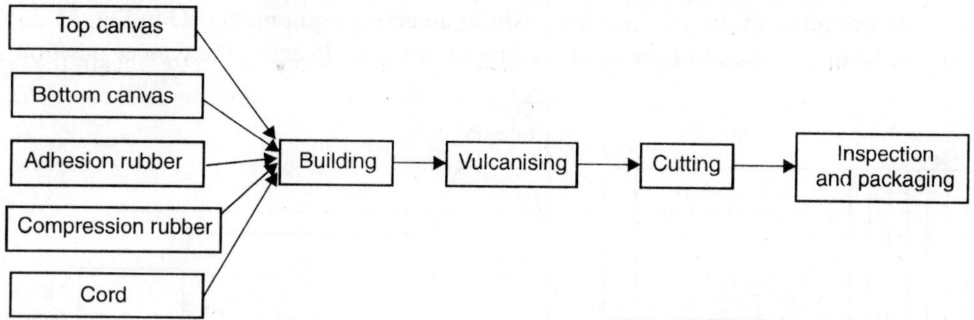

Fig. 13.17. Raw-edged belt manufacturing process.

Building

Building is basically the same as for wrapped V-belts. The main difference is that the rubber sheet, cord, and canvas specified for each belt length are wound directly on the mandrel and then vulcanised. Therefore, there is no need for an expanding mandrel, nor are the processes of skiving and wrapping necessary, as in the case of wrapped belts. One of the most important functions in the building stage is the control of cord tension, which is described below.

The tension applied to the cord is an important factor in belt quality. During the building process, the tensioned cord is wound on the rubber sheet material at a specified cord pitch. The rubber sheet material has been made and finished to a certain thickness in the sheeting stage and rolled onto the mandrel as the first step of the building phase. The following points are important:

1. If the tension fluctuates during winding, it will affect cord alignment and result in belt vibration and/or reduced belt life.
2. Belt length is affected by the amount of cord tension. Higher tension makes the finished belt shorter and lower tension makes it longer.
3. Overtensioning the cord will affect the post-vulcanisation demoulding process.
4. Tension deterioration during belt running is caused by improper cord treatment and/or incorrect tensioning during the building stage.

The optimal belt tensions are determined by belt type. Winding the cord in a stable manner with the determined tension is one of the most important factors in ensuring belt quality.

As shown in Fig. 13.18, cord tension tends to be higher at the start of cord winding (accelerating period) and lower at the end (decelerating period). This effect is attributed to a mechanical loss in the pulleys used in the cord feeding system. For this reason, and also because of the heat characteristics at, and insufficient flow of rubber to, both ends of the mould, both ends of the sleeve are, in fact, removed as scrap. In other words, reducing the accelerating and decelerating time and keeping constant tension on the cord are two very important points for maintaining quality and also productivity.

The schematic drawing in Fig. 13.19 and the control block diagram in Fig. 13.20 can be used to explain a typical tension control system. The cord is wound on the rubber, which has already been rolled onto the mandrel attached to the main shaft. In recent years, as AC servomotor has been used to control accelerating/decelerating time.

Cord tension is arrived at by the amount of weight placed on an accumulator (referred to as a "dancer"). The tension is changed by changing either the weight or the air pressure, using a precision air cylinder. Of the two methods, changing the weight is the more popular. If main shaft rpm increases for any reason, the dancer goes up. In this case the position detecting equipment (SG in Fig. 13.20) locates the dancer to accelerate the speed of the cord feeding motor, thus keeping the dancer position fixed.

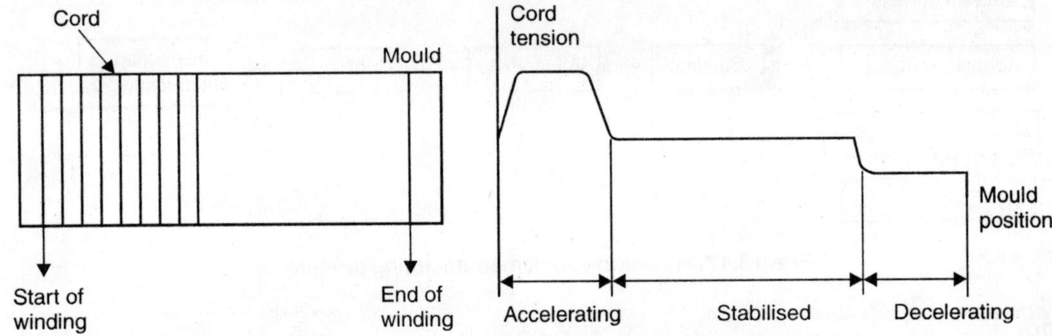

Fig. 13.18. Fluctuation of cord winding tension.

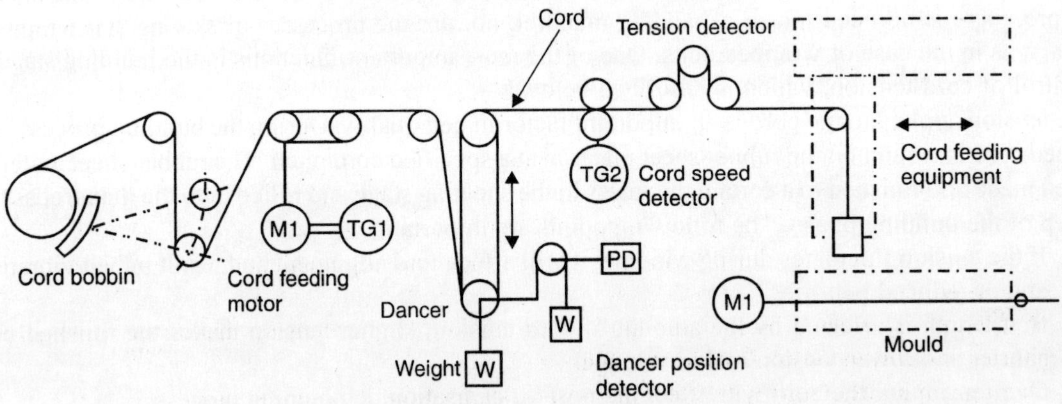

Fig. 13.19. Cord tensioning system.

With a dancer, a special calculator is provided to control the various conditions. An amplifier connected to the calculator provides synchronisation between the electrical and mechanical systems to prevent hunting of the dancer, which affects the cord tension consistency. There is also a cord tension detector in the cord feeding equipment. Since mechanical loss in the cord feeding equipment (Fig. 13.19) moves the dancer up and down, it is important to minimise such mechanical loss. Proper control of cord tension is a key factor not only in raw-edged belts but also in making quality wrapped V-belts and V-ribbed belts.

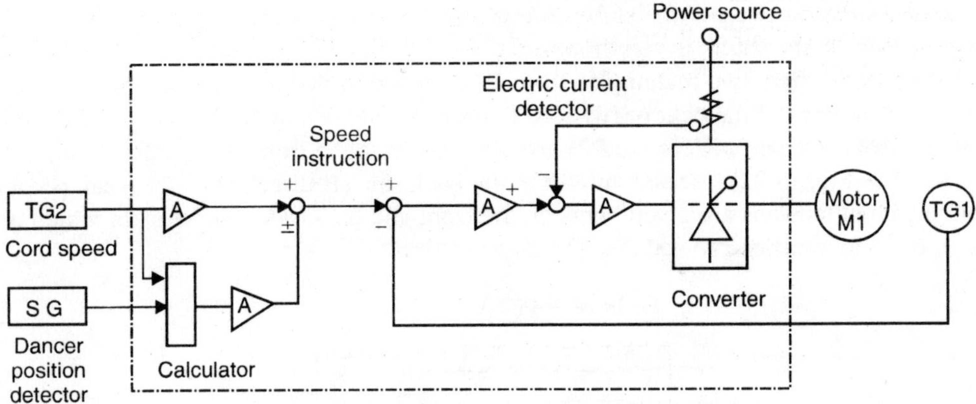

Fig. 13.20. Block diagram of tension control.

VULCANISATION

Vulcanisation is similar to the process for wrapped V-belts described above.

Cutting

After vulcanisation, the mould is cooled and the slab is demoulded to be cut into the designated belt width. A description of the cutting process follows. Fig. 13.21 outlines the raw-edged V-belt cutting classification.

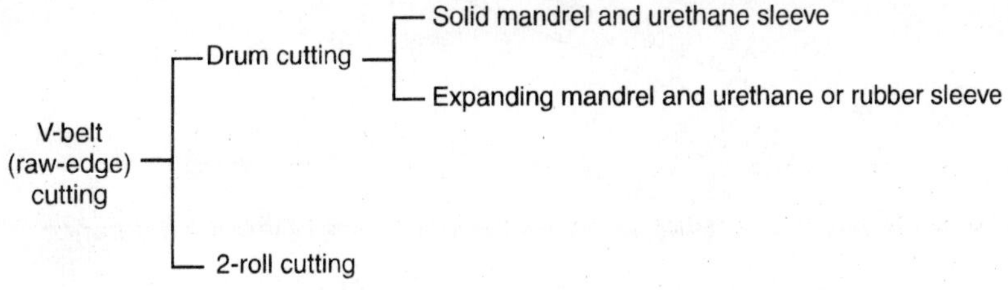

Fig. 13.21. Raw-edged V-belt cutting classification.

Drum cutter

In the make-up of a drum cutter, the slab is mounted on a vulcanised rubber (or urethane) sleeve, which is mounted on a mandrel (solid or expanding) so that the slab is given tension and securely set. The mandrel is driven by the main shaft motor. Two cutters move forward to cut the belt at the angle specified, with a time gap between cuts. When cutting is finished, the equipment senses the cylinder's receding movement and the cutter base is pushed forward by the specified pitch for further cutting.

Two-roll cutting

A two-roll cutter consists of a main roll, a tension roll, and two round cutting blades. The already vulcanised slab is put onto the main and tension rolls, and tension is applied. Then the slab is cut by the two round cutting blades, which have been set at a specified angle. A load cell (side pressure) or a differential transformer (position) is located at the slab end to detect and control lateral slab movement for cutting to the specified width.

In the control system for two-roll the slab is moved by the deflection of one shaft. The technique of repositioning the slab on the shafts is very important.

The two-roll system offers two techniques to properly position the slab. These entail using: (i) the difference in circumference of the belt; or (ii) cord tension. In the case of cord tension, when tension is deflected with its front side up (a in Fig. 13.22), the slab moves to the front side (A direction). With the front side down (b in Fig. 13.22), the slab moves to the back side (B direction). The weak point of this system is that different tensions are given between the front and back sides of the slab, which result in slab slackness, making it difficult to stabilise the angle setting.

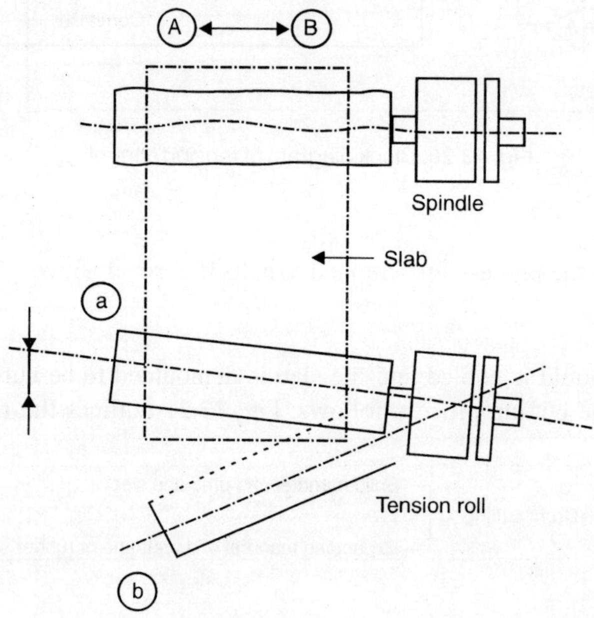

Fig. 13.22. Changing slab position on two-roll cutter by use of tension difference.

Detection of slab position and feedback control

The detection of either the position of the slab edge or the unbalanced pressure given on the slab edge activates feedback control to determine slab position. Fig. 13.23 shows slab position control achieved by use of slab travel distance.

Belt cutting conditions

In drum cutting and two-roll cutting, belt cutting accuracy greatly depends on the difference between the speed of the mandrel and the speed of the cutting blade (a function of their relative circumferences) and the feed-in speed of the cutting blade into the belt. Productivity will be affected by low speed operation, and heat and accuracy problems can be expected in high speed operation. Experience dictates the use of a cutting blade like the one shown in Fig. 13.24 and the cutting conditions listed in Table 13.8.

Table 13.8. Cutting conditions.

Peripheral speed of main shaft mandrel	200–1000 m/min
Peripheral speed of cutting blade	50–120 m/min
cutting blade feed-in speed	0.2–0.8 mm/rev

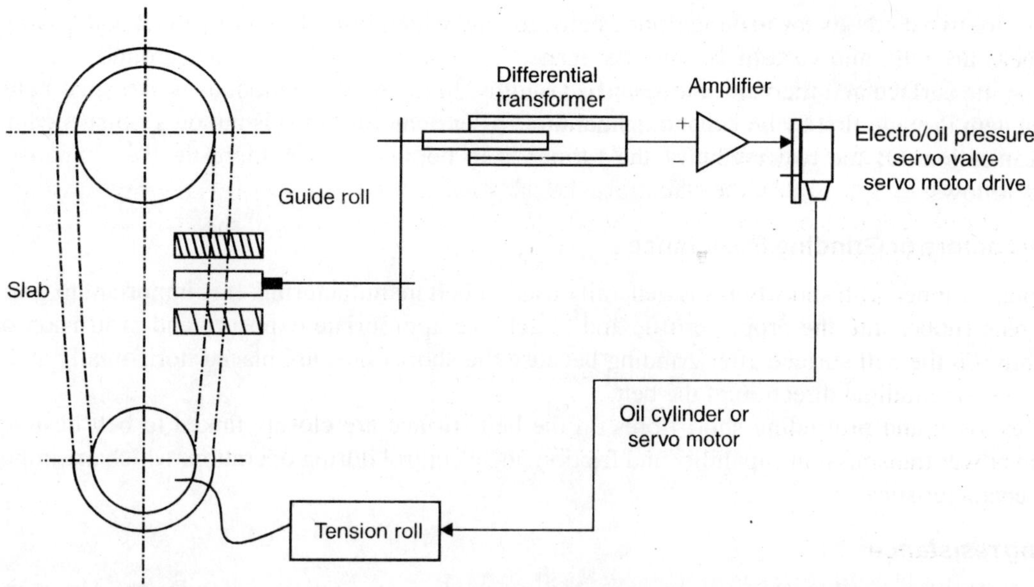

Fig. 13.23. Control of slab position on two-roll cutter by use of slab travel distance.

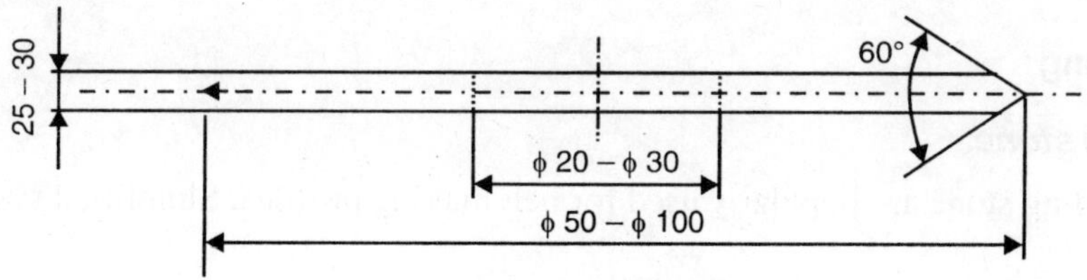

Fig. 13.24. Cutting blade.

METHOD FOR PROCESSING V-RIBBED BELTS

The process flow for ribbed belt production is shown in Fig. 13.25. The remarkable process is grinding; the other processes are much the same as for wrapped V-belts and raw-edged V-belts.

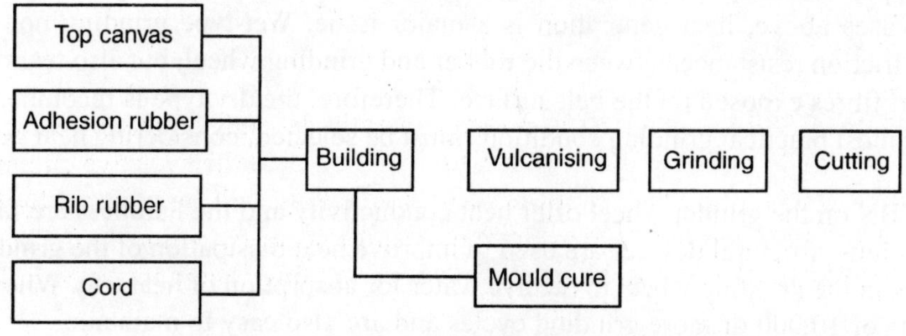

Fig. 13.25. V-ribbed belt manufacturing process.

There are two methods for making ribbed belts. In one, which resembles raw-edged belt production, ribbed belts are built and vulcanised on a flat mould. After demoulding, they are ground to give a rib profile on the surface and then cut into specified widths. In the other method, as in wrapped belts, the moulded item is made first by an expanding mandrel. After demoulding, it is put on a rib-shaped mould for vulcanisation. Of the two methods, the former is in popular use. A more detailed expansion of grinding follows.

Control Factors of Grinding Resistance

Chloroprene rubber with short fibres is generally used in belt manufacturing. It is important to grind the chloroprene rubber into the proper profile and to achieve appropriate exposure and protrusion of the short fibres on the belt surface after grinding because the short fibres are placed horizontally at a right angle to the longitudinal direction of the belt.

The exposed and protruding short fibres on the belt surface are closely linked to belt design with regard to power transmission capability and friction noise control during operation, which are important quality characteristics.

Grinding resistance

Unlike metal grinding, the rubber will become concaved during grinding, following the Hertz contact theory, resulting in friction resistance, deformation, loss resistance, etc. Grinding resistance refers to the sum of the above-mentioned theoretical resistances and the force needed for grinding.

Practice of Grinding

Types of grinding stone

Three types of grinding stone are popularly used for belt making profiles. Simplified systems are shown in Fig. 13.26.

Their features can be briefly described as follows:

1. Grinding grains are glued on the profile core metal, where grain wear or deterioration of the belt surface finish is regarded as ending the life of the grinding stone (e.g., grinding wheel with electrodeposited diamond grains).
2. When grinding performance becomes impaired, or the grinding stone loses its profile, it can be re-shaped by tooling and dressing (e.g., conventional GC, WA, etc. grains).
3. Between-type, with easy maintenance (e.g., cubic boron nitride (CBN) grinding stone).

In any of the cases above, heat generation is a major issue. Wet-type grinding not only greatly increases surfaces friction resistance between the rubber and grinding wheel, but also tears a substantial amount of the short fibres exposed on the belt surface. Therefore, the dry type is recommended. In any event, the best and most practical grinding conditions must be selected, considering heat generation and productivity.

Diamonds or CBN on the grinder wheel offer heat conductivity and the hardness are far superior to other materials. Various structural devices are used to improve heat dissipation of the grinding wheel or to provide a hollow in the grinding wheel to receive water for absorption of heat, etc. Wheels made this way have a lifetime of 10,000 or more grinding cycles and are also easy to maintain.

Fig. 13.27 shows an example of grinding conditions or a diamond wheel that also employs a cooling structure. The temperature of the grinding stone and belt during grinding can be maintained at 60°C or lower. Incidentally, 60°C is a safe temperature, which does not damage the belt.

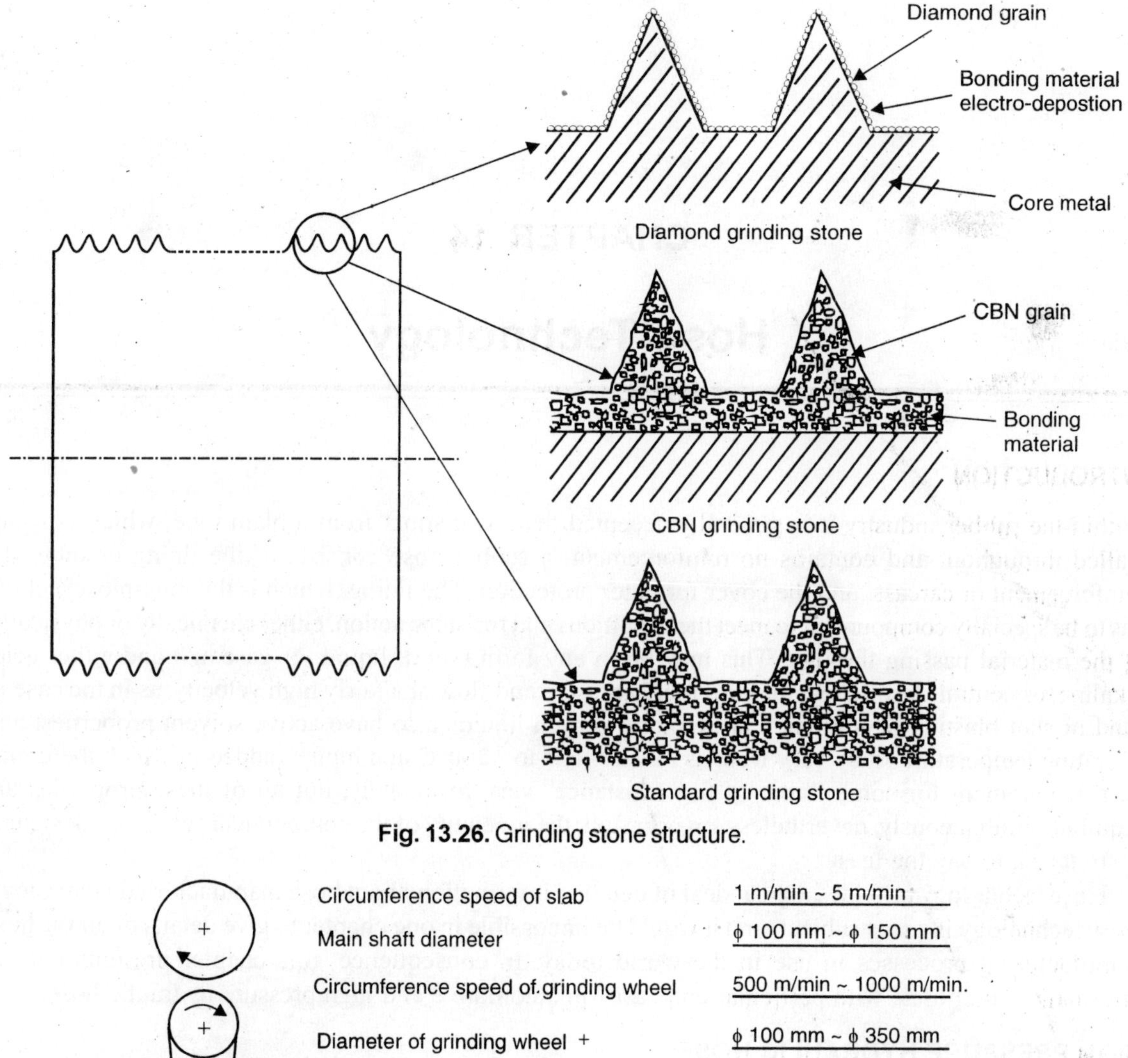

Diamond grain
Bonding material electro-depostion
Core metal
Diamond grinding stone
CBN grain
Bonding material
CBN grinding stone
Standard grinding stone

Fig. 13.26. Grinding stone structure.

Slab

Circumference speed of slab	1 m/min ~ 5 m/min
Main shaft diameter	ϕ 100 mm ~ ϕ 150 mm
Circumference speed of grinding wheel	500 m/min ~ 1000 m/min.
Diameter of grinding wheel +	ϕ 100 mm ~ ϕ 350 mm

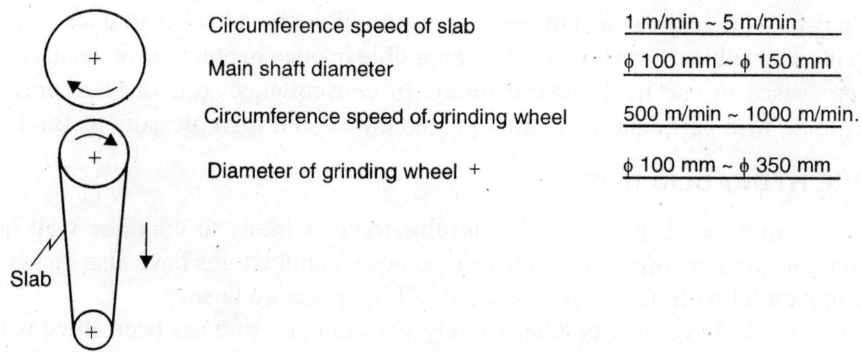

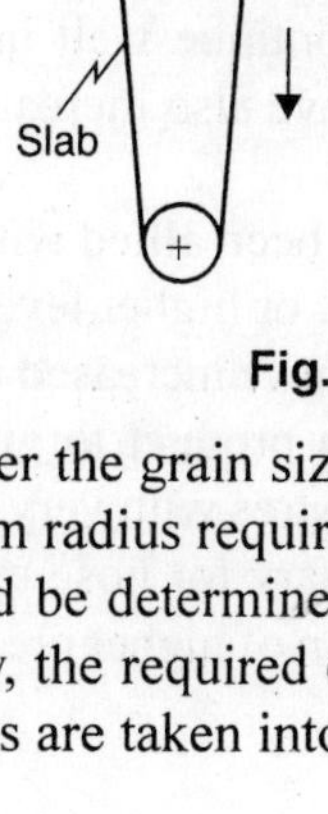

Fig. 13.27. Example of grinding condition with diamond wheel.

The larger the grain size of the grinding stone, the smaller the amount of heat generation, provided the minimum radius required by the item to be ground is accommodated. The grain size of the grinding stone should be determined also by the surface coarseness, as well as the state of short fibre cutting. Accordingly, the required characteristics of the belt should be the determining factors. When all these requirements are taken into consideration, #100–#170 grain sizes are generally used.

CHAPTER 14

Hose Technology

INTRODUCTION

Within the rubber industry it is generally accepted that, as distinct from a plain tube, which is solid-walled throughout and contains no reinforcement, a rubber hose consists of the lining or tube, the reinforcement or carcass, and the cover for outer protection. The lining, which is the innermost section, has to be specially compounded to meet the conditions and resist the action, either chemically or physically, of the material passing through. This may be in any form (solid, liquid, or gaseous) and either acid, alkaline, or neutral. A solid may be extremely abrasive and flow at a fairly high velocity, as in the case of sand or shot blasting operations. Furthermore, certain liquids also have active solvent properties, and operating temperatures may vary from as low as –65° to +200°C and higher, added to which there may be a requirement for some flame or fire resistance. Very fortunately, not all of these properties are required simultaneously, nevertheless, on occasion, the ingenuity of the compounders and hose designers is stretched, to say the least !

Hose technology provides a great deal of detailed information about hose manufacturing technology. Hose technology is a vast subject, and it would be impossible in one chapter to give details of all the hose manufacturing processes in use in the world today. In consequence, this chapter presents current information on hoses, with particular emphasis on automotive and high pressure hydraulic hose.

HIGH PRESSURE HYDRAULIC HOSE

In recent years hydraulics has grown steadily, and this trend is likely to continue well into the next century. The technological demands made on hydraulic hose manufactures have also increased and will intensify further in parallel with market requirement of high pressure hoses.

In many cases pressures have risen or, alternatively, constant pressure has been allied with increased bore size, both with a view to giving faster response times, greater work rates, or higher levels of power transmission at point of need. Pressure levels available for high pressure hoses have increased inordinately with improved design, and the wire industry has provided a higher strength product to give the hose designer the necessary material to facilitate these advances. Larger diameter wires with very high tensile strengths are now available which have the processing characteristics necessary for hose manufacture. Aramid textile fibre has also played a significant role in allowing the production of higher pressure hoses, particularly for the offshore oil industry.

Hoses have become slimmer in wall, lighter in weight, and more flexible; thus they are able to work at lower bend radii and with increased working pressure capability and, because of improvements in compound design and the availability of newer elastomers, at much higher temperatures.

The distinction between rubbers and plastics has become quite blurred, with some thermoplastic having properties that make them virtually indistinguishable from traditional vulcanised rubber, and such materials have found a place in the hose manufacturing industry.

Hoses have been combined with electric cables as an integral product to allow the transmission in a single flexible conduit of electric power or communication signals and hydraulic pressure.

Methods of manufacture of rubber hydraulic hoses have not changed markedly over the past 20 years. Hoses from 5 to 32 mm bore are usually made on flexible rubber or thermoplastic mandrels. The inner liner is crosshead extruded onto the lubricated mandrel from a cold feed rubber extruder. A short freezing section just prior to the braider ensures that the inner liner is hard enough to withstand the stresses of braiding. After braiding the hose is covered on a cold feed crosshead extruder; frequently the hose is branded at this stage immediately downstream of the crosshead. Antitack, if required, is applied, and after a period of maturing the hose is either passed through a lead extruder (where a lead sheat slightly smaller than the outside diameter of the hose is applied) or wrapped with nylon fabric on an orbital or concentric wrapping machine.

Whether wrapped or leaded, the hoses are coiled onto drums, which are placed in an autoclave to effect cure. The time and temperature of vulcanisation depends on several factors such as the mandrel types, but it is important in hydraulic hose manufacture to obtain a tight cure to minimise compression set, an important property for good end fitting retention. It is also important to have compounds that cure at similar rates, to ensure good bonding between compounds.

After cure and cooling, the lead is stripped off or the wrap is taken off and the mandrel should be loose inside the vulcanised hose. The mandrel loosens because of the significant thermal expansion differential between the mandrel (and the inner liner material) and the metal reinforcement. During heating in the autoclave, the mandrel and the inner liner compounds swell relative to the wire reinforcement. The inner liner material is forced outward into the braid interstices, and bonding to the wire reinforcement takes place. Upon cooling, the mandrel shrinks back to its original size, leaving the inner liner bonded to the wire, and a gap develops between the mandrel and the inner liner.

It is possible to blow mandrel lengths up to 500 m out of hose in favourable circumstances, but a more practical length is 200–300 m for smaller sizes and perhaps 150–200 m for 25 and 32 mm bore sizes. Hoses are usually proof tested at this point to pressures recommended in the relevant SAE (Society of Automotive Engineers: U.S.) or DIN (German standard) specifications. The water is blown out by air pressure, and the finished hose is ready for inspection and coiling prior to despatch.

Significant advances in manufacturing equipment in the past few years have helped the hose manufacturer to make hydraulic hose of greater consistency in terms of diameter and performance, to increase productivity, and to reduce incidence of faults with resultant improved average length mix.

Cold feed extruders with mixing screws properly fed with consistent and dry compounds can extrude the inner liner onto the mandrel with a good level of accuracy (± 0.1 – ± 0.2 mm, dependent on bore size), especially if the system is controlled with the help of a laser or optical micrometer. Covers can be extruded over the wire reinforcement layer with equipment similar to that used for liner extrusion.

Ultrasonic concentricity/eccentricity equipment, which can monitor the concentricity of the inner liner on the mandrel or the cover on the reinforcement and allow adjustments to be made if concentricity is not perfect, can now be placed just downstream of extruder die.

Braiders have become significantly faster with the introduction rotary wire braider. This braider not only operates at approximately twice the speed of the preceding generation of high speed wire braiders but also gives a very high quality braid with very few wire crossovers–provided wire spooling quality is good.

Lead extrusion is slowly being superseded by fabric wrapping hose, utilising high speed concentric wrapping machines; an even more recent development is the use of a thermoplastic sheath in place of the lead. The thermoplastic sheath gives a smooth finish to the hose similar to that given by the lead extruder. Hydraulic hoses between 38 and 63 mm inside diameter are usually manufactured on steel mandrels, although some producers make up to 50 mm wire braided hose on flexible mandrels. Production on steel mandrels is usually carried out on mandrels between 15 and 40 m long.

Spirally reinforced hoses of all sizes (from 6 to 50 mm) are normally manufactured on steel mandrels, but some manufacturers have adopted the flexible mandrel process for the production of spiral hoses up to 25 mm internal diameter.

Hydraulic fluids are very diverse. Water with 5% oil emulsion is widely used in mining applications; water–glycol mixtures with lubrication agents are used in the offshore oil industry. Low viscosity mineral-based oils are used widely in the earth moving industry and in general hydraulics. Oil spillage from a leaking hydraulic system can cause serious environmental problems, and consequently the use of biodegradable oils, such as rapeseed oil, is gaining favour.

The multiplicity of hydraulic fluids presents a constant challenge to the hose engineers, particularly in the compound design.

Wire Braid Hoses

One and two wire braid hoses are the most widely used hoses in general hydraulic applications such as the backhoe digger.

The lighter weight hoses with smaller bend radii have become much more popular—although such hoses have been a single-braid hose with two wire braid performance.

One wire braid hoses conforming to SAE 100 R1 A (thick cover), SAE 100 R1 AT (thin cover), DIN 20022 1-ST (thick cover), and DIN 20022 1SN (thin cover) are usually manufactured incorporating an inner liner of nitrile (NBR) to give maximum resistance to the hydraulic fluids.

Polychloroprene rubber (CR) is also used as the inner liner in some instances. Although its oil resistance is not as good as that of NBR, it is quite suitable for mineral-based hydraulic oils.

For high temperature applications, the inner liner is manufactured form a specialty high temperature compound based on chlorinated polyethylene (CM) or hydrogenated NBR (HNBR).

The cover material is normally polychloroprene or blends of polychloroprene and styrene butadiene rubber (SBR) for good abrasion, and weather resistance, although blends of NBR and polyvinyl chloride (PVC) and NBR and SBR are also being used. Recently the use of SBR alone for the cover of hydraulic hoses has become widespread despite the need to improve the weather resistance by incorporation of chemical antiozonants and antioxidants.

For higher temperature applications the cover is normally manufactured with a compound based on CM or its sulphonated form (CSM), the latter being employed particularly when the cover is coloured, since compounds based on CSM are not prone to loss of colour quality upon exposure to light. Recently one wire braid hoses, which are slimmer and ligher in weight, have been finding applications in some hydraulic fields. Such hoses are constructed with a thin inner liner, a layer of braid usually composed of a smaller gauge wire (e.g., 0.25 mm) to keep the braid layer as thin as possible, and a thin cover. The hoses are ligher and more flexible than conventional one wire braid hoses and can thus be used at considerably smaller bend radii. The main disadvantage of the slim hose is that it is less robust than a conventional hose. More care is also required in the production of the inner liner because the presence of any contamination could give rise to a weakness under service conditions that could lead to pinholing and failure.

Hoses with two wire braids are produced by techniques and from materials similar to those used for one wire braid hoses but the wires are separated by an interply (also known as insulation or friction layer), which is designed to fill all the voids in the wire braided layers during vulcanisation. The thickness of the interply layer varies according to the gauge of wire used in the braiding layers. Typically the gage of interply is 0.30 mm for 0.25 mm wire in the braid, increasing to 0.6 mm for 0.38 mm wire in the braid.

The compound used for the interply must be capable of bonding to the wire reinforcement, and it is highly desirable that it bonds to the inner liner and cover materials.

It is usual to apply the interply layer on wire braid hoses longitudinally with a slight overlap or seam, which flows and disappears during vulcanisation to leave an even layer of rubber between the wire braids.

Slim two-braid hoses have become quite popular as the lighter weight and greater flexibility have permitted the use of shorter lengths of hose in some cases where a tighter bend radius is possible. The shorter length of hose again reduces weight compared to the conventional hose and also helps to keep costs down.

The extra care required in the production of the inner liner is just as necessary for slim two-braid hoses as it is for slim one-braid hoses. Contaminating particles or any voids or weaknesses in the inner liner will cause failure of the product at proof test or in service. Any eccentricity or undergauge portions of inner liner will also create weak spots, which can lead to failure in service.

Slim three-braid and four-braid hoses are now produced to give higher pressure capability or a bigger safety factor, which may be necessary if cover damage is likely and corrosion of the wire by ingress of water could cause a loss of burst pressure capability and increased risk of bursting in service. For very difficult operating conditions such as in coal mining, a combination of high carbon steel and stainless steel has been used to improve the life of hoses and to improve safety. A hose with either one braided layer of Hipac, carbon steel wires, or two slim braided layers made with high carbon steel to provide the majority of the burst strength is overbraided with stainless steel to protect the inner carbon steel braided layers in the event that the cover is damaged. It is now possible to obtain stainless steel wires with reasonable high strength (2050 N/mm^2) and also wires with brass coating to enable direct bonding of the compounds to the stainless steel braided layer. This modification enhances performance of the hose in a very hostile environment. Stainless steel is also sometimes used over the outside cover of a hydraulic hose to provide physical protection from mechanical damage to the hose. In the latter case the stainless steel does not contribute to the strength and burst performance of the hose, whereas when the stainless steel is an integral part of the hose construction, it does contribute to the bursting strength of the hose.

A recent innovation in the technology of two wire braid hose is the concept of applying the wires in each layer asymmetrically. The two-braid structure is made up of multistrands of a variable number of monofilament wires braided asymmetrically; that is, there are fewer strands in one direction than the other.

The construction is claimed to allow optimum coverage of wire in each braided layer, leading to additional burst strength, better impulse life, longer service life, and increased flexibility. These advantages are achieved by a more favourable stress distribution on the reinforcing layers, which reduces the extreme deformation of the individual wires at the crossover points.

Single wire braid hoses are available which have textile braid over the tube, then the single-wire braid, then a textile braided layer over the wire braid layer. Such hoses conform to SAE J517 100 R5 for hydraulic applications. End fittings for this type of hose are usually the re-usable type which can be assembled with a pair of wrenches and do not require the use of a swaging press. There is no rubber cover, which would have to be removed, and with this type of hose and fitting an assembly can be made up in the field without the need for sophisticated equipment.

Wire braid hoses with considerably higher working and burst pressures have been developed over the past decade, and as pressures in hydraulic circuits have become higher, so the use of such hoses has grown. The higher pressure capability is achieved by the use of larger diameter mono-filament wires of high strength.

Specialty one and two wire braid hoses have been produced for hydraulic applications with special requirements. Some fire-resistant hydraulic fluids are based on phosphate esters, and special compounds are necessary to cope with this type of fluid. Compounds based on ethylene propylene diene monomer (EPDM), butyl, or chlorobutyl rubber as the inner liner are commonly used to contain phosphate esters, since compounds based on NBR and polychloroprene swell and soften excessively in this type of fluid. A thermoplastic inner liner has also been used for hoses used with phosphate esters, polyamide being the most commonly selected polymer. The polyamide inner layer is bonded to a rubber layer during vulcanisation; the remaining hose construction is made up in the same way as an all-rubber hose.

Another special type of wire braid hose is the low temperature hose. The ability of the hose to remain flexible, and therefore serviceable, at temperatures of –50°C or even –55°C lies in the flexibility of the rubber components of the hose. NBR or polychloroprene is the usual choice for the inner liner, but special low temperature plasticisers are necessary to improve low temperature flexibility. In the case of NBR, a lower acrylonitrile polymer than that required for hoses operating at temperatures no lower than –40°C is necessary.

Flame-resistant, or more correctly flame-retardant, hoses can be produced through the design of special compounds. Polychloroprene, CM, and CSM are naturally flame retardant because each has a chlorine atom in the polymer main chain. The natural flame-retardant properties of these polymers can be further enhanced by the use of special fillers such as aluminium trihydrate, low flammable plasticisers such as trixylyl phosphate, or chlorinated paraffin and special additives such as antimony trioxide. With all the flame-retardant materials added to a polychloroprene cover compound, it is possible to render the hose completely self-extinguishing in the absence of an outside source of flame.

The main disadvantage of the chlorinated polymers is that under combustion conditions they give off copious quantities of hydrogen chloride gas, which is extremely acidic and choking and can attack delicate electronic equipment. This highly undesirable possibility has led to the development of hoses that are very flame retardant but, upon combustion, do not give off acid gases; smoke generation is also reduced.

Spiral Hoses

The design of very high pressure hoses involves the need to use spiraling, rather than braiding, techniques.

The efficiency of wire braiding falls away quite markedly as wire diameter increases. If efficiency of braid using 0.25 mm diameter wire is 100%, then for 0.30 mm diameter wire the efficiency is 90%, and for 0.38 mm diameter wire the efficiency falls to 70%. A further problem with larger diameter wire is that high coverage of wire cannot be obtained because the higher bending modulus of the wire requires a larger "gate" to pass under the opposing band of wire.

In consequence, there is an upper limit to the wire diameter that can be braided efficiently—the limit is about 0.4 mm. If higher pressure hoses are to be designed, the only remaining options are to have multiple layers of wire braid or to use higher strength wires. These options can produce some benefit but they are limited in as much as stress transfer between wire braid layers is not perfect and higher tensile strength wires, which are harder, produce very little benefit.

The only real way to achieve significantly higher burst and working pressure levels is to apply the wires in discrete spiral layers, each layer being separated from the next by a thin layer of high bonding vulcanised rubber compound.

Four or more layers of spiral wires can be applied over a hard inner liner compound. The inner liner can be made up of an extruded compound with fabric or textile braid between the liner, and the first wire layer or the inner liner can be made from layers of coated fabric. To build up the hose, each wire spiral layer is wound in the opposite direction to the preceding layer, an interply layer of rubber is applied in a spiral fashion. Four or six layers are normally used for hydraulic applications, wire diameters up to approximately 1 mm can be applied. Efficiency of reinforcement is very high, provided good consolidation is achieved during vulcanisation. Spiral hoses not only offer the advantage of efficient use of the wire reinforcement but have the added benefit that the fatigue life of the hose under dynamic impulse conditions is vastly superior to that of a wire braid hose. The reason for this is quite obvious when one considers that in a wire braided hose, since the bands of wire are woven to form the braid, the wires are by definition in contact with wires in the same layer. Wires at the edge of each band are very prone to notched fatigue failure under dynamic impulsing conditions. When one or two wires at the edge of a band of wires has broken, there is a loss of diaphragm strength in the tube and failure can occur by pinholing, or the remaining wires may break to produce a catastrophic failure.

In spiral hose the wires are not interwoven but are separated from one another by a thin layer of interply rubber, which is bonded to the wires. Wire-to-wire contact does not take place until the layers of rubber between the wires have been displaced by the action of heat and pressure. In a well-designed spiral hose the interply rubber will resist displacement, and so the dynamic impulse life of the produce will be longer—more than 1 million impulse cycles at 133% of maximum working pressure, which translates into many million cycles at working pressure.

The main disadvantage of spiral hose compared with wire braided hose is that it is more difficult to attach end fittings successfully without creating high stress levels at the junction of the hose and fitting.

Over the years many novel fitting designs have emerged for spiral hoses, but two main techniques for obtaining good fitting retention exist. The first is to remove a section of inner liner as well as cover and to swage the fitting directly onto the wire sandwich, using a separate section of the fitting to give a hydraulic sealing area. The other main technique is to crimp the wire/rubber sandwich severely at one end of the fitting and form a seal at the other end of the fitting.

Hot Water Washing Hose

The market for hoses for hot water washing has grown very strongly with a high proportion of garages and filling stations in many parts of the world providing the facility to wash vehicles in a few minutes with a mixture of hot water and detergents at moderately high pressures—typically 200 bars. The use of hoses for cleaning in abattoirs has led to the growth of similar hoses but with coloured cover.

Hoses are normally single wire braid rubber hoses similar to those used for low pressure hydraulic applications. The inner liner material is usually a compounded butadiene acrylonitrile rubber (NBR), which has quite good resistance to hot water at the temperatures used and perhaps, more important, has good resistance to the wide range of chemicals and detergents used in hot water washing.

The hoses are usually thin covered to allow the application of a swaged fitting without the need to remove the cover. The cover material is usually polychloroprene, chlorosulphonated polyethylene, chlorinated polyethylene, or styrene butadiene rubber.

Hoses for the car wash market are usually black, to give maximum weather and abrasion resistance; for the water wash requirements of the food industry, a smooth coloured cover is preferred because it does not mark the area being cleaned, as a black hose can. The smooth cover is preferred in the food processing industry because the hose covers must remain clean, and this is difficult to achieve if the hoses have a wrapped finish.

The water wash industry also uses some two wire braid hoses where pressures are higher or an improved safety factor is required, and there has been extensive use of "one equals two" braid hoses, which are based on the Hipac principle.

Water Blast Hose

The cleaning power of water has led to the development of water blasting as a major industry. Here cold water without additives is used to blast away unwanted surface contamination from any surface. A typical application is the removal of barnacles and other marine growth from the hull of a ship before repainting. Another application is the cleaning of drains and sewage pipes. In this type of application the water jets are arranged in such a way that the water power is used not only to clean the pipework but also to propel the cleaning head along the pipe at the same time.

To achieve a satisfactory cleaning action and cleaning rate, the hoses must work at very high pressures with high flow rates of water passing through. Typically pressure is around 700 bars (10,000 psi), and the most popular size of hose is 12.7 mm diameter. The minimum safety factor between minimum burst pressure and operating pressure is 2.5:1 but it is more usual to have a higher safety factor for around 3:1.

This safety factor takes the pressure requirements of the hose beyond what can be achieved with wire braiding alone, and it is usually necessary for the wire to be applied by spiraling in discrete pairs of layers. Hoses for this application are usually four layers of spiral wire or two layers of spiral wire with a braided layer over the spiral layers.

The pressure requirements and water volume requirements of the water blast industry have risen over the past few years and no doubt will go on rising. This trend has led to the development of hoses with higher pressure capability and/or larger bore hoses. Hoses with four spiral layers of heavy gauge, very high tensile strength wires plus a heavy gauge wire braid, have become common, and recently hoses with eight spiral wire layers with a braid over the outer layer were developed to give very high pressure capability and high water throughput rates.

Water at 700–1400 bars (10,000–20,000 psi) is dangerous if not under good control, and thus it is important to understand the requirements of very high pressure water blast hoses and hose assemblies. The choice of wire-reinforced rubber hoses for this application offers certain advantages. The product must be flexible, for ease in handling the gun that contains the nozzle. The hose must not fail catastrophically (e.g., because of sudden burst or the loss of an end fitting), and the hose must be robust and able to withstand rough physical treatment, whether it is used with a hand-held gun or propelled along a drain piper or sewer.

These objectives are achieved by the use of a very hard rubber hose liner, over which is applied a fabric layer or layers, followed by wire reinforcement layers applied over the rubber-fabric inner layer. Each layer of wire is separated from the next by a thin layer of vulcanisable rubber, which bonds chemically to the wire reinforcement layers to produce a strong rubber-to-metal bond. The outer cover is also bonded to the outer layer of wire reinforcement. The bonding is achieved during vulcanisation of the hose, when consolidation of the product also takes place.

Consolidation of the product involves the exclusion of all voids and air pockets that have formed during the hose building process. If all voids are not removed during vulcanisation, poor stress transfer and poor levels of chemical bonding will result in a product that does not meet the design criteria for burst pressure and could be dangerous in service.

End fittings for water blast hoses are designed so that they do not blow off in service: after removal of the rubber cover and a section of the inner liner, a part of the fitting is clamped directly onto the wire. A separate section of the fitting is used to achieve the hydraulic seal between the hose and the fitting.

A well-manufactured hose assembly will be safe in service if it is regularly inspected for physical damage, which could allow ingress of water and corrosion of the wire reinforcement. Proof testing of the hose will determine whether corrosion of the wire reinforcement has taken place.

Wire-reinforced rubber hoses are unlikely to fail catastrophically or to develop dangerous pinhole leakage and, with good end fitting design and application, they do not suffer from end fitting blow-offs.

Thermoplastic Hydraulic Hoses

The most commonly used thermoplastic hydraulic hose is manufactured with an inner liner of copolyester, a single braid of polyester or polyamide fibre, and a copolyester or polyurethane outer cover. Plasticised polyamides 11 to 12 are sometimes used for the inner liner or outer cover. The hoses are not manufactured with an internal mandrel and thus can be manufactured in very long lengths. There is normally no chemical adhesion between the inner liner and the reinforcement or the reinforcement and the outer cover—although some mechanical adhesion is achieved between the braided reinforcement and the latter during the extrusion of the outer cover. Adhesives are sometimes used with thermoplastic textile-reinforced hoses.

Thermoplastic hoses can be manufactured with very high strength fibres such as aramids, and the burst and working pressure levels attainable can be very high. A burst pressure of 3450 bars (50,000 psi) can be achieved on a hose with inner diameter of 6 mm. Although thermoplastic hoses can be manufactured in very long lengths (several kilometers), they are not very robust and would be prone to mechanical damage or kinking if used in a typical industrial application such as on a mechanical digger. In practice the hoses are incorporated into hydraulic or electrohydraulic umbilicals, which often are used to operate control systems in the offshore oil industry.

Textile-reinforced thermoplastic hoses are generally non-conductive unless designed to be conductive by the incorporation of antistatic wires. The non-conductive properties may be an advantage where conductivity of the hose would create a hazard. The lack of inherent good conductivity of textile-reinforced thermoplastic hoses can also create a hazard when the creation of a spark from static buildup could cause an explosion in the presence of an explosive gas or a fire in the presence of flammable solvents.

If an antistatic thermoplastic textile-reinforced hose is required, a soft copper wire can be applied over the outer textile reinforcement layer either by spiral application of the wire around the hose at approximately the same pitch as the reinforcement or by application longitudinally of the wire after it has been heavily crimped. In both cases the wire can be connected to the metallic end fitting when the hose assembly is made up.

Wire-reinforced thermoplastic hose has been around for many years in the form of stainless steel reinforced products lines with polytetrafluorothylene (PTFE). These hoses do not have an outer cover; this would be superfluous, since the reinforcement is not prone to corrosion. The introduction of rubbery thermoplastics has led to the development of a new generation of thermoplastic wire-reinforced hoses that are much more like the traditional rubber type.

The choice of inner liner material is usually a copolyester or a plasticised polyamide. The liner is pressurised, and one or two wire braid layers are applied on a typical wire braiding machine used for rubber hoses. An outer cover of a rubbery thermoplastic such as a softer grade of copolyester or a thermoplastic polyurethane or a thermoplastic elastomer such as Santoprene is applied on a crosshead extruder.

This type of hose lacks some of the robustness of the vulcanised rubber wire-reinforced hose, but it is more robust and less prone to kinking than the textile reinforced thermoplastic hose, and it is electrically conductive. This set of properties has resulted in the use of such products in some hydraulic applications, particularly where difficult chemicals or solvents are conveyed inside the hose. An example of such an application can be found in very high pressure paint spraying which uses solvents that would attack a rubber hose.

Thermoplastic hoses with spiral reinforcement are also available, and for hoses with a small bore size the pressure capability is very high indeed. The minimum burst pressure of a six-spiral hose with a bore diameter of 5 mm is 6200 bars (90,000 psi). As the pressure requirements of the hose increase, the inner liner material must become harder to obtain the necessary diaphragm strength. Harder thermoplastic materials, which have the necessary fluid resistance and high modulus, include POM or unplasticised polyamides.

AUTOMOTIVE HOSE

The use of hose in automotive applications has grown tremendously over the past few years, in line with the general increase in the number of cars and heavy vehicles. The widespread adoption of power steering, air conditioning, and cruise control systems has also increased the amount of hose in each vehicle, but this trend has been offset somewhat by the reduction in the amount of hose used for servo brake systems, where thermoplastic tubing has generally taken over from textile-reinforced rubber hose.

Space is at a premium in modern vehicles, leading to compact engine compartments. This, together with the need to have fuel-efficient engines and go aerodynamics for the sake of combining low fuel consumption with higher performance levels, has resulted in engines running hotter in confined spaces. As a result, there is a need for improved hoses, many of which are expected to last the life of the vehicle. The space premium has also resulted in many automotive hoses being moulded into complex shapes to accommodate the space restriction.

Fuel injection systems are good for performance and economic fuel consumption, but some fuel is recirculated to the tank, and this may result in buildup in the fuel of peroxides and hydroperoxides, which are extremely aggressive toward many polymeric hoses.

Fuel permeation through the wall of a fuel hose can lead to contamination of the atmosphere and in some parts of the world the amount of fuel that may be lost by permeation to the atmosphere is severely restricted on ecological grounds.

Lead in fuel has been shown to produce unacceptable levels of lead in the atmosphere, and lead-free fuel has gained widespread use, encouraged by tax concessions from governments. This concern for the environment has resulted in fuels having higher proportions of aromatics to keep the octane ratings at acceptable levels. Other additives to fuel include ethers and alcohols.

The result of all these changes is that the automotive hose designer has faced a considerable challenge to meet the demands of the automotive industry.

Coolant Hoses

The main polymer used for compounds for hoses that carry coolant is EPDM (because of its outstandingly good resistance to heat and hot water, as well as ethylene glycol, which is the most widely used antifreeze). EPDM also has very good weather resistance, especially with regard to attack by ozone, which causes cracking in many unsaturated polymers. Compression set of EPDM is satisfactory, provided the polymer is carefully chosen and a tight cure is achieved.

The technique for manufacturing coolant hoses range from fibre reinforcement of the rubber to textile knitting, braiding, and spiralling.

Fibre reinforcement of rubber can be achieved by a process in which the fibre Santoweb is composed of short cellulose fibres treated to bond to the hose polymer during vulcanisation. Extrusion of the fibre-reinforced compound takes place through a conventional cold feed extruder, but in the region of the die the extrudate is expanded diametrically very rapidly, to give a significant amount of radial orientation to the fibre reinforcement as well as an element of axial fibre orientation. This leads to a reasonable balance between radial and axial reinforcement, and adequate working pressure capability is achieved in the product.

Hoses can be produced straight or, by more recent innovation, the production of curved hoses is effected by moving the die in the head of the extruder. Further development of this idea has resulted in the die of the extruder being moved in various directions by the use of a programmable computer to move the die by fixed amounts in the desired direction. The system is known as the Iddon Computahose.

It is possible to produce hoses of various shapes from fibre-reinforced compound by the Computahose system simply by installing the appropriate programme. This is certainly a novel and ingenious way of producing curved hoses.

Most coolant hoses are at present still produced by the older techniques of extruding an inner liner, passing this through a textile knitting machine, then extuding a cover over the knitted reinforcement. The unvulcanised hose is chopped to length, then pushed onto lubricated shaped mandrels, vulcanised in high pressure steam, stripped, and trimmed. Any branch or T-pieces required in the coolant hoses can be built in before vulcanisation.

Smaller bore coolant hoses, such as from radiator to header tank, are usually produced on a flexible mandrel in very long lengths, which are vulcanised by continuous vulcanisation techniques in steam, salt, or fluidised bed or pan cured in steam. If the hose is not shaped, vulcanisation is completed; if the hose is to be shaped, the hose is partially vulcanised, the mandrel removed, and the hose put onto a lubricated shaped mandrel, where vulcanisation is completed and the desired shape achieve.

For heavy vehicles in which the coolant hose is expected to last for several hundred thousand kilometers, the hoses can be made of silicone rubber reinforced by aramid textile. This combination gives very good service life—normally the life of the vehicle.

Aramid textile reinforcement of coolant hoses for cars has increased steadily: in Europe, 50% of all coolant hoses are aramid reinforced.

Power Steering Hoses

Power streering hoses are composed of three main types: high volumetric expansion textile-reinforced hoses, a low volumetric expansion textile braid hose, and a return hose that is textile reinforced. There is also a wire braided hose.

The main types of high volumetric expansion hose are based on SAE J188 types 1 and 2. These textile-reinforced rubber hoses have high volumetric expansion characteristics that damp pump pressure pulsations. Type 1 requires a volumetric expansion of 9.8–26.2 cm^3/m and type 2 a volumetric expansion of 26.2–55.8 cm^3/m, both at a working pressure of 90 bars (1300 psi). Both type 1 and type 2 have a maximum working pressure of 103 bars (1500 psi) and a minimum burst pressure of 414 bars (6000 psi).

These hoses are mandrel built—normally on a flexible rubber or thermoplastic mandrel by the process used to make wire-reinforced hydraulic hose. They consist of an inner liner of polychloroprene, NBR, or (more recently, for higher temperature applications) hydrogenated NBR. The reinforcement for type 2 is usually polyamide or polyester, the yarn of a tyre cord type construction with high levels of twist giving high extension characteristics to permit the high volumetric expansion required. Type 1 reinforcement is also polyester or polyamide but not twisted to the same degree. In both types of hose the yarns are treated to give strong dynamic bonding levels to the rubber. The outer cover is usually polychlorprene, CSM, or CM. A low expansion power steering hose based on SAE J191, which has a maximum working pressure of 90 bars (1300 psi), is sometimes used where pressure pulsations created by the pump do not require to be damped. Power steering return hose is usually based on SAE J189. Both SAE J191 and J191 are usually manufactured with NBR inner liner and polychloroprene outer cover; the reinforcement is usually rayon or polyvinyl alcohol, which is braided.

A one wire braid power steering hose conforming to SAE J190 is also sometimes used where pressure pulsations from the pump do not require to be damped.

Fuel Hoses

Fuel hoses can range from a section of polyurethane tube with no reinforcement to a quite sophisticated reinforced hose with an inner liner composed of a veneer of fluoroelastomer inside another fuel-resistant polymer, a textile reinforcement layer, and an outer cover. Some fuel hoses are completely immersed in the fuel inside the fuel tank all the time, and therefore the other cover must exhibit as much resistance to fuel as the inner liner. Many fuel hoses have no outer cover and consist of an inner liner and a braided reinforcement only. There is an increasing trend for fuel hoses to be shaped.

Fuel hoses have traditionally been manufactured with an NBR or NBR/PVC inner liner, reinforced by cotton, rayon, polyamide, or polyester. If a cover was required, it was usually polychloroprene or CSM. The majority of fuel hoses today are manufactured of the same materials. More sophisticated fuels, fuel injection, higher under-the-hood temperatures, and the requirement for low permeation of fuel through the wall of the hose have led to the use of other polymers.

Fuel hose for the more difficult conditions may have an inner liner of NBR, HNBR, CO, or ECO or may have a veneer of fluoroelastomer with NBR or ECO as the second polymer, specially compounded to bond to the fluoroelastomer compound and the textile reinforcement layer. A recent introduction is the use of a thermoplastic (polyamide) inner veneer bonded to an outer rubber layer.

Most fuel hoses are produced on mandrels, although significant amounts are made without mandrels. Mandrei-made hose processes usually utilise flexible rubber or plastic mandrels and are either pan-cured or continuously vulcanised.

Air Brake and Vacuum Brake Hose

Normally hoses for air and vacuum brakes are made to the SAE 1402 and 1403 standards, respectively, and consist of an inner liner of NBR, CR, or NBR/SBR blend, a textile reinforcement of polyester, polyamide, rayon, or polyvinyl alcohol; and an outer cover of CR or CSM. Continuous vulcanisation or pan curing is the usual method of manufacture.

Brake Hoses

A more modern technique for producing this type of hose is by flexible mandrel and continuous vulcanisation–either steam or pressurised liquid salt.

Turbocharger Hose

Turbocharging of cars to give extra performance has grown quite markedly. Hoses connected to the turbocharger must withstand high temperatures and are usually made with silicone rubber, polyacrylates, or ethylene acrylic rubbers and reinforced with aramid yarn: such hoses are capable of operation at 150°C for a substantial length of time, with occasional excursions up to 170 or 180°C. The hoses must be resistant to oil as well as hot air.

Oil Cooler Hoses

Oil coolers are frequently used on heavy vehicles and also on some larger cars, particularly those with automatic transmissions, where temperatures of around 150°C are encountered.

Oil cooler hoses for cars are usually textile reinforced with end fittings attached. Chlorinated polyethylene or ethylene acrylic is the usual choice of polymer, but if oil temperatures will frequently rise to 150°C or more, a fluoroelastomer must be used as the inner liner.

INDUSTRIAL HOSE

"Industrial hose" is a broad term that generally refers to hoses for air, water, gas welding, filling station pump delivery, steam, and so on.

Air, Water, and Welding Hose

Air and water hoses were made by the moulded long length process; these were either braided or spiraled, and a high proportion of today's air and water hoses are still manufactured by this process–although now spiral reinforcement is normally used.

Briefly, the process manufactures textile spiral-reinforced hose without a mandrel, in a continuous process using a horizontal fluidised bed to vulcanise the hose. Production rate is very rapid, but long production runs are necessary if the process is to be economical. The process can manufacture hoses over the size range 6.3–38 mm for air, water, welding (including twin hoses), and filling station pump hoses as well as some types of automotive hose such as heater hose.

Another new method for the continuous manufacture of air and water hose is, that the unusual feature of this product is that the inner liner is vulcanised rubber while the cover is thermoplastic. The rubber inner liner is extruded and vulcanised continuously, and a plastisol of PVC and plasticisers is applied over the vulcanised inner liner. The plastisol is compounded in such a way that it bonds to the inner liner. Polyester, polyamide, or stronger yarns such as aramid are applied together with more plastisol. The plastisol is gelled by the action of heat, whereupon a PVC cover is applied, which bonds strongly to the plastisol surrounding the yarns, and the hose is complete.

Compared to rubber hose, the thermoplastic cover gives a more attractive appearance, with coloured covers and contrasting stripe quite easily achieved.

Steam Hose

Rubber steam hoses are normally reinforced with wire, although textile-reinforced hoses are available for low pressure steam applications. The internationally recognised safety factor for steam hose is 10:1 minimum burst pressure to maximum working pressure. Steam is available to 17 bars (250 psi) pressure in many factories, and for a steam hose that can withstand 232°C and have an adequate safety factor on burst to working pressure, wire is the natural choice for reinforcement.

The methods of manufacture of steam hose are very similar to those used for wire reinforced hydraulic hoses. Sizes up to 25 or 32 mm are usually manufactured on flexible rubber or thermoplastic mandrels, and larger sizes are made on steel mandrels.

The compounds for steam hose play a large part in product longevity, since the mode of failure of steam hose may be leakage at the fittings due to excessive compression set of the compounds, or the cover may become brittle and flexing of the hose may cause cracking. The inner liner may in some circumstances suffer from "popcorning," a phenomenon similar to explosive decompression failure in high pressure gas hoses. Leakage of the hose inevitably accompanies popcorning.

Fittings for steam hose are usually of a clamp type, which can be gradually tightened during service if leakage occurs. There is a limit to the degree of tightening that can be accommodated, however.

EPDM or butyl rubbers are the usual choice for both the liner and outer cover of steam hose, as well as the interply layer of rubber when this is required between the two wire braided layers. The compounds must be designed in such a way that the maximum resistance to steam and heat is achieved and the compression set resistance is maximised. Relatively low levels of filler are incorporated into the compounds and low levels of platsiciser are used, because the latter could be extracted by the steam. Close attention must be given to ensuring that all materials are very well dispersed in the inner liner compound; any badly dispersed particles can act as a nucleus for popcorning.

Popcorning can occur when the inner liner becomes fully saturated with steam and the pressure inside the hose is reduced rather quickly. The dissolved steam in the wall of the inner liner will gradually diffuse out of the wall of the liner. If, however, there are any weaknesses in the wall of the hose (e.g., a particle of grit or a small air bubble), such anomalies act as nuclei for the steam to escape quickly from the wall of the liner by rapidly expanding and blowing a small disk of rubber out of the inner lining. Popcorning weakens the lining and makes it thinner, and failure can occur upon repressurisation.

Hose technology continues to advance as it has done over the past quarter-century. The fundamental methods of manufacture of most hose types have not, however, undergone any major changes in manufacturing technology.

CHAPTER 15

Rubber Sports Goods Manufacturing Technology

INTRODUCTION

This chapter is primarily concerned with balls for golf, tennis, and squash which, with badminton, are the most popular competitive games played by individual players. Table tennis is also referred to briefly because, while the balls are not made of rubber, the surface of a modern table tennis bat is, and the game has a close affinity with court tennis.

The discussion of the properties, specifications, and manufacture of sports equipment is generally straightforward, but certain necessary equipment qualities are intangible and difficult to define in scientific terms, yet are extremely important to a player's full and satisfying appreciation of the game. Examples are the "click" of a golf ball when contacted by a club, the "response" and "control" of a ball on the strings of a racket, and the tactile texture of a tennis ball cloth. Such qualities determine whether an otherwise satisfactory product will be successful commercially, and so manufacturers spend considerable time and energy developing their products based on often secretive "know-how," acquired over many years.

GOLF BALLS

Golf balls originally were made by stuffing feathers under high pressure into a stitched leather container, which was subsequently hammered into a spherical shape. But later it was found that *gutta-percha* could be easily softened in hot water and fashioned to make a golf ball that generally performed better than the so-called feathery. In addition, it was found that the surface cuts and indentations inevitably produced on the *gutta-percha* ball by the club head actually improved the flight of the ball. Thus the science of golf ball aerodynamics was born.

The significant details in the development of the wound ball over the years since its inception include the development of a liquid centre for the windings improved rubber thread for greater resilience, plus fast and efficient winding machines. Synthetic balata (*trans*-polyisoprene) and "ionomer" cover materials, better paint, and attention to aerodynamics have also enhanced ball quality and performance.

Specification

Two separate authorities control equipment used in the game of golf throughout the world, the U.S. Golf Association (USGA) in the United States and Canada, and the Royal and Ancient Golf Club of St.

Andrews (R & A) elsewhere, but they adopt identical ball specifications :

Ball weight: not greater than 45.93 g (1.62 oz.)

Ball diameter: not less than 42.67 mm (1.68 in.)

The ball is required to be "spherically symmetrical," to avoid any asymmetry, which might give directional flight properties to ball judiciously placed on the tee.

Maximum ball resilience is also specified in terms of the speed of projection of the ball from a specially designed apparatus in which the ball is struck by a hammerhead moving at a controlled speed. On this machine the ball is restricted to a maximum velocity of 77.2 m/s (255 ft/s) at a temperature of 24°C (75°F).

Only maximum ball weight and minimum ball size are referred to in the specification because, other things being equal, the smaller and heavier the ball (within limits), the farther it can be hit. For ultimate performance, therefore, manufacturers are encouraged to produce balls to the maximum allowed weight and minimum allowed size.

The compression of a golf ball (i.e., its resistance to deformation under static load) is not specified, and it is left to manufacturers to produce balls to meet players' requirements. The most widely accepted format for expressing compression is referred to an arbitrary scale related to a ball's deflection under a load of weight 90.68 kg (200 lb). A deflection of 0.254 cm (0.1 in.) is counted as 100 and deflection of 0.508 cm (0.2 in.) and greater as zero. On this scale, balls favoured by professional golfers are in the 90–100 range. In general, balls of harder compression have a higher resilience, and manufactures attempt to produce a larger percentage of balls of harder compression. In the case of wound balls, this effort is limited by the tension at which the thread can be wound.

Wound Golf Balls

Wound golf balls are made from tightly stretched rubber thread wound on to a "centre," which is necessary to provide additional weight, to initiate the winding operation, and to allow the free distortion of the core upon impact so that high resilience can be generated. Until recently, the best golf balls incorporated "liquid" centres almost exclusively because such centres can distort readily upon impact with minimum loss of energy. The ball is completed by moulding on a protective cover, which incorporates the surface pattern.

Golf ball centres

The original liquid centres consisted of a variety of fluids contained in a latex bag at its neck and wound with wide rubber tape prior to the winding operation. A modern method for liquid centre manufacture is as follows:

1. A suspension or "paste" of barytes and bentonite clay in water and glycerine is prepared and dispensed into a series of hemispherical cavities in a rubber mould. Two such moulds are located together so that spheres of paste are formed, and the moulds are suitably jigged and passed through a freezer.
2. Upon leaving the freezer, the frozen spheres are demoulded and encapsulated in unvulcanised rubber by a die-forming operation.
3. The encapsulated spheres are allowed to thaw and are then subjected to a compression moulding process to vulcanise the covering.
4. Before winding can commence, the centres are rotated in a barrel at a low temperature to freeze the paste again and to make the spheres perfectly spherical for the winding operation.

With the advent of *cis*-polybutadiene rubber highly resilient solid centres could be made by a simple compression moulding process, and such centres have been progressively replacing liquid centres. Because of their high resilience, the new centres cause little energy loss in the ball and, since their resilience is additive to that of the windings, balls of increased resilience can be produced.

The spherical shape must be preserved, and usually this means that the solid centre must be frozen prior to winding, although with certain winding machines this is unnecessary. A typical formulation for the solid centre is as follows:

Ingredients	*Parts by weight*
cis-Polybutadiene	100
Barytes	100
Dicumyl peroxide (40% active)	5
Cure time at 150°C 15 min	

Thread and winding

Golf ball thread can be made from suitably compounded latex, but the greatest volume of thread is made by the established dry rubber processes of mill mixing, calendering, drum curing, and multiknife slitting. Conventionally, the thread is made tc dimensions of 0.5×1.5 mm^2 (0.020×0.060 in.^{2}) or similar. The thread is compounded to give an elongation of about 1000%, with high tensile strength and good chafe resistance, to facilitate handling on the winding machine. A typical thread formulation (parts by weight) is as follows:

Ingredients	*Parts by weight*
cis-Polyisoprene	100
Aldehyde/amine	0.5
Zinc oxide	3
4,4'-Dithiodimopholine	1
Sulphur	1
Cure time at 135°C: 2h	

A winding machine consists basically of two driven rollers on which the core sits. The rollers not only rotate, they oscillate axially in antiphase, so that the core is oscillated about a vertical axis as well as being rotated about a horizontal one. Two idler rollers press down onto the core under the action of a pneumatic cylinder, which is programmed to release pressure on the core periodically.

Thread is fed to the core from a spool or box, and it is tensioned by passing around a rubber-covered roller controlled by an electromagnetic brake, the drag of which can be adjusted to stretch the thread to an appropriate degree.

The action of the machine is to rotate the core in a random manner under the action of the driven rollers, the idler rollers, and the tension exerted by the thread. In practice, the thread is always wound onto the smallest diameter at any instant, and the core can reorient itself whenever the pressure exerted by the idler rollers is reduced. A microswitch stops the machine when the core has reached the correct size.

Cover moulding

Until the 1960s, golf balls were almost exclusively covered with *gutta-percha/balata compositions* because these materials could be moulded at the temperatures that did not degrade the highly treated thread (85–96°C). Synthetic *balata* (*trans*-polyisoprene), which became available later, is now used to the exclusion of the natural material in certain balls. At about the same time a new thermoplastic copolymer was developed and was given the generic name "ionomer." It was characterised "by having ionic cross-links in the solid state which are removable on fusion and restored on reversion to the solid state". The copolymer, named *Surlyn* (Du Pont registered trademark), was based on ethylene and a carboxylic acid and offered for the first time a man-made thermoplastic cover material that could be processed at reasonable temperatures to produce a good, tough ball cover.

The moulding processes for balls covered with *trans*-polyisoprene and with ionomer are distinctly different. Because of the higher processing temperatures of ionomer (160°C), special moulding procedures are adopted as given below.

Trans-polyisoprene. Half-shells, formed by compression moulding from suitably compounded *trans*-polyisoprene material are assembled onto the wound core. The assembly is compression moulded at 90°C followed by cooling by means of refrigerated water to aid extraction from the mould. The equator spue line is carefully ground, and the ball cover is subjected to vulcanising. Usually, cover vulcanisation is carried out by exposing the balls to carbon disulphide vapour for several hours. The carbon disulphide reacts with a secondary amine and zinc oxide in the cover compound to produce zinc dibenzyl-dithiocarbamate, and vulcanisation is completed by oven postcuring the balls for 48 hours at 32°C.

Ionomer. Several different grades of Surlyn are available, and cover compositions having desirable characteristics have been evolved by blending two or more different types together or by blending with other thermoplastic materials. The half-shells, produced by injection moulding, are linked together for ease of handling.

Because ionomers require higher moulding temperatures, a special moulding procedure is necessary to avoid thread degradation in the wound core. The core/cover assemblies are put into a hot mould, the mould is closed rapidly, and cooling is applied immediately. Special mould coatings are necessary for good release properties. The equator spue line is accurately trimmed after moulding. Because the cross-links are thermolabile, no subsequent cross-linking is necessary, as in the case of balata compounds.

Finishing

Polyurethane paints developed specifically for golf balls have been subject to continuing improvement. Chlorination is necessary to obtain good paint adhesion to balls covered with *trans*-polyisoprene, but for Surlyn balls specially developed treatments have been devised involving abrasive treatment and/or flame plasma treatment combined with the use of special paint primers based on epoxy or urethane in petroleum or water carriers, respectively. Ball marking is normally carried out by foil printing, but pad printing and transfer methods are also used. The surface is protected by a final coat of polyurethane lacquer.

Nonwound Golf Balls

Solid balls

Since the demise of the solid gutty ball, attempts have been made to produce improved nonwound balls. Such attempts were generally unsuccessful, however, until developments in polymer technology provided the appropriate materials. A good nonwound ball requires a hard, highly resilient material that is tough enough to withstand the repeated impacts to which a golf ball is subjected. Such compositions which claimed a homogeneous moulded golf ball "...predominantly composed of a filled elastomer highly cross-linked into a three-dimension network with long flexible cross-links formed from a polymerised cross-linking monomer, said elastomer being one capable of being so cross-linked by said cross-linking monomer". The most appropriate monomers are esters of unsaturated acids, and those preferred are acrylates and methacrylates. These are used in conjuction with *cis*-polybutadiene.

A typical formulation for a 42.67 mm (1.68 in.) ball is shown below:

Ingredients	*Parts by weight*
cis-Polybutadiene	100
Butylene glycol dimethyacrylate	62.5
Silica	62.5
Dicumyl peroxide (40% active)	3.13
Cure time at 150°C: 10 min	

While Bartsch-type balls were superior to earlier solid balls, they did not have the resilience of a wound ball, being 10% short in distance. They have found application in driving ranges, where their durability is attractive and distance less important.

Two piece balls

Further developments in nonwound ball technology resulted in the advent of balls with nonwound cores (i.e., solid cores) requiring the protection of a separate cover because of their limited toughness, termed two-piece balls to distinguish them from simple (one-piece) solid balls and conventional (three-piece) wound balls.

The technology is again based on the use of cross-linking monomer, the preferred monomers being metal salts of polymerisable organic acids. Particularly suitable materials are zinc diacrylate and zinc dimethyacrylate. These may be formed either *in situ* (i.e., during the actual compounding operation) or in a separate operation in which they are partially polymerised, then added in the normal compounding procedure.

Generally the "metal-containing polymerisable monomers" produce cores of higher resilience then Bartsch-type monomers but tend to be more brittle. For this reason they need the extra protection afforded by a separate "cover," and so the two-piece format was devised. A core formulation is as follows:

Ingredients	*Parts by weight*
cis-Polybutadiene	100
Zinc diacrylate	35
Zinc oxide	18
Dicumyl peroxide (40% active)	5
Cure time at 160°C: 20 min	

The availability of ionomers allowed the possibility of directly injecting the cover onto the core—a process basically untenable for a wound core because of the high melt temperature necessary. The direct cover injection method requires the accurate location of the core in the centre of the mould by means of a method that uses retracting supporting pins in the ball mould. Typically, this is done with a vertical clamp injection moulding machine.

Four cavities are installed in the mould, which shuttles in and out of the injection position for ease of core insertion and ball extraction. Surface imperfections due to the core support pins and injection points are removed by vibrating the balls in abrasive medium. Otherwise the finishing procedure is the same as for an ionomer-covered wound ball. Two-piece ball manufacture lends itself to a high degree of automation: cores are placed in the moulds and moulded balls removed robotically. A manufacturing cell incorporating a core dispenser, two injection moulding machines, and a deflashing press, all serviced by a single robot. In this case the robot collects eight cores at a time from a storage hopper, loads them into the mould cavities, removes the balls after moulding, and places them in the deflashing press shown in the foreground. This cell makes 4200 balls per hour.

The performance of a two-piece ball is characterised by a flight trajectory lower than that of a wound ball, as a result of the higher moment of inertia of the ball around its centre point compared with a wound ball and different mode of deformation. Less spin is imparted by a lofted club, resulting in a flatter trajectory and a less acute angle of impact, which gives the two-piece ball greater running ability when it strikes the ground. Generally, the characteristics of the ball are more suited to the longer handicap golfer, the better golfer preferring the stopping effect of a wound ball (arising from its greater spin) on and around the green.

Golf Ball Moulds

The significant improvement in ball performance in recent years is due not only to material and process improvements but also the accuracy with which golf ball moulds can be made. Improvement in ball aerodynamics has resulted, determined by surface pattern uniformity: for instance, variations of 0.001 in. in dimple depth on the ball are significant. Various procedures have been adopted for manufacturing moulds, but the most common techniques are based on hobbing or otherwise pressing the desired surface profile into the mould surface so that a large number of moulds can be accurately reproduced. Spark erosion techniques are often used for producing the hobs or moulds directly.

TENNIS BALLS

The game of tennis (more properly, lawn tennis) are similar, being based on one or a pair of players on opposite sides of a net hitting ball back and forth, but the type and size of the court, the method of scoring, and so on are very different. Real tennis balls have been traditionally made from a spherical stitched envelope of leather or cloth stuffed with rags, horsehair, or similar material, while tennis balls have always been based on rubber.

Originally, tennis balls were made solely of natural rubber, but the wearing and playing properties of the balls were improved by covering them with flannel stitched around the rubber "core." An early development was to pressurise a hollow core with gas. Originally, core manufacture was based on the "cloverleaf" principle whereby uncured rubber sheet was stamped into a shape resembling a three-leaf clover and this was assembled into a roughly spherical shape by machinery adapted for the purpose. Chemicals generating pressurising gas were added before the assembly was closed, and these were activated upon moulding the core to a spherical shape in heated cavities. This process was used for many years until the precision of the game demanded a higher degree of uniformity (especially wall thickness) than could be obtained with the cloverleaf method. Now, normally two separate half-shells are compression moulded and then assembled together to produce a core. The original flannel cloth was replaced by special "melton" cloth made specifically for the purpose, and the stitching has been replaced by a vulcanised rubber seam.

Specification

The ball specification, set by the International Tennis Federation, covers ball weight, size, rebound, and "compression" tested at 20°C and 60% relative humidity are given below :

Weight: 56.7–58.5 g (2–2.062 oz.)
Diameter: 6.54–6.86 cm (2.575–2.700 in.) (go/no go ring gauge)
Rebound from 254 cm (100 in.) onto concrete: 135–147 cm (53–58 in.)
Compression:
 Forward deformation: 0.56–0.74 cm (0.220–0.290 in.)
 Return deformation: 0.89–1.08 cm (0.350–0.425 in.)

The compression test procedure requires the use of a modified Stevens machine, which is designed to measure the deformation of a ball under a load of 8.165 kg (18 lb), when the load is initially applied (forward deformation) and upon recovery (return deformation), after the ball has been subject to deformation of 2.54 cm (1 in.). For a perfectly elastic ball, the forward and return readings should be the same; thus any difference between the readings gives a measure of the hysteresis loss of the ball.

Prior to compression testing, it is stipulated that the ball must be deflected through 2.54 cm (1 in.) on each of three diameters at right angles. This series of deflections is designed to remove from the ball any "set," which can occur after prolonged storage.

Modern Manufacturing Technology

The manufacturing operations consist basically the following steps :

1. Half-shell moulding.
2. Core assembly.
3. Melton preparation.
4. Core covering.
5. Covered ball moulding.
6. Testing and marking.

Half-shell moulding

Normally half-shell moulding is carried out by a simple compression moulding operation using accurately prepared blanks, as prepared by a Barwell preformer. For high volume operation it is normal to load the press from preloaded jigs, which enable a large number of cavities to be loaded simultaneously. Again, to aid handling, it is normal for the flash from the individual cavities to be joined so that the press can be quickly stripped. Material wastage (flash) is more than compensated for by higher operational speed. A blanking-out operation removes flash from individual half-shells.

Core manufacture

The edges of the half-shells are lightly abraded where they are to be jointed, and vulcanising rubber solution (normally based on natural rubber with a fast curing system) is applied. The process then differs according to whether the chemical or compassed-air method is to be used for inflation.

The chemical method is older and has been largely superseded, but it has the advantage that the gas pressure (hence core properties) can be accurately controlled from the quantity of chemicals used and because inflation occurs after the core has been assembled. Under certain conditions of storage, however, products of the reaction can crystallise out and cause "rattlers." Simple compression moulding is performed, using mating hemispherical cavities. The inflation chemicals are normally sodium nitrite and ammonium chloride, which produce nitrogen during the moulding process.

The compressed-air inflation method uses a complex clamshell press arrangement in which the individual edge-solutioned half-shells are placed separately in opposite mating hemispherical cavities of the press. When the press is closed, the half-shells opposite each other are brought together to form a core. However, the closing operation takes place in stages, as follows:

1. The press closes until the edges of the half-shells are almost but not quite touching.
2. In this position, the internal platen area is isolated from the atmosphere by a rubber sealing ring around the periphery of the platens.
3. Compressed air is introduced to the platen area at the required pressure.

4. The press platens are brought together, thus trapping the compressed air between the half-shells and forming a pressurised core.
5. The platens are heated, to vulcanise the rubber cement, and cooled prior to core extraction.

The mould and press arrangement for compressed-air inflation involves a considerably larger mass of metal than is needed for chemical inflation, and process times tend to be longer. Also, great care is necessary to ensure that the cores are equally pressurised over the whole platen area. Thus good equipment maintenance is necessary. The inflation method is also applicable to other gases.

Core formulations for pressurised balls are normally based on natural rubber containing high loading of fine particle filler for low gas permeability. A formulation is given below :

Ingredients	*Parts by weight*
Natural rubber	100
GPF black	30
Clay (fine particle size)	31
Zinc oxide	9
Sulphur	3.5
Diphenyl guanidine	2
Cyclohexyl benzthiazylsulphenamide	1

Cure times at 150°C

Half-shell: 2.5 min

Core

Chemical inflation: 4 min

Air inflation: 12 min

Cores are normally pressurised to around 0.8 bar above atmospheric pressure, but the pressure is gradually lost because rubber compounds are permeable to gas. Balls, therefore, gradually lose rebound and compression, and after about 4 months in the atmosphere they may no longer meet specification. If longer storage periods are anticipated, the balls must be packed in pressurised cans.

Pressureless (nonpressurised) balls have been developed and have sold in some markets for many years. The basic technology of making a pressureless ball is that of producing highly resilient, high modulus compounds of lower than normal density. These materials can be used to produce cores with walls of greater than normal density. These materials can be used to produce cores with walls of greater than normal thickness, which provide unaided the necessary compression and rebound properties. Otherwise, manufacturing procedure is the same as for pressurised balls.

Compounding methods for pressureless balls are varied and involve the use of high styrene resin, cellulosic filler, aminoplast resin, and copolymer of an olefin and an unsaturated monocarboxylic acid copolymer. The various formulations attempt to produce appropriate qualities of ball performance and player satisfaction as well as fulfilling the specification requirements. In practice, pressureless balls have had only limited commercial success. Less than 10% of balls currently produced in the world are of the pressureless variety.

Since the days of the cloverleaf core, attempts have been made to produce a precision core by avoiding the separate operations on half-shell and core moulding. Methods of producing uncured half-shells that can be assembled into pressurised cores in only one vulcanising operation. The main problem is that of producing a strong joint between the halves of the core while minimising local distortion from the pressure that must be applied to effect jointing of the unvulcanised material.

Melton cloth

Special melton cloths have been developed for tennis balls. Best quality tennis balls use melton with a high wool content (50–65%) for three primary reasons :

1. To enable the nap of the cloth to be satisfactorily consolidated by fulling (or felting).
2. To allow easy removal of moulding lines after the cover moulding operation.
3. To give low soiling properties in play.

For optimum wearing properties, the balance of the fibre content is made up by nylon. No synthetic fibre has challenged nylon in this respect. The melton cloth is produced by the long-established processes of weaving, raising, and fulling.

The basic cloth is produced by weaving blended wool/nylon weft yarns into a light cotton warp, where the warp weight normally constitutes only about 15% of the final cloth weight. A "sateen" weaving pattern is used in which the weft lies predominantly on one side of the warp, and this area becomes the wearing surface of the cloth. Cloth weight for a best quality ball is about 0.7 kg/m^2.

The "face" of the raised cloth is then subjected to a *raising* operation in which the fibres of the weft are carefully teased to produce a fluffy texture without destroying the basic integrity of the yarns.

In *fulling*, the raised cloth is worked in water containing soap, with the result that the raised wool fibres become felted together to produce a consolidated surface, which ultimately becomes the wearing surface of the ball.

Considerable skill is exercised by the cloth manufacturers, particularly in the raising and fulling processes, so that, in play, the cloth surface generates and maintains as long as possible the appropriate surface texture for desirable ball flight qualities and end wear. In achieving this objective, account must be taken of the widely differing surfaces on which tennis is now played, ranging from grass, clay, shale, and many synthetic "carpet" surfaces to highly abrasive "all-weather" surfaces.

While melton cloths are used predominantly for top quality balls, so-called needle cloths have also been used, but mainly for lower grades. Here a process of "needle punching" produces a consolidated fabric with the general appearance of a melton directly from staple fibres by avoiding the stages of yarn preparation, weaving, raising, and fulling. In the needle process, a mat of randomly oriented fibres (sometimes assembled on a light "scrim" cloth) is passed through a needle loom, on which a multitude of barbed needles are made to reciprocate up and down. The barb on the needles point down, and as they pass into the mat of fibres, they pick up individual fibres, which are pushed down, where they remain as the needle withdraws. The action of hundreds of needles punching and retracting in this way gradually consolidates the fibres into a continuous interlocking structure which depends for its integrity on interfibre friction.

Such needle cloth is cheaper to produce than melton, but they tend to be less uniform in thickness and to lose surface nap more quickly. Because fabric consolidation does not depend on the felting properties of wool, high synthetic fibre contents can be used, but mould line removal after ball moulding then becomes the controlling factor.

The cloth is normally coated on its reverse side with a vulcanising cement by means of one or more spreading operations. Dumbbell shaped blanks are then cut from several layers of cloth at once by means of an automatically indexing stamping press. The axis of the blank is aligned at approximately 45° to the warp direction, to permit the dumbbell to stretch axially during the covering operation.

The edges of the blanks are next coated with a vulcanising cement, which subsequently forms the seams of the ball. This is normally done by first putting the blanks into a stack, compressing the blanks tightly together into a jig, and immersing the assembly in a vat of vulcanising rubber solution. The immersion is normally carried out automatically: jigged stacks are taken by overhead conveyors to one more dipping and draining stations.

Core covering

Prior to the covering operation, the cores are subjected to surface abrasion. Next they are coated with rubber solution in a barreling operation in which cores, plus a measured volume of rubber solution, are tumbled together, with the result that each core is coated with a uniform layer of solution. Rubber latex is often used instead of rubber solution.

The covering operation is carried out by hand or by machine using two dumbbell blanks for each core. The difficulty of the operation can be appreciated by recalling that the cloth can stretch, the core is deformable, and the tackiness of the rubber cement on core and cloth produces contact adhesion.

Covered ball moulding

After covering, the balls undergo cover moulding to vulcanise cloth to core and the joint between the blanks. This operation is carried out at a temperature that avoids the heat discoloration of the cloth and is followed by a cooling process, as in the case of all inflated products.

Testing and marking

The covered ball moulding operation compresses the nap of the cloth, and the split line of the mould produces a crease; both effects are removed by a steaming operation in which the balls are tumbled in wet steam. After stabilising, the balls are tested for compression and printed by a hot transfer method.

Future developments

While developments involving both core and cloth have occurred over the years, the inflation medium has received little attention since the development of the compressed air method. However, it is known that certain gases, because of their molecular complexity, permeate through rubber at a significantly lower rate than air or nitrogen, and such gases could have application in tennis ball cores. One such gas is sulphur hexafluoride, which has been tried experimentally but not adopted commercially.

SQUASH BALLS

The game of squash rackets (squash for short) is thought to have been developed at Harrow School. Boys awaiting their turn on a rackets court are reported to have played with a rackets ball in an adjacent room or corridor, but having broken windows with the hard ball, they resorted to using a rubber ball of similar size. The modified game spread to other public schools and then to the armed services. The Squash Rackets Association was formed in England in 1929, when the rules were standardised, and an International Squash Rackets Federation (ISRF) was formed in 1967.

The game has developed differently in the United States, where, among other differences, harder balls are specified. The term "hardball" has come to denote the American game, which is controlled by the U.S. Squash Rackets Association (USSRA), while the game controlled by the ISRF is known as "softball" in the United States.

Specifications

The diameters of the ISRF and USSRA balls are identical, but weight, compression, and rebound are different, as indicated below. In addition, different methods are used to test for compression. The "speed" of the ball in play is determined by its rebound and compression, where higher rebound and high compression (i.e., lower deformation) gives higher speed. Balls of different speeds are identified by means of coloured dots moulded into recesses in the surface of the ball.

ISRF ball (softball)

The ISRF has published the following specifications:

Diameter: 39.6–41.4 mm (1.56–1.63 in.)

Weight: 23.3–24.6 g (0.82–0.87 oz.)

Compression: change in diameter under a 2.44 kg (5.29 lb) load: 3.0–7.00 mm (0.12–0.28 in.) all balls.

Rebound from 254 cm (100 in.) onto concrete: 45.7–66.0 cm (18–26 in.)

Segregated into: 5.08 cm (2 in.) bandwidths identified by yellow, white, red, and blue dots, respectively.

USSRA ball (hardball)

The USSRA specifications are as follows :

Diameter: 39.6–41.4 mm (1.56–1.63 in.)

Weight: 19.3–21.8 g (0.68–0.77 oz.)

Compression: change in diameter under 16.78 kg (37 lb) load:

(a) white and blue dot balls: 21.6–26.7 mm (0.85–1.05 in.)

(b) triple white dot: 11.4–15.2 mm (0.45–0.60 in.)

Rebound from 254 cm (100 in.) onto concrete: 31.1–95.3 cm (15.0–37.5 in)

Segregated into: 10 cm (4 in.) bandwidths starting at 38.1 cm (15.0 in.), 53.3 cm (21.0 in.), and 82.6 cm (32.5 in.), identified by white, blue, and triple white dots, respectively.

The appropriate playing qualities of a squash ball cannot be completely and adequately described by a specification. The skill in producing a good squash ball lies in the compound formulation, which provides (in addition to the specified requirements), satisfactory rebound characteristics over ranges of impact speeds and playing temperatures. In addition, appropriate surface friction is necessary to produce acceptable rebound angles, and minimum change in this surface friction over a reasonable playing life is desirable. The necessity for good durability is particularly important, in view of the extremely high deformation the ball undergoes while in play. This requirement has particular significance in terms of the strength and durability of the joint between the half-shells from which the ball is made.

Squash Ball Manufacture

Squash ball formulations are based on butyl rubber and are jealously guarded by manufacturers, who develop their own manufacturing expertise and know-how. Heat buildup in the ball causes it to speed-up in play, and this effect must be controlled by judicious compound. Also wall marking proprieties must be minimised. Ball manufacture is similar to that of a tennis ball core, with the separate stages of half-shell production, ball assembly, and final vulcanisation. Appropriate surface abrasion is produced by high speed rotation in an abrasive barrel followed by washing. Surface marking is applied by a pad printer to a recess in the ball surface, normally before surface abrasion, to permit the removal of any excess.

TABLE TENNIS BAT RUBBERS

The origins of table tennis are obscure, but the game seems to have been developed at about the same time as lawn tennis. The bats were originally made solely of wood, but "sandpaper" was later used on the blade to improve the grip on the (celluloid) ball.

Rubber covering for the blades was introduced in the of "pimpled sheet" (i.e., rubber sheet with moulded projections, or pimples, to increase the grip of the bat surface on the ball. During the late 1950s, sponge rubber was introduced as an alternative bat surface, but pimpled sheet and sponge rubber were later combined. Originally, the sponge was used as a base for the pimpled sheet so that the pimples contacted the ball (the "pimples out" orientation); later, however, the orientation was reversed so that the smooth back of the pimpled sheet contacted the ball ("pimples reversed"). Compound developments for the reversed pimpled sheet enabled tacky surfaces to be produced, which gave improved ball grip (hence spin), and sponges of differing resilience were developed. These are now made of high resilience for attacking play and lower resilience for defensive play, but the compound formulations are not revealed by manufacturers.

The thickness of pimpled sheet and so-called sandwich rubber (pimpled sheet and sponge) has been specified by the International Table Tennis Association to not exceed 2 and 4 mm, respectively, and the number of pimples must be between 10 and 50 per square centimeters.

Pimpled sheet is made from calendered sheet, compression moulded against perforated metal plates supported on an appropriate fabric with good release properties, which imparts a textured impression to the pimple extremities. The textured surface, in turn, produces good ball grip and aids in bonding to the rubber sponge when in reversed form. Great care must be taken to ensure moulding uniformity from the viewpoints of both thickness and weight.

The sponge rubber is made from natural rubber and/or polybutadience rubber. A blowing agent giving fine pore size is used to enable it to be cut accurately to a 2 mm thickness by band sawing from a moulded block. Considerable development work has been devoted to sponge, with particular reference to improving its resilience for producing high ball speeds in attacking play.

Rubber-based adhesives are used to bond the components of the sandwich together and to bond the pimpled sheet (or sandwich) to the bat blade.

CHAPTER 16

Cable Technology

INTRODUCTION

Power cables are the most useful and practical media (agents) used for conveying electricity from generating stations to factories, domestic premises, and other locations, as well as to equipment, tools, and appliances of all types. The variety of uses for cable and the need to co-ordinate service requirements have resulted in a large number of designs, constructions, and material systems.

CONSTRUCTIONAL ELEMENTS OF POLYMER-INSULATED CABLES

Polymer-based electrical cable essentially consists of metallic low resistance conductors (to carry the current) covered with a polymeric insulation, which isolates the conductors from each other and from the surroundings, and a sheath or jacket formulated primarily with regards to its mechanical properties and environmental conditions. Other major components may include a conductor and insulation shield and a metal screen (mainly to obtain a radial uniform field), a reinforcing metal wire or belt armouring for mechanical protection, water blocking tape, and a metal sheath to prevent moisture ingress. Fig. 16.1 shows the construction of a typical polymer-insulated cable.

Metallic Conductors

Economy and versatility favour copper (electrical resistivity at 20°C; 1.724–1.727 $\Omega . m \times 10^{-8}$) and aluminium (electrical resistivity at 20°C: 2.803–2.826 $\Omega . m \times 10^{-8}$) for conductor materials in power cables, although in flexible elastomeric cables only copper conductor is used. To compensate for the difference in conductivity, the conductor cross-sectional area of aluminium cable must be increased by a factor of almost 1.6 over that of copper. The international electrotechnical commission standards IEC 28 and IEC 111 specify the electrical characteristics of copper and aluminium, respectively. To prevent the adverse chemical reaction of copper with the polymer, the metal is coated with tin, particularly in the case of elastomer-insulated cables. For high temperature applications, tinned copper may be replaced by nickel-coated copper or silver-coated copper. Similarly, if the strength of the conductor is important, copper-cadmium alloy may be used in place of copper with a little sacrifice in conductivity.

Three forms of conductors are generally used: solid, stranded, and flexible (bunch). Solid copper conductors are preferred up to 16 mm^2 cross-sectional area, stranded copper conductors for 25 mm^2 and above. Generally, circular solid and stranded aluminium conductors from 25 mm^2 upward and sector-shaped conductors from 50 mm^2 above are used, in accordance with the German standard DIN VDE 0295. Cable flexibility can be achieved by stranding conductors and using a large number of smaller diameter wires. "Concentric lay" refers to the conventional stranding practice in which a single wire or a group of wires in the centre forms the core, and successive complete layers of wires are applied, one on top of the other. Normally the direction of application is reversed between layers, but in some special cases, it may be the same (i.e., unidirectional stranding). In bunched conductors the individual wires are merely twisted together, all in the same direction and with the same length of lay. Since there are limitations to the total number of wires that can be twisted together to form a single satisfactory bunch, for the larger conductors small individual bunches are stranded together in layers (i.e., multiple bunch stranded).

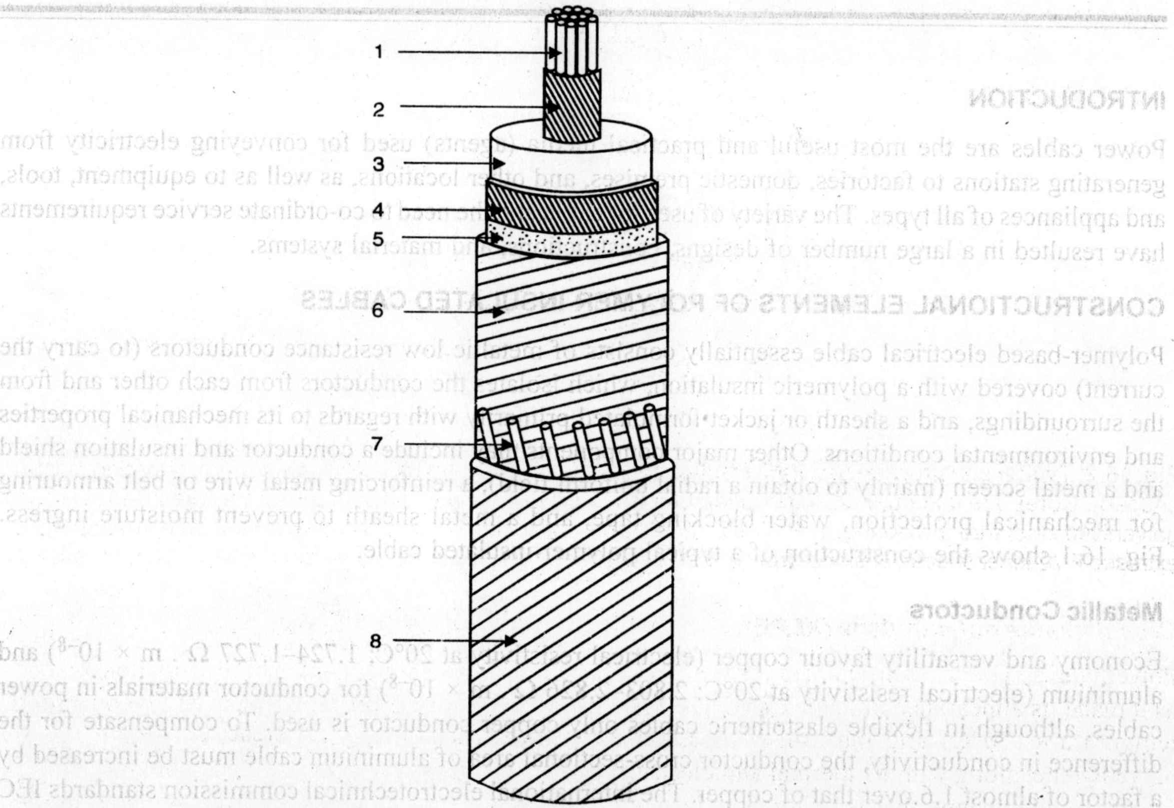

Fig. 16.1. Typical construction of a high voltage extruded polymer insulated cable: 1. stranded conductor; 2. semi-conducting conductor shield; 3. extruded insulation; 4. semiconducting dielectric shield; 5. shielding copper tape; 6. inner sheath; 7. wire armouring; and 8. outer sheath.

Polymeric Materials for Cable Insulation

Polyvinyl chloride (PVC)

PVC has assumed particular significance as an insulating material in low and medium voltage cables because of its low compound cost, ease of processibility, and recyclability. For cable insulating PVC having a *K*-value of 65–70 is generally recommended. PVC is compounded by the incorporation of stabilisers like basic lead sulphate and phthalate, plasticisers like dioctyl phthalate (DOP) and trioctyl trimelliate (TOTM), fillers like calcium carbonate, and so on. PVC-insulated cables are widely used in the 1.1–3.3 kV range and have found limited applications in 6.6 and 11 kV cables. A British standard (BS 6346) specifies the requirements for PVC compounds for cables up to 3.3 kV. IEC 502 specifies these by stipulating PVC/A up to 3.3 kV and PVC/B for higher voltages. Fig. 16.2 shows typical construction for a 300 mm^2, 3.3 kV, stranded sector-shaped aluminium conductor, PVC-insulated cable.

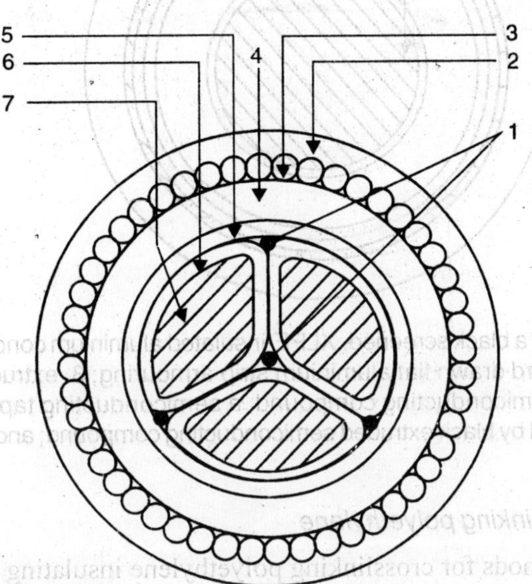

Fig. 16.2. Typical construction of a PVC-insulated aluminium conductor (300 mm^2, three-core, stranded, sector shaped). Cores are laid up; PVC inner sheathed is extruded. Cable is black and has a single layer of round galvanised steel wire armouring. 1. PVC filler; 2. PVC outer sheath; 3. galvanised round steel wire armouring; 4. extruded PVC inner sheath; 5. binder tape; 6. PVC insulation; and 7. stranded, sector-shaped aluminium conductor.

Crosslinkable polyethylene (XLPE)

Polyethylene, which has unique dielectric properties, found wide applications in cable insulation as soon as it made its appearance on the market. The main problem in straight polyethylene is its low operating temperature (70°C) and maximum short-circuit temperature (160°C). These two properties of polyethylene, along with its thermal and electrical ageing, are greatly improved by crosslinking.

Thus polyethylene that is to serve as insulating material is almost always crosslinked; as a result, maximum operating and short-circuit temperatures have been enhanced to 90 and 250°C, respectively. Cross-linked polyethylene insulated power cable is the most popular power cable in the world, and the demand for XLPE cables is expanding rapidly as users perceive its basic advantages, such as a simple transmission system, lower maintenance requirements, low permittivity, and low dielectric loss.

The development of a superclean tree-retardant grade has led to successful applications up to 275 kV cables, and an attempt is being made to use XLPE in 500 kV cables in Japan. Fig. 16.3 shows typical construction of a 500 mm^2, single-core, 33 kV, aluminium conductor cable, insulated with XLPE.

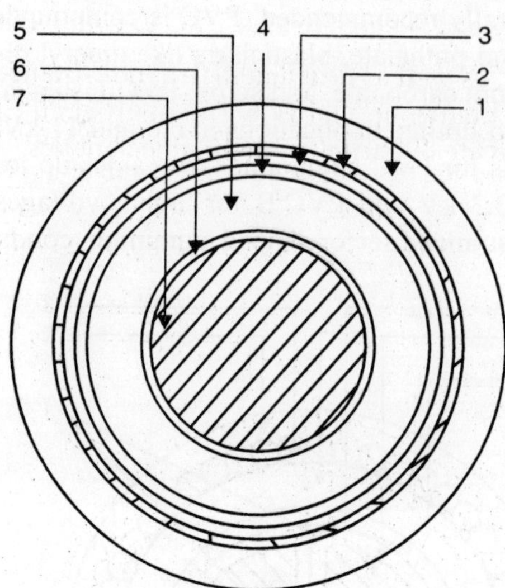

Fig. 16.3. Typical construction of a black screened, XLPE-insulated aluminium conductor (500 mm^2, single core, 33 kV). 1. PVC outer sheath; 2. hard drawn flat aluminium strip armouring; 3. extruded PVC inner sheath; 4. core screened by black extruded semiconducting compound, a semiconducting tape, and a copper tape; 5. XLPE insulation; 6, conductor screened by black extruded semiconducting compound; and 7. stranded compacted circular aluminium conductor.

Principle methods for cross-linking polyethylene

There are three principal methods for crosslinking polyethylene insulating materials.

1. *Cross-linking by peroxides:* Peroxide mixed with polyethylene decomposes when exposed to high temperature (heat) and pressure, generating radicals that abstract the hydrogen radicals (H·) from the polyethylene, producing polyethylene macro-radicals. The polyethylene macroradicals combine to form crosslinks. This technique is employed to produce cables for high and extra high voltage.
2. *Cross-linking by electron irradiation:* The polymer chains are cross-linked directly by high energy electron beams. Here, the high energy electrons generate the polyethylene macroradicals by abstracting the hydrogen radicals. Generally 1.1 kV XLPE cables are produced by this technique.
3. *Silane cross-linking:* Polyethylene is also successfully crosslinked through peroxide-initiated grafting of alkoxysilane followed by hydrolysis and condensation in the presence of moisture and a condensation catalyst. Cables up to 33 kV are manufactured by this technique.

For long life and reliable service performance antioxidants and sometimes voltage stabilisers and tree-retarding additives are incorporated in XLPE compounds.

Ethylene propylene rubber (EPR)

Regarding such insulating properties as dielectric strength and loss factor, ethylene propylene copolymer (EPM) and ethylene propylene diene terpolymer (EPDM) are surpassed only by XLPE. Other outstanding characteristics of these elastomers are resistance to ozone, oxygen, and ionisation, good flexibility at low temperatures, and high resistance to weather and light. Peroxide-cross-linkable PE and EPDM are suitable for insulation at voltages up to 100 kV.

Silicone rubber

Excellent electrical insulation properties, as well as heat and radiation resistance, make silicone rubber usable continuously at conductor temperatures up to 180°C (for short-circuit periods, up to 250°C). Silicone rubbers are crosslinked by peroxide by means of infrared techniques.

Protective Sheaths

Sheaths or protective coverings provide external protection and prevent the ingress of moisture, mechanical damage, chemical or electrochemical attack, and destruction by fire or any other harmful influences that are detrimental to the cable itself.

Thermoplastic sheaths

Polyvinyl chloride

PVC compounds are predominantly used as sheathing materials for power cables because they combine high tensile strength and elongation, moderate resistance to heat deformation and chemicals, and especially resistance to ageing and flame.

Polyethylene

High strength polyethylene sheath is increasingly preferred for mechanical protection, especially for medium and high voltage cables with XLPE insulation. A PE sheath is recommended in the new German specification DIN VDE 0273/.87 for XLPE cables laid in the ground. High density polyethylene (HDPE), which meets the specifications, is particularly advantageous for conditions involving high ambient temperature (e.g., the tropical and subtropical countries).

Polyamide (PA) and polyurethane (PU)

Polyamide and polyurethane possess outstanding mechanical properties and good resistance to oils; these two materials are used as protective coverings for flexible cables. Polyamide protective coverings are applied to, among others, flexible and wiring cables for use in mineral oil extraction and in aircraft.

Elastomer sheath

When cables are designed to operate under adverse conditions of temperature, pressure, humidity, shock, dynamic flexible conditions, vibration, hazardous atmosphere, flames, fires, etc., the obvious choice is elastomeric sheaths. Finally the major factor that placed elastomeric cables in such an important position is their special-purpose applications. Special cables are mainly used for mines, railways, aircrafts, ships, submarines, electric equipment, nuclear power stations, and instrumentation.

Polychloroprene (PCP)

Vulcanised PCP sheaths possess good resistance to heat, oil, weathering, flames, and mechanical stress. Added advantages over PVC are higher flexibility and resistance to tear extension and abrasion. These sheaths are particularly suited for flexible cables in underground mining applications and in locations plagued by fire hazards.

Chlorosulphonated polyethylene (CSP)

Vulcanised CSP sheaths possess even better mechanical properties, than PCP, as well as superior heat, oil, and weather resistance. CSP also is chosen for the same range of applications for which PCP is suitable.

Nitrile-butadiene rubber (NBR) and PVC-NBR blends

Nitrile rubber shows excellent oil resistance and is blended with PVC to improve its mechanical strength and ozone resistance. Cables that are immersed in oil or come in direct contact with oil have sheaths of NBR or PVC-NBR blends. The physical and electrical properties of the polymers used for insulation and sheathing are given in Tables 16.1 and 16.2, respectively.

Table 16.1. Physical properties of polymers used for insulation and sheathing.

Polymer	Tensile strength (N/mm^2)	Elongation at break (%)	Specific gravity	Resistance to oil and solvent	Resistance to environment[a] Weather	Heat	Flame
PVC	20	200	1.4	VG	G	F/G	E
XLPE	15-20	300-500	0.92	F	G	VG	P
EPR/EPDM	6-8	300-500	0.89	P	VG	VG	P
Silicone rubber	9	200-300	1.15-1.5	G/E	E/O	O	E
PCP	15-18	300-500	1.23	G	E	V	VG
CSP	20	300-500	1.2-1.28	G	E	E	VG
PU	22	300	1.1	G	G	VG	P/E
Fluoropolymers	15	300	1.8	G/E	E/O	O	E

[a] P, poor; G, good; VG, very good; E, excellent; O. outstanding.

Table 16.2. Electrical properties of polymers.

Polymer	Volume resistivity at 20°C (Ω . m)	Permitivity at 50 Hz	tan δ at 50 Hz	Maximum continuous operating temperature (°C)
PVC	1×10^{11}	6-7	0.08-1.0	70
XLPE	1×10^{15}	2.3-5.0	0.0003-0.0004	90
EPR/EPDM	1×10^{13} to 1×10^{14}	3.2-4.0	0.004-0.01	90
Silicone rubber	2×10^{12}	2.8-3.6	0.002-0.025	150
Fluoropolymers	1×10^{15} to 1×10^{16}	2.0-2.3	0.0003-0.0005	250

Armour

Armour provides the cable with mechanical strength and protection. It also serves as an electrical screening. Steel wires or helically wound steel tapes are generally used for armouring. Copper or aluminium wires are used where electrical screening is very important.

Electrical Screening

Semiconducting shields

In high voltage (HV) and extra high voltage (EHV) cables (⩾ 6.6 kV), semiconducting layers are applied as conductor shields (also called inner semicon) and insulation shields (outer semicon). They fulfill the following functions :

1. Provide a uniform electric field around the insulation by reducing the potential gradient over the surface of the stranded conductors and inside the metal shielding.
2. Prevent corona discharge at the surface of the stranded conductors and the insulation by maintaining close contacts between the inner and outer surfaces of the insulation.
3. Provide protection during short-circuit against damage caused by the heating of the conductor and the copper shielding tape.

Semiconducting compounds are made by incorporating conducting carbon black with such ethylene copolymers as ethylene vinyl acetate (EVA) and ethylene ethyl acrylate (EEA). Volume resistivity of these compounds usually ranges between 10 and 100 Ω . cm and should not be exceed 10^4 Ω . cm. Insulation shields are usually fabricated with either bonded extruded semiconducting layers or strippable shield layers.

Metallic screening

Metallic screening provides the path for transmission of the discharge currents in the longitudinal direction of the cable toward the ground. Metallic screening is usually constructed of a layer of copper wires over a helix of copper tape or tapes or a layer of copper tape over each individual core, which may be each screened with conducting layers.

DESIGNATION OF CABLES

Cable designations usually include the following informations :

1. Abbreviated description of the cable design and its component materials.
2. Number of cores by nominal cross-sectional area of conductor (mm^2).
3. Indications of shape and type of conductor.
4. Rated voltage (kv).

Voltage Designation

The design voltages for cables are expressed in the form of U_0/U, where U_0 is the power frequency voltage between conductor and the ground and U is the power frequency voltage between conductors for which the cable is designed. Both U_0 and U are rms values. The operating voltage is the voltage between conductors and ground that is present in the power installation.

Additional Symbols for Cables

Special characteristics are indicated as follows:

1. HOFR, heat, oil and fire resistant.
2. FRLS, fire resistant, low smoke.
3. FS, fire survival.

Ships and aircrafts are among the applications of such cables.

CABLE DESIGN

For cable design, it is necessary to take account of both the electrical stresses that may be experienced during service and the ambient conditions where the cables are laid. Primarily the cable design depends on the following factors:

1. Voltage grade.
2. Current-carrying capacity (ampacity), both continuous and cyclic.

3. Short-circuit level of the system.
4. Types of operation, conditions of installation, and ambient conditions.
5. Mechanical stress.

Current-Carrying Capacity and Selection of Conductor Cross-sectional Area

The cross-sectional area of the conductor is quoted basically not as the geometrical but as the electrically effective cross-sectional area, that is, the cross-sectional area as determined by the resistance of the conductor. The current-carrying capacity and the temperature rise of a cable depend on the type of construction, the characteristics of the materials used, and also operating conditions. To achieve a safe design and a full service life for a cable, the conductor cross-sectional area must be chosen such that current-carrying capacity is higher than current loading for both the normal and shot-circuit conditions. This design ensures that the cable will not be heated above the rated maximum permissible operating and short-circuit temperatures. The minimum number of wires, the diameter of the wires, and the resistance of the conductor are laid down in international standards (IEC 228 and DIN VDE 0295).

Electrical Stress and Insulation Thickness

Dielectric strength, volume resistivity, permittivity, dielectric loss, and corona resistance are the major electrical requirements of cable insulation. Insulation thickness is determined by recognising that the power cable must withstand not only the steady state A/c voltage but also transient lightning impulses and switching surge voltages. The choice of the design stress of a cable can be determined by either the impulse or the A/c voltage requirement of the system in which the cable is installed. The value is usually determined from the basic insulation level (BIL), which is meant to represent the maximum transient voltage that can occur, plus a suitable safety factor.

Two formulas are available to determine the insulation thickness of extruded cables. One is based on impulse breakdown voltage and the other on A/c breakdown voltage, as follows:

$$t_1 = \frac{\text{BIL} \times \alpha_d \times \beta_t}{E_1} + t_0$$

where

t_1 = insulation thickness, to be determined from the impulse breakdown voltage.
BIL = basic impulse level
α_d = degradation factor
β_t = temperature factor
t_0 = thickness of the conductor shield
E_1 = mean electrical stress for the impulse breakdown voltage
and

$$t_2 = \frac{U_0 / \sqrt{3} \times \gamma_d}{E_2} + t_0$$

where

t_2 = insulation thickness, to be determined from the A/c breakdown voltage
U_0 = system voltage
γ_d = degradation factor
t_0 = thickness of the conductor shield
E_2 = mean electrical stress for the A/c breakdown voltage

Dielectric Properties

Charging current and losses depend on the dielectric properties of the insulation. Losses determine the power rating of a cable (i.e., its transmission capability). As the operating voltage increases, dielectric power loss also increases according to the following formula :

$$\text{dielectric loss} = U_0^2 \omega c \tan \delta$$

where

U_0 = operating voltage
c = electrostatic capacitance
ω = 2π times frequency
δ = dielectric loss angle

It is seen that the losses are proportional to the square of the operating voltage. However, this relationship is partly mitigated by the increase in insulation thickness with the increase in operating voltage, which reduces the cable capacitance. Typical values of relative permittivity and tan δ of three major polymers used for insulation are as follows:

Polymer	*Permittivity at 50 Hz*	*tan δ*
PVC	6-8	0.08
XLPE	2.3	0.0003
EPR	3.5	0.004

Compound Design

Polymers in their virgin state are not capable of meeting all the requirements of insulation and sheathing. Thus other ingredients such as antioxidants, fillers, process aids, and curing agents must be mixed in, to impart the desirable product properties after processing. Antioxidants are added to increase the service life by preventing polymer ageing and degradation. Fillers are mainly used to improve the mechanical properties of the elastomers and to some extent for reducing the cost of the compounds. Process aids, as the name implies, help to ensure smooth processing (i.e., mixing and extrusion). Finally, curing agents induce crosslinks between the polymer chains to form three-dimensional networks, thus increasing the dimensional stability, mechanical properties, continuous operating and short-circuit temperatures, and service life of the cable. Elastomer compounding is commonly done by internal mixers and two-roll mixing mills. XLPE and PVC compounding is done by co-rotating and counter rotating twin screw extruder, respectively. Because of the important and stringent requirement for good dispersion, particular attention is paid to the level and range of fillers as well as to their particle sizes.

MANUFACTURING TECHNIQUES

As a consequence of the range and variety of cables, a large number of processes are involved in their manufacture. These can be grouped as follows :

1. Conductor forming, that is, drawing and annealing of wires, coating (tinning and plating) of wires, bunching, and stranding.
2. Insulating and sheathing, which consist of various steps involving mixing and compounding, extrusion, and curing, as appropriate to the characteristics of the cable being manufactured.
3. Assembling, including laying up of cores, taping, braiding, and armouring.

Fig. 16.4. is a flowchart for manufacturing processes and quality control check.

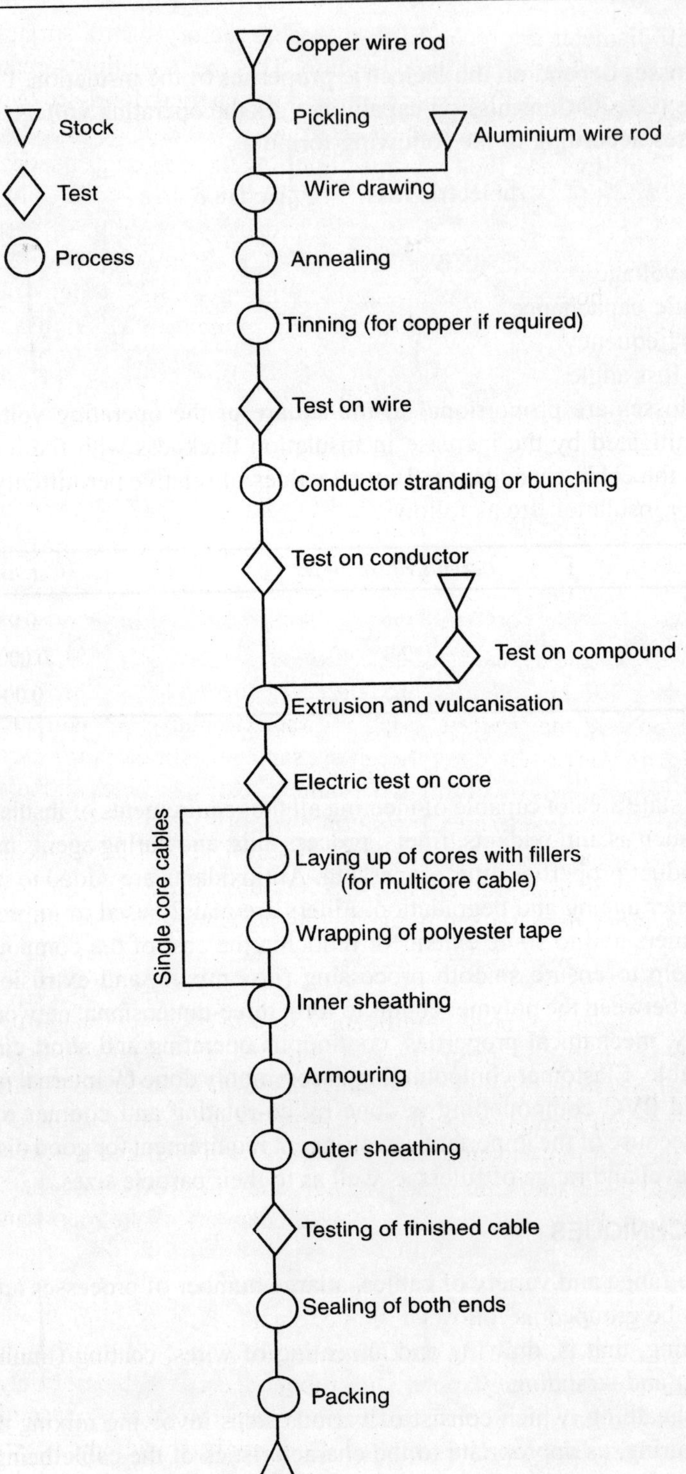

Fig. 16.4. Flowchart for processes and quality control checks in the manufacture of polymer-insulated cables.

Conductor Forming: Copper Wire Drawing and Annealing

Copper rods of 1/4 inch diameter are received in the cable factory with a surface coating of copper oxide, resulting from the preceding hot rolling operation. This oxide coating is removed by "pickling"; that is, the rods are placed in a bath of dilute sulphuric acid for a period, followed by washing, and finally neutralising in an alkaline bath. The rods are then electrically butt welded and continuously fed into a wire drawing machine. "Drawing" the rods consists of pulling them through a series of dies, the orifices of which are gradually reduced in diameter. The hard drawn wire is then annealed to restore the ductility, elongation, and electrical conductivity necessary for power cable. "Annealing" is carried out in annealing furnace under a nitrogen atmosphere at about 400–550°C for 1–1.5 hours, followed by slow cooling to ambient temperature for 6–8 hours. Conductor stranding increases the flexibility of the cables. It can also increase the maximum electrical stress by 20%. The number of individual wire or strands in a stranded conductor N_s is usually a multiple of 6, plus 1:

$$N_s = 1 + 6n + 6(n-2) + 6(n-3) + \ldots + 1 + 3n(n+1)$$

where n is the number of layers of wires.

Extrusion

Insulation and sheathing involve various techniques of extrusion appropriate to the characteristics of the materials used and the curing process. Extrusion of insulation and sheathing is carried out using conventional machines fitted with crossheads. The conductor to be covered (or the core or cable core to be sheathed, as the case may be) passes through a core tube supported coaxially within the head and located concentrically with the die by a tapered extension to the core tube—the core point. HV and EHV cables based on EPR and XLPE are manufactured either by the tandem or the triple common extrusion process. In the tandem process the inner semicon is first extruded on the conductor followed by the coextrusion of insulation and outer semicon through the same crosshead. In triple common extrusion all three layers are extruded simultaneously through the same crosshead. Better interlayer consolidation and elimination of irregularities can be achieved by adoption of one of these processes. Computer-aided design techniques of screws and dies based on the theoretical and experimental analysis of flow characteristics of the polymers and programmed temperature distribution provide good control over molecular packing and arrangement, thus minimising the morphological defects in the bulk insulation.

Curing Processes

The technology of crosslinking has been significantly improved in the last two decades. Numerous procedures and process variants were proposed and in many cases also tried out. The technology has passed through several stages in this period, and ultimately the efficiency and the limits of the individual processes became known well enough to permit a comprehensive evaluation.

Any assessment of the different vulcanisation processes must take into account the size and voltage spectrum of the cables to be manufactured, the design and construction of the cables, the technical requirements, the quality of the product, and the overall economy of the plants and processes.

Batch curing

In the batch or discontinuous method, the insulation or sheathing is cured in an autoclave in steam at a pressure corresponding to 130–170°C; curing cycles vary between 15 and 90 minutes. The cable or core in its uncured state is spirally overlapped with rubberised cotton tape or polyester film for better consolidation, then coiled in trays or reeled on drums. Where maximum consolidation and improved surface finish are required, a metal sheath is applied. Commonly, a lead tube is extruded onto the cable by either a ram or a screw-type press. The lead is stripped off after the cure and reused.

Continuous curing

Conventionally, a continuous vulcanisation (CV) plant consists of a curing tube attached to the die face of the extruder head. There are three types of continuous vulcanisation plant: horizontal continuous vulcanisation (HCV), catenary continuous vulcanisation (CCV), and vertical continuous vulcanisation (VCV). The HCV plant has an obvious disadvantage, namely for reasonable tube diameter and with practical tension, the cable must touch the bottom of the tube. This problem is more critical with cable having a core diameter above 15 mm. For cables larger than this, the CCV and VCV arrangements are more appropriate. The vulcanising tube of a CCV line is generally maintained at an angle between 12 and 25°. With a tension appropriate to its weight per unit length, the cable can be made to conform to the longitudinal axis of the tube. The catenary sensor is set at a point before which the crosslinking reaction of the cable surface is completed; thus any possibility of surface damage or scratch marks is eliminated. After passing through the heating zone, upon completion of the crosslinking reactions, the cable enters the cooling tube, and after cooling it is wound up. The total length of the tube (including the catenary part, the straight part, and the cooling part) has gradually been increased to 70–130 m. Indeed, lately equipment with a length of 150 m has come into wide use. However, when the outer diameter of cable exceeds 80 mm or the wall thickness reaches 20 mm, tension control in the catenary becomes critical, and cable extruded with insulation is brought to deformation by gravity before the crosslinking reaction takes place (i.e., during heating in the tube). Therefore manufacture of XLPE cable with diameters exceeding 80 mm is usually carried out by VCV. Since VCV has a vertical tube, the problem of deformation by gravity does not arise, even in the case of large diameter cable. Processing is thus easier than with CCV. One of the disadvantages of VCV, however, is that a special building having a high tower is required for the installation of the vertical equipment. In the case of both CCV and VCV, the heat required to raise the temperature in the curing zone of the tube can be obtained from steam, high temperature nitrogen, or radiant heaters fixed on the outside of the tube. Particularly for HV cables, however, the use of steam is avoided because it generates macrovoids in the insulation. The heat provided to the insulation causes the peroxide to decompose, generating volatiles such as acetophenone, methane, water vapour, and α-methyl styrene. To keep to an acceptable limit, the size of the voids formed by these gases in the insulation, the tube is pressurised to about 1 MPa.

Mitsubishi Dainichi continuous vulcanisation (MDCV) long land die process

In MDCV process the equipment is horizontal and the die of the extruder is extended for several meters, forming a vulcanising tube to maintain the pressure and prevent void formation. The uncrosslinked extruded core exiting the die enters the tube, and its surface is coated with a lubricant. The core is then slid and moved through the tube and heated, to complete the crosslinking reaction.

Liquid curing

The attempts so far to achieve a liquid cure on commercial scale can be classified into two groups. One is the process using silicone oil and the other is the crosslinking process known as the liquid curing medium (LCM) or salt bath process, which uses sodium, potassium, and a molten eutectic mixture as the heating medium.

Infrared curing

Another important crosslinking method used for the cable manufacture is infrared curing. By this technique, the elastomers are continuously crosslinked by the infrared energy source after extrusion at ambient pressure and at relatively low temperature. Silicone rubber cables, especially, are made by applying this technique.

SPECIAL-PURPOSE ELASTOMERIC CABLES

Fire Survival (FS) Cables

During a fire certain circuits, such as those for emergency lighting and alarms, for the safe shutdown of equipment, and for rescue operations, must remain operative. This demanding service requires special cables with fire resistance and fire survival properties, and such products have been developed.

Silicone rubber insulated FS cables

The insulation for FS cables is a composite of silicone rubber and glass fibre braid (up to 16 mm^2) or silicone rubber coated glass fibre fabric tapes (25–300 mm^2). These cables are used for power distribution and signaling services and will meet the requirements of the fire resistance test of IEC 331 at 750°C for periods in excess of 3 hours.

Mica/glass and EPR insulated FS cables

The cable construction to IEC 92-3 is designed to withstand the IEC 331 fire resistance test at 1000°C for 3 hours. The design features a special inner sheath, combining low smoke, flame-retardant, and halogen-free properties, as well as galvanised steel wire armours (helical or braid) and a special low smoke sheath. The CSP outer sheath is designed to limit the emission of hydrogen chloride gas when decomposed at 800°C to 5%. This sheath also meets the physical requirements of BS 6899 for heat, oil, and fire resistance. Typical construction of such a cable is shown in Fig. 16.5.

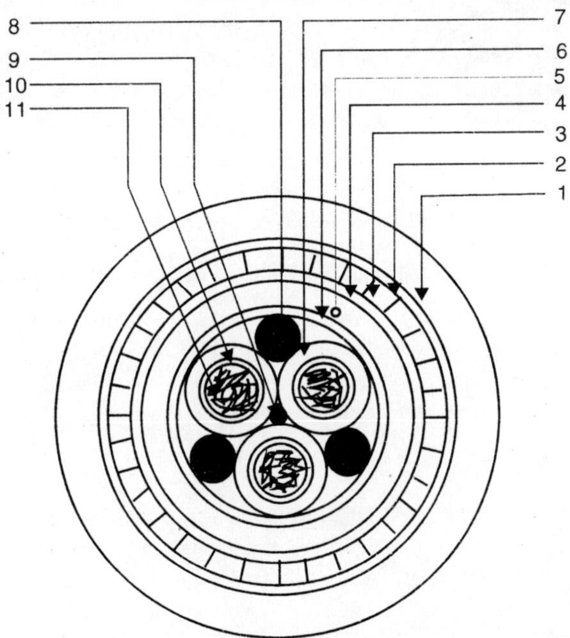

Fig. 16.5. Typical construction of a three core, 1.1 kV fire survival cable 10 mm^2 (7/1.35 mm). 1. outer sheath of black low smoke, low halogen elastomeric compound; 2. layer of glass type; 3. layer of a flat galvanised steel strip armour; 4. layer of glass tape; 5. extruded inner sheath of elastomeric compound; 6. layer of glass tape; 7. heat-resistant elastomeric materiai. 8. asbestos fillers. 9. asbestos centre; 10. two layers of glass mica tape; and 11. stranded tinned annealed copper conductor.

Special constructions complying with BS 6883 and IEC 502, as appropriate, are available, which in addition to withstanding a 1000°C fire resistance test will also withstand impact and water resistance tests.

Radiation-Resistant Cables

For applications in power plants, silicone rubber cables that are resistant to heat and radiation have been manufactured by adopting infrared curing techniques. The chemical structure of silicone rubber imparts a forbearance to radiation and also offers several advantages: long service life under normal conditions, improved circuit integrity when cables are exposed to fire, and no evolution of potentially corrosive halogen- or sulphur-containing gases as by-products of combustion.

CHAPTER 17

Rubber-to-Metal Bonding Components

INTRODUCTION

The unique utility of Rubber to Metal (R to M) assemblies began to be appreciated around the turn of the century through applications that helped cushion motion and deaden noise. An early example was the rubber-clad carriage wheel. And it was not long before rubber received attention as a means of alleviating the severe vibration of vehicles powered by external and internal combustion engines.

Eventually, bonded R to M composites came to be employed in the mechanical world to not only minimise the damaging effects of vibration and to protect against shock, but also to seal against leakage of oil and other fluids, and to simplify assembly by providing a tolerance for misalignment.

The motor vehicle industry is by far the biggest user of bonded elastomer components. Typical applications such as engine mounts, suspension bushings (bushes), transmission and axle seals, couplings, and body mounts are engineered largely to each vehicle's requirements. Manufacture usually involves moulding of the elastomer to shape, vulcanisation, and bonding, all in a single-stage press operation.

In the aerospace sector, R to M mounts and bushings fulfill similar and at least as critical functions. Well-engineered assemblies are vital for reliable sealing of the fuel systems in rocketry and missiles. Sophisticated mounts are necessary for aircraft engine isolation. Less esoteric but no less important to aircraft operation are a wide variety of bonded avionics mounts. For helicopters, advanced R-to-M technology has resulted in such highly specialised devices as low maintenance rotor bearings, embodying alternating layers of rubber and metal in special geometric configurations.

R to M assemblies are further necessary in their aforementioned functions for the nonaeronautical vehicles of war (tanks, submarines, etc.) and for rail and rapid transit rolling stock, as well as for trackbeds and rail crossings. Bonded parts are essential for materials handling machinery, shipping containers, construction and agricultural equipment, leisure devices, air systems, and business machines of various types.

On a more prosaic plane, rubber roll of all sizes with bonded metal cores are used widely for processing rice and other food and in the production of paper and continuous metal sheet stock. Finally, an atypical but significant specialty area of bonding is that of tank and chute lining (with relatively thin sheets). Rubber-lined vessels or conduits are required for mining and chemical processing operations and also for scrubbers in the utilities industry.

RUBBER TO METAL ASSEMBLIES

Rubber-metal composites for shock and vibration isolation are sometimes referred to as bonded rubber springs. The basic designs of such R to M mounts have not change greatly over the years, particularly for the less demanding applications such as isolation or damping of industrial machinery. There has been substantial progress but more in an evolutionary rather then revolutionary vein.

There have been interesting geometric refinements, such as in the design of marine dock fenders. And there have been a few rather new applications, such as the architectural mounts that extrapolate from the concept of rubber bridge bearings–to accommodate expansion and contraction–that have been around for 100 years or more. Architectural mounts were originally developed about 30 years ago to isolate residential and office buildings from the rumble and noise of nearby or underground rail transit systems. But more sophisticated architectural mounts began to be designed to protect buildings from earthquake damage–the so-called seismic mounts (Fig. 17.1). These multiple pad assemblies are required to cope with strong shear forces. On balance, however, the more striking or more obvious progress has been with assemblies for dynamic applications in motor vehicles and in R to M composites for the aerospace and defense sectors.

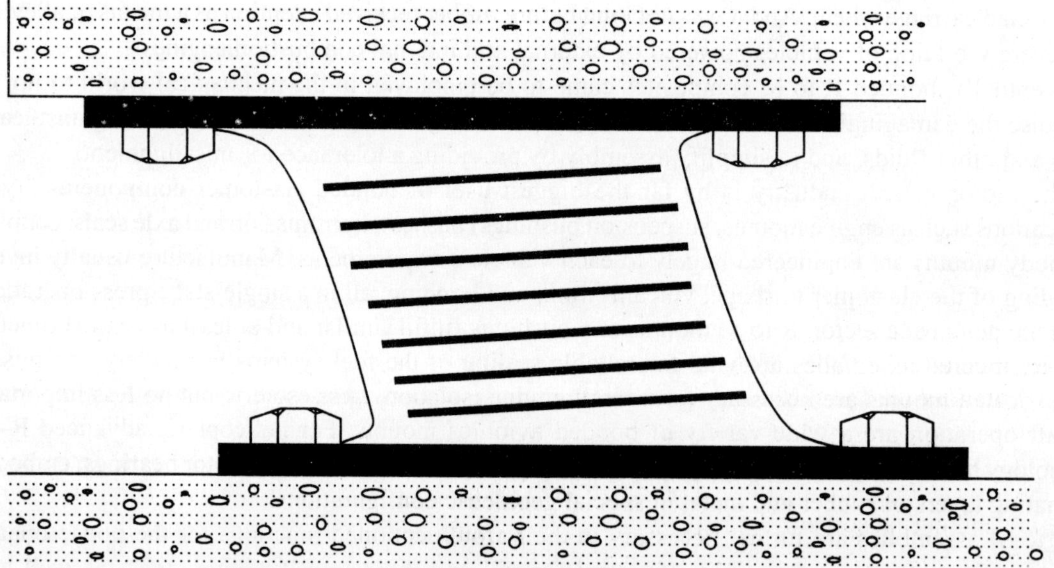

Fig. 17.1. Schematic of seismic mount in shear deformation.

In some instances, thermal degradation of the rubber after years of service (and as aggravated by the softness of the compounds) resulted in catastrophic failure of the mountings such that the engine could break loose from the chassis. Many engine mount designs were then changed to embody a fail-safe cage-type assembly that caused the engine to be retained even in the event of total rubber loss.

To facilitate the manufacture of cage mounts, two-step bonding came to be developed. By this method union was established to a first metal member during normal vulcanisation and to the second metal part by post-vulcanisation bonding. This procedure placed new demands on adhesive systems.

To maintain or upgrade car handling and ride quality, tighter specifications became necessary for such critical properties of the rubber as change in dynamic modulus with temperature. These also began a clear trend to increased numbers of mounts and bushings for vehicle drive trains and suspensions.

The most advanced automobile engine mounts widely used are the so called hydraulic or fluid mounts. These assemblies not only have the traditional bonded rubber springs but also embody the fluid dashpot principle. Confined liquids such as mixtures of ethylene glycol and water are able to move back and forth through an orifice or through baffles. A secondary mechanism of damping thereby imposed results in a more effective control of vibration. Such dual action mounts are being designed not only for motor vehicle engines but also as isolators (e.g., for truck tractor cabs) and now for suspension bushings, as well.

Even more sophisticated mounts have been or are being developed that combine electronic feedback with dual-action control and damping. "Hydromounts" now in production with electromagnetic switches are being further engineered to make use of the electrorheological properties of confined liquids. In effect, the viscosity of this new type of fluid can be changed by application of an electrical field so that its viscoelastic damping is even more closely attuned to changes in the engine's operation occurring in a fraction of a second. For the aerospace and defence sectors, there have been advancements not only in mechanical design but also in materials. Demands of extremely high operating temperatures have resulted in mounts and isolators with elastomers other than natural rubber (NR) or the early synthetic diene types.

After the broad categories of assemblies for control of vibration or shock, seals probably comprise the next most important class of elastomer-metal composites. Seals are of course used in engines pumps, compressors, and axles to prevent leakage of fluids or gases. The terms "radial seals," "shaft seals," and "bore seals" describe the main types, which are generally circular with open centers to accommodate rotating shafts. The sealing function necessitates the use of elastomers which maintain resistance to permanent set in service.

Advancements in seals for transmissions or differential housings have been less striking in regard to mechanical design than in regard to rubber usage and adhesive application. The availability of ultrahigh performance specialty elastomers has also made possible sealing elements of new types (e.g., for fuel injection systems and for engine valve stems).

Valve designs of the 1970s were the earliest to exploit durable fluoroelastomer sealing by mechanical attachment, but bonding now prevails. These valve stem seals are radial shaft seals in the broadest sense but are atypical in being very small and very exacting of manufacturing tolerances. They embody, for example, minute coiled spring steel retainer washers. The valve stem moves through the cylindrical seal's opening by up-and-down motion rather than by rotation. The typical modern automobile may contain as many as 80 separate seals.

With motor vehicle applications comprising the largest outlet for R-to-M parts, it is instructive to group the assemblies according to the principal vehicular functions. Here is a typical categorisation.

Engine: mounts, absorbers or dampers, oil seals (for bearings, valves, and sumps), fan hub couplings.

Transmission and drive train: seals, couplings, bushings, supports, dampers.

Suspension: bushings, macpherson struts, mounts, seals, seatings, bellows.

Steering: couplings, bushings, seals, bellows.

Exhaust: supports, hangers.

Body: mounts, dampers, overload pads, weatherstrips, window channeling.

MATERIALS

Elastomers

As indicated by its position at the top of the list in Table 17.1, natural rubber is still by far the principal elastomer for motor vehicle engine mounts. The *Hevea brasiliensis* polymer–and to some extent its almost identical synthetic cousin, *cis*-polyisoprene–is still unsurpassed in offering dynamic characteristics that are so well suited for a broad range of dynamic engineering applications. In bonded springs such as anti-vibration mounts and bushings, design engineers continue to capitalise on NR's high fatigue resistances, good strength, and stiffness in more than one dimension.

There have been noteworthy advances over the past 25 years in compounding of all the general-purpose diene elastomers but especially in NR. The niceties of dynamic properties as a function of filler type and amount have undergone considerable refinement. Compounding for much improved reversion resistance is now possible with "efficient" and "semiefficient" vulcanisation systems.

Table 17.1. Elastomer for R to M assemblies.

Mounts, bushings, miscellaneous	*Seals*
Natural rubber	Nitrile-butadiene
cis-Polyisoprene (synthetic)	Hydrogenated NBR
Styrene-butadiene	Silicone
Polychloroprene	Polyacrylic
Butyl rubber	Fluoroelastomer
Ethylene-propylene terpolymer	Fluorosilicone
Nitrile-butadiene	Ethylene-acrylic
Epichlorohydrin interpolymers	

Two unusual classes of synthetic elastomer, both embodying polyether linkages in their main chains, have been tested extensively in mounts and bushings because of attractive combinations of performance characteristics. Copolymers of propylene oxide and allyl glycidyl ether submit good dynamic properties over a wide range of operating temperature along with good resistance to thermal degradation and ozone attack. However the "GCO" elastomers have been specified for relatively few bonded applications in the motor vehicle sector.

Comprising the second class of vulcanisable polyether elastomers, also with good dynamic properties, are the epichlorohydrin interpolymer. Elastomers of this broad type are now beginning to be used in small volume in automotive mounts or bushings.

Other elastomers that have found use in dynamic R to M applications include, more or less respectively, polychloroprene rubber (CR), styrene–butadiene rubber (SBR), and nitrile rubber (NBR). In many cases the synthetic is blended with NR or less often with another synthetic, for compromises in performance and /or economics. Silicones are used to some extent in vibration control isolators, particularly for aerospace/defence applications. These provide exceptionally constant performance–resonant frequency and transmissibility–over an extremely broad temperature range.

Butyl rubber (IIR) finds use in mounts requiring high hysteresis loss such as certain dampers and isolators (e.g., cradle mounts for front-wheel drive automobiles). It is also used in truck body mounts, in bonded overload pads, and in elastic damper hang-straps.

EPDM, like butyl rubber, is important in applications calling for excellent ageing characteristics such as bonded weatherstrips, window channeling, boots, bellows, and body mounts–that is, where dynamic properties are less critical. Its capacity to be heavily oil extended with no great sacrifice in elastomeric qualities is advantageous in selected bonded applications.

Other elastomers used in R to M parts, especially where shock control is targeted and ease of fabrication is a consideration, include polyurethanes (particularly the reactive "castable" sort) and a few of the newer elastomeric thermoplastics (or "melt-processible" elastomers).

The thermoplastics have been among the more significant developments in specialty rubbers. Important commercially available elastomeric thermoplastics include various formulations of elasticised polyvinyl chloride, urethanes, styrenics, polyesters, and a broad range of thermoplastic polyolefin (TPO) types. The advantages of thermoplastics for manufacturing operations are obvious: reuse of scrap or trim waste, and absence of requirements for capital–intensive compounding, milling, and curing facilities (i.e., the ability to be fabricated into products using plastics processing machinery.

Thermoplastic elastomers are obviously not practical for bonded R to M assemblies that will do service at very high operating temperatures. But they are beginning to receive attention in applications having reactively modest performance demands, and adhesives are available to bond most types to metal. Ultrahigh performance vulcanisable elastomers for seals have been successful in meeting the more and more stringent demands of the aerospace (and automotive) engineers. The polyacrylates found application, as did NBR, in automatic transmission seals and in seals for crankshafts, pinions, and oil pans. The NBR elastomers, like NR for mounts and bushings, have not been superseded for many motor vehicle seal applications. This remarkable steadfastness is due to the superb oil resistance imparted by the nitrile group. Carboxylated NBRs are not uncommon for seal applications. Offering even greater durability under seal service conditions are the more recently developed hydrogenated nitrile rubbers (HNBR). Ethylene-acrylate ester elastomer is also used in bonded seals because of good temperature and oil resistance, along with reasonable cost.

During past the most demanding seal requirements were being met by new synthetic polymers of very high fluorine content. Breakthroughs in new methods of curing enabled fluoroelastomers to be offered with very low permanent set for extended service under the worst conditions of high temperature and contact with corrosive gases, fuels, oils, and hydraulic fluids.

Metals and Other Rigid Substrates

The rigid members for bonded rubber springs continue to be predominantly steel, particularly for the high volume assemblies of the motor vehicle industry. Castings are used as well as stampings, including a few castings fabricated via the power metallurgy route. For seals, low carbon cold-rolled steel is often the framework of choice. More and more of these various steel parts are today chemically treated (i.e., phosphatised) to minimise corrosion of the entire metal piece.

Lead is specified occasionally for bonded parts requiring a substantially concentrated center of mass and is sometimes a minor constituent of steel alloys to facilitate forming.

Very little aluminum has been used over the years for automobile R-to-M assemblies. But there are reports recently of aluminum beginning to be specified for mounts or bushings in Japan because of improvements over steel in dynamic performance.

Treated aluminum alloys are certainly important in aerospace applications (e.g., for avionics mounts), as are stainless steel and even titanium for highly sophisticated applications.

There has been considerable interest in recent years in reducing the weight of bonded rubber springs by replacing metals with rigid engineering plastics. For example, glass fibre reinforced thermoplastic polyamides are beginning to be used in a small way for automotive bushings. Manufacture in many cases is by regular bonding in the press, but some firms are using postvulcanisation bonding.

One other type of bonded elastomer assembly that has recently assumed importance in the automotive sector makes use of glass rather than metal as the rigid member. Polyurethane elastomers, thermoplastic olefinic rubbers, and elasticised polyvinyl chloride are employed as the gasketing for prefabricated automobile windows, the so called modular window assemblies.

MANUFACTURING METHODS

As noted in the introduction, most R to M assemblies are produced by single-stage press operations wherein shaping, vulcanisation, and bonding all occur more or less simultaneously. The sequence of manufacturing operations is straightforward. The major changes of the past 30–40 years have been in the areas of metal preparation and adhesive application and in the development of advanced injection moulding techniques.

Metal Preparation

It is axiomatic for any type of industrial bonding operation that substrates should be carefully prepared and cleaned. Even autogenous metal surface contaminants such as heavily hydrated iron or aluminum oxides can interfere with establishment of the necessary adhesive bond, one that not only will be strong initially but will withstand various adverse agencies in its end-use service environments.

In most cases, metal preparation involves at least scouring or solvent degreasing, often in combination with grit blasting or some other type of abrasive roughening.

The better two-coat adhesive systems require only that steel parts be sufficiently clean for good primary (initial) bond formation and for robust resistance to service environments. But chemical pre-treatment can lessen not only the corrosion tendency of exposed metal surfaces of the final assembly but also the demands on the adhesive system (i.e., in ensuring the durability of the bond for the lifetime of the part).

A typical metal pre-treatment sequence involves mechanical descaling (grit blasting), solvent degreasing or alkaline scouring, rinsing, phosphate treatment, rinsing, chromate sealing, and drying. The chromate sealing step is not entirely desirable in light of today's environmental concerns–sewage effluent contamination or disposal problems–but it does provide the highest order of service protection of the final R to M part.

Phosphatisation prior to bonding must be carefully controlled. Very thick layers of inorganic phosphates–iron phosphate structures or, in the case of the applied zinc type, mixed iron–zinc composites–can have much less cohesive integrity than the R to M adhesive. In other words, improperly applied inorganic phosphates can act as weak boundary layers. In such cases, the strength of the bonded assembly is limited to the force necessary to start delamination within the phosphate conversion coating layer.

Adhesive Application

One or two-coat adhesive systems are applied to metal parts by spraying, dipping, bushing, and even by roll-coating. Spraying predominates, particularly for high volume motor vehicle assemblies. But whatever the method, film thickness control is essential for satisfactory adhesion and service performance.

Advances in spray application have been notable over the past 10–20 years. Electrostatic spraying, the first advance held great promise for considerably improved economy of use of adhesive systems. However, the electrostatic method places unusual demands on solvent selection and manipulation, since solvent polarity must be within defined limits. And the spray tips easily become clogged. Certain adhesives therefore lend themselves better than others to electrostatic application.

The availability of chain-on-edge machines with automated spraying stations has made possible the most significant advance in this method of adhesive application. The chain on edge machine is very well suited for small and not particularly complex metal inserts that can be hand-arrayed, continuously conveyed to stations where electronic trip mechanisms result in spray coating of just the right amount of primer or adhesive, and dried before removal to bins.

Roller-coating, also by hand, is the most practical technique for tank lining, whereas adhesive application to coil metal (open width) is typically done by reverse roll coating.

The frames or inserts for seals generally involve methods of adhesive application different from those used for the metal parts for mounts or bushing. In one technique the parts to be immersed are contained in a revolving basket at the end of a metal treatment line. Another more complicated method spirals the metals through a reservoir of adhesive and from there to a conveyor line drying station.

MOULDING

The three main types of press continue to be used for bonded assemblies–compression, accounting, transfer moulding.

Injection moulding of course enables faster throughputs, with vulcanisation cycles of 1 or 2 minutes at 200° C$^+$ being feasible for many synthetic elastomer assemblies. But with NR's continuing dominance for mounts and bushings, milder cure cycles of 4–6 minutes at 175°C are nót atypical.

For seal production there continues to be considerable compression moulding of NBR assemblies. Injection molding is increasingly being used for ultrahigh performance fluoroelastomer seals.

It should be noted that the mechanics and conditions of moulding place special demand on adhesive systems. As molten rubber enters mould cavities in advance of vulcanisation, it may move laterally across adhesive-coated metal surfaces. The tendency of the adhesive to become thermoplastic under these conditions can cause it to be wiped or swept along by the rubber–to the detriment of the final adhesive bond. Some commercially available adhesives are much more resistant to this mould "sweeping" phenomenon than others.

Also important is the ability of the adhesive to withstand exposure to near vulcanisation temperatures before the coated metal is in actual contact with the vulcanising rubber. Since most adhesives are reactive chemical systems, "prebake tolerance" is an important practical requirement.

Another mould-related problem that sometimes arises also is attributable to the reactive constituents of adhesive systems. Certain ingredients may volatilise or sublimate from the heated metal surface prior to rubber contact, thus causing fumes at the press and occasionally mould fouling.

Postvulcanisation bonding requires manufacturing setups different from normal assembly fabrication. The best quality bonds, for example of vulcanised natural rubber, are achieved by using the same metal primers employed for normal vulcanisation bonding. Appropriate adhesive covercoats are applied either to the primed metal or to the surface of the vulcanised rubber article. The assembly is then jigged under compression and heated in an oven at a temperature high enough to establish strong forces of adhesion–primer to metal, covercoat to primer, and covercoat to vulcanised rubber. But for less demanding applications there is also some use of instant-setting cyanoacrylate adhesive and of two-part urethane or epoxy cements that cure at room temperature.

Postvulcanisation bonding is nevertheless a practical means of simplifying the assembly of struts or mounts with two or more bonded interfaces (i.e., as designs become more complex).

ADHESIVES

Much if not most bonding is still being done with sulphur-vulcanising natural rubber, which adhered spontaneously to metal surfaces that had been suitably plated with copper alloys.

Polyisocyantes, particularly triphenylmethane triioscyanate, had been effective in press-bonding steel to several of the synthetic diene rubbers. It was hypothesized that the isocyanate group combined during vulcanisation with hydrated metal oxides to form polyureas and triazine ring structures that were strongly adsorbed on the metal surface. The vulcanising synthetic rubbers were assumed to offer various active hydrogen sites that were also able to react chemically with this type of adhesive.

Aromatic isocyanates, probably because they do form true chemical bonds across the interface, can provide R to M assemblies with fairly good resistance to underbond corrosion and attack by water or organic solvents. In a practical sense they are difficult to use under less then ideal conditions because of their great reactivity with adventitious water or even moisture in the air. If metal parts are coated with triphenylmethane triisocyanate and laid over prior to bonding under even moderately humid conditions, much of the adhesive activity is lost.

Special adhesives had to be developed for the newer non-sulphur-vulcanising specialty elastomers. More significantly, broadly versatile two-coat adhesives emerged that were able to bond all the important high volume diene rubbers over a wide range of curing conditions. They were not sensitive to normal variations in humidity. Nor did these newer adhesive systems restrict the part designer's choice of metals. Steel, phosphatised steel, galvanised steel, stainless steel, treated aluminium, brass, titanium–all were able to be successfully bonded as long as the parts were clean.

Broad-Purpose Adhesives

Nitrile rubbers (NBR) and polychloroprene (CR) were observed to be less demanding of the adhesive composition than SBR, NR, or IIR–elastomers that are significantly less polar. Also very important in determining how easy an elastomer may be to bond is its total stock recipe: vulcanisation system (including sulphur level and accelerators), types and amounts of fillers, and antidegradants.

This variation in "bondability" defined the challenge that was met so successfully by adhesive formulators. Because they do bond a wide variety of elastomer types as well as different stock recipes, broad-purpose R to M adhesives tend to be complex and highly proprietary in their makeup. Unlike many industrial structural adhesives, they cannot be categorised generically as "epoxy," "phenolic," "urethane," or "acrylic." The versatile R to M formulations can contain as many as six to eight components in addition to the solvents. Certain of these ingredients are able to become chemically active at elevated curing temperatures. Thus versatility and ability to perform under a variety of vulcanisation conditions appear to be due to more than one mechanism of bonding and, in particular, to one or more processes of chemical reaction across the rubber-adhesive interface. This type of directed chemical reaction can be described as crossbridging. Thus, depending on the elastomer in question and its particular vulcanisation system, one or more of the adhesive's multiple bonding mechanisms will generally be operative to ensure the formation of strong chemical bonds.

The significance of crossbridging is emphasized in the highly schematic diagram of Fig. 17.2. Other important bonding processes are wetting and chemisorption (at the metal end), and adsorption and interdiffusion, which probably occur interfacially to at least some extent.

The widespread use of various halogen-containing polymers in primers and covercoats for vulcanisation bonding is gaining importance. Highly chlorinated resins are known to wet metals efficiently and also to provide good barriers to chemicals that may attack and undermine the adhesive-to-metal bond. Thermosetting resins or reactants bolster performance, possibly by formation of interpenetrating polymer networks. For interaction with vulcanising rubbers, there has been speculation about the ability of chlorinated resins to act as Friedel-Crafts reagents and to thus bond to aromatic sites across the interface. It has further been conjectured that dehydrohalogenation may play a mechanistic role, particularly if sites are thereby created in the adhesive phase for attachment by polysulphidic chains extending across from the vulcanising rubber side.

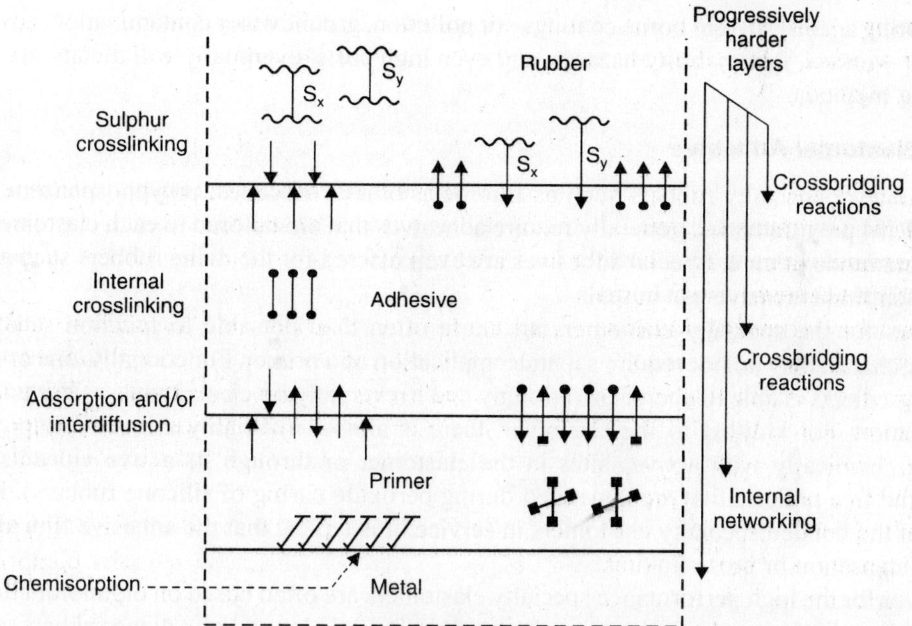

Fig. 17.2. Schematic of vulcanisation bonding processes.

Polyfunctional chemicals (nonisocyanato) that are not in themselves adhesive film formers but do appear to function as crossbridging agents are also disclosed to R-to-M adhesive patents. By analogy with what is known of diene rubber cross-linking, it is likely that the polyfunctional compounds react across the interfaces in the fashion of many nitrogenous "ene" reagents.

Most R-to-M bonding for motor vehicle assemblies is done with robust two-coat systems (i.e, using both a primer and a covercoat). The primer is designed to wet and chemisorb strongly on clean or chemically treated metal surfaces. Its chemistry, again, is such as to achieve high cross-link density during the vulcanisation cycle and to become strongly knitted to the covercoat that is applied over it.

Also available are single-coat adhesives that enable high quality bonded parts to be manufactured without a separate application of a primer. They are often used on metals that have been both phosphatised and chromate-sealed.

Some of the newest developments in general-purpose adhesives are in formulations that offer superior resistance to very specific service conditions or to environments that are not encountered by most R to M assemblies. For fluid engine mounts (i.e., those embodying contained fluids as well as rubber springs), two coat adhesives are now available that better withstand interfacial attack by the contained liquids at high operating temperatures. Very hot mixtures of ethylene glycol and water, for example, can slowly disrupt the R-to-M bond provided by some of the older commercial adhesives.

Also developed in recent years are adhesives for assemblies that will do service under conditions of cathodic disbonding: for example, rubber bonded to metals for under water service, where the metals are protected from corrosion by the use of sacrificial anodes.

Probably the most important adhesive development efforts of recent years have been directed to water-based products for the high volume diene elastomers. The organic solvents that are used for virtually all bonding agents today are becoming less and less acceptable in the industrial world. The same forces militating against solvent borne coatings–air pollution, groundwater contamination, adverse health exposure for workers, flammability hazards, and even high costs–eventually will dictate use of solvent-free bonding methods.

Specialty Elastomer Adhesive

High performance specialty rubbers such as fluoroelastomers, silicones, polyphosphazene types, and castable or RIM polyurethanes, generally require adhesives that are tailored to each elastomer's specific chemistry and mode of cure. Special adhesives are even offered for the diene rubbers such as NBR and HNBR, which find extensive use in seals.

Adhesives for the specialty elastomers are more often than not able to function satisfactorily as single coats; that is, they do not require separate application of a primer. Functionally, one or more of the adhesive ingredients is able to chemisorb strongly and irreversibly on clean metals as driven by the heat of vulcanisation. For knitting to the elastomer, there is almost invariably a reactive moiety that can crossbridge chemically with active sites in the elastomer or through its active vulcanising agency (e.g., with the free radicals that are generated during peroxide curing of silicone rubbers). For suitable durability of the bonded specialty elastomers in service, it is typical that the adhesive film also undergo internal condensation or cross-linking.

Adhesives for the high performance specialty elastomers are often based on organofunctional silanes or reactive phenolic resins. As in the case of adhesives for the high volume diene rubbers, a few of the newer bonding agent prototypes for specialty elastomer make use of water as the carrier rather than organic solvents.

Thermoplastic Elastomer Adhesives

As already discussed, thermoplastic elastomers of various types have become increasingly important in recent years. Adhesives are available to bond most of the thermoplastic rubbers, but they generally function rather differently from vulcanisation bonding agents. Physical chemistry must be relied on through processes of diffusion and adsorption–as set apart from cross-bridging or interfacial chemical reaction–to achieve bonds strong enough for durable practical composites. For the manufacture of metal assemblies with thermoplastic elastomers, very limited heat is available to drive the physical or chemical processes that are believed to be important for good bond formation. Even so, some of the thermoplastic elastomer adhesives are formulated as two-part systems, for cross-linking internal to the adhesive. This enables bonds having better resistance to service environments (e.g., moisture) than would otherwise be the case.

The more nonpolar thermoplastic elastomers generally require two-coat adhesive systems. Others–polyurethanes and elasticised polyvinyl chloride, for example–are in some cases effectively bonded to metal or to glass by single-coat formulations.

TESTING

Primary Bond

The primary R-to-M bond can be arbitrarily defined as that which can be characterised or measured on completion of the assembly operation. It can sometimes be distinguished from the bond that is (or is not) retained after long-term service or after environmental exposure. Test protocols for primary bonds have been recognised by the American Society for Testing and Materials for many years. As delineated under Adhesion to Rigid Substrates (D42-81), there are five standard methods:

1. Rubber part assembled between two parallel metal plates.
2. 90° stripping test–rubber part assembled to one metal plate.
3. Measuring adhesion of rubber-to-metal with a conical specimen.
4. Adhesion test–postvulcanisation bonding of rubber to metal.
5. 90° stripping test–for rubber tank lining, assembled to one metal plate.

Test method 1 is often referred to as the butt joint test; the force of rupture is straightforward tension. Method 2 involves peeling forces, whereas the conical test (method 3) is most discriminating of bond quality in concentrating shear forces within a very narrow zone of the bonded surfaces (Fig. 17.3). The standard method for postvulcanisation bonding also involves preparation of butt joint test pieces. For tank lining, bonded sheets are arranged vertically and peeled in the same direction.

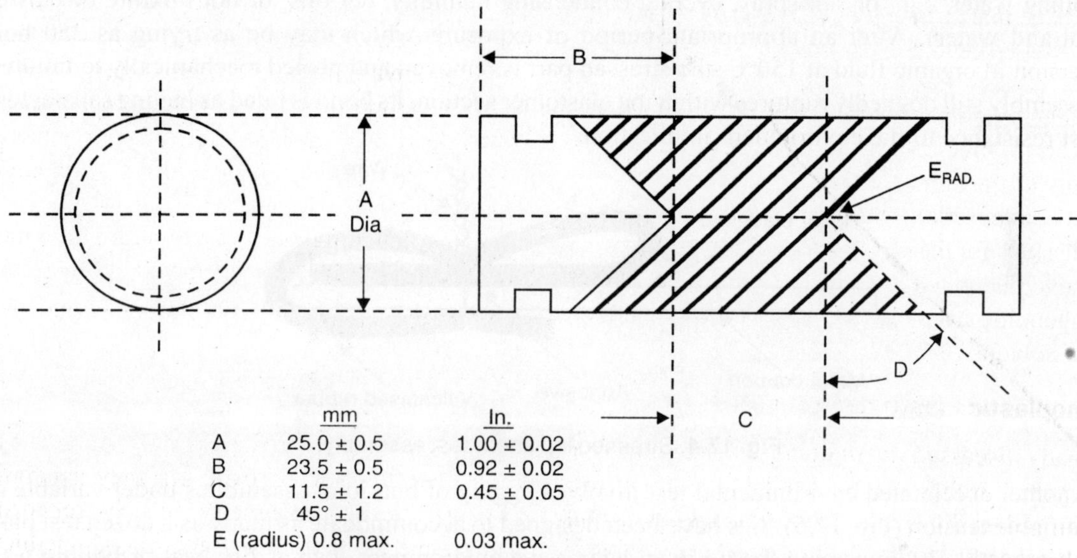

Fig. 17.3. Drawing of conical test assembly.

These tests are valuable to adhesive suppliers for purposes of new product development and technical service (i.e.,. for comparative measurements).

As important as static tests to the fabricators of end products such as engine mounts are special dynamic fatigue tests, which often are tailored to very particular applications and service conditions. The most advanced dynamic testing of highly engineered R-to-M assemblies involves servo-controlled electrohydraulic test trains. These devices make use of advanced electronic controls and are of course highly computerised. Dynamic properties can now be logged over a wide range of frequencies and service conditions in a fraction of the time required for a single frequency –response curve with the test equipment.

It is interesting to note that design, production, and quality engineers often view R-to-M bond testing as a yes/no proposition –the assembly either fails within the rubber (good) or interfacially (bad). Yet changes in the forces of rupture together with changes in the nature of the failure can convey information that is of great value in the course of systemic analyses or experimentation–for example, in development or improvement of manufacturing processes.

Environmental Resistance

An excellent primary bond is of little or no value if it does not hold up in service. Accordingly several accelerated tests have evolved over the years that focus on the resistance of the bonded assembly to various service environments. Most of these tests involve exposure to heat, to salt spray (resistance to underbond corrosion), to boiling water, or to organic fluids. The latter category subdivides into lubricating oils, hydraulic fluids (several types), fuels, and antifreeze mixtures.

One test that is useful for screening purposes is a modification of the ASTM 90° stripping test. For exposure to aggressive environments, the rubber strip is tensioned back on itself as in Fig. 17.4. The highly stressed rubber is tightly tied down (e.g., with a stout section of wire) and "scored" or cut directly at the exposed interface with razor blade. It is then subjected to the service condition (immersion in boiling water, e.g., or salt spray, cycling condensing humidity, hot oils, or hot mixture of ethylene glycol and water). After an appropriate period of exposure–which may be as trying as 250 hours immersion in organic fluid at 150°C–the stressed part is removed and peeled mechanically to failure. If the assembly still doggedly ruptures within the elastomer section, its bond is rated as having satisfactorily robust resistance to the environment in question.

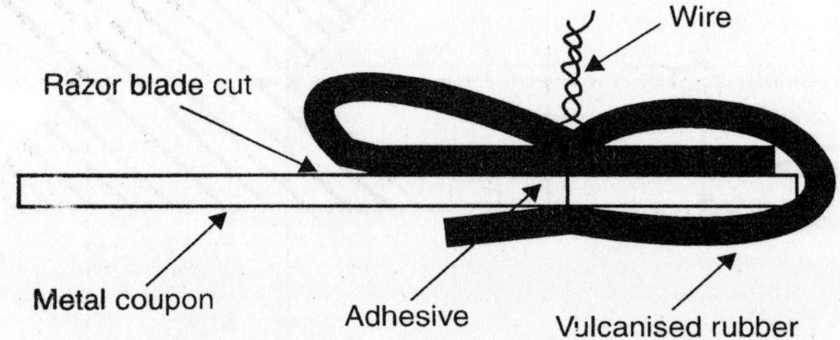

Fig. 17.4. Stressed lap shear test assembly.

Another accelerated environmental test involves jigging of butt joint assemblies under variable and quantifiable tension (Fig. 17.5). Jigs have been designed to accommodate as many as a dozen test pieces simultaneously. By immersing the jig in an aggressive environment such as dry heat or boiling water, resistance can be measured quantitatively in many instances, simply as a function of time to failure. (With appropriate levels of tension, the parts generally fail catastrophically–as a result of attack on the bond–or not at all.

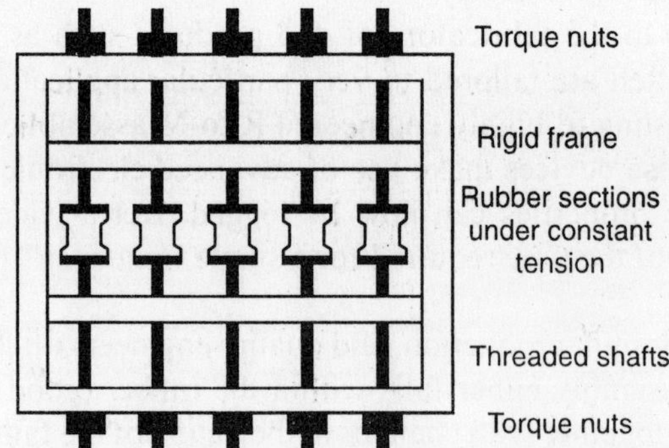

Fig. 17.5. Drawing of test jig with butt joints in tension.

CHAPTER 18

Vibration Isolators and Mounts

INTRODUCTION

Vibration is what one feels upon putting one's hand on the hood of a car with the engine running, or on the base of a running motor. As a simple illustration, consider the case of a compressed spring (Fig.18.1). When released, assuming negligible resistance, such a spring exhibits wavy and repetitive motion, which is a function of time. These are the essential features of mechanical vibration.

Vibration is a magnitude (force, displacement, or acceleration) that oscillates about a mean reference. The maximum displacement from the mean is called the "amplitude" of the vibration. The interval of time within which the motion sequence repeats itself is called a cycle or period, and the number of cycles executed in a unit of time (e.g., second or minute) is called "frequency," Any kind of vibration is commonly characterised by "frequency," in cycles per second or hertz (Hz), and "amplitude."

Shock is defined as a motion in which there is a sharp, sudden change in velocity. Shock usually entails a single impulse of energy of short duration and large acceleration. Examples are blows of a hammer or a box falling to the ground from certain height. Shock is generally characterised by g: meters per second per second.

PRINCIPLES OF ISOLATION

Vibrations can be isolated but shocks need damping; hence the two are treated separately. Many applications involve both vibration and shock in varying proportions, however, and here a judicious blend of the two principles becomes necessary.

The task of a rubber technologist is to design a suitable compound and maintain modulus at a level that will satisfy the load–deflection requirement of a given design. If no damping is involved, the best elastomer will be natural rubber, which can be compounded to ensure that optimal elastic properties will be maintained over a long period of service. Table 18.1 gives a typical formula and the typical application.

Rubber mounts are not normally effective for very low frequency applications because of the impractically high frequency ratio. In such cases the natural frequency of the isolator must be lowered, by changing either the mass of the isolator (since it is inversely proportional to natural frequency) or its stiffness. In many cases it is impractical to change the mass of the isolator. Thus one changes the other directly proportional quantity (i.e., the stiffness) by, for example, adding fillers or cork powder (which incidentally also improves damping by hysteresis), or by using reinforcing wires or inserts. Such cases always require a special design.

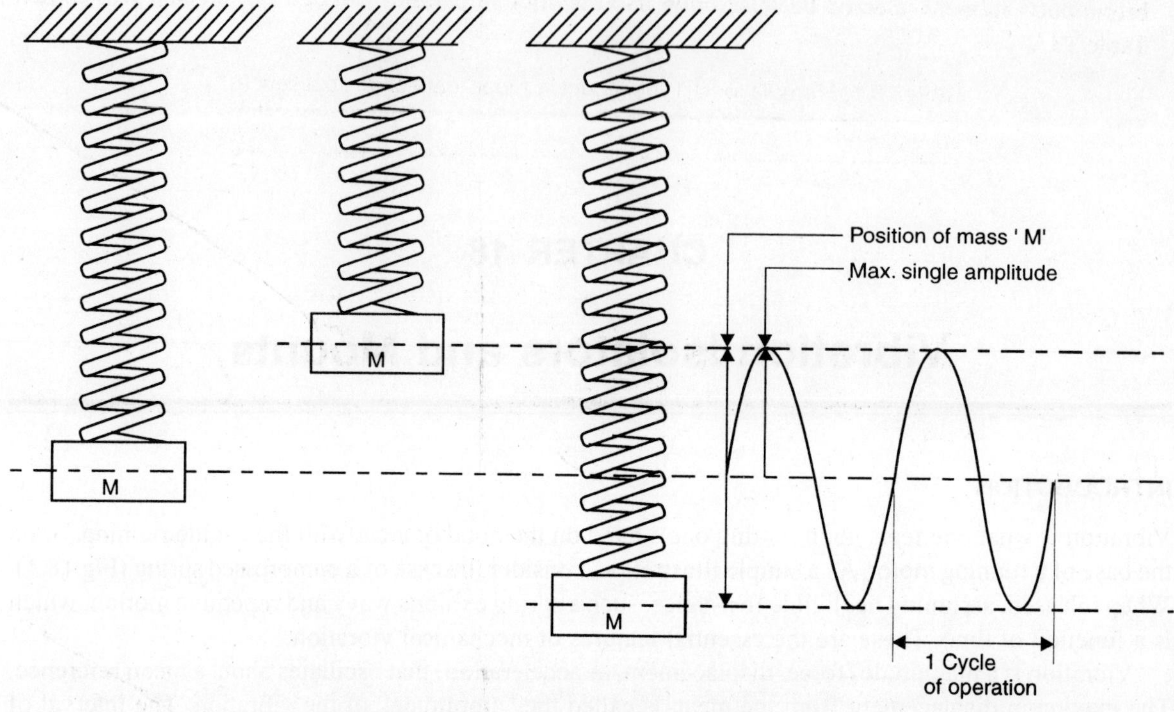

Fig. 18.1. Essential features of mechanical vibration.

While an elastic composition is recommended for vibration isolation, a certain amount of damping is wished for in some applications, to depress resonance rise or to cause rapid die-out in case of shock. In such cases the same compounding principles used for low frequency applications will apply for particular elastomer. But elastomers themselves are characterised by differing degrees of transmissibility at different ratios of disturbing frequency to natural frequency.

At a frequency ratio of 1 (i.e., under resonating conditions), the descending order of transmissibility is as follows :

1. Natural rubber/EPDM.
2. Chloroprene rubber.
3. Styrene butadiene rubber.
4. Polyisoprene rubber.
5. Nitrile rubber.
6. Butyl rubber.

As the frequency ratio increases and crosses into the magnification area the order reverses itself, the least magnification being shown by butyl rubber.

Similarly, compounding with carbon black, fillers, and so on lowers transmissibility by inducing an increase in damping. Since this also increases modulus, the frequency of resonance is also shifted toward a higher level. It must be determined for each application whether the resonance needs to be shifted toward the higher side or the lower side, and the decision depends on the polymer and compounding system. For example, in a particular application involving a typical "O"-type mount (Fig. 18.2), the natural compound given in Table 18.2, causes resonance at a frequency of 28 Hz, whereas resonance was tolerable only well below 20 Hz, which could be achieved by the blend of natural rubber and bromobutyl rubber (selected because butyl itself is not compatible with natural rubber), also shown in Table 18.2.

Table 18.1. Formula for U-Type Mount in a load-deflection application.

Ingredients[a]	*phr*
Natural rubber	100
Zinc oxide	5
Stearic acid	1.5
PBNA	0.8
IPPD	0.8
Black FEF	20.0
Oil	5.0
CBS	1.5
Sulphur	1.5
Physical properties	
Shore hardness	50°
Tensile strength	20.77 MPa
Modulus at 3000%E	11.61 MPa
% Elongation	600

[a] The following abbreviations are used in Tables 18.1–18.5: PBNA beta naphthyl amine; IPPD: N-Isopropyl. N phenyl-p-phenylene diamine; FEF: fast extrusion furnace; TMT: tetramethyl thiuram disulphide; CBS: N-cyclohexyl-2-benzothiaxole sulphenamide.

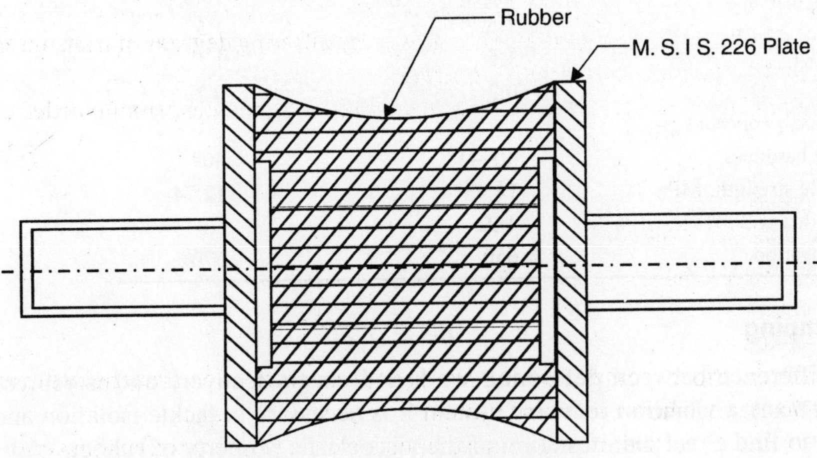

Fig. 18.2. O-type mount.

Similarly, in the case of a cup-type mount (Fig. 18.3), an initial attempt to restrict resonance amplification featured the use of white factice in a natural rubber formulation Table 18.3. The successful formulation, however, was a predominant blend of bromobutyl rubber also given in Table 18.3.

Fig. 18.3. Cup-type mount.

Table 18.2. Formulas for an O-Type mount.

	Phr	
Ingredients	*Natural compound*	*Bromobutyl compound*
Natural rubber	100	25
Bromobutyl	–	75
Zinc oxide	5.0	5.0
Stearic acid	1.5	1.5
PBNA	1.0	1.0
IPPD	1.0	1.0
Black FEF	5.0	5.0
Pine tar	10.0	10.0
White factice	10.0	10.0
CBS	1.5	1.5
Sulphur	1.5	1.5
TMT	0.2	0.2
Physical properties		
Shore hardness	40–41°	40°
Tensile strength, MPa	14.43	12.74
Modulus at 300% E, MPa	4.92	5.0
% Elogation	650	670

Principles of Damping

Damping is the difference between deforming work and elastic recovery and is represented by the hysteresis loop. In many a vibration reducing system it is necessary to tackle isolation and damping at the same time and to find a real manifestation of the viscoelastic property of rubber.

Rubber, in dynamic condition, can be conceived of as a hybrid of steel spring and liquid-filled dashpot: the former represents the fully elastic three-dimensional network of vulcanisate having a modulus, dependent on amount but not rate of strain, that is retained for long periods without relaxation and without heat buildup, and the latter represents raw rubber having a modulus dependent on rate but not on amount of strain, which relaxes with time and loses heat. In vulcanised rubber both these characteristics are present in varying proportions, depending on the amount of cross-linking and network formation. Formula for machine mount is given in Table 18.4.

Table 18.3. Formulas for a Cup-type mount.

Ingredients	*Phr* *Natural compound*	*Bromobutyl compound*
Natural rubber	100	25
Bromobutyl	–	75
Renacit- 7	0.2	–
Zinc oxide	5.0	5.0
Stearic acid	1.5	1.5
PBNA	1.0	1.0
IPPD	1.0	1.0
Black FEF	5.0	5.0
Pine tar	10.0	10.0
White factice	20.0	20.0
CBS	1.5	1.5
Sulphur	1.5	1.5
TMT	0.2	0.2
Physical properties		
Shore hardness	35°	34–35°
Tensile strength, MPa	13.35	11.12
Modulus at 300% E, MPa	4.57	4.73
% Elongation	650	675
Amplification at resonance	8.5	3.5

Table 18.4. Formula for machine mount.

Ingredients	*phr*
Natural rubber	100
Zinc oxide	5
Stearic acid	1.5
PBNA	0.8
IPPD	0.8
Black FEF	25.0
Oil	5.0
CBS	1.5
Sulphur	1.5
Physical properties	
Shore hardness	52°
Tensile strength	21.19 MPa
Modulus at 300% E	12.04 MPa
% Elongation	550

For a given mount design such complex requirements will be governed by the static stiffness to correspond to the desired natural frequency required to isolate the disturbing frequency of the engine and also minimum loss factor (tan δ) required to absorb the shock. Compounding features that can bring about changes include increasing the loading of carbon black, as shown in Table 18.5.

Table 18.5. Formulas for mount.

	phr		
Ingredients	*Compound 1*	*Compound 2*	*Compound 3*
Natural rubber	100	100	100
Renacit-7	0.2	0.2	0.2
Zinc oxide	10.0	10.0	10.0
Stearic acid	1.5	1.5	1.5
PBNA	1.5	1.5	1.5
Black FEF	5.0	10.0	15.0
Pine tar	10.0	10.0	10.0
CBS	1.5	1.5	1.5
TMT	1.5	1.5	1.5
Physical properties			
Shore hardness	35°	40°	45°
Tensile strength, MPa	15.14	16.97	17.67
Modulus at 300% E, MPa	7.39	8.66	9.01
% Elongation	650	630	600
Static stiffness, N/mm	654	879	1159
Loss factor	0.104	0.137	0.268

Designing and Compounding for Vibration Isolation and Shock Absorption

Designing and compounding for vibration isolation and shock absorption call for a fair estimate of actual requirements. In addition, the exact dynamic properties must be worked out within practical limits of variations. In most other applications one tends to exceed the minimum requirement by a safe margin to be sure to conform to the specification, as well as to compensate for certain unforeseen conditions. For example, if the tread rubber of a tyre is to be designed for abrasion resistance, one might design and formulate the material for use on concrete road surfaces so that even on tarred road it will perform better. But in the case of vibration management, no such safety factor is relevant at all: one must be precise within certain limits, nothing less, nothing more. A spring designed for a truck obviously will not work for a car, and vice versa.

MANUFACTURING TECHNOLOGY

The manufacture of vibration isolators and shock mounts is much simpler than their design and formulation. Mainly it involves moulding by compression, transfer, or injecting, depending on the design and the volume of production. A vast majority of these devices are metal-to-rubber bonded products. Some are partly bonded and partly assembled (Fig. 18.4), and some are simple assemblies (Fig. 18.5).

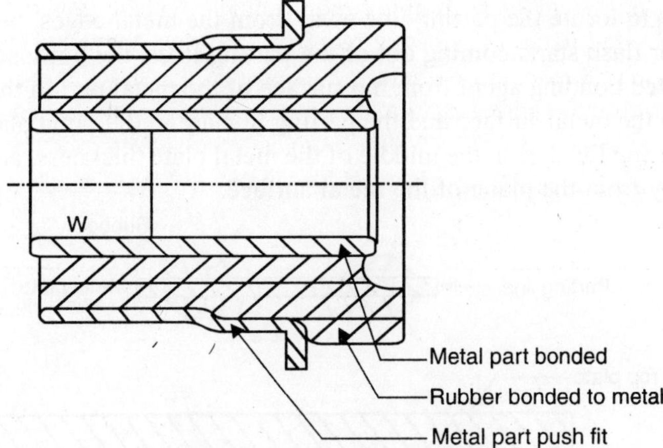

Fig. 18.4. Partly bonded mount.

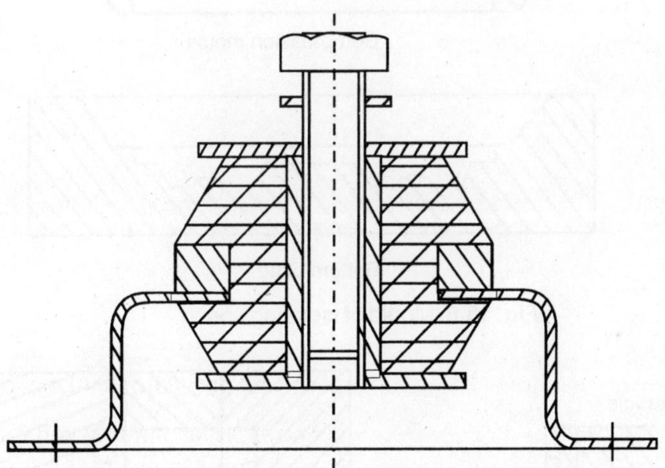

Fig. 18.5. Assembled mount.

In the case of metal bonding, surface preparation of the substrate is of vital importance. Ideally this is done by blasting of chilled angular iron grits, normally preceded by degreasing with chlorinated solvents and followed by application of chemical bonding agents, usually proprietary formulations of polymeric materials, different resins, reinforcing fillers, and so on, in a solvent base. Metallic inserts so prepared must be handled carefully to avoid contamination before taking up for moulding alongwith the rubber compound. Take-up should be completed within 24 hours of metal preparation.

Normally such products are manufactured by a moulding process: by compression or transfer moulding (injection moulding in a compression moulding press) or injection moulding. Injection moulding is normally used only in the case of mass-produced articles like shaft seals used in automotive transmissions. Some shaft seals are still moulded by compression and some are moulded by transfer methods, but the trend has been toward injection moulding, particularly when very large quantities are involved.

Figs. 18.6. and 18.7 present typical examples of metal-bonded vibration isolating springs and their moulds meant for use in the compression mode.

Care must be taken to locate the parting line away from the metal edges, because otherwise during compression, as rubber flash starts coming out of the parting lines, the hot viscous rubber will tend to sweep the already coated bonding agent from the surface of the metal part in the vicinity of the parting line, particularly when the metal surface and the parting line are in the same plane.

In Fig. 18.6, the parting line lies in the middle of the metal plate thickness, and in Fig. 18.7, there are two parting lines, away from the plane of the metal surface.

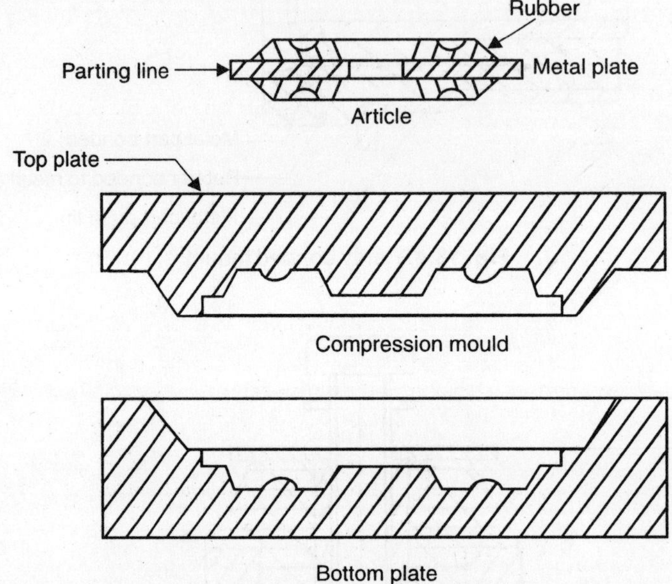

Fig. 18.6. Mould of isolating spring.

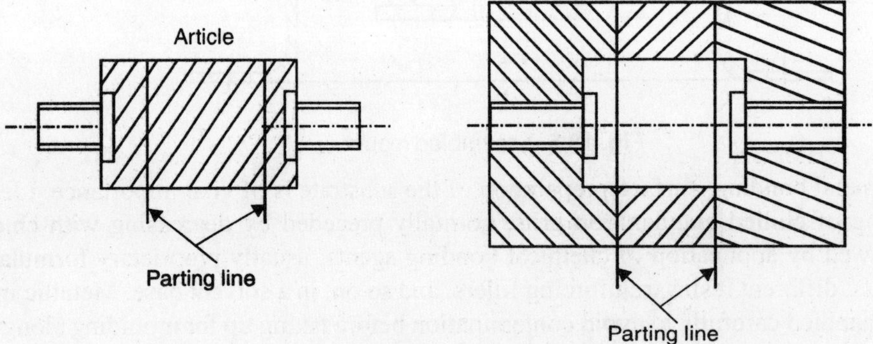

Fig. 18.7. Compression mould of O-type mount.

In compression moulding great care is necessary to keep the rubber blank uncontaminated by dust and foreign matter. Preforming to the nearest dimension is necessary. Transfer moulding is a kind of injection moulding done in a compression moulding press. Typical articles and their mould designs are shown in Figs. 18.8 and 18.9. Here also parting lines must be located away from the edges of the metallic parts to avoid flushing away the bonding agent. The metallic parts are located in a sort of a groove, which almost locks the rubber and prevents the bonding agent from flowing out.

The advantage of transfer moulding is that because a freshly injection surface of rubber is made to adhere to a metallic surface, incidence of weak spots is negligible. This defect is not uncommon in compression moulding because soiling of either the rubber blank or the painted surface of the metal parts is unavoidable.

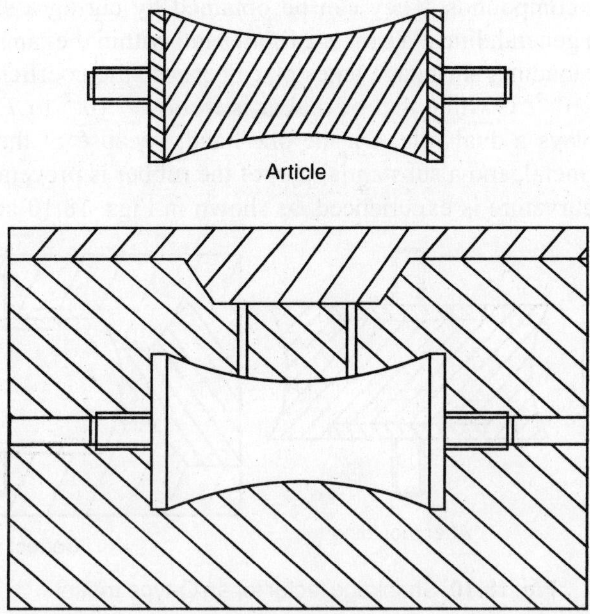

Fig. 18.8. Injection mould of O-type mount.

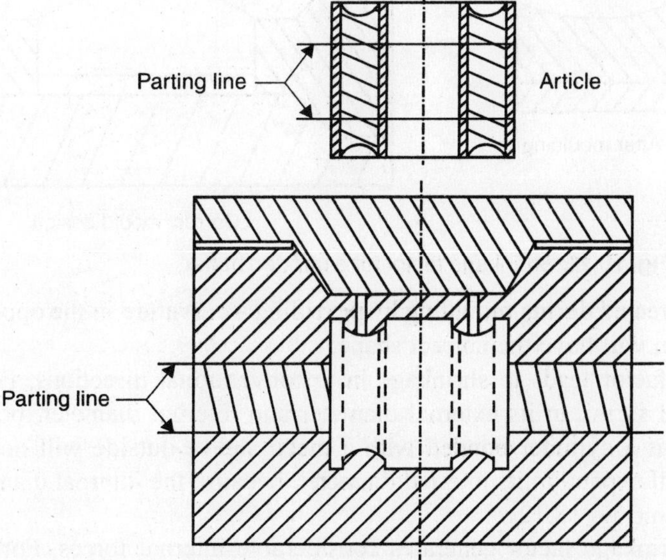

Fig. 18.9. Mould of bonded silent block.

Similar moulds are used in injection moulding and in transfer moulding. The other vital aspect of mould design for metal-to-rubber bonded products is the shrinkage factor. "Shrinkage" is the term used to describe the difference in the dimensions of the mould and the article produced from it when both are measured at ambient temperature. If strict dimensional accuracy is required, shrinkage factors must be determined for individual compounds. They can be obtained by curing a standard test specimen at production temperature. In general, linear shrinkage figures fall within the range of 1.5–3.0% depending on polymer type and filler loading; shrinkage takes place because the coefficient of thermal expansion (volume) of steel is 0.3×10^{-4}/°C while that of rubber is from 4×10^{-4} to 7×10^{-4}/°C.

The shrinkage factor plays a dual role. On the one hand, because of the difference in shrinkage (rubber shrinks more than metal, and a substantial part of the rubber is prevented from shrinking by the bonded metal), a kind of curvature is experienced, as shown in Figs. 18.10 and 18.11.

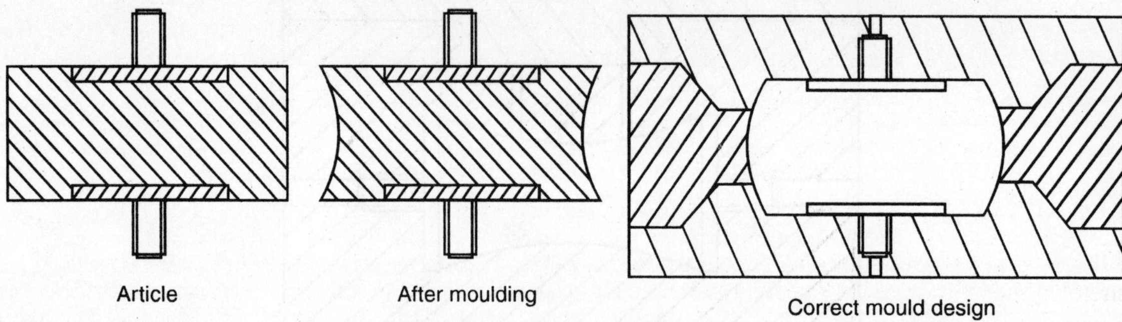

Fig. 18.10. Shrinkage factor for an O-type mount.

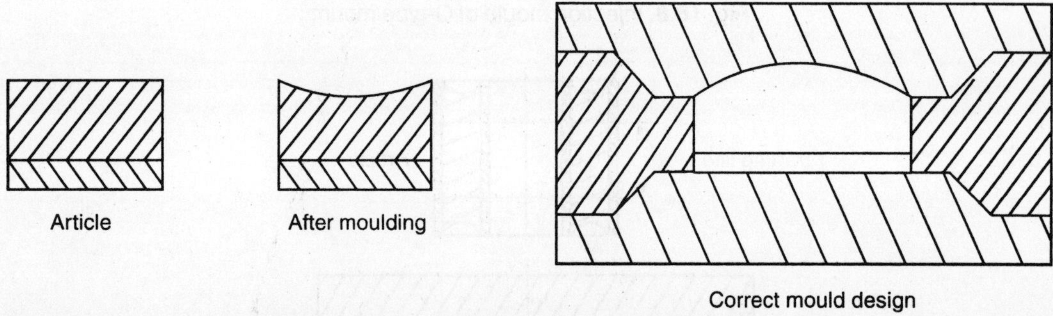

Fig. 18.11. Shrinkage factor for a bonded buffer.

This curvature must be corrected during moulding by providing a curvature in the opposite direction so that after shrinkage the item will have the correct shape.

Sometimes the shrinkage factor leads to shrinkage in unconventional directions. For example, a simple cylinder of rubber will shrink in its external diameter and internal diameter, both toward the centre of the cylinder, whereas a cylinder bonded with a metal sleeve outside will not shrink in its external diameter at all but will expand in its internal diameter, because the internal diameter is free to shrink in the opposite direction (Fig. 18.12).

On the other hand, the shrinkage factor generates considerable internal forces. For example in a cylindrical article bonded both externally and internally with a metallic sleeve (Fig. 18.12), rubber is not allowed to physically shrink either toward the internal diameter or toward the external diameter; therefore stresses are generated inside rubber portion, and the bond is thereby placed under some strain.

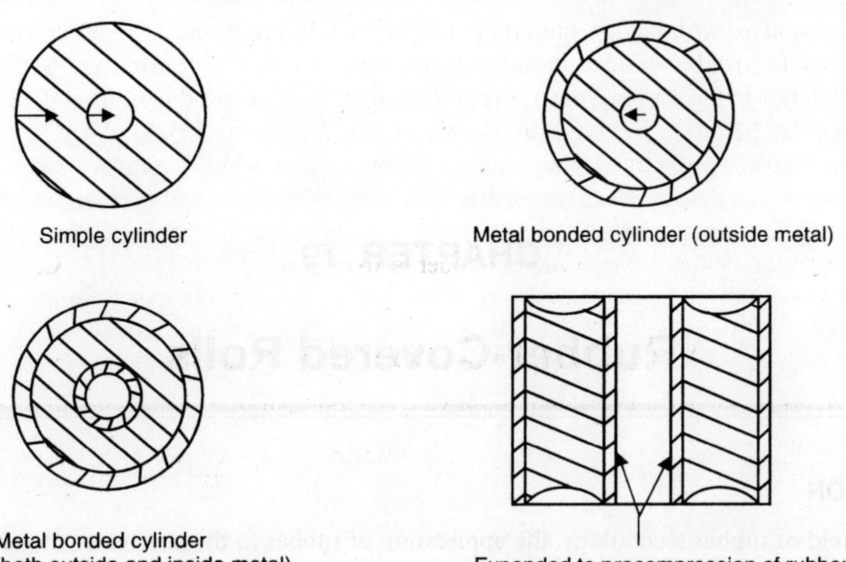

Fig. 18.12. Shrinkage factor for a bushing.

This is sometimes compensated by precompression of the rubber–for example, by expanding the inner metal of the cylindrical composite before the part is put into service. This counteracting of shrinkage by precompression is a bonus effect, since the main virtue of precompression is that in service the rubber will not pass through zero strain and will thereby gain a many fold increase in its fatigue life.

CHAPTER 19

Rubber-Covered Rolls

INTRODUCTION

In the broad field of rubber technology, the application of rubber to the surface of metal cores to make rubber-covered rolls, generally referred to as rubber rolls, is a highly specialised area comprised of unique methods, practices, and standards using a well balanced combination of the rigidity of the metal core and the elasticity of rubber.

Although rubber rolls may vary in importance, their versatility ensures that they are used almost in all sectors of industry for a wide variety of purposes and on almost all kinds of machines–from tiny rolls for making envelopes and movie film to giant rolls used in paper making.

The functions of rubber rolls in various applications are diverse, demanding different properties from the rubber that is used. This, in turn, means that rubber rolls must be made for specific applications, which makes standardisation difficult.

One of the most extensive uses of rubber rolls is in the control of liquids by pressing or squeezing, as exemplified in the manufacture of paper and in the processing of metals, textiles, plastics, and leather. Such typical devices as the press roll, the touch roll (paper making), the padding roll (textile), and the transferring roll (printing) all take advantage of the elasticity of rubber.

Rubber rolls are used for many conveying purposes, such as feed rolls and guide rolls of various kinds, as used in the paper (table rolls), steel (sink rolls, deflector rolls, tension rolls), textile (guide rolls, tension rolls, expander rolls), and plastics industries. The properties of rubber that are important for such applications include elasticity, friction, abrasion resistance, and chemical resistance. In addition, resistance to solvents and plasticisers as well as resistance to heat are desired in applications such as laminating, embossing, and coating (textile, plastics, paper, and printing).

DESIGN OF RUBBER ROLLS AND CHOICE OF MATERIALS

The design of rubber rolls involves consideration of general factors such as the work place, the working conditions, the desired performance, the required durability.

The core is designed depending on factors such as the load, the speed, and the required precision. The cores may be built of seamless, welded, or cast pipe and may be made from mild steel, cast iron, stainless steel, bronze, aluminium, etc. However, the use of drawn steel pipes and mild steel as the core material is desirable because of the ease with which these materials can be bonded with rubber.

It is almost impossible to vulcanise soft rubber satisfactorily to a cast iron core because during vulcanisation, heat drives gases out of the porous iron, causing blisters in the covering and resulting in early service failure. To overcome this problem, a layer of ebonite is applied to the core and before the soft rubber cover.

Most cores are hollow cylinders, although solid cores are not unusual. The hollow cylinder has heads in each end, and in these heads are installed journals or shafts to support the bearings on which the roll turns.

In the design of cores, some allowance is required, consideration of the rubber covering process and reprocessing as well as the conditions of use. Even though the load is light, a slender core deforms as a result of the forming pressure during vulcanisation or bends because of pressure during scraping or grinding, resulting in lack of accuracy. Besides, it may be that the new metal surface of the core cannot be prepared for repeated covering of rubber.

Large cores require vent holes, which allow the vulcanising medium to heat the shell from the inside as well as from the outside, resulting in an increase in the efficiency of vulcanisation.

For rolls that are to operate under heavy loading, high speeds, or both, it is customary to design the core so that water cooling can be used to carry away the heat of hysteresis that is a generated when the rubber goes through the nip.

In addition to the conventional type of roll body, there are types that have unique designs for specific industrial applications. For example, the suction press roll used in paper making is generally made of bronze or stainless steel and has holes drilled through the surface of the metal core and, later, through the rubber covering. A stationary suction box is provided inside the roll to aid in the removal of water by drawing it into the rubber cover.

In another example of a specially designed roll body, internal oil pressure is used to ensure uniform nip pressure across the face of the roll, which permits crown adjustments while the roll is running. These rolls consist of a smooth bore in a cylindrical shell, a stationary centre shaft, and accurately machined hydraulic components. In such applications the roll covering manufacturer generally bolts "dummy heads" onto the shell, thus providing support so that the roll can be covered.

After the core has been made ready, the rubber covering is applied on its surface. It is important to consider what is expected of the rubber covering. The following areas are significant: (i) chemical resistance; (ii) heat resistance; (iii) physical properties (iv) hardness; (v) thickness of the rubber; and (vi) economics.

Chemical Resistance

The proper type of rubber selection depends on careful study of the chemicals and solvents that will come in contact during service. There has not been and possibly never will be an all-purpose rubber. The correct choice of the rubber to be used for coverings that will offer optimum chemical resistance is sometimes very difficult because so many factors, especially under actual service conditions, can affect the behaviour of the rubber. These factors include the nature of the chemicals and solvents, their concentration, the degree of purity and nature of suspended solids, the extent and continuity of contact with rubber coverings, the effect of chemical absorption and swelling, the effect of temperature, the effects on the rubber compounding ingredients, and the effects of pressure and stress.

Basically the polymers can be attacked in two ways: (i) by solvent swelling; and (ii) by chemical attack.

Solvent swelling

The "three solvency indices," namely solubility parameter, hydrogen bonding, and dipole moment, help in predicting a polymer's resistance to swelling actions on theoretical grounds are well known. Although such theoretical treatment is of great use in preliminary sorting, in the final analysis the practical roll covering compounder cannot be satisfied with such results. It is certain that for optimum serviceability, measurement of the effect of the working environment on physical properties is a must.

Chemical attack

If chemical attack causes degradation, the application of "solvency indices" is clearly of little significance. Degradation implies an alteration of the chemical structure of the polymer, either by simple chain scission or by more complex alteration of molecular structure, and in such cases, testing is essential. At elevated temperatures the chemical attack is accelerated, and other complicating features come into the picture (e.g., oxidation, cross-linking, chain scission). The accuracy of accelerated immersion tests as carried out in industrial laboratories is open to question, owing to the complexity of the degradation processes entailed and the complications produced by variations in sample thickness.

In the case of solvent swelling, the test temperature can affect both the penetration rate and the equilibrium swelling value.

Additionally, the effects of pressure and stress during prolonged work contact with chemical media cannot be ignored. Obviously, it is not possible to gain information on this point from normal immersion tests. However, the study of chemical resistance under static and dynamic oscillating loads may be helpful.

The compounding and curing of rubber to be used in roll covering may introduce some variables that must be taken into consideration if no better material is available. When a rubber is known to be perfectly resistant to attack by a chemical, compounding exerts relatively little influence. However, the presence of certain carbonates (e.g., calcium carbonate in hot acidic conditions and silicious fillers in hot alkali) should be avoided because of their susceptibility to attack.

From the examination of the behaviour of stressed vulcanisates in corrosive media, it has been observed that by using fillers that increase the strength and stability of the vulcanisate, a higher stress is required to cause breakage at a given time.

In the case of swelling by solvents, it is evident that reduction of available rubber by an increase in the compounding ingredients will in itself reduce the volume swelling. A maximum number of cross-links, to obtain the tightest possible cure, generally is good practice to keep swelling and chemical attack to a minimum. Where resistance to heat is a prime requirement, a suitable curing system is the answer.

Plasticisers used in compounding frequently are leached out when in contact with organic fluids. This loss of plasticisers may or may not alter the hardness of the rubber covering, depending on the amount of plasticiser leached and the swelling action of the fluid. However, the serviceability of the roll covering may be affected as it physically becomes poorer. A good practice is to use non-extractable plasticisers. An assessment of volume or weight change together with any alteration in hardness is a good guide for roll coverings that are not required to withstand stresses, although additional tensile tests are of great use as supporting evidence. It is important to carry out the swelling tests until equilibrium is attained, while defining the ratio of surface area to volume and the testing temperature.

A thick rubber covering will withstand both chemical attack and swelling better than a thin one. Similarly, attack will be slower if just the surface comes in contact (as opposed to dipping the whole unit in a bath).

Usually, chemical attack is reduced by dilution of the aggressive reagent with an inert liquid. Similarly, the diffusion coefficient and the penetration rate of many organic fluids depend on concentration. In practice, it is quite unsafe to assume that a known swelling agent may be tolerated because it is present in small quantities in an otherwise inert medium. The behaviour of some rubbers in binary solvent mixtures, as examined, and by relating the calculated cohesive energy density of the mixture to the swelling produced, revealed that swelling maxima do not take place at cohesive energy density values characteristic of the polymer. The use of thicker coverings will only delay the ultimate failure, and the best answer is to use other materials or to alter the design of the system.

Heat Resistance

The temperature encountered during service must be considered in selecting the type of rubber. The roll service temperature is raised not only by the ambient high temperature or by enhancement from the hot web, but also by heat buildup in the rubber due to high speed rotation of the roll under a high load. Because of the poor heat conductivity of rubberlike materials, the surface temperature of roll covers cannot be used as a criterion of heat buildup inside the rubber covering. Studies have shown that under certain conditions, inside temperatures are more than 40°C higher than that at the surface of the roll covering.

Physical Properties

The rubber, selected from the standpoint of chemical resistance and heat resistance, must be tested for ability to withstand the working conditions (load, shock, shearing, abrasion, etc.) For example, silicone rubber cannot be used in many cases because it is inferior to other rubbers in its physical properties. Furthermore, it has been necessary to replace general-purpose rubbers like natural rubber (NR) or styrene butadiene (SBR) by nitrile or carboxylated nitrile or even polyurethane rubber, which have better loadbearing capacity in applications where very high nip pressure are encountered.

Hardness and Thickness of Rubber

Since the compression deflection characteristics of rubber are utilised in rubber rolls of all types either to control liquids or to absorb shock loads, the hardness and thickness of rubber is one of the most important criteria and should be determined only after due consideration.

The load-deflection characteristics of a rubber roll cannot be predicted from simple hardness values, since these are determined by the ratio of cover thickness and finished roll diameter along with Young's modulus. Thus two rubber rolls of the same core and finished diameter and with the same hardness can show different load-deflection characteristics depending on different Young's modulus values. However, the normal rubber modulus values at definite strains such as 100 to 200% alongwith elongation at break values, provide some indication; for example, lower deflection corresponds to higher modulus values and vice versa. Again the ratio of rubber covering thickness to finished roll diameter is very critical in determining the load-deflection characteristics. For example, if the same rubber covering compound having a definite Young's modulus value (or normal rubber modulus value at 100 to 200% strain) is used for a cover thickness of 10 mm and an overall diameter of 200 mm, the compression may be 2.5 mm, whereas for a cover thickness of 20 mm and an overall diameter of 200 mm, the compassion might be as high as 6 mm. A higher compression deflection characteristic is advantageous in certain applications where the web may break and enter the nip, an area in which the chances of damaging the rubber covering are greatly reduced. However, proper hardness (coupled with modulus) must be maintained to prevent excessive deflection, which could result in permanent set and waviness of the rubber covering.

Much emphasis is placed on specifying the hardness of rubber coverings, which indeed is merely a measure of the degree of resistance to penetration of the indenter and thus largely dependent on thickness, making a difference between the apparent hardness and the true hardness. As the rubber covering thickness is reduced by grinding, the apparent hardness increases but the true hardness may remain unaltered. Thicknesses of 10 and 6.5 mm are considered to be essential for measuring rubber covering hardness up to 50 Shore A and 50–100 Shore. A, respectively.

The hardness of the roll is the middle reading of three taken 120° apart around the centreline. The allowable limit of any specified hardness usually is ± 5 Shore A. As a normal practice, the limit for hardness variation within a roll is taken as 4 Shore A. The hardness variation arises not only from dispersion defects but also from non-uniformity in heating during vulcanisation. In the heavy rolls (e.g., paper mill rolls), the crown will not hold with satisfactory uniformity if there is excessive variation in hardness. Again the softer areas may cause wet streaks, while the harder spots give a dry surface. Similar anomalies may be observed in printing rolls, particularly ink transfer rolls or spreader rolls, wherein softer areas may cause more deposition while the harder spot gives a finish which is not distinct.

It is quite normal to set the hardness of rubber between 30 and 85 Shore A and the thickness between 10 and 35 mm, but there are exceptions such as foam rolls for printing (with a hardness of 15 Shore A) and platen rolls for typewriters (with a hardness higher than 90 Shore A).

Economic Factors

When several rubber materials meet the requirements, such factors as cost, life, grinding interval, and loss due to the exchange of roll should be taken into consideration.

PROCESS OF RUBBER ROLL MANUFACTURING

The process of rubber roll manufacturing consists of the following major steps :

1. Preparation of core and rubber-to-metal bonding.
2. Formation of rubber covering.
3. Curing.
4. Grinding and finishing.

Preparation of Core and Rubber-to-Metal Bonding

Excellent adhesion between rubber and core is decisively important to the satisfactory performance of most rubber rolls. Irrespective of the method of bonding, the core must be suitably prepared. The first step of core preparation is to expose the cores to steam at about 150°C for several hours. This is required to remove the anticorrosive materials applied on new cores, which are normally rough machined. Although light metal cores are not protected by an anticorrosive agent, their surfaces are slightly more porous than steel cores and should also be cleaned with steam. Any deeply cut grooves should be machined off the core before the steam treatment; this is to avoid the increased risk of trapping air between the core and the rubber covering when the latter is applied. In the case of used metal rolls, the old rubber covers are removed on a lathe or, less frequently, by burning off.

After this preliminary cleaning, the cores are sandblasted to produce a virgin metal surface that is free from rust and scale and also to increase the surface area for better bonding. Finally the cores are cleaned with a solvent to remove any trace of oil and grease.

The prepared core is applied with bonding agent as soon as possible, to avoid reoxidation of the surfaces. In the manufacture of rolls, good rubber-to-metal bonding is ensured by two methods: chemical bonding agents and use of an ebonite base layer.

Chemical bonding agents offer short curing times (unlike ebonite) and give high adhesion values even at elevated temperatures (contrary to ebonite bonding, which begins to soften at a relatively low temperature: ≈ 70–80°C). Chemical bonding agents are, however, somewhat sensitive to mechanical influences. If rolls, particularly those having a relatively thick rubber cover, are subjected to dynamic stresses, the increasing shear forces may cause the cover to break loose in the course of time.

Ebonite tie stock is sometimes preferred, especially when the rolls have a thick rubber covering and are subjected to heavy dynamic stresses. Nowadays small and medium-sized rolls are almost exclusively produced with chemical bonding agents. Ebonite offers very high separation strength and the rubber covering undergoes little deflection, since up to a quarter of it may consist of ebonite, which has little tendency to be deformed. When using an ebonite bonding system, that the soft rubber cover should not be allowed to overcure (since the cure time of ebonite is long), and care should be taken to ensure that the ebonite is cured to the optimum (since adhesion is unsatisfactory when a sufficiently high degree of cure is not attained). Furthermore, pronounced sulphur migration occurs between the ebonite and the soft rubber in the case of overcure, leading to the formation of a thin layer of semihard rubber at the interface between the ebonite and the soft rubber; such semihard rubber has unsatisfactory mechanical properties and frequently results in failure between the two materials. An additional tie stock is generally used to absorb the migrating sulphur without impairing its mechanical properties, to avoid such adhesion failure between ebonite and soft rubber layers.

Chemical bonding agents are applied uniformly as a thin film, usually by brushing. The build up of the roll can be started as soon as the film of bonding agent has dried. Sometimes, one or two additional coats of a specially prepared assembly solution are applied for easier buildup, since most of the chemical bonding agents are without intrinsic tack.

In the ebonite bonding system, the core is given two or three coats of an ebonite solution, each coat being allowed to dry before the next is applied. This series is followed by covering with a layer of ebonite approximately 1 mm thick.

Forming of Rubber Covering

The following methods are normally used for covering the metal cores that have been prepared and treated with bonding agent :

1. Plying of calendered sheeting.
2. Plying of extruded sheeting.
3. Covering with extruded profile.
4. Covering with extruded hose.

Rubber cover compound 1.0–1.5 mm thick taken from a calender is passed through cooling drums, then wound with a liner (normally polyethylene sheet). Such thin sheets are preferred because they can be removed from the calender without forming air bubbles. The cover sheet is next applied over the core, mostly using hand rollers, until the built-up diameter is roughly 4–6 mm greater than the desired finished diameter of the rubber roll. During application of the sheet, some solvent wiping is carried out in case the compound does not have sufficient building tack. Sufficient care is taken to ensure that solvent is fully evaporated before plying up. Alternatively, such sheets can be warmed by passing over hot platens or by heating with an infrared bulb for reviving the tack of the compound.

In the course of mechanisation, the hand building process has been replaced by the three-roll machine and special buildup machines. The three-roll machines are provided with a variable speed drive, which particularly facilitates wrapping down. The rolls of such machines generally have a rubber covering (Fig. 19.1).

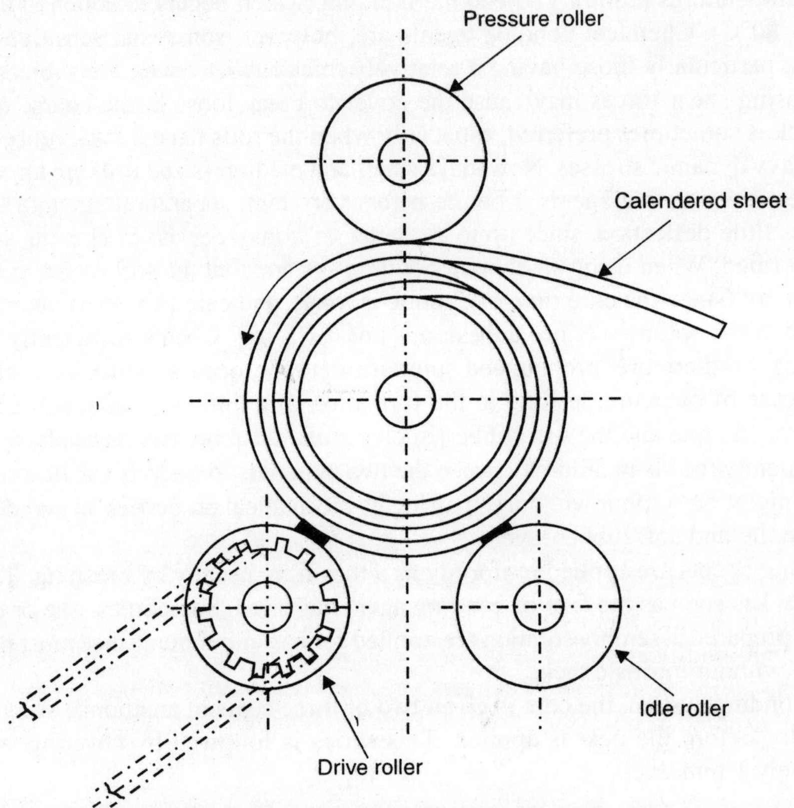

Fig. 19.1. Three-roll machine for roll covering using calendered rubber sheet.

To ensure that the sheeting is rolled on satisfactorily, the pressure roll usually has a somewhat softer covering; in most cases, the pressure is adjusted hydraulically. Such machines permit good productivity, especially for the covering of small, light rolls up to about 2 m long. By reason of their design, such machines have not proved suitable for building up heavy rolls.

In another machine using calendered sheet, the core is placed between the dead centres or between a dead centre and a jaw chuck. The drive system is of variable speed, and generally the roll is turned continuously and very slowly, while at the same time the sheeting is smoothed by pneumatic pressure or smoothing rolls. In one refinement of this process, the roll remains stationary while the sheeting is being smoothed and is turned only when the direction of the smoothing rolls changes. This system is particularly suitable for building larger rolls; the machines are connected to heatable bench, which warms the sheet, thus improving its tack.

Plying of extruded sheeting is better than plying of calendered sheeting in some respects because the sheet coming out of the extruder is free from blisters. Such blister-free sheets can be used much thicker (usually in the range of 4–8 mm), thereby reducing the building time. Furthermore the hot sheet is used directly in the plying-up operation. The core is rotated in such a way that its surface speed is always similar to that of the extrudate, which is squeezed over the rotating core by an adjustable press roll placed vertically above the core. Normally 20–30 cm extruders are used with fishtail head and roller head dies. This process is quite fast, but only comparatively short rollers can be covered because of the machine limitation.

In the relatively new method of covering cores with extruded profile, the ply is laid down at an angle to the core. In effect, the width of the extruded strip is greater than the thickness of the covering, and only one pass is required to cover a roll of average thickness. The ply interfaces are not concentric to the metal core, since they are in conventionally covered rolls, either calendered or extruded. The most important effect of this method is that there is no generation of differential pressure when the roll is stretched in a nip or swollen by chemicals, as in conventionally piled-up rolls, where we find blistering or ply separation. Rather, the roll acts like a monolithic piece of the rubber, eliminating such problems.

This method has further advantages. For example, air trapping between plies, which occurs with conventional methods using extruded or calendered sheets, is eliminated because the width of each strip in contact with the base is narrow and there is no tendency to bridge over surface irregularities. Again, conventional building machines use a pressure roll that bears straight down on the sheet. The pressure roll in the new machine presses the profiles sideways against each other and also forces the profile radially against the core. A chamber at the bottom edge of the roller provides a slight milling action, which intimately welds the rubber being used at the hard base, which was applied by conventional methods, thus providing maximum adhesion of the cover to the roll. The new method also screens the rubber immediately before it is applied to the roll, eliminating the possibility of contamination of the roll by foreign matter. Finally, the new method has the advantage of flexibility, both in the variety of rolls that can be covered and in the materials that can be applied, along with the fact that a comparatively small extruder can be used. In another method of roll covering, an uncured rubber tube, prepared by extrusion, is put on the core. In a modification of this process, cured tubes of correct outer diameter are then put on the core, using a special type of bonding agent (postvulacanisation bonding), which eliminates the need for any further grinding or finishing. The process is now limited to small rollers, where pressure on the rolls is not high (Fig. 19.2).

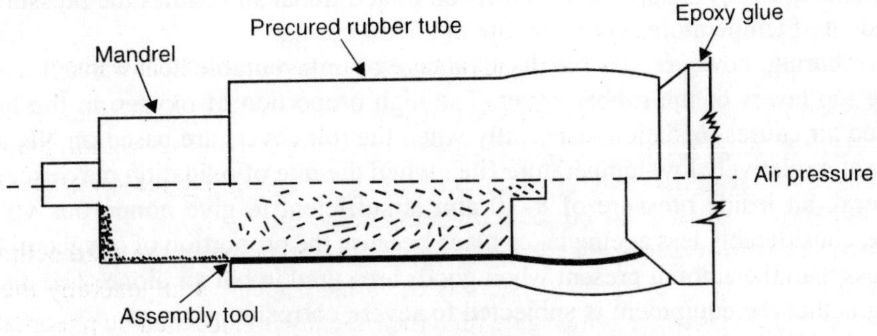

Fig. 19.2. Precured tube building.

Whatever method is employed for building, the built-up cover is made slightly larger than the final design dimensions, to compensate for the grinding operation to follow. Furthermore, the thickness at the middle (where wrapping is commenced) and also at the ends is made a few centimeters higher, which serves to compensate for the flow of rubber during wrapping.

Finally, the material protruding at the ends of the roll is cut off to measure and, if necessary, the side surfaces of the rolls are also covered. To prevent the compound from flowing away at the ends of the roll, blocking disks are normally fixed at the ends of the roll.

The built-up roll is next wrapped, either in a lathe or by the build-up machine, to prevent sagging and porosity. The wrapping is wound as tightly as possible from the centre to one of the ends, back to the other end, and then to the centre again. To ensure the necessary pressure, a particularly narrow fabric strip (~ 25 mm wide) is used for the first wrapping. Then several more wrappings are made as tightly as possible, this time with a particularly broad tape.

Cotton is the most widely used material for tapes. The pressure exerted on the cover by a cotton tape can be increased by wetting the cotton tape and winding it on in the wet state, to take advantage of the shrinkage that occurs as the tape dries. Currently nylon and polyester tapes are popular because of the number of times they can be reused.

Curing

Roll covers are usually open-cured in horizontal autoclaves, the curing media being steam, hot air, a combination of steam and hot air, or water. Except for small rolls that must be made in large numbers, curing in the mould generally is not practical for three reasons: (i) it is uneconomical to prepare moulds for the production of small quantities; (ii) the inset of the metal core, which has large heat capacity, makes it very difficult to cure the rubber in a short time; and (iii) since the rubber layer is thick, slow curing is desirable.

For the most part, saturated steam is used in directly heatable curing autoclaves because of simplicity of operation. The heat transfer of saturated steam, compared with that of other gaseous media, is the most favourable for the curing process. Steam curing offers the desired uniformity of the temperature with considerably shorter curing time. However, the dependence of the steam pressure on temperature and vice versa is often felt to be a considerable disadvantage of steam curing: particularly for large rolls, which must be cured at the lowest possible temperature to prevent reversion, the pressure corresponding to the temperature might not be high enough to prevent porosity reliably. Steam-air curing, which combines the advantages of steam and hot air curing, is practiced for large rolls. Apart from the efficient heat transfer brought about by steam, the introduction of additional air enables the pressure to be raised at will, independent of temperature, as obtainable in hot air curing.

Simple hot air curing, however, has the disadvantage of unfavourable heat transfer associated with oxidation of the top layers of the rubber cover. The high proportion of oxygen in the big autoclaves using compressed air causes oxidation, especially when the roll covers are based on NR and are cured for long periods at a relatively low temperature (i.e., when the rate of oxidation may exceed the rate of curing). In general, an inside pressure of 8–10 atm is sufficient to give nonporous vulcanisates. In steam-air curing, considerably less ageing takes place because the proportion of oxygen in the autoclave atmosphere is less than the amount present when goods are cured in hot air alone. The disadvantage of steam-air curing is that the equipment is subjected to severe corrosion.

Water curing has limited applications when a combination of low curing temperatures and high pressure is desired, as in the production of thick ebonite linings. In water curing the wrapping process can be eliminated, and there is hardly any premature ageing induced by oxygen. The main disadvantage of this method is that it makes heavy demands on space and time because the low curing temperatures necessitate very long curing times.

The following factors determine vulcanising time and temperature: (i) type of rubber; (ii) thickness of cover; (iii) bonding system employed; and (iv) size and condition of the core, particularly whether it is solid or hollow.

The curing time for rubber rolls can be divided into four stages or steps, namely rising time, preheating time, curing time, and evacuation time.

A step cure is necessary for large rolls, medium-sized rolls, and also small solid cores or undrilled rolls. It is not normally necessary for pipe rolls and small rolls with thin cover.

The purpose of the step cure is threefold. First, it serves to heat the interior portions of the roll and the rubber cover by increasing the temperature without curing the cover, and second, by heating the compound without curing it, viscosity is reduced, allowing entrapped or generated gases to escape. Thus the softened ply surfaces begin to bond thoroughly before the vulcanisation reaction commences.

The step cure usually begins at about 100°C and progresses at intervals of 10–20°C, taking from half an hour to 2 hours for each step, depending on roll size and the maximum curing temperature used. The total time to reach curing temperature may vary from about 2 hours for small rolls to 20 hours for very large rolls. Actual curing temperatures usually range from 135 to 150° C. With large undrilled or solid rolls it is recommended to cut a centre band approximately 30 mm wide, recover with new compound using 2 mm thick sheets with separating agent, and run a test cure.

Rising time is the time taken to raise the temperature in the autoclave from room temperature to that required for curing. The rising time is characterised by a continuous and even increase of temperature and depends on such factors as the size and initial temperature of autoclave, the thermal diffusivity of rubber covering, the method of cure, and the facilities for the rapid introduction of steam or hot air. Rising time differs from case to case: with relatively large rolls, for example, it must be extended some what to include an additional preheating time in the curing cycle, to allow the whole mass of rubber to be heated. The temperature corresponding to the preheating time period should not be high enough for cure to take place. Since the full pressure is not available throughout the entire rising time, including the preheating time, porosity might develop if cure were to commence at this stage. It is more advantageous to use compounds having a comparatively short scorch time and a long cure plateau than to use compounds with delayed action and peaky cure, particularly when large cores are employed.

The thickness of rubber covering is important in general because of the poor heat transfer properties of elastomeric materials; it is particularly important with large undrilled or solid core rolls. The amount of heat that reaches the cover from the inside of the roll is determined by (i) the amount of heat circulated from the core; (ii) the condition of the interior surface; and (iii) the thickness of the shell itself.

Norman considered a model roll covering with rubber 2.5 cm thick of a core 45 cm in diameter, 30 cm long, and 27 cm in internal diameter with end plugs 50 cm long. It was observed that conduction of heat through the rubber covering alone could not produce a useful rate of temperature rise in the metal core and that conduction of heat from the ends was insufficient to affect the centre of the roller. Thus it was concluded that for satisfactory heating of the rubber in contact with the centre section, an adequate internal steam supply is essential.

Generally speaking, the rate of cure of the rubber compound can be neglected for large rolls because the overall cure time is far in excess of the optimum cure time of the compound. Thus rubber covering compounds must be designed to withstand large degrees of overcure without impairing their physical properties. Again, the thermal diffusivity of a rubber compound is more important than the optimum cure time. In a typical example, a core of 2 cm, diameter was covered with 5 cm thick rubber using two different compounds: compound A, with heat diffusivity of 0.0015 cm^2/s, having an optimum cure time of 20 minutes at 140°C, and compound B, with heat diffusivity of 0.00085 cm^2/s, having an optimum cure time of 15 minutes at 140°C. After 90 minutes of cure time, it was observed that the roller using compound A was fully cured, while that with compound B showed porosity and sulphur bloom.

The largest influence in curing is exerted by the core. If the rubber cover has been applied on a hollow or tubular core, the roll can be heated both from the cover and from the core, and under such conditions it is generally unnecessary to extend the preheating time unless the rubber cover is very thick. Where the mass of metal is relatively small, axial conduction can produce the effect of heating the rubber from the inside. When the core is larger in diameter, the metal acts like a "heat sink" and the inside heating effect is found to be negligible. This is exemplified in two rolls using the same rubber compound, 5 cm thick, with cores 20 cm long; one core diameter was 2 cm and the other 10 cm. Under identical conditions, the roll with a core diameter of 2 cm cured satisfactorily while the other when cut out, showed distinct porosity and sulphur bloom in the midregion of the roll cross section. When the metal is considerably thicker than the rubber, radial heating has the most effect, and axial heating of the metal does not cause temperatures at the rubber-metal junction higher than in the centre of the rubber. When the core is solid, the heat flows from the cover to the core and from the ends of the core to the centre. For open steam curing near 140°C, the following rules have been suggested for exceeding the optimum cure time.

Allowance for core

(i) For hollow core, nil, and overall cure between 50 and 60% of that for a light core; (ii) for light core (hollow tube with closed ends), approximately 1 hour; and (iii) for solid core, nil up to 75 mm diameter, ranging up to four steps of 1 hour each for 150 mm diameter.

Allowance for ebonite

1 hour in excess of normal cure, except for hollow cores, where no allowance is made if the cure cycle exceeds 2 hours.

Allowance for rubber covering

In addition to any allowances made for the core and ebonite, 5 minutes for each 0.75 mm up to 75 mm and 8 minutes for each 0.75 mm up to 75–100 mm covering.

The evacuation time is normally kept to a minimum to avoid continued post-heating. However, in some cases it is necessary to evacuate the autoclave slowly: for example, in the case of thick ebonite covering temperature and pressure are dropped gradually to avoid blow-out and cracking. This requirement also applies to rolls whose covers are based on polymers of very low gas permeability such as butyl or nitrile rubber. The extended evacuation prevents damage to the covers by expansion of the gas due to a sudden reduction of pressure. Finally, after evacuation, the rolls are allowed to cool at room temperature and the blocking dies and tapes are removed.

Grinding and Finishing

The diverse uses of rubber rolls in different industries call for finishes suitable to the production methods and the service conditions in those industries. The following finishes are standard:

Type 1 (wheel or belt finish). Customarily the roll surface is ground in a single operation, using a grinding wheel or power-driven abrasive belt, and one size of grit. This finish usually shows grinding and chatter marks.

Type 2 (rough grind). A refinement of type 1; grinding marks are visible, but chatter marks are not.

Type 3 (sanded smooth finish). After a type 1 finish, further finishing operations are performed, using power-driven or hand-applied grit or girts of a finer type than those used earlier. Grinding and chatter marks are removed.

Type 4 (polished finishes). Very fine finishes can be obtained on some rubbers of specific hardness. They are usually referred to by the finishing grades of abrasive paper or cloth used to produce the desired surface. There are no scratches on the surface.

Type 5 (creep finish). This rough finish is produced with a rasp or saw.

Type 6 (tooled finish). Light thread marks are visible around the roll.

Types 1,2 and 3 are the most common finishes required for paper making, textile processing, steel processing, etc. Type 4 is the usual finish for graphic arts rollers and is sometimes required in other industries. Types 5 and 6 are specialities and are used by all industries as and when needed.

The surface finish and tolerance of a roll covering must satisfy very strict requirements. A precision grinding machine, as encountered in the machine tool and engineering industries, is suitable, although special cylindrical belt grinding machines have also been developed which grind the entire roll length at once. The factors that affect the grinding operation are the abrasion resistance and hardness of the rubber, alongwith its tear strength and modulus. The size and peripheral speed of the grindstone, its grit size, the grinding depth, the traversing speed, and the rpm of the roll are contributing factors, as well.

The grindstone chosen greatly affects both the quality of the surface finish and the speed of the grinding operation. A fine stone does not necessarily give a high finish in all covers. When the rubber cover is soft, the pores of a fine stone very soon become clogged, and in consequence much heat is formed, resulting loss in speed and an impaired finish. On the other hand, various ebonite stocks can be really finely ground only with a fine stone. In most cases, therefore, the grindstone should be coarse and sharp–a vitrified high quality carborundum stone of high porosity comes nearest to being universally applicable. When the grain structure is considerably coarser than the particles of the rubber grindings, it is easy for the latter to fall away from the stone, and smearing is prevented. A blunt stone fails to cut the rubber and removes it by friction, resulting in development of heat with the risk of stone cracking and an unsatisfactory finish. The grinding stone is selected on the basis of the hardness of the rubber cover:

Hard rubber (roughing and finishing), carborundum C36–J5–VG

Medium-hard rubber (roughing and finishing), carborundum C36–L5–BR

Soft rubber (roughing), carborundum C36–L5–BR

Soft rubber (finishing), carborundum C60–M5–BR

where C is silicon carbide: 46, 36, 60 indicate grit; J, L, M signify grade; VG is vitrified; and BR is resinoid.

A wheel diameter of 25–45 cm having 2.5–5 cm width and a wheel speed of 1500–3700 rpm is used. Roll rotation is in the same direction as the wheel, the rpm of the roll being 12–20, with carriage feed of 10 to 25 per minute.

The depth of the rough cut varies with the hardness of the roll cover:

For hard rolls, 0.38–0.50 mm

For medium rolls, 0.75–1.0 mm

For soft rolls, < 0.75 mm

The carriage feed or traverse speed varies with the roll diameter; a lower traverse speed will be used for a bigger roll diameter and a higher speed for a lower diameter. In all finishing grinding, the wheel is not fed into the work but is locked at a position where it just makes contact with all parts of the roll surface and is fed back and forth until the desired finish is produced.

In most cases rubber rolls are ground dry, and when the roll surface is sticky, loading of the wheel is prevented by dusting the rubber with soapstone. However, wet grinding is also practiced. This procedure is 10–20% faster, resulting in significant reduction in secondary finishing and polishing operations and giving accuracy within 0.05 mm.

For a high finish, the roll is polished with emery or other abrasive cloth while it is driven by a high speed lathe; uniform pressure should be maintained during the polishing to prevent ridges and low places. Number 1 coated abrasive may be used for preliminary polishing, followed by 0 for final finishing.

Surface treatment

As a practical matter, there are three methods by which the surface of a roll can be treated: heat treatment, irradiation (e.g., UV), and chemical treatment (e.g., chlorination).

Surface treatments usually are intended to reduce the adhesion of material that comes into contact with the roll. For example, the UV irradiation of spinning rolls is intended to reduced the yarn's tendency to wind up, and the chlorination of printing rolls reduces the absorption of printing ink.

SPECIAL COMMENTS

Dimensional Standards

The dimensional standards required of rubber rolls are quite stringent and vary from one application to the other.

For paper mill and industrial rolls, the tolerance on diameter is ±0.38 mm for 200 mm diameter and an additional ±0.05 mm per 25 mm of diameter from 200 to 750 mm; for diameters exceeding 750 mm, it is ± 1.5 mm. TIR is 0.05 mm, and finished face length tolerance is 6.0 mm.

For steel mill rolls, the tolerance on diameter is ±0.2 mm up to 125 mm diameter, with an additional ± 0.05 mm per 25 mm of diameter from 125 to 750 mm; for diameters exceeding 750 mm, it is ± 1.5 mm. TIR is 0.13 mm, and finished face length tolerance is 6.0 mm.

For graphic art rolls, the following tolerances apply.

	Letter press	*Offset ink*	*Offset water*	*Rotogravure newspaper*	*Aniline (Flexo)*
Tolerance on diameter, mm	+ 0.75	+ 1.13	+ 0.63	+ 2.4	+ 0.25
TIR, mm	0.13	0.08	0.08	0.13	0.08
Finished face length tolerance, mm	± 1.5	± 1.5	± 1.5	± 1.5	± 1.5

Crown

Crown is the difference in diameter between the centre of a roll cover and reference points at or near the ends of the face. Some bigger rolls, particularly paper machine rolls working under heavy stress, require crowning for three principle reasons: (i) to compensate for the deflection of the roll body under operating condition; (ii) to cause the strip or web being processed to track properly; or (iii) to compensate for the contour of the material being handled by the roll. An improper crown will adversely affect the operation of the machine, the quality of the material processed, and the performance of the roll covering.

The deflection to be compensated for is the sum of the natural deflection of the roll due to its own weight and the deflection due to external loading. This defection is determined by the design and physical characteristic of the roll body and the amount of loading imposed on it. Since each different load requires its own crown, if a roll is to be used with more than one load, a compromise must be reached regarding which crown is to be used. The type, thickness, finish, amount of crown, and hardness of rubber cover affect the accuracy to which crown can be produced; the standard tolerance for crowns used to compensate for roll deflection is as follows :

Crown (mm)	*Tolerance*
⩽0.25	±0.025 mm
0.26–1.0	±0.05 mm
> 1.0	±5% of total specified crown

For a given lever weight, a soft roll requires slightly greater crown than a harder cover.

Roll Balance

Vibration from unbalance shortens roll covering life, damages bearings, and other parts of the machine, and has seriously adverse effects on the product.

Because of high rpm, all rolls must be balanced to reduce vibration. If the weight of the roll is doubled, the centrifugal force is double; if the velocity is doubled, however, the centrifugal force is quadrupled, thus showing the importance of balancing at high speeds.

Roll balancing is the manipulation of the centre of gravity so that it falls along the axis of revolution.

Static balance

If a roll supported without friction remains at rest wherever the rotation is stopped, all the local irregularies are self-compensating and static balance is said to exist. If a roll is not in static balance, sufficient weight must be added to the "light" side to bring it into balance. Usually this is accomplished by drilling hole and adding lead before the rubber covering. Static balance is used chiefly for rolls operated at a surface speed of less than 180 m/mm. However, rolls that are balanced for one speed are not necessarily in balance for other speeds.

Dynamic balance

Even if a roll is statically balanced, one end of it may have an eccentric centre of gravity, with the result that correction has been applied to the other end. This change cannot be detected by simple static balance and will become noticeable only when the roll is speeded up. The resulting oscillation, known as dynamic imbalance or couple or moment imbalance, requires investigation to determine whether one end or both requires counter-weighting.

In such a situation each end might require its own particular counterbalance, with the sum of the two again producing static balance. The rubber cover of a press roll that is out of balance may develop a flat spot across the face or may separate from the core. Separation may also occur if the top roll is out of balance.

Kinematic balance

When a roll is made from a pipe, the walls vary in thickness along the face of the roll and also around the circumference. When such a roll is statically and dynamically balanced and then speeded up, local conflicts in balance along the face appear, with the heavy spots in the shell being slung farther from the centre. When the shape of the roll is deformed by its own weight irregularities, the resulting condition is called kinematic imbalance. To avoid this problem, the use of a machined core (both inside and outside), balanced at operating speed before covering, is recommended.

Static unbalance is expressed in kilograms at the working surface, while dynamic unbalance may be expressed in kilogram-centimeters and is the measure of one or more forces that result in vibration when the roll is revolved. This vibration is measured in linear movement at the ends of the roll. Specifications for dynamic balance should be expressed in terms of the amplitude of vibration at the ends of the roll for a given speed, as measured by a recording vibrometer. For rolls on which dynamic balance is specified, the minimum amplitude is 0.13 mm at the ends of the roll twice of the centre of the roll for a given speed.

Speciality rubbers

A wide variety of rubbers are used for making rubber coverings (NR, SBR, IIR, CR, CSPE, polysulphide, PU, etc.). Among these, the polyurethanes (PUs) need special mention because of their extensive use lately in making rubber coverings–for example, in printing, laminating, coating, and paper and steel mill operations, and in applications such as ink transfer, ink application, squeezing, press roll, tension roll, and nip roll.

This breadth of use reflects the wide range of hardness of these materials, coupled with excellent solvent resistance, excellent abrasion resistance and very good load-bearing capacity. Two different fabrication methods are used depending on the type of PU, namely, polyurethane elastomers and castable polyurethane. The elastomer grades are processed in much the same way as other conventional solid rubbers. These are polyester-or polyether-based polymers cross-linked either at the end groups or at reactive sites in the chain; the cross-linking agents include polyisocyanates, diamines, sulphur, and peroxides.

In the casting system, liquid or semiliquid starting components are converted into the roll covering directly in the mould, where the core is also positioned. The starting raw materials for casting systems are liquid polymeric diol, isocyanate (bifunctional or polyfunctional), and chain-extending agents (monomeric, diol, triol, diamine, etc.). In this process, low molecular weight intermediate products containing terminal isocyanate groups (prepared *in situ* from polyol and diisocyanate or as prepolymers) are chain extended with monomeric diol or diamine. In the final step, the terminal isocyanate groups react with active hydrogen of the urethane or ureas groups, giving rise to three-dimensional cross-links.

Casting may be carried out in two ways: (i) a two step procedure using higher temperature, in which a prepolymer is produced first; (ii) a one -step procedure, in which the temperature employed is near room temperature.

The products include both linear and partially cross-linked materials, the former being chemically and physically similar to thermoplastic urethane (TPU). Many high quality castable polyurethanes are still made by a two-stage process, although there is increasing use of one-component systems based on methyl diisocynate (MDI) variants. One-pack casting systems, employing blocked diisocyanates that are activated by heating, are also in use; here the properties are limited by the presence of the blocking agent.

In the two-step method, the polymeric diol and a diisocynate (e.g., NDI, TDI, or MDI) are heated in an inert atmosphere and a viscous prepolymer is formed, terminated by isocynate groups. This compound is degassed in vacuum at the temperature at which the system is to be cast (~ 105–120°C). The monomeric diol or diamine, which acts as a chain extender, is then added with stirring; frequently a catalyst (e.g., tin octoate) is also used. The pot life of the system is limited, and pouring into the preheated mould containing the core should be completed as quickly as possible, taking care to avoid air entrapment. The core may remain in the mould from several minutes to several hours (involving chain extension, branching, and cross-linking) at a temperature of 100–150°C. Curing in the mould is stopped as soon as the material has attained sufficient strength to be demoulded without damage. Postcuring is thereafter carried out at 100°C for several hours or at room temperature for about a week. Products with low hardness may be obtained by reaction of polymeric diol with a polyfunctional isocyanate. Monomeric diols, used as chain extenders, also give lower hardness than aromatic amines.

In the one-step method, a prepolymer is not prepared; rather, individual components–polymeric diol, diisocyanate, and extender (diamine)–are mixed together and react to form the product. The relatively lower viscosity of one-component system enables casting to be carried out at lower temperatures, which is advantageous because heating of the vessels and tubing become unnecessary. The best known application of one-shot cast polyurethanes is for soft printer rollers from lightly branched poly(diethylene adipate) and TDI, where hardnesses from 15 to 35 Shore A are achieved by varying the ratio of polyester diol to TDI.

INDUSTRIAL ROLLS

Now that the present-day manufacture of rubber rolls has been summarised, we turn to the characteristics and compounding of rubber rolls related to a few important industrial applications.

Paper Mill Rolls

Rubber rolls used in the paper industry must satisfy a great variety of requirements, the specific requirements to be fulfilled depending on the function of the individual roll. The basic principles of operation are essentially the same for both the types of machine used for paper making, namely, the Fourdinier machine and the cylinder machine. In each case a sheet is formed on a travelling wire screen or a cylinder, dewatered under rollers, dried by steam-heated rolls, and finished by calender rolls.

The chief function of paper mill rubber rolls is to remove water from the sheet of pulp that is being converted to paper. The pulp at the feed end of the wire screen where table rolls and suction rolls (both rubber-covered) are used is roughly 96% water; at the exit point from the wire screen, water content is reduced to about 83%. From the wire screen, the paper is transferred to the first felt blanket, which carries it through a series of rubber-covered press rolls, where more water is removed; the water content is now about 65%. The paper then passes through smoothing rolls (steel and rubber-covered) and is picked up by the second felt blanket through a series of press rolls (rubber-covered), followed by a series of drying rolls internally heated by steam. Upon exiting the second series of press rolls, the pulp has a water content of approximately 15%, which is further reduced (to ~ 6%) by the drying rolls. Finally, the paper enters the calender, where the finish is imparted.

General requirements of paper mill rolls

The general requirements of paper mill rolls are as follows :

1. High mechanical stability and good dynamic property.
2. Uniform hardness.
3. Resistance to the chemicals used in paper making (acidic and alkaline solutions).
4. Resistance to the temperatures likely to be encountered.
5. Resistance to scratching and marking.
6. Good grinding characteristics.

Almost all paper machines use at least one suction roll, whose main purpose is to draw water from the wet paper web. A suction roll consists of a perforated metal cylinder, either bronze or steel, with a wall thickness of 25–50 mm. A rubber cover is applied to this perforated cylinder, and after vulcanisation the rubber is drilled so that the preforations in the rubber lead into the holes in the metal shell. A suction box, 10–15 cm wide, is connected to a source of low pressure and extends throughout the length of the roll on the inside. The suction rolls operate at speeds from 10 to 350 m/min and are subjected to pressures varying from a few kilograms up to as much as 70 kg/cm^2 in localised areas.

A suction roll and a press roll will be deflected during operation, the actual amount depending on the type of metal used, the wall thickness, the diameter and length of the roll, the pressure applied, and the speed. The speed has a direct bearing on the number of bending or stress reversals that take place in any unit of time; for example, with an average size suction press roll 5.25 m long, this deflection may be of the order of the 2.5 mm. To ensure that a uniform line of pressure is exerted on the paper web, a camber or crown is made into the rubber surface. Thus, the rubber cover must flex and yield to overcome the basic bending or flexing of the metal core.

However, roll deflection is only one small factor to consider in the case of drying. The metal drying, 3–4 m in diameter, is under steam pressure and can bow out in the centre. The large heads, however, restrict the ends. Because of their mass, these heads change the surface configuration of the dryer, making it almost impossible to crown or camber a rubber-covered roll so that the two will match uniformly from end to end. It is the ability of the rubber cover to crepe and flow that assumes predominant importance in solving this problem of camber.

A freshly ground roll, almost irrespective of its camber, quickly assumes the surface qualities of the dryer and continues to operate satisfactorily until the creep becomes so great that the stresses in the rubber cause it to shear. Thus periodic grinding is needed to prevent rubber cover failure due to excessive creep.

As indicated above, the speed at which the rolls move is an important factor, bearing directly on the number of bending or stress reversals that occur in any unit of time. At each passage through the nip between two rolls, the rubber is deformed. This deformation must be recoverable before the entire revolution of the roll is completed. Otherwise additional stress is placed on the non-recovered rubber resulting in corrugations, cracks, and ultimate failure of the rubber cover.

It is desirable that the table rolls be as light as possible, to decrease their inertia. Usually, table rolls are made of thin-walled, lightweight seamless steel tubing covered with ebonite. They are balanced dynamically at the service speed, and the rubber cover is ground and polished to make them touch the wire evenly. The rubber compound used in table rolls is less easily wet by water than other kinds of roll surface. At high roll speeds, this lower affinity causes water to be thrown off the rubber at a lower tangent. Consequently, less water is carried back up to the wire, and water is removed from the pulp sheet more rapidly and thoroughly.

To permit great pressure, paper mill press rolls are made quite heavy. The important function of the press rolls is to obtain high dehydration without damaging the quality of the paper. In the stages where the paper contains much water, soft rubber rolls are used under low pressure, and the hardness of the rubber rolls and the pressure are gradually increased. It costs about 12 times more to remove a kilogram of water from paper with steam than with rubber-covered rolls.

Sometimes the total drying cost can be varied 40% or so simply by adjusting the hardness of the press rolls. If the cost of steam consumption is abnormally high, the press rolls probably are not doing their jobs satisfactorily. If the steam cost is high but the felts wear longer than would normally be expected, the press rolls probably are too soft.

An important characteristic of rubber-covered press rolls as used in paper mills is their ability to absorb high impact loads in localised areas. Under normal operating conditions, the applied load is spread more or less uniformly across the entire length of the roll. When a break in the web occurs, or the paper wraps around the roll or a wad goes through the nip between the rolls, the total applied load is transmitted to a small area of the rubber cover. Under normal conditions, the applied load may be 13 kg/cm^2, whereas under any of the conditions mentioned above, the actual load may reach a maximum of 65–75 kg/cm^2.

Speedy recovery from deformation is also an important characteristic of rubber rolls. At each pressure through the nip between the two rolls, the rubber is deformed. However, if the rubber cover does not recover before the entire revolution of the roll is completed, additional stress is placed on the non-recovered rubber and the result, as noted earlier, is corrugations, cracks, and ultimate failure of the rubber cover.

Finally, there is the problem with respect to the release, or the lack thereof, of paper fibres. A perfect wet press roll that exhibited absolutely no affinity for the wet paper web would be as unsuitable as one to which all the fibres adhered strongly. This is truc bccause the paper web must adhere to the rubber roll to a certain degree if it is to be separated from its carrier–a wool felt.

It is expected then that the roll surface will attract and hold the paper web for a definite distance beyond the point of contact with its engaging roll, then releasing the web completely without leaving adhering fibres to collect on the doctor blade or scraper that runs against the roll surface to remove such adherents. The rubber technologist can only make a careful, searching study of each individual mill and its own operating conditions, and on this basis formulate and fabricate rubber rolls for satisfactory performance.

Since paper mill rolls must be reground occasionally, the ability to furnish a smooth, unblemished or ground surface is important. Regardless of other desirable physical and chemical properties, rubber covering compounds become unsatisfactory if, under the heat and pressure of the grinding wheel, they develop strings or bells of rubber, leaving a creep surface on the roll. Equally unsatisfactory is a rubber compound that permits the grinding wheel to pick out small pieces of rubber, leaving a pock-marked or dimpled surface.

Natural rubber is still the preferred polymer for paper mill roll covers. Blends of polybutadiene and natural rubbers also offer interesting opportunities. Some roll covers are now being made with chloroprene rubber. Cover stocks based on nitrile rubber are used in special cases only (e.g, for drying cylinders).

Approximately 30–40 ppr zinc oxide is used for good thermal diffusion and dynamic properties. Blends of carbon black with precipitated silica are favoured for obtaining higher values of elongation at break; coupled with higher tear strength. Sulphur is used in the range of 3–5, along with an accelerator system, which enables the roll covers to be cured for the usual long periods without risk of reversion. Addition of polystyrene sometimes improves the release characteristics. Once again, the need to make a careful, searching study of each individual roll and operating conditions is stressed as a prerequisite to producing rubber rolls that will perform satisfactorily.

Steel Mill Rolls

Rubber rolls used in steel mills, in continuous electrolyte tin plating machines, and in galvanising plants are not crowned but are ground to a uniform diameter throughout. Rubber compounds used for covering rolls that handle metal sheets and strips must be designed to resist the cutting and tearing action of rough, sharp edges. Most steel mill rolls are covered with polychloroprene rubber and some with NR in the hardness range of 45–80 Shore A. In the rugged service of sheet steel rolling mills, oiling rollers made from nitrile rubber are used for lubrication of finished sheet metal.

In a strip steel mill, rubber rolls are used as conveyors and hold-down rolls in acid pickling tanks and as wringer rolls at tank ends, for removing excess pickling solution from the strip. In stainless steel manufacture, rubber roll covering prevents the stainless alloy from touching ordinary steel or iron and eliminates scratching and other damage. To prevent marking, light coloured rubber stock is used for handling stainless steel.

Continuous electrolytic machines use rubber rolls of three types: (i) tension rolls, to pull the sheet through the line; (ii) wringer rolls, to remove water and chemical solution from the plated surface; and (iii) guide rolls which act as deflectors in changing the direction of travel of the sheet. The rubber stock must not mark or discolour the tinned surface. Such rolls are made from SBR and CSPE.

In galvanising plants, rubber rolls remove excess water from the surfaces of the iron sheets, after they have cleaned, to hasten drying with hot air.

Printing Rolls

The function of printing rolls is to apply ink to a selected and prepared surface at a controlled rate, the quality of printing depending very much on the quality of the roll cover. The specific requirements for the rubber coverings of printing rolls depend on the printing process (letterpress, lithography, offset, photogravure printing, duplication process, etc.). Offset printing uses an inking roll, which transfers ink to the plate, and a watering roll, which supplies water to the nonprinting area of the plate, to repel ink.

The following factors must be considered for such rolls: strength, resistance to oils and solvents, resistance to glaze, hardness, and dimensional accuracy.

Appropriate strength

Printing rolls rotate at high speed, which means that they must have appropriate strength. Reinforcing fillers are needed to increase the strength, and since the rolls are of low hardness, the amount of such reinforcing fillers must be limited. A proper balance of reinforcing fillers and plasticisers is required. Reinforcing fillers also improve the modulus and tear strength of the compound, properties that are essential for good grinding characteristics.

Resistance to oils and solvents

Depending on the process of printing, the inks employed may be based on vegetable oil (letterpress), mineral oil (rotary press), or aromatic solvents. The vehicles for offset printing ink are composed of linseed oil and mineral oil.

Printing rolls are constantly exposed to ink during printing and are cleaned with kerosene afterward. Thus resistance to such oils and solvents is an important property. The problem of oil/solvent resistance is intensified because today's fast-drying printing inks possess highly varying chemical compositions. Accordingly, the ink mixtures exert very dissimilar effects on rubber, and the roller coverings must be designed to ensure resistance to swelling from a series of chemical substances. If the tendency to swell caused by these substances is not stringently limited, the dimensions as well as the hardness of the roller will be seriously affected, which of course will be reflected in the printed image. Moreover, a swollen rubber roll may result in an extensively intensive contact with printing plate, thereby causing damage to the roll surface due to incision of the type.

Nitrile rubber is used mainly, with a large amount of plasticisers, since hardness required is low (20–40 Shore A). Since most plasticisers are easily extracted by the solvents or ink, with consequent decrease in volume, nonextractable plasticisers should be selected.

Resistance to glaze

Offset printing inks contain phenol-alkyd resin to impart gloss and are mostly of the quick-drying type, suitable for use with high speed printing machines. Linseed oil based inks containing metallic soaps as dryers tend to form a dried, oxidised film of ink on the rubber rolls, which is known as glazing. Glazed rubber rolls become slippery and less effective for the transfer of ink; subsequently the glazed film may develop cracks and pit marks on the rubber surface.

Correct hardness

The hardness of rubber printing rolls greatly affects their efficiency in spreading ink, in accommodating ink transfer, and in ensuring the ability of the machine to absorb shocks. Thus hardness is one of the most important factors.

Correct dimensional accuracy

For a correct and uniform ink transfer, smooth interaction of different rollers with extremely narrow tolerances is of decisive importance. This is particularly applicable for rubber inking roller, which must maintain an exact clearance between the last distributor roller and the impression cylinder. Only a minute change of the diameter would cause the contact to become too intensive or too weak, resulting in an excessively light or bold type face.

Tanning Industry Rolls

All tanning industry rolls require nonstaining cover compounds. Degraining, buffing, and splitting machine rolls are mostly covered with natural rubber, whereas in fleshing and dehairing machines, rolls with nitrile rubber covering are used. The hardness of splitting and fleshing machine roll covers usually is 45–50 Shore A, while those of the buffing and degraining machines are in the range of 60–70 Shore A. The roll covers used on fleshing machines to separate the flesh from leather are required to satisfy rather exact demands with respect to tensile strength, compression set, tear propagation strength, and resistance to swelling.

Rolls for Plastic Industry

Polyethylene extrusion and lamination

In the making of polyethylene film with a ring die, known as the inflation process the extrudate is always inflated by air that caused to flow into the cylinder. This process is very effective in the production of extremely thin films as well as in the manufacture of bags. Rubber rolls work as tight rolls to retain the air fed from the centre of the ring die and at the same time to pinch the folded tube-shaped film and to feed it onto the winding machine. When the film touches the roll, it is already cooled. The functional requirement of such pinch rolls is excellent elasticity, so that air can be sufficiently retained. Rubber rolls with hardness of roughly 60 Shore A made of natural, chloroprene, and nitrile rubbers are frequently used. In the case of thick films, nitrile rubber with a superior abrasion and heat resistance is employed.

In applications calling for embossing or lamination of the extruded sheet, the T-die process is used. Here, fused polyethylene flowing from the T-die is received first by the cooling roll and is embossed or laminated with the web in a half-fused condition. Hence the rubber rolls should have tear resistance and non-tackiness.

Since normally polyethylene film is made wider than paper in the paper lamination process and the excessive thick polyethylene film is cut off at the edges, the fused polyethylene is likely to stick to the surface of the rubber rolls. Under such conditions, silicone rubber rolls are preferred. When the working width of the rubber roll is restricted to slightly smaller than the width of the paper, the polyethylene of both edges will not stick to the roll surface, and the rubber rolls made of chloroprene or nitrile rubber may be employed.

Polyethylene printing

The polyethylene film surface is oxidised by corona discharge to overcome defects due to nonpolarity, and the adhesive property and printing ability of the material are thus improved. In the corona discharge process, rubber rolls work as the negative electrode and CSPE covering (2.5–5.0 mm thick) is used because it has superior resistance to high voltage and to ozone. Careful attention must be given to the selection and mixing of the compounding ingredients, since poor dispersion causes a short discharge, resulting in damage to the cover.

PVC processing

The rubber rolls employed of the calendering of polyvinyl chloride (PVC) consist of touch rolls (lamination rolls), embossing rolls, sand rolls, and so on, where properties such as heat resistance, plasticiser resistance, and abrasion resistance are required. Butyl and EPDM rubbers are commonly employed as the covering material for touch rolls with hardness to 65–80 Shore A and embossing rolls with hardness of 55–85 Shore A.

Sand rolls are embossing rolls used for matting the backside of the film and sheet, to prevent them from blocking. Sand rolls are prepared by compounding a large amount of emery sand in SBR or EPDM, with special attention to dispersion and the prevention of splitting off of the emery particles.

Textile Mill Rolls

Rubber rolls are used in the textile industry for different purposes to withstand widely different conditions involved in desizing, scouring, dyeing, mercerising, and finishing. The media used in the textile processing include not only water and acidic, and alkaline solutions but also oxidising agents, softeners, resins, and such organic solvents as gasoline, chlorinated hydrocarbons, and aromatics.

Coverings with a natural rubber base are widely used in the removal of water-soluble sizes and for rolls needed for some bleaching processes (using cold hypochlorite, chlorite, and peroxide liquors), mercerising, dyeing (with any dyestuffs that do not attack NR, and finishing). Roll covers based on nitrile rubber are particularly suitable for use in desizing with fet dissolving soaps, mercerising (e.g. in neutralisation compartments), special dyeing processes e.g., pigment padding, finishing and some types of special finishing (e.g., flame proofing). The rubber covering is made in the range of 65–80 Shore A. Guide rollers of jiggers are usually covered with ebonite, while the pair of squeeze rollers pressing on the dyestuff (rinsing agents, etc.) must be covered with soft rubber.

Rolls for Foodstuff Industry

Rollers covered with blends of natural and polybutadiene rubbers, nitrile rubber, and carboxylated nitrile rubber are used in machines for the dehusking of rice. The covers of these rollers must be quite hard (~ 90 Shore A), yet show sufficient elasticity and low compression set, as well as high abrasion resistance.

Dairies used heated dying cylinders with nitrile or silicone rubber covers for the evaporation of water in the production of powdered milk.

Rollers covered with nitrile and silicone rubber are also used in bread, cake, and cookie factories–for example, in the application of fat-containing cookie or cracker, fillings to the products and in pressing the baked top layer onto the filling. In this operation, the flavour of the products cannot be influenced by the rollers.

Other Applications

In the coating industry rubber rolls are used either in spreaders using a doctor knife or in roller coaters; solvent resistance and non-stickiness are important requirements of such rolls. Another interesting application, employed in various textile processing industries, is as expander rolls for controlling the tension across the width of textiles. In the plywood industry rubber rolls play an important function in wide belt sanders.

The end users of the rubber rolls are located in an extremely wide range of industries, and information from all raw materials suppliers, including the manufacturers of synthetic rubber and compounding ingredients, together with interaction with the end users of rubber rolls, are valuable to the rubber roll manufacturer for the necessary future development of this important product.

CHAPTER 20

Sealing Technology

INTRODUCTION

A successful seal is the solution of an engineering problem (leak prevention) with an engineering material (typically an elastomer). Elastomer seals made from elastomers are attractive because they have a minimum volume change under a uniform hydrostatic compression (that of steel is 150 times greater). This property is due to the elastomer's Poisson ratio of 0.499, compared to 0.3 for steel.

Further, the force required to develop an equivalent strain in rubber is 10^{-5} times the value for steel. Of course we are not considering replacing steel with rubber, nor are we considering replacing rubber with steel; these are only material strength comparisons. Specially synthetic rubbers normally utilised for seals are by their nature more expensive (up to 2000 times greater than the cost of natural rubber). Their usage is economically justified, however, because of one or more unique properties required for some specific and demanding function. These compounds probably account for less than 0.5% of the synthetic rubber volume, but they are indispensable in many critical markets. The failure ramifications need no elaboration. From a design standpoint, the elastomer formulation can be compounded for service under stress (gaskets), strain (O-rings), or load (shaft seals).

PHYSICAL AND CHEMICAL CONSIDERATIONS

Any elastomeric seal, whether it is designed to perform under stress, strain, or load, is evaluated on the basis of numerous overlapping basic physical and chemical considerations. Although we briefly review them, a broader understanding requires additional study. Seals, sealing concepts, and the sealing mission will become more focused subjects. Logically, the first consideration is the compatibility of the elastomer with the sealing environment: that is, elastomer selection coupled with the elastomer formulation. Second is a consideration of the seal configuration and an understanding of sealing principles. Finally, of the effects of the operating environment, which can range from temperature extremes to ageing—both physical and chemical, we must evaluate dynamic properties and liquid—gas interactions.

Selection of Elastomer

The basic elastomers should be considered in light of several basic statements.

1. They belong to a physical category referred to as "glass-forming liquid"; that is, at some arbitrary low temperature, their physical state is a glass having a glassy modulus and subject to glass-type fractures.

2. They are also referred to as "supercooled liquids", because of their ability to change from the "rubbery" physical state to a glassy state (100-fold increase) through a very narrow temperature excursion (7–25°C). Fig. 20.1 illustrates this in greater detail.
3. They should also be considered to be "supercondensed gases," since they are typically developed from C_2 and C_4 monomers, normally in a gaseous state.

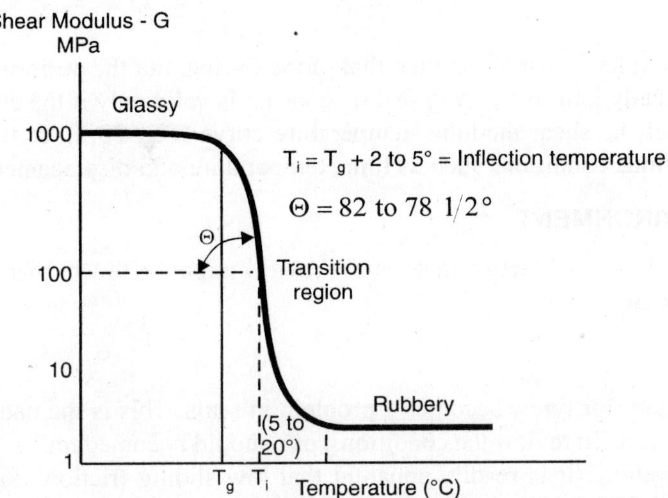

Fig. 20.1. Shear modulus versus temperature (idealised) for a typical crosslinked elastomer having a hardness of 75 shore A.

This 1000-fold increase in viscosity from a gas to the elastomeric state gives us a basis for accepting thermal pressure and other thermodynamic manifestations. The actual elastomer selection process is increasingly driven by such external constraints as costs, which might in turn dictate a marginal elastomer or often a less-desirable moulding process. Possibly the desired operating parameters cannot be reached by any elastomer.

A typical rubber formulation normally consists of the following ingredients :

1. Elastomer.
2. Filler (reinforcing agent and/or diluent type).
3. Plasticiser (for improvement of low temperature properties).
4. Crosslinking system (vulcanising agents).
5. Age resisters (antioxidants, antiozonants).
6. Process aids.

These are not firm guidelines, of course; there can be variations. For example, logically saturated elastomers (butyl rubber, ethylene propylene copolymers and terpolymers, etc.) have minimal antiozonant requirements. Similarly, high temperature fluorocarbon elastomers require no age resisters; metal oxides often serve this purpose. Fluorocarbon elastomer formulations normally do not contain plasticisers, which are typically too volatile to be retained at the normal higher operating temperatures to which fluorocarbon rubbers are subjected. We will use a simple model compound to illustrate both physical properties and mechanical response to time-temperature shifts. Antioxidants, plasticisers, and process aids are omitted in the interest of simplicity. The cross-linking chemistry principally develops the more mechanically mobile polysulphidic cross-link.

SEALING CONFIGURATIONS

Configurations are numerous, but we will limit the discussion to two basic seals: O-rings and lip seals. Fig. 20.2 illustrates the most common gland configurations utilised for O-ring seals. The subtleties of gland design allow us to maximise elastomer performance under marginal operating conditions, which are detailed below.

O-ring

One cannot envision a simpler construction than that of the O-ring, but the performance characteristics are astounding. The basically intuitive O-ring sealing concept is valid only if the elastomer is operating on the rubbery portion of the shear modulus–temperature curve (Fig. 20.1). In the Fig. 20.1 O-rings dynamic response to various conditions such as time, temperature and displacement are highlighted.

THE OPERATING ENVIRONMENT

This is the real world, and external factors that require review might be few or many. Some of the more obvious are discussed below.

Friction

Friction, which is necessary for tyres, becomes a problem in seals. This is the natural consequence of using a vis-coelastic material. To review the conditions, one should become familiar with the differences between adhesion and wetting. It is readily apparent that low sliding friction requires a lubricant or surface coating. Maximum life and low friction result when the lubricant wets both the seal and the moving member.

"Wetting" Phenomenon

Possibly one of the least appreciated aspects of sealing technology, wetting is a phenomenon that must be clearly understood. In the simplest application, the lubricant should be wetting both the seal and the shaft. It should be noted that the lubricant boundary between elastomer and the solid surface must possess *interfacial free energy* to be stable.

This stability is governed by both thermodynamics and kinetics, so what we really have is normally a metastable (i.e., time-dependent) system. Thermodynamics, which is essential irresistible, eventually wins. "Breakout friction" is a good example. Additionally again it should be noted that "solid polymer surfaces, therefore, are inherently non-equilibrium structures and exhibit a variety of time and condition-dependent properties that may change dramatically with those variables. This is specifically to the energetic differences between physical adsorption and chemisorption. The exact nature of molecular interactions in these interfacial regions is still somewhat imprecise, as evidenced by ongoing annual conferences on adhesion promoted by several specific groups.

There is general agreement however that an adequate interface first requires adsorption of the fluid on the surface. This is followed by a multilayer adsorption of the fluid through self-association, which creates a combined film thickness that may range from 3 to 150 nm.

The shearing of this film is the prime source of dynamic or running friction and the subsequent heat buildup in a seal. The ability to "wet" the surface occurs when the physical attraction of the surface is equal to or greater than the surface tension of the lubricant. The extremes might be illustrated by water on cellulose versus water on polytetrafluoroethylene (Teflon).

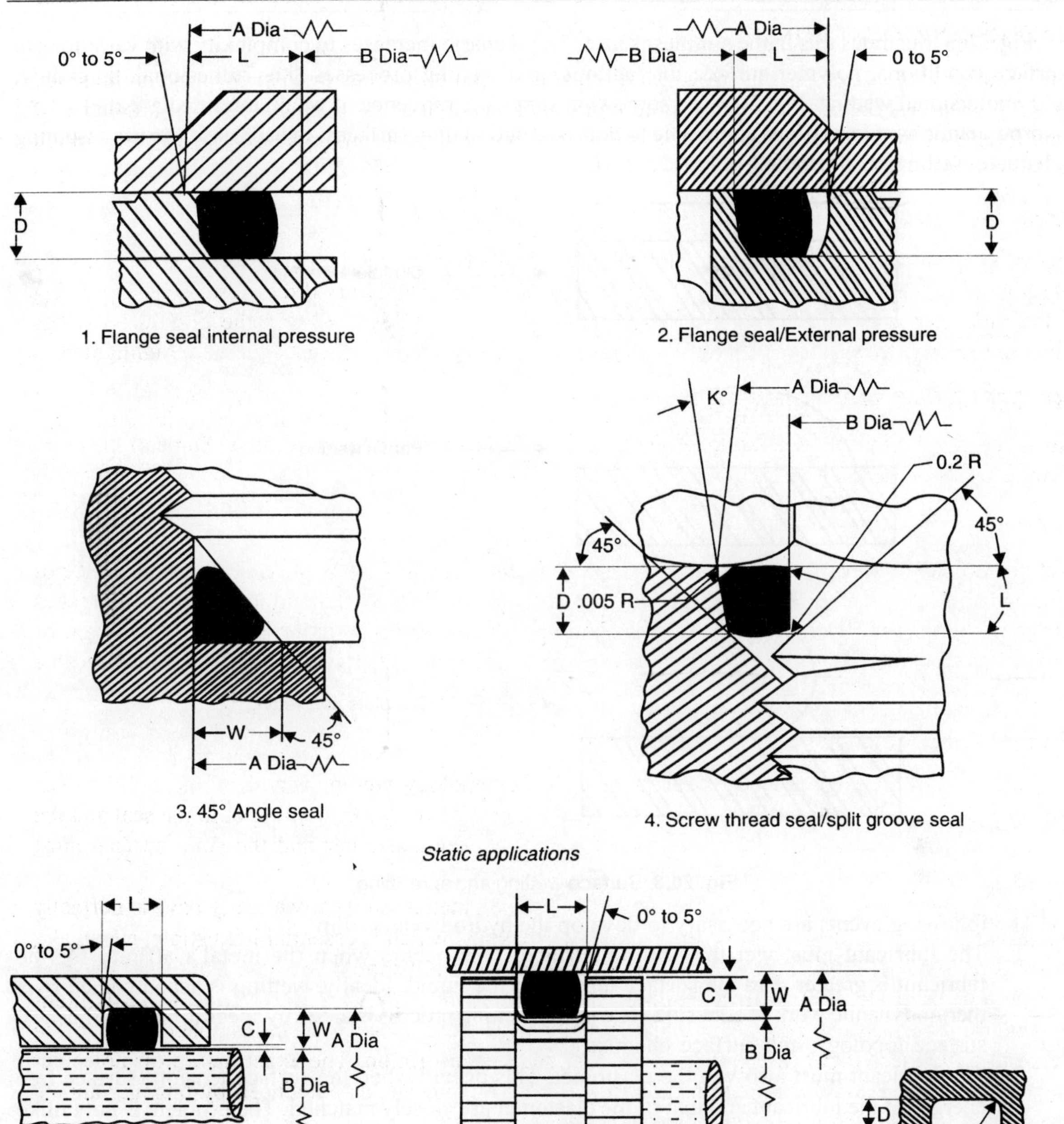

Fig. 20.2. Common O-ring groove designs.

Fig. 20.3 illustrates this in the simplest sense. The scenario increases in complexity with variations in surface conditions, polymer surface interactions, and wetting processes (i.e., adhesional, spreading, and immersional wetting. Fig. 20.4 is a somewhat simplistic sequence of weak, intermediate, and strong hydrodynamic film formations applicable to both seal and sealing surfaces. In each case, we are assuming adequate wetting, as illustrated in Fig. 20.3(a).

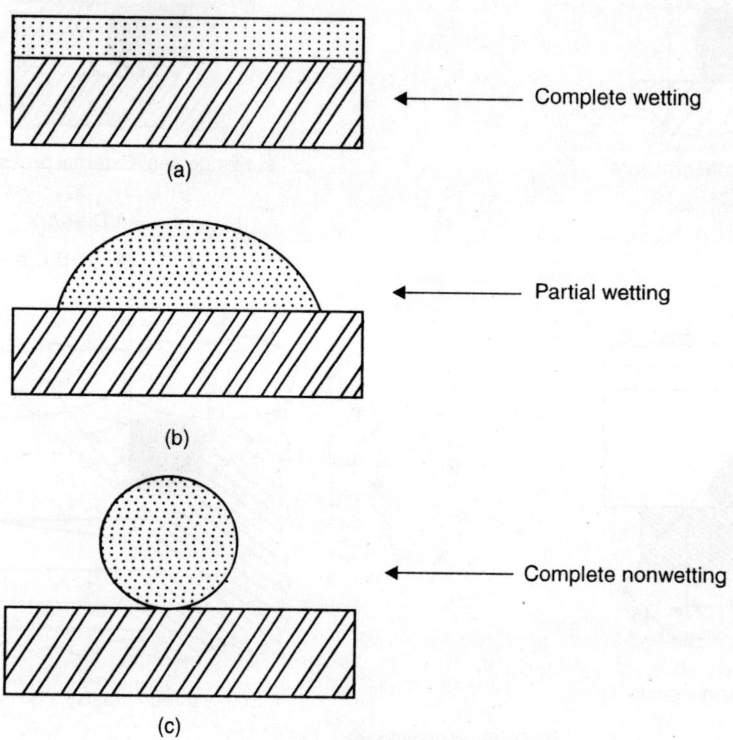

Fig. 20.3. Surface wetting and spreading.

The following events are necessary to develop the hydrodynamic film :

1. The lubricant must wet the sealing surface. This occurs when the metal's affinity for the lubricant is greater than the surface tension of the liquid. Ideally, wetting is accompanied by a thermodynamic work of adhesion. The initial wetting process is aided by specific surface finishes, surface topology, and surface chemistry.
2. The lubricant must also wet the elastomer. This occurs when the surface tension (surface free energy) of the lubricant and that of the elastomer are closely matched. The problem is not unlike that of coating and ink technology.

Lewis Acid–Base Considerations

The elastomers can also be roughly characterised in a Lewis acid-base sense as having the characteristics of Lewis acid, non-specific, amphoteric, or Lewis base. Lewis acid-base context, in a more recent discussion on the swelling of nitrile rubber in various solvents, describes the swell response in Lewis acid-base considerations. The acid-base relationships for elastomers have not been legitimately defined to date, so the free surface energy concept for elastomer surface characterisation is at the moment a more practical consideration. The effects of fillers, plasticisers, and process aids can alter the apparent free surface energy of the base elastomer in a rubber formulation; this, of course, is a Lewis acid-base manifestation.

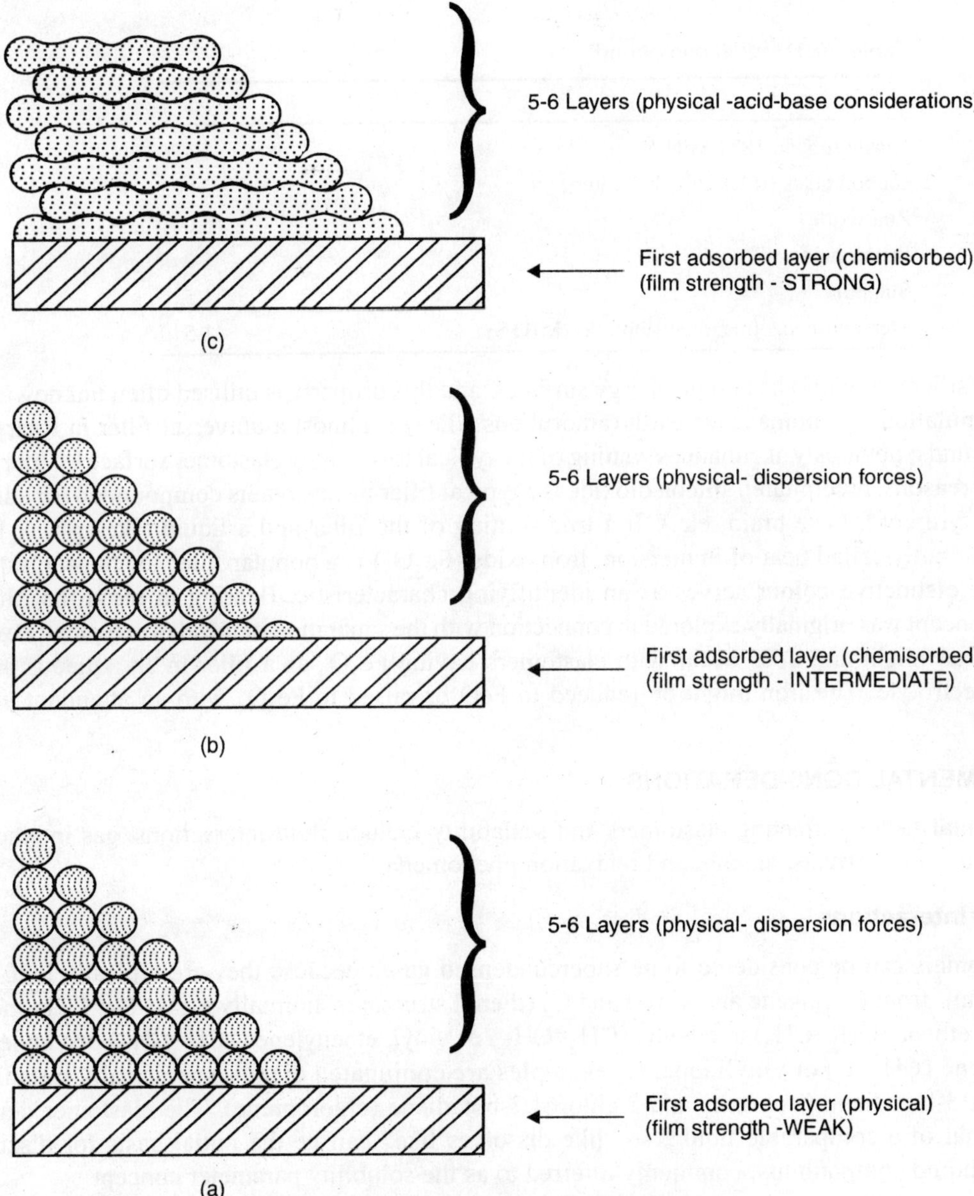

Fig. 20.4. Hydrodynamic film formation.

FILLERS

Fillers serve numerous functions, but common to all events is their effect brought about by increasing the viscosity of the formulation. As noted earlier, fillers increase viscosity. It is known that carbon blacks behave differently in similar elastomers even though they might contain equal volumes of N990 or N330. Table 20.1 illustrates a typical carbon black filler study using a 38% acrylonitrile (ACN) rubber in a model compound when compounded with 28% by volume of decreasing particle size carbon blacks.

Table 20.1. Model compound.

Ingredients	*phr*
Nitrile rubber, 38% ACN	100.0
Carbon black (filler: 28% by volume)	50.0
Zinc oxide	5.0
Stearic acid	1.0
Sulphur	1.5
Mercaptobenzothiazoledisulphide (MBTS)	1.5

Silicas (silicon dioxide) have high energy surfaces, and this property is utilised often unknowingly in rubber formulations. Diatomaceous earth (amorphous silica) is almost a universal filler in rotary shaft seal compounds; obviously, it enhances wetting of the typical low energy elastomer surface by lubricants. For similar reasons, precipitated silicon dioxide is a typical filler in elastomers compounded for adhesion to fabrics (tyre cord, hose braid, etc.). If a true wetting of the filler and a liquid takes place, heat is evolved, normally called heat of immersion. Iron oxide (Fe_2O_3) is a popular filler for fluoroelastomers because its distinctive colour serves as an identifiying characteristic. Bearing in mind that the zeta potential concept was originally explored in connection with the separation of metal ore (typically oxides), consider what might and does occur with elastomers having Fe_2O_3 as a filler in an aqueous or non-aqueous electrolyte. The iron might be reduced to FeO or raised to Fe_3O_4 with an attendant volume increase.

ENVIRONMENTAL CONSIDERATIONS

Environmental factors affecting elastomers and sealability include fluid interactions, gas interactions, heat, cold, dynamic effects, ageing, and relaxation phenomena.

Elastomer Interactions

Most elastomers can be considered to be supercondensed gases because they are typically long-chain polymers built from C_2 (alkene and vinyl) and C_4 (diene) structures, normally a gas. The C_2 monomers range from ethene ($CH_2{=}CH_2$) to ethenyl ($CH_2{=}CH{-}$) or vinyl, ethenylene ($-CH{=}CH-$) or vinylene, or ethenylidiene ($CH_2{=}C$) or vinylidene. C_4 examples are conjugated dienes, specifically butadiene, 2-methyl-1, 3-butadiene (isoprene) and 2-chloro1,3-butadiene (chloroprene). The elastomer densities approach that of a comparable liquid, so "like dissolves like" can be the initial basis for discussing elastomer–liquid compatibility, commonly referred to as the solubility parameter concept.

Liquid-Elastomer Interactions

The solubility parameter concept, on discovery, appears to offer a rationalised approach for a simplified elastomer selection in various solvents. In summary, the classic solubility parameter concepts are valid for non-electrolytes. Electrolytes, however, are less predictable. Self-association of molecules (by H-bonding, dipole-dipole association, etc.) can create problems in solubility estimations. Again, if the solvent-solute conditions are not in thermodynamic equilibrium, the solubility parameter concept is not a valid predictor.

Gas-Elastomer Interactions

The interactions between gases and elastomers are more complex. *Perfect* gases (oxygen, fluorine, etc.) achieve high densities at higher pressures; thus almost any elastomer can be utilised to seal them at low pressures, as long as consideration is given to such effects as oxidation. An indication of the potential swelling effects of any gas can be determined by considering the critical pressure and critical density.

Heat

Elastomers extended under a constant load contract when heated (Gough-Joule effect). The stress in rubber at a constant extension increases with rising temperature under normal conditions and is approximately proportional to the absolute temperature. The generation of heat when rubber is stretched under adiabatic conditions is thermodynamically related. Shaft seals should be designed with these considerations in mind: that is, the inner diameter of the moulded seal might be larger than the shaft to be sealed, and the garter spring were to have sufficient force to reduce seal bore sufficiently to create initial seal interference. As the shaft seal temperature increases, the seal contact stress remains relatively constant. In less demanding applications, the phenomenon described as thermoelastic inversion can also be utilised.

Cold

It was earlier stated that elastomers should be considered to be a supercondensed gas. This statement is true as long as the elastomeric material is in the rubbery state. As an elastomer cools, it contracts as a result of reduced molecular activity (third law of thermodynamics). This reduction of molecular motion coupled with the reduced volume diminishes the physical space for the elastomer chain motions (Fig. 20.5). The basic physical state progresses from rubbery to leathery (or transition) to glassy. At the glassy state, there is an abrupt change in sope`of the curve of volume change versus temperature.

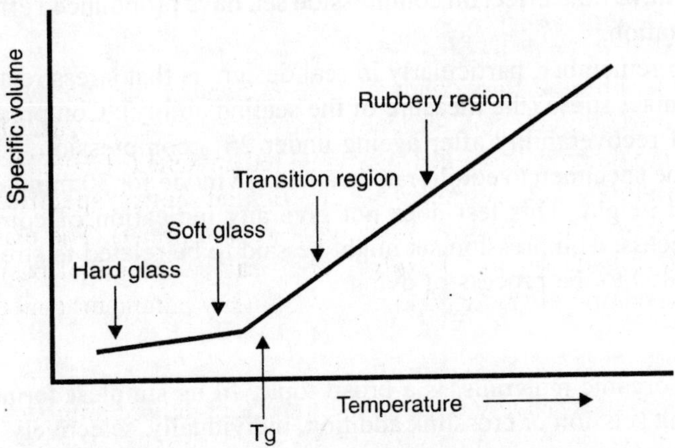

Fig. 20.5. Specific volume versus temperature of a typical glass-forming liquid.

Dynamic Mechanical Properties

When the term "dynamic mechanical properties" is used we instinctively think of tyres, motor mounts, or shock absorbers, but expand this to include shaft seals, static and dynamic O-ring applications, lip seals, and so forth. Most engineering applications of rubber are under some degree of deformation or load. Basic to this discussion is that the dynamic mechanical properties of rubber within reasonable deformations (< 25%) are *independent* of the magnitude of deformation. Conversely, they are strongly dependent on rubber's molecular architecture, temperature, time (frequency) and, to some extent, compounding.

Relaxation Processes

Elastomers, as noted earlier, are characterised by their viscoelastic properties. The viscous component, typically modeled with a dashpot, is unrecoverable energy. The elastomeric component, modelled with a spring, is stored energy. Elastomeric components subjected to a constant force (motor mounts) undergo strain relaxation (crepe). A component under a constant deformation (O-ring) has a decrease in stress over time as a result of stress relaxation.

It has been stated that: "Stress relaxation and crepe are intimately related, and one may be calculated from the other if stress-strain behavior of the rubber is known." It is generally agreed that elastomeric components operating within conservative temperature limits have relaxation processes that are proportional to the logarithm of time. Excessive temperatures cause relaxation processes that are proportional to linear time. Relaxation rates are strongly influenced by the base elastomer's molecular weight, molecular weight distribution, and branching.

It has been found that compounding variables, such as fillers, crosslink type and density, plasticiser, and process aids, also affect relaxation rates.

Compression Set

Compression set is a simple test that can be performed under constant stress (ASTM D395, Method A) or constant strain (ASTM D395, Method B). If the test is run at normal temperatures, it is measuring physical (primary) relaxation processes. At elevated temperatures, it is measuring chemical (secondary) relaxation processes. The results are difficult to correlate in terms of stress-strain relaxation. For example, some reinforcing fillers, which have little effect on compression set, have pronounced effects (increasing the rates) on stress-strain relaxation.

The important difference to remember, particularly in seal design, is that stress relaxation gives an actual value indicating seal contact stress (the measure of the sealing ability). Compression set, on the other hand, is an indication of recoverability after ageing under 25% compression. The actual value reported is based on allowing the specimen to equilibrate in a relaxed mode for 30 minutes before taking the measurement of recovered height. This test does not give any indication of compressive stress available during the ageing process. Compression set might be said to be related to stress relaxation in the same way that death is related to the process of dying.

Ageing

Ageing, characteristic of most organic materials, is a broad topic. In its simplest form, the elastomer might undergo chain or crosslink scission or crosslink addition, individually, selectively, or collectively.

THE MOULDING PROCESS

As already mentioned above the costs typically dictate the moulding technique, with the transfer or injection processes offering the tightest manufacturing controls. Consider now the nature of the compounded elastomer: a high molecular weight, high viscosity, semisolid material. The moulding process typically has to create a part with complex geometry, often bonded to a metal substrate. One might readily envision a relatively stress-free moulded component with a rheological response approximately that of an isotropic solid (i.e., a material that swells equally in all directions). Actually, there is a strong tendency based on mould design, that normally develops some degree of anisotrophy. The typical elastomeric state, which might best be categorised as a *single phase* with a dictated state of maximum disorder, now has ordered domains as a result of the moulding process. This new physical state, a combination of disorder (normal) and order (chain orientation), has pronounced anisotropic tendencies.

A natural consequence of anistropy in a moulded seal is erratic sealing at different time-temperature conditions. The condition is often apparent by a simple swelling test: nitrile rubber aged 24 hours at room temperature in a solvent such as methyl ethyl ketone. Typically, for the test to be effective swelling should be in excess of 150%.

CHAPTER 21

Mechanical and Miscellaneous Products

INTRODUCTION

In this chapter, attention is turned to the large number and variety of other products consuming relatively little rubber because of their small size, their small quantity usage, or their low content of new polymer. Many of them are commercially very remunerative to the manufacturer; others, because of their technical importance for the engineering applications in, for example, the aircraft, electrical, and electronic industries, receive a great deal of effort in government and co-operative research laboratories, as well as in individual manufacturer's laboratories.

Because of their variety, it is difficult to group or classify them; furthermore, it is not the intention to attempt to discuss or to list every product. Many are produced by one or other of the methods with no particular technique or technology being involved.

It is proposed to consider the more important and interesting items under the headings of the five processes: (i) spreading; (ii) calendering; (iii) extruding; (iv) moulding; and (v) handbuilding. It is, however, desirable to draw attention to the several methods that may be adopted to produce certain of them, and so some overlapping and repetition are unavoidable.

PRODUCTS BASED ON SPREAD FABRICS

Spreading of fabrics is carried out as a preliminary to their incorporation in the products. In addition, final products are made from fabrics, usually spread with natural or synthetic rubber on one side only, and used for their waterproof quality, e.g., macintoshes, groundsheets, and tarpaulin-type covers.

The mechanical, electrical, automobile, and aircraft engineering industries have requirements for a great variety of gaskets and diaphragms which for strength and stiffness must have fabric reinforcement. These are frequently made by spreading fabrics of cotton, rayon, nylon, polyester, or glass fibre with rubber, all types of elastomer being used.

The shape is punched from the fully cured sheet. Surface finish is important in many applications and determines the method of vulcanising the spread fabric. The finest, smoothest surface is obtained by moulding, but often an acceptable finish can be achieved by wrapping the fabric on itself on a large drum with an interleaf of fine textured cloth, special paper, or plastics film.

Flat punched diaphragms can be post-formed to simple convoluted forms by a short heat treatment when pressed between a metal and a rubber shaped jig (Fig. 21.1).

PRODUCTS MADE BY THE CALENDERING PROCESS

Unsupported Sheeting

For quantity production of unsupported rubber sheet, i.e., without fabric insertion, the calendering operation is most economic, assuming that the vulcanisate requirements can be met by a mix which will calender smoothly. Unsupported sheeting is required for hospital use, for sticky-back repair patches, for gaskets and diaphragms, and for cutting into thread for garment and golf ball manufacture. A calendered sheet is also required for built-up products, moulding, etc., e.g. the sheet may be wrapped on a mandrel to form a tube for subsequent parting off into washers.

Supported Sheeting

Products referred above can equally well be produced by means of the calender. The choice of method depends on many factors—the ability of the rubber mix to friction or calender smoothly, the degree of adhesion required and obtainable by the various methods, the thickness required, and the uniformity of thickness needed. The vulcanising methods available for the cure of calender produced sheeting and fabric-reinforced sheeting are the same as for spread fabric.

EXTRUSIONS

Continuous or Long Lengths

It is hardly necessary to do more than mention the variety—tubing, sealing and trimming strips for automobiles and domestic appliances, shoe welting, etc., supplied in long lengths to other industries and users to cut up as required. The vulcanising methods are either batch or continuous and both solid and cellular rubber sections are manufactured. Trade marks, names, sizes, and other details may be imprinted on the extrusion by allowing it to pass over an engraved or embossed roller as it leaves the machine.

Precision Extrusions

The compounder's and the diemaker's arts are usually adequate to ensure that extrusions meet the designer's requirements dimensionally and shapewise, at least in the smaller size range. There are, however, large section extrusions where collapse and distortion occurring during handling and cure, demand special treatment. Tubing may need to be blown onto a mandrel; other sections require support on specially made jigs. For fine dimensional tolerances or when the extruded surface is not perfect enough, grinding the outside of the tubing is commonly practised. In other instances, bends and other shapes are made by putting the uncured extrusion onto bent or shaped mandrels before cure.

Products Made from Cured Extrusions

Scarf-joining

Vulcanised or semi-vulcanised solid or hollow extrusions are scarf-or butt-joined to produce rings for such items as seals and driving bands. The techniques vary. The freshly cut surfaces are coated with a solution of the same mix, with the addition of a bonding agent in the case of solvent-resistant polymers. A further aid to a satisfactory bonding is to insert a 0.1-0.2 mm calendered film between the two solutioned faces of the joint. After making the joint, it is clamped under light pressure in a sample two-piece mould, heated electrically or placed between heated platens. A short cure is given to the joint, followed frequently by an open steam or oven cure to complete the vulcanisation of the whole component.

Parted-off components from extrusions

Uncured tubing or any other extrusion is cut into suitable lengths for use as moulding blanks, by parting-off in a lathe, often by automatic mechanisms. The extrusion is supported on mandrels coated with rubber or other non-metallic material to protect the knife and avoid contamination of the blanks. Fully vulcanised tubing is similarly cut into washers and gaskets. If dimensional tolerances are tight on the internal and external diameters, the extrusion will need to have been cured on a mandrel of the correct size, allowing for shrinkage and the outside diameter ground before the parting-off operation. This is only one way of producing a rectangular section washer; punching or trepanning from calendered or moulded sheet and moulding are alternatives. Stationers' bands are made from extruded tube allowed to flatten, and then vulcanised, and cut to the required width by guillotine.

Endless shaped seals from extrusions

For automobile and aircraft window and domestic appliance door seals, accurately shaped corners may be required. A widely used technique, to save the cost of complicated and large moulds which would be needed to make them in one piece, is the corner moulding of extrusions. Semi-vulcanised extrusion is prepared and cut to length, the cut ends being coated with solution. Two lengths are placed accurately in a simple corner mould made to the shape of the corner with a section corresponding to that of the extrusion. The mould is clamped between the heated platens of a press, and rubber is transferred to the cavity to form a corner piece integral with the two lengths of extrusion. The operation is repeated with the necessary other lengths of extrusion to produce an endless seal shaped to fit the window or door. Alternative methods include the scarf-joining of the extrusion into an endless ring and then by means of suitable mandrels or formers (which may be of a meltable or soluble substance) setting the corners to shape by means of a heat treatment. In this case, the extrusion must be in such a state of cure that it will assume the required permanent set.

MOULDED PRODUCTS

The large number of general mechanical goods, as they are termed in the industry, manufactured by moulding, usually in compression-type moulds, do not call for any particular comment. There are a few special items which will be briefly discussed.

Solid Tyres

Rubber-tyred wheels, casters, and drive pulleys are manufactured in a variety of sizes and sections, and also solid rubber tyres for perambulators, push-chairs, etc., as separate items for fixing to a grooved wheel. The rubber-tyred product, in which the rubber is moulded and bonded to the metal or other rigid wheel, is made by applying an adhesive or ebonite layer to the metal surface followed by an extruded section or calendered strip; the assembled unit is then placed in a simple compression two- or three-piece mould and vulcanised under heat and pressure.

Solid rubber tyres for application to a wheel are moulded with an endless metal helix and/or fabric core on the inner diameter to give it the rigidity and stiffness to hold on the rim—sufficient stretch is present to permit assembly on a grooved wheel.

Hollow Articles

Inner tubes for tyres of all sizes are nowadays moulded. A pre-moulded rubber-based valve stem is cemented to an extruded tube, the ends of which are cut, usually with a hot knife, and butt-joined. The tube is then inflated and moulded in watchcase-type presses. The previous method, still in use for cycle tubes, used a vulcanised extruded tube with overlapped joint, made by hand.

Under this heading are also included play balls, which may be moulded on the lines of tennis balls, and hand-built and open steam or oven cured products such as bladders, fuel tanks, liquid storage bags, and dinghies. In addition, there are the moulded hollow components and articles—boots and gaiters for engineering joints, cell sacks, handle grips, and hot water bottles—for which the mould contains a core rigidly positioned and located but separate, so that after cure the product can be removed. Careful blank sizing and shaping is necessary to ensure uniform and consistent wall thickness. If compression moulding is employed, the compound formulation and condition of the blank must be such that good knitting of the separate blanks occurs as they flow together. The vulcanisate, in the hot state as it leaves the mould, must have sufficient elongation to permit the extraction of the core, which, in the case of a hot water bottle, for example, is considerable. Transfer or injection moulding is necessary for many products of this type to eliminate uneven wall thickness, retracted spew (backrinding), air trapping, and other faults, and to save material.

Diaphragms

Reference has already been made to the production of simple flat or slightly convoluted diaphragms from unsupported and supported spread or calender proofed material. There are diaphragms with peripheral and/or centre beads and bosses for assembling and sealing purposes, and with deep convolutions which must be moulded (Fig. 21.1). No special techniques are called for, apart from accurate blank preparation for the reinforced diaphragms to ensure that during moulding the fabric is neither distorted, strained, nor damaged by the flow of rubber. Simple two-plate compression moulds with accurate dowelling or taper location are commonly used.

Table 21.1. Burst strengths of diaphragm fabrics: mullens test. 31.7 mm diameter disc.

Fabric	*Burst strength (kgf/cm²)*			
	60 g/m² fabric	*85 g/m² fabric*	*135 g/m² fabric*	*240 g/m² fabric*
Nylon polyester	10	17.5	31	60
Glass	5.3	11	17.5	25

Top hat diaphragms, which are designed to roll in use, giving long strokes with minimal resistance to movement, are the subject of patents. The majority of such products have a fabric reinforcement which may be square woven or knitted. The essential of the technique is the pre-forming of a top hat from the fabric and then moulding around it a rubber coating on one side of the fabric with the necessary beads or rims to provide sealing when assembled in the metal components.

It is common practice to produce diaphragms with the rubber on one side of the fabric only, the side which faces the pressure. This ensures that no separation of fabric and rubber takes place as is liable to occur if the fabric side faces the pressure.

Seals

Rubber seals are used to seal anything from a wristwatch to the freight door of an aircraft. The former may be an O-ring of bore 0.8 mm and cross-sectional diameter 0.5 mm; the latter may have a section more than 40 mm across and a circumferential length of 15 metre. Fig. 21.2 is a chart of seal types, not all of which are made wholly or partly of rubber. Some of these types have already been discussed as being manufactured by techniques other than moulding.

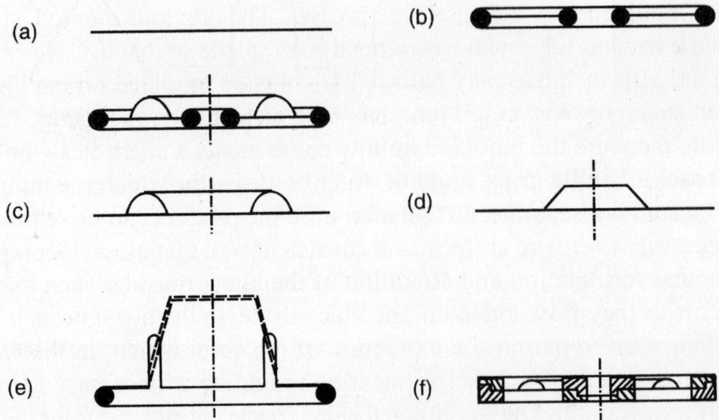

Fig. 21.1. Types of diaphragm. (a) flat (simple punchings from moulded or calendered sheet; the surface can be polished, cloth marked, or sueded); (b) beaded (essentially flat, but moulded with a bead on the inner and/or outer periphery to provide better sealing); (c) convoluted (formed from flat sheet or moulded with a bead); (d) dished (moulded with or without beads); (e) rolling type (usually made as a 'top hat' and assembled inverted to achieve true rolling action); (f) metal bonded (an elastomeric diaphragm of any shape is bonded or adhered to metal mating members to provide still better sealing and stability).

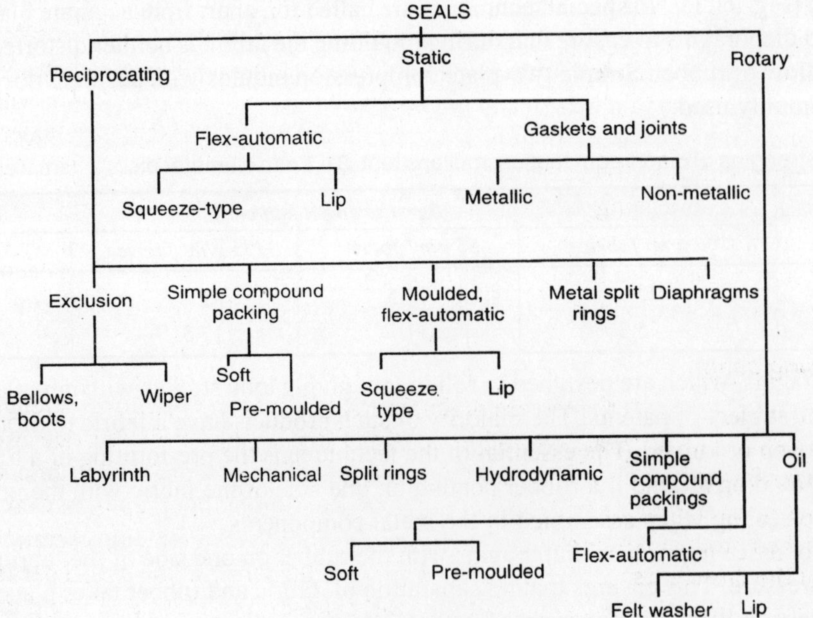

Fig. 21.2. Seal types.

From the manufacturing point of view, those produced by moulding can be considered under three headings. Squeeze-type seals and lip seals are in common use, and within each type there are several section shapes (Figs. 21.3 and 21.4). The former operate by distortion under radial or axial compressive strain, as discussed below. The O-ring is very widely used, though some designers prefer the rectangular, D, or delta sections. The T-seal has the advantage that it can be clamped on assembly and does not, therefore, become easily displaced or twisted.

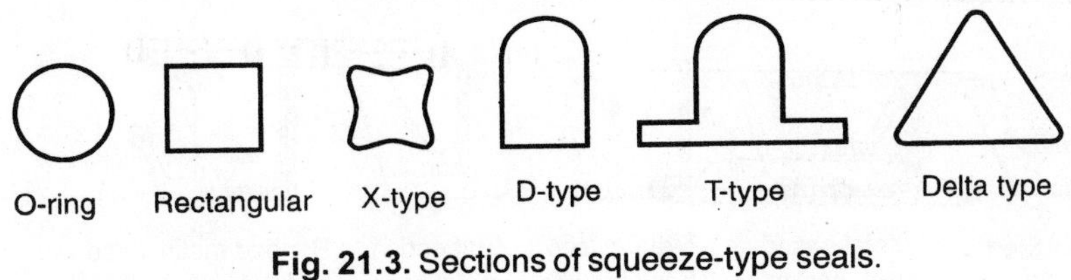

Fig. 21.3. Sections of squeeze-type seals.

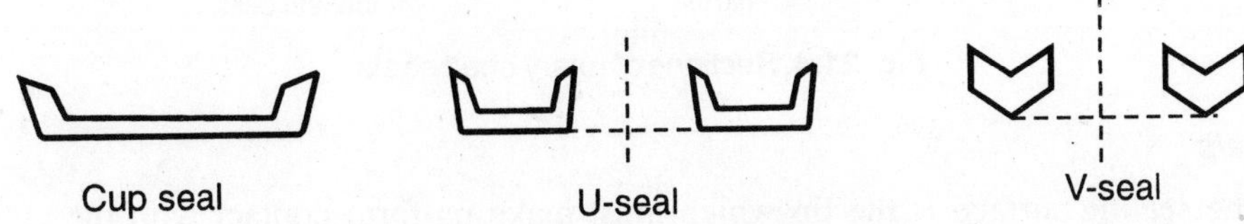

Fig. 21.4. Sections of lip seals.

In the case of the lip seals, sealing is achieved by distortion of the lip assisted by the pressure in the system.

All-rubber seals

The majority of squeeze-type seals are all-rubber, and the same applies to a few of the simpler U-seals and cup-seals for light duty. Moulding is the method of production adopted for all except the smaller rectangular-section seals required in large quantities, which are made by punching or trepanning from sheet or by parting-off extrusion, and the large bore diameter ones, which are made by scarf-joining.

The moulds are of the two-plate compression type with single or multiple cavities. A substantial part of the technique lies in the mould design and manufacture to ensure perfect closure with thin flash and accurate match under the high-pressure moulding conditions that the product and material require. The blanks used are either cubes for small sizes or rectangular or cord section rings, prepared by parting-off from the extruded tube, punching from sheet, or joining the extruded cord. Their shape and size is important because most of the moulding faults, such as air trapping, retracted spew (back-rinding), flow marks, and 'not made', arise from incorrect blanks. Considerable attention has been paid to transfer and injection of O-rings, and in particular to the production of flashless or 'less flash' products.

Fabric-containing seals

Fabrics are incorporated in seals to increase modulus and stiffness and to provide strength and wear and tear resistance. Fabrics of all available fibres are used depending on the service use, in particular, temperature. Manufacture involves the thorough impregnation of the textile fabric with the rubber compound, either by spreading or frictioning, and then a careful building upto, as near as possible, the shape of the finished product before placing in the mould. During the moulding operation, minimal flow and distortion of the fabric must take place and trapping of the fabric at the mould parting lines must be avoided as it detracts from the appearance of the seal, exposes it to ingress of liquid and other foreign matter, and generally reduces its service life.

Metal-containing seals

Many seals for rotary shaft applications contain metal stiffening, usually as a punched and formed cup to which the rubber is bonded (Fig. 21.4). Attention needs to be paid to the location of the metal during the moulding operation and choice of blank shape to ensure satisfactory and consistent products.

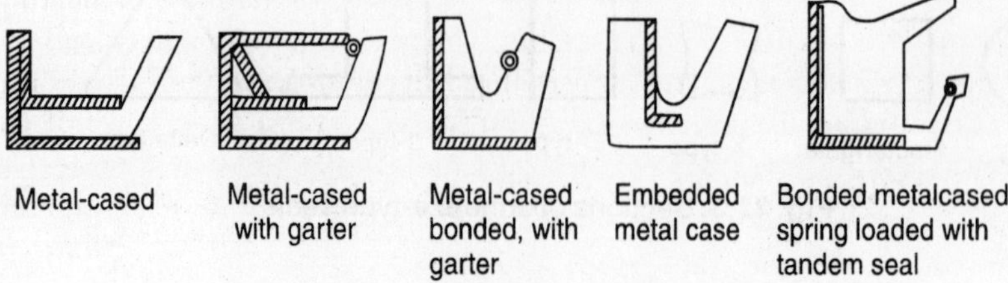

Fig. 21.4. Sections of rotary shaft seals.

Finishing seals

In lip seals, the sealing surface is the lip which must make uniform contact with the cylinder walls, against which it operates. Because it is not easy to mould this lip without a parting line near to its edge, it is common practice to make the mould parting line at the lip, which is made longer than required and knifed off to dimension in a lathe as a subsequent finishing operation. Automatic machines are often used to do this. Flash removal from O-rings and moulded rectangular and similar section seals is carried out by high-speed felt and cotton mops and abrasive wheels.

Seal design and material selection

The development and selection of rubber mixes for seals is dictated usually by three factors. First, the fluid with which the seal will be in contact; secondly, the temperature range of the environment in which it will operate; and thirdly, the mechanical environment, including such factors as pressure, stressing, and abrasion. The fluid and temperature will restrict the choice of the base polymer, usually to one or two; compounding will be directed to achieving the correct hardness and abrasion resistance and to adjusting the swelling behaviour of the rubber in the liquid. The designer will want to know the likely dimensional changes in the seal material due to uptake of liquid and differential thermal expansion.

Important in seal applications is the elastic property of the rubber vulcanisate. The distorted rubber exerts a pressure on the contacting surface to maintain the seal. Unfortunately, no rubber is perfectly elastic, and the stress in the rubber decays or relaxes with time. This stress relaxation can be measured directly or more frequently, the degree of permanent compression set the rubber has taken up on removal of the stress is determined.

There are two factors of importance in compression set testing. First, at the rubber-metal clamp interface, with the cylindrical test-piece of BS 903, there can be any degree of slippage, and the amount of slippage affects the result. It is probably easiest to achieve the no-slip condition, but the 'recovered' test-piece in this case has concave top and bottom faces, which lead to measurement difficulties. This complication can be overcome by the use of an O-ring, or portion thereof, as a test-piece. There is, however, some evidence of a size effect here, probably related to the rate of cooling during the recovery step in the test, which is the second factor to influence the result. At the conclusion of the heating period, the clamps and test-pieces can be allowed to cool before the strain is released; alternatively, the test-pieces can be removed from the clamps immediately and allowed to recover for the specified period of 30 minutes, either at the test temperature or at room temperature. The spread of results obtained by such variations of procedure depends on the magnitude of the viscous component of the elastomer's viscoelasticity. If high, not only is the spread high but one arbitrarily chosen method may give misleading and unreliable indications of the elastomer's behaviour as a seal. It is in such circumstances that stress relaxation observations are valuable as giving a more complete picture of what is taking place.

Other factors related to those mentioned above are the change of hardness on heating the rubber and on immersion in the liquid which the rubber is sealing. Too great a softening may lead to extrusion of the seal between the clearances, with ultimate rapid failure. Strain in the rubber affects liquid uptake, and diffusion rates of various liquids in rubbers of different types cover a wide range of values.

Nitrile rubber mixes are the workhorses of the seal industry, because of their resistance to mineral and ester-based lubricants and to hydraulic liquids. The mineral oils offered vary in composition and in the swelling they produce in nitrile rubbers. It is important that the rubber and oil are matched.

HAND-BUILT PRODUCTS

The broad principles of the manufacture of most important and typical products in this group have already been described. All types of rubber are used, with a preference for natural rubber, if it meets the vulcanisate properties required because of its building tack. Butyl rubber has become established as a reservoir-lining rubber and also for other water-storage containers which involve handwork. Nitrile, polyurethane, and chloroprene rubbers are used for fuel cells and storage containers on a base of nylon fabric.

MOULDED RUBBER GOODS

Moulded rubber goods find extensive use in railways, automobiles, bicycles and also in many industrial and domestic appliances. The product range varies from small O-ring play balls, bushes, shock absorbers, rubber rollers, battery containers, etc. This industry can also work as an ancillary one to many large scale units. In fact, almost all the large scale manufacturers of automobiles, scooters, moped, cycles, machines depend largely on small scale units for their entire range of rubber products.

Raw Material Required

Raw materials required are raw rubber, reclaimed rubber, sulphur, MBT, MBTS, TMT, powder, carbon black, China clay, whiting powder, zinc oxide, stearate, lubricants, binding materials, etc.

Process of Manufacture

All the rubber chemicals are mixed with raw rubber after proper mastication in a two-roll mixing mill. Depending upon the nature of rubber used, it might be necessary sometimes to pass steam through the rollers. After compounding is over, it is the usual practice to extrude the same to form slabs and cut pieces; after weighing they are fed into moulds and cured either with steam or electricity in presses (hand-operated). These are then taken out, screened, tested and packed.

Composition of rubber compound

Items	*Quantity*	*Items*	*Quantity*
Raw rubber	4 kg	MBTS	100 gm
China clay	5-18 kg	Zinc oxide	300 gm
Sulphur	100 gm	Stearate	150 gm
TMT powder	100 gm	Carbon black	100 gm

Machinery and equipments required

1. Two-roll mixing mills, with 25 H.P. motor reduction gear and other accessories.
2. Extruder; steam heated presses; baby boiler; weighing machine; miscellenous equipments and tools.

DIPPED RUBBER GOODS (DRG)

Dipped goods from rubber latex like surgical gloves, finger stalls, catheters and veterinary appliances are finding increasing use. Some of these are: (i) industrial gloves; (ii) finger stalls; and (iii) transfusion tubes. The raw materials required are: (i) latex; (ii) DRG; and (iii) zinc oxide, sulphur, casein, caustic soda, accelerator, antioxidants, calcium nitrate, distilled water, packing material. etc.

Process of Manufacture

Rubber chemicals like sulphur, zinc oxide, accelerators, antioxidants, etc., are dispersed with Dispersol L.N. in a pot mill in suitable proportion separately. The dispersed solution is mixed with 60% DRC centrifugal latex in a Hobart type mixer. This is taken in a tank and moulds of the required products made of either wood, aluminium, or porcelain are dipped in the tank. These are then vulcanised with the moulds in air ovens, removed from the moulds and dusted with talc. No solid material should be fed into the mixer to avoid coagulation.

Machinery and equipment

Hobart type mixer; pot mill; thermostatically controlled air oven; coagulation tanks; moulds; weighing machine.

RUBBER MATTINGS

Rubber mattings mainly find applications in automobiles. These are made by coating rubber compound over Hessian cloth either by spreading or calendering technique. Due to constant use, these need early replacement. There is a substantial demand for rubber mattings of different types for cars, scooters and household requirements. The raw material required are: Hessian cloth, raw rubber, sulphur, rubber chemicals, solvents, coal dyes and colours, etc.

Process of Manufacture

Raw rubber is masticated and is mixed with all other ingredients in a two-roll mixing mill. The rubber compound so obtained is sheeted out thinly and dropped into suitable solvent for 10-12 hours. The rubber absorbs the solvent and becomes soft and plastic. The mass is then taken into mixing churners where little more solvent is added and mixed well to get a dough like consistency. It is then spread on Hessian cloth in a spreading machine by the Doctor Blade Technique. To get designs, the sheets are passed through embossed rolls before curing. It is then cured in a vulcaniser.

Machinery and equipment

Two roll mixing mills; spreading machine; Churner; Vulcaniser; Baby boiler; and miscellaneous equipments and tools.

RUBBER BUSHES FOR AUTOMOBILES

Rubber bushes are the important auto ancillaries which play an important role. Rubber bushes are also known as resilient bushes. The raw material required are : natural rubber; zinc oxide; stearic acid; carbon black; process oil; PBN; MBT; and sulphur.

Process of Manufacture

The rubber, with other ingredients is compounded in a mixing mill. After compounding, the next step is extrusion to form slabs and cut pieces. After weighing these are fed into moulds and cured in a hydraulic press. Now the bushes from the hydraulic press are ready for packing.

Plant and machinery

Rubber mixing mill; hydraulic press; extruder; baby boiler; moulds; weighing machine; and testing apparatus.

RUBBER TUBING

Rubber tubings are flexible rubber pipes of low diameter and used in the transfer of fluids from one place to another. These are generally known as delivery tubes. They are characterised by inertness to corrosive chemicals and even gases for burners. Rubber tubings are most popular because of their low cost as well as inertness to most of the materials which are conveyed. The raw materials for compounding are: rubber; petroleum; antioxidants; zinc oxide; white clay; dibenzothiazol disulphide; sulphur. These are vulcanised by steam at 50 psi pressure.

Manufacturing Process

Rubber is compounded in a mixing mill after the addition of various ingredients. The mixed mass is then extruded through a suitable extruder in the form of tube in length of 50' to 100' which largely depends upon the diameter of the tube. The extruded tubes are continuously cooled with talc on a pan stick and then transferred to vertical vulcaniser or autoclave where vulcanisation is done in open steam under pressure of 48-59 lbs per square inch.

Plant and machinery required

Mixing mill with motor; extruder; vulcaniser; boiler (oil fired); tray tables; weighing machines; testing equipments etc.

THE DESIGN AND FORMULATION OF SYNTHETIC BUTADIENE-ACRYLONITRILE COPOLYMER LATICES FOR GLOVE APPLICATIONS

Protective gloves are manufactured from a wide range of polymers both natural and synthetic in origin. Increasingly synthetic latices have been selected, as glove users, working in areas as diverse as the pharmaceutical and food industry, demand ever improved and specialised properties.

This section explains the nature and chemistry of synthetic butadiene-acrylonitrile copolymers, and how to formulate and understand the impact of factors such as the composition of the nitrile latex and the cure formulation. In particular this study will demonstrate the importance of the acrylonitrile, the carboxylic acid content and zinc oxide cure on the mechanical and solvent resistance properties of dipped films.

Synthetic butadiene-acrylonitrile copolymers (nitrile) are increasingly used in the manufacture of many types of protective gloves, including thin wall, industrial heavyweight flock lined and fabric supported. They offer advantages in terms of processability and product performance compares very favourably with that of other materials synthetic and natural in origin.

Nitrile is manufactured by emulsion polymerisation. Monomers butadiene, acrylonitrile and a carboxylic acid typically methacrylic acid are employed in the presence of an anionic surfactant, a free radical generating species and a chain transfer agent. The resultant latex is a stable, aqueous dispersion with fine particle size, typically in the range 100-200 nm, with a narrow particle size distribution. Prior to use for the manufacture of gloves, the nitrile latex must be first compounded with various ingredients in order to define the properties of the final product. Additives include zinc oxide, sulphur and accelerators. Subsequent curing enables a cross-linking reaction to take place between the polymer chains leading to tailor-made physical properties.

Experimental

Synthesis of latices

A number of experimental nitrile latices were produced using semi-continuous and also batch technology in main and pilot production facilities in Kluang, Malaysia and Stallingborough, United Kingdom. These samples differed in the relative ratios of butadiene (BD), acrylonitrile (AN) and methacrylic acid (MAA) that were charged to the reactor. Latex properties are given in Table 21.2.

Table 21.2. Experimental latex properties.

Ingredients	*AN (wt%)*	*Particle Size (nm)*	*Gel (%)*	*Surface tension (nN/m)*	*Tg°C*
Exp. Latex A	39	139	36	29	–10
Exp. Latex B	39	144	44	31	–12
Exp. Latex C	26	129	55	32	–25

Compounding and film formation

Latices were compounded under a range of conditions, as detailed in the relevant results sections. Polymer films were obtained by dipping ceramic formers into concentrated electrolyte solution, drying and then dipping into the compounded nitrile latex sample. The film was then dried and leached in demineralised water, followed by further drying then curing in a high temperature oven at 120°C. The film was allowed to cool and then placed in a controlled humidity/temperature environment prior to testing.

Tensile testing

Tensile strength, moduli and elongation measurements were determined using a Hounsfield tensiometer. Ten dumb-bells were cut from each sample of dipped film and a statistical median of the results was obtained.

Solvent resistance

Resistance to solvent was measured using flame ionisation detection and using nitrogen as a carrier gas. Four discs were cut from each polymer film and an average breakthrough time calculated.

Results and Discussion

Influence of acrylonitrile and carboxylic acid on tensile properties

Three experimental latices, compounded at a zinc level of 1.0 wt per cent, accelerator level of 0.5 wt per cent and a sulphur level of 0.5 wt per cent were evaluated for tensile strength at break and modulus. The results are shown in Table 21.3.

Table 21.3. Influence of acrylonitrile and carboxylic acid on tensile properties.

Ingredients	*AN (wt%)*	*MAA (wt%)*	*Tensile Strength (MPa)*	*M300 (MPa)*
Exp. Latex A	39	6	36	10.6
Exp. Latex B	39	3	22	4.4
Exp. Latex C	26	6	33	3.9

The high acrylonitrile, high methacrylic acid latex (A) demonstrates the highest overall tensile strength and modulus values. However, on decreasing the acrylonitrile level (latex C) a large reduction in modulus and a small fall-off in tensile strength is observed.

Furthermore, when compared to latex A, the lower methacrylic acid level in latex B results in low modulus together with lower tensile strength. These results clearly indicate that the methacrylic acid level dominates the tensile strength behaviour whereas the acrylonitrile level has only a relatively minor effect. The modulus values are seen to depend on both the acrylonitrile and the methacrylic acid level however.

The three latices were also compounded at two different levels of zinc oxide and sulphur and the modulus results are shown in Table 21.4.

Table 21.4. Influence of zinc oxide and sulphur levels on film modulus.

ZnO level (wt%)	*Sulphur level (wt%)*	*Exp. latex A M300 (MPa)*	*Exp. latex B M300 (MPa)*	*Exp. latex C M300 (MPa)*
1.0	0.5	10.6	4.4	3.9
3.0	0.5	12.8	4.5	9.8
1.0	2.0	13.1	4.9	5.0
3.0	2.0	17.6	5.5	13.2

The results indicate that the modulus of latex A responds to changes in both zinc oxide and sulphur level. In contrast the lower methacrylic acid latex (B) demonstrates very little variation in modulus as the ratio of compound ingredients are changed. Latex C (medium AN, high MAA) exhibits similar behaviour to latex A and accordingly there is a strong dependence on both zinc oxide and sulphur level. Interestingly, the difference in modulus between latices A and C narrows at the higher zinc oxide levels.

Influence of acrylonitrile and carboxylic acid on solvent resistance

Three experimental nitrile latices were compounded using a formulation of 3.0 wt per cent zinc oxide and 0.5 wt per cent sulphur and the resistance to toluene permeation time was measured for the resultant films. Results are shown in Table 21.5.

Table 21.5. Influence of acrylonitrile and carboxylic acid on solvent resistance.

Ingredients	*AN (wt%)*	*MAA (wt%)*	*Toluene breakthrough time (min)*
Exp. Latex A	39	6	14.2
Exp. Latex B	39	3	13.5
Exp. Latex C	26	6	6.1

The results indicate that the acrylonitrile level of the latex has the most dominant effect with respect to solvent resistance. Toluene resistance times are highest for both of the latices containing 39 per cent acrylonitrile, irrespective of the methacrylic acid level. At a lower acrylonitrile level (latex C), a large reduction in solvent resistance is observed.

The high acrylonitrile, high methacylic acid latex was further compounded at zinc oxide levels of 1.0 and 3.0 wt per cent and sulphur levels of 0.5 and 2.0 wt per cent. Resistance to toluene permeation time was measured for films dipped from these compound variants. The results are shown in Table 21.6.

Table 21.6. Influence of zinc oxide and sulphur level on solvent resistance.

	Exp. Latex A – High AN, High MAA		
ZnO level (wt%)	*Sulphur level (wt%)*	*Toluene breakthrough Time (min)*	*M300 (MPa)*
1.0	0.5	13.1	10.6
3.0	0.5	14.2	12.8
1.0	2.0	13.2	13.1
3.0	2.0	15.3	17.6

The above results indicate that the effect of zinc oxide and sulphur on solvent resistance appears to be minimal when the acrylonitrile level of the latex is high. There is only a slight increase in toluene resistance time when the combination of high zinc oxide and high sulphur is used. The same experiment using latex C (medium AN, high MAA) was performed and the results are shown in Table 21.10.

Table 21.7. Influence of zinc oxide and sulphur level on solvent resistance.

	Exp. Latex C – High AN, High MAA		
ZnO Level (wt%)	*Sulphur Level (wt%)*	*Toluene breakthrough Time (min)*	*M300 (MPa)*
1.0	0.5	3.9	3.9
3.0	0.5	6.1	9.8
1.0	2.0	7.1	5.0
3.0	2.0	6.8	13.2

The results indicate that both the zinc oxide and the sulphur level are contributing to help improve the solvent resistance. However, whereas a large increase in modulus results from an increased zinc oxide level, this is not the case when the sulphur level is increased. Therefore our study indicates that it is possible to optimise the sulphur level in order to achieve increased solvent resistance without compromising the modulus value.

The influence of zinc oxide on tensile properties

Further work on the role and interaction of zinc oxide, was carried out using experimental latices with varying levels of methacrylic acid.

It is well known that of the typical cure ingredients added (sulphur, accelerator, etc), zinc oxide has the major contribution to the observed tensile properties. Fig. 21.6 shows the effect of methacrylic acid level on tensile strength at zinc oxide levels ranging from 0 to 10 wt per cent.

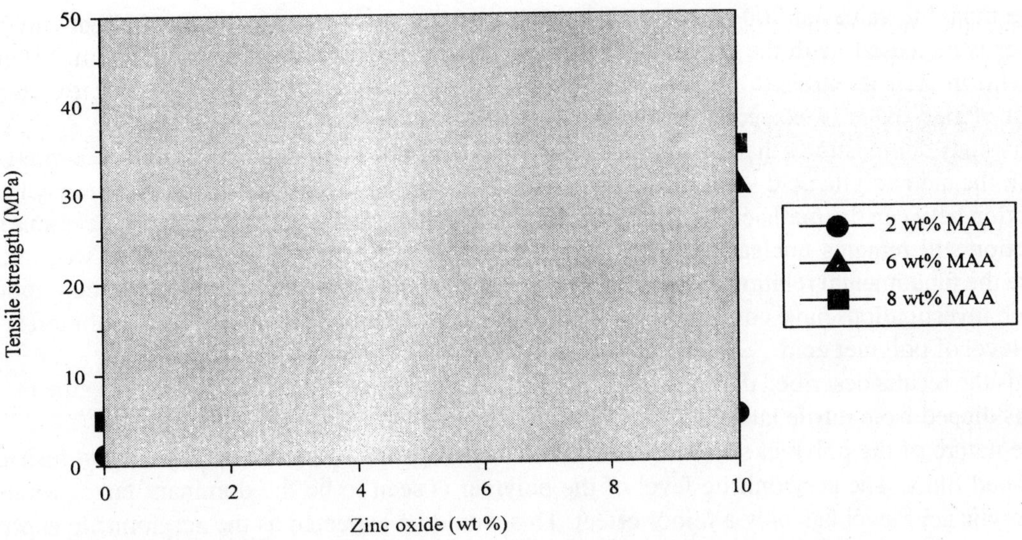

Fig. 21.6. Influence of methacrylic acid on tensile strength at zinc oxide levels 0 to 10.

The tensile strength of the latex containing 2 per cent wt of MAA is seen to be extremely low and, significantly, found to be independent of the zinc oxide level above 1 per cent wt. However, when we consider that the stoichiometric level of zinc oxide for this latex is 0.95 per cent wt (assuming that one mole of zinc oxide interacts with two moles of methacrylic), this relationship and behaviour is not surprising. The tensile strength behaviour of the latices containing 6 per cent wt of MAA and 8 per cent wt of MAA both show a similar dependence to the zinc oxide level. The tensile strength initially rises steeply as zinc oxide is increased and then plateaus off as the stoichiometric amount is exceeded. This occurs at approximately 3 per cent wt of zinc oxide for the 6 per cent wt of MAA latex and 4 per cent wt of zinc oxide for the 8 per cent wt of MAA latex, corresponding to stoichiometric levels of 2.85 and 3.80 per cent wt respectively. Fig. 21.7 shows the influence of methacryliac acid on modulus at zinc oxide levels 0 to 10.

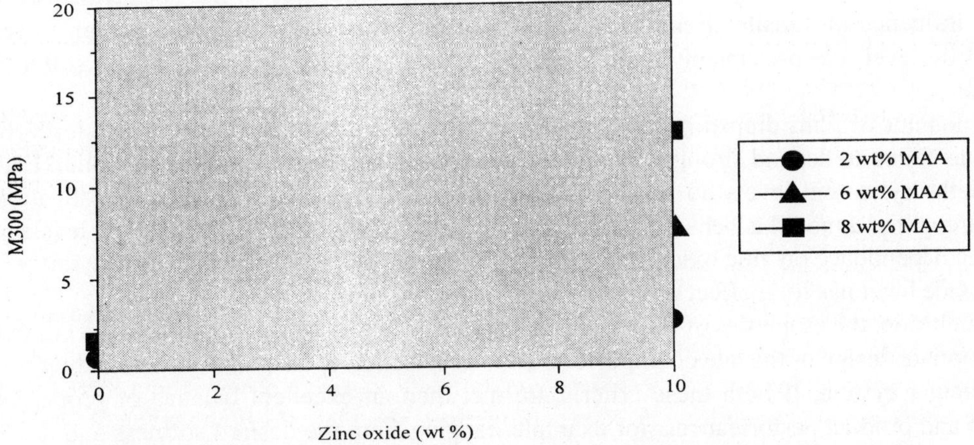

Fig. 21.7. Influence of methacrylic acid on modulus at zinc oxide levels 0 to 10.

The modulus values at 300 per cent elongation are seen to increase as the methacrylic level in the polymer is increased, with the greatest difference being evident at high zinc oxide levels. Similarly as seen with the tensile strength values, the modulus values are seen to level off as the stoichiometric amount of zinc oxide is exceeded for the three systems.

This study demonstrates that the tensile strength and modulus of dipped nitrile films, depends strongly on both the methacrylic acid level of the latex and on the compounded level of zinc oxide. Clearly the interaction between the methacrylic acid and zinc oxide is hugely influential, but the exact nature of this interaction still remains unclear and is of significant interest to our research team, as we continue to explore the fundamental relationships of nitrile latex design and cure compound formulation. Indeed our research investigations have confirmed that zinc oxide dissolves in the latex and its dissolution is related to the level of polymer acid.

Thus the results described demonstrate a number of fundamental concepts concerning the properties of films dipped from nitrile latex.

The nature of the polymer structure is found to have a strong influence on the solvent resistance of the dipped films. The acrylonitrile level of the polymer is seen to be the dominant factor whereas the methacrylic acid level has only a minor effect. This would be expected as the acrylonitrile is present at a much higher overall level than the methacrylic acid and so its relative effect will be greater. The distribution of the two monomers in the dipped films may also have an influence on solvent resistance. It is recognised that acrylonitrile will be distributed throughout the polymer chain, whereas it is speculated that the methacrylic acid may exist locailised in small ionic domains.

Consequently the effect from the relatively polar cyano groups in the acrylonitrile will dominate the resistance to non-polar solvents. It has also been shown that the solvent resistance of a latex with a lower acrylonitrile level may be improved by increasing the amount of zinc oxide or sulphur compounded into the nitrile latex. The more favourable and recommended approach is to increase the sulphur level, as this does not have such a detrimental effect on the measured modulus and increase the stiffness of the film.

The tensile strength and modulus values of films dipped from nitrile latices, are also found to depend strongly on the relative monomer ratios used during the polymerisation process. The methacrylic acid level is seen to influence the tensile and modulus behaviour indicating that the ionic interaction between zinc and the carboxylic acid groups is a key factor. The acrylonitrile level has been shown to have only a minor influence on tensile strength. However, the modulus of the polymer is seen to respond to acrylonitrile level due presumably to its inherent stiffening effect compared to the softer monomer butadiene.

The modulus of films dipped from nitrile latices produced at high methacrylic acid level (6 per cent by wt) was found to depend strongly on the level of compounded zinc oxide and sulphur. Latices with lower methacrylic acid levels do not demonstrate the same trend in modulus. Investigations of the tensile strength and modulus behaviour over a wider range of compounded zinc oxide levels highlights the strong dependence on zinc oxide until the stoichiometric amount is exceeded. Any further increase in zinc oxide level has little effect on tensile strength or modulus.

In conclusion, the properties of films dipped from nitrile latices can be optimised by a combination of the appropriate design of the latex composition and the selection of a suitable compounding package for that particular system. If both these criteria are met then an excellent balance of properties can be achieved and product performances, for example, ranging from the desired softness and dexterity of a medical glove or a stiffer more robust feel for an industrial glove, may be engineered simply and efficiently.

DEVELOPMENT OF EPDM RUBBER BASED ENGINE MOUNTINGS

Mechanical vibration and shock are present in varying degrees in virtually all moving equipment. Therefore vibration control techniques in the form of shock and vibration isolators have been devised to provide dynamic protection to all types of equipment. The direction of force transmission is from the equipment to the support structure. Alternatively if support structure is the source of the vibration and / or shocks, the purpose of the isolator is to reduce the dynamic motion transmitted from the support to the equipment. In either case principle of isolation is the same. The isolator is a resilient element and frequently rubber is used.

Engine moutings are used in automobiles to mount the engine to the base structure. Natural rubber is used as resilient material because of its excellent properties, such as excellent dynamic resilience, flexing and fatigue resistance etc., which are essential for this type of application. However natural rubber has poorer ozone resistance and heat resistance, which can deteriorate the life of engine mounts. Present day automobiles run at much higher temperature and improved heat resistance of rubber is important for prolonged life of rubber mountings.

Ethylene – Propylene- Diene terpolymer (EPDM) has excellent heat and ozone resistance, excellent dynamic properties over wide service temperature. By proper selection of EPDM grades and by proper compounding, important physical properties required for engine mountings such as dynamic resilience, fatigue resistance and durability at high temperature can be achieved.

Another important area is bonding of EPDM rubber to metal part, which can be now achieved by use of proprietary bonding agents.

EPDM RUBBER BASED ENGINE MOUNTING

EPDM rubber based engine mounts for passenger car is presented here.

Experimental

1. Compounding of EPDM for Engine mountings.
   It was recognised that Ultra high molecular weight EPDM namely oil extended grades are required to get the required low hardness and physical properties close to that of natural rubber (NR). Herlene ® EPDM grade H-645 is used. H-645 is an oil extended grade containing 75 phr low volatile paraffinic oil. Properties of raw H-645 are given in Table 21.8. It was compounded in a laboratory 1.5 litre capacity internal mixer as per recipe in Table 21.9 and evaluated for various physical and ageing properties. Tricure system consisting of accelerated sulphur and peroxide was found to give best combination of properties in combination with co-agents HVA-2® and SARET-633®. These compounds are used for moulding engine mountings. Proprietary natural rubber based compound was also studied.
2. Transfer Moulding of engine mountings.
   Metal parts were cleaned by degreasing and sand blasting. Proprietary adhesive primer Chemlok ® 205 was applied and dried for 30 minutes at room temperature. Thereafter cover coats Chemlok ® 236 A was applied and coating dried for 30 minutes. The metal parts were placed in transfer mould and EPDM compound made from Herlene –645 was transfer moulded at 180°C for 30 minutes to get the engine mounting. After ageing overnight, the samples were tested for dynamic properties.

Results and Discussion

Raw polymers properties of H-645 EPDM are given in Table 21.8. Formulation used in the study are given in Table 21.9. Proprietary co-agents namely HVA-2 and Saret-633 were used in the study as these are recommended for improving cross-link density as well as rubber to metal adhesion. It can be seen from the results of physical properties, that H-645 based EPDM compounds exhibited excellent physical and heat ageing properties.

Table 21.8. Raw polymer properties of H-645 oil extended EPDM.

Property	*Typical value*
Mooney Viscosity ML_{1+4} @ 125°C	52
Ethylene to Propylene ratio	67/33
ENB Diene content, %	8.5
Volatile matter, %	0.7 Max
Oil content, phr	75

Table 21.9. Enginge mounting formulation based on Herlene® EPDM.

Ingredients	*Compound with HVA-2*	*Compound with SARET-633*
EPDM Herlene-645	175	175
Zinc oxide	5	5
Zinc stearate	1	1
N-650 black	70	70
Antioxidant-MBI	2	2
Antioxidant - TDQ	1	1
Bis (t-butyl peroxy) di-isopropyl benzene	2.2	2.2
Coagent - HVA-2	1	—
Coagent - SARET-633	—	1
CBS	1	1
Sulphur	0.8	0.8
Total	259	259
Rheomertric @ 175°C, for 30 minutes. 1°C ARC.		
ML, LB-IN	6.17	6.36
MH, LB-IN	24.96	23.24
TS2, M.M	2.22	2.03
T90, M.M	13.38	12.43
Test slab cured @ 175°C, for 20 minutes.		
Hardness, Shore-A	47	48
Modulus @ 100%, Kg/cm²	10	9
Modulus @ 300%, Kg/cm²	40	32
Tensile srength Kg/cm²	193	190
Elongation @ break, %	750	800

(Contd ...)

Ingredients	*Compound with HVA-2*	*Compound with SARET-633*
Hot air ageing @ 135°C, for 70 hrs.		
Change in hardness	+5	+5
Change in tensile %	–26	–24
Change in elongation %	–25	–22

Tan delta studies were conducted at frequency of 15HZ, with two formulations based on H-645 and compared against proprietary NR based compound used for engine mountings. The results are given in Table 21.10. The results indicated that high molecular weight EPDM (i.e. oil extended grade) are reasonably comparable to NR compounds.

Table 21.10. Tan delta and dynamic propertis.

Test conditions:

Frequency : 15 Hz.
Pre load : 20 kg
Peak amplitude : 0.5 mm.

Ingredients	*NR proprietary compound*	*HVA-2*	*SARET-633*
K – dyn	26.39	27.49	29.51
C- dyn	0.0204	0.049	0.052
Tan delta	0.086	0.15	0.14
Shear stiffness (kg/mm) @ 130 kg	17 – 19	18 – 19.28	18 –18.86
Bond strength (in kg)	1200	1060	1215
Bashore resilience, %	50	48	50

These compounds were used for transfer moulded engine mountings for passenger cars and various dynamic properties were studied. Two type of engine mountings were moulded:

1. One for studying spring rate at compression load deflection.
2. For studying axial deflection spring rate at 45° angle.

The results of compression load deflection is given in Table 21.11 and graphically in Fig. 21.12. It can be seen that H-645 formulation containing Saret-633 gave marginally better properties as compared to formulation containing HVA-2. Spring rate was determined for both H-645 based EPDM compounds and is given in Table 21.12 as well as in Fig. 21.9.

Table 21.11. Compression load deflection properties.

Load in Kg	*Deflection of H-645 with HVA-2*	*Deflection of H-645 with SARET-633*
50	2.31	2.19
100	5.04	4.70
150	7.38	6.98
200	9.82	9.12

Load in Kg	*Deflection of H-645 with HVA-2*	*Deflection of H-645 with SARET-633*
250	12.20	11.24
300	14.63	13.30
350	16.83	15.30
400	18.89	17.06
450	20.56	18.80
500	22.08	20.34

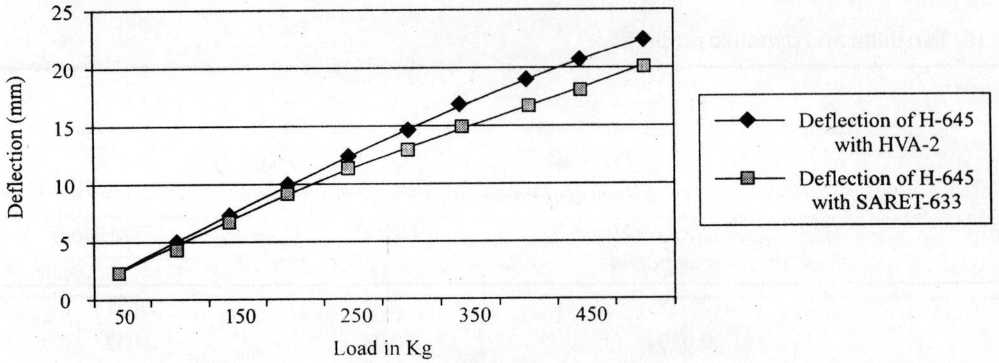

Fig. 21.8. Effect of co-agents on deflection.

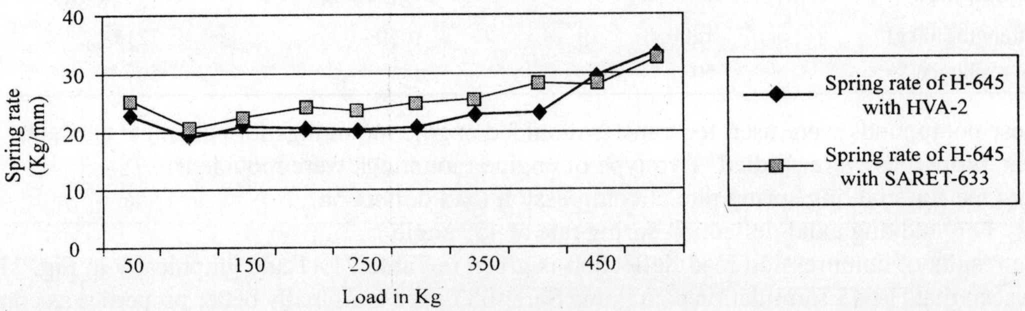

Fig. 21.9. Effect of co-agent on spring rate.

Bond strength between rubber and metal was determined by Pull test and it was found that by using Chemlok 205 primer and Chemlok 236A top coat, very good bond strength can be achieved as seen in results in Table 21.10. The bond strength with H-645 EPDM containing Saret 633 is comparable with NR.

Comparision of NR based proprietary formulation for engine mountings was carried out against H-645 based EPDM compounds in axial load deflection studies carried out at 45° angle. The results are given in Table 21.13 and Table 21.14 and Fig. 21.10 and Fig. 21.11. It can be seen from the results that compounds based on oil extended Herlene H-645 is quite comparable to NR, with respect to load deflection and spring rate.

Table 21.12. Spring rate of H-645 compounds.

Load in Kg	*Deflection of H-645 with HVA-2*	*Deflection of H-645 with SARET-633*
50	22.76	24.79
100	18.32	19.92
150	21.37	21.93
200	20.49	23.36
250	21.01	23.58
300	20.58	24.27
350	22.73	25.00
400	24.27	28.41
450	29.94	28.74
500	32.89	32.47

Table 21.13. Axial load deflection (In 45°C angle natural rubber v/s oil extended EPDM rubber H-645.

		Deflection in mm	
Load in Kg	*Natural rubber*	*HVA 2 compound*	*SARET 633 compound*
50	1.24	1.29	1.29
100	2.39	2.39	2.54
150	3.68	3.54	3.73
200	4.98	4.69	5.07
250	6.27	5.79	6.27
300	7.42	6.94	7.42
350	8.52	7.99	8.66
400	9.67	9.14	9.76
450	10.82	10.15	10.91
500	11.82	11.05	11.96

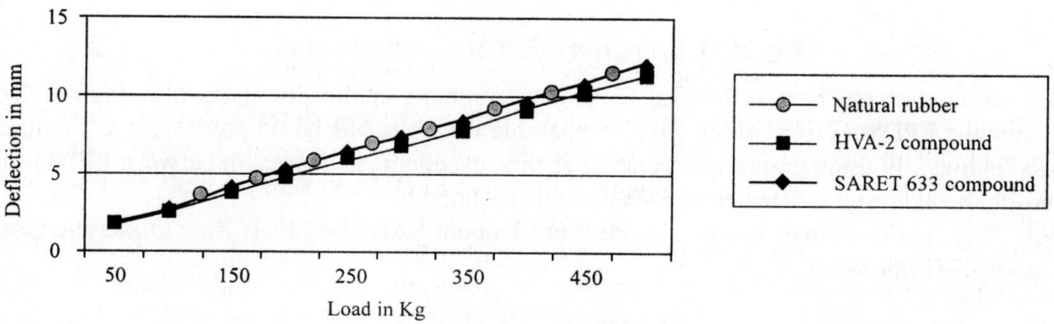

Fig. 21.10. Axial load deflection 45°.

Table 21.14. Axial deflection spring rate (In 45° angle) natural rubber v/s oil extended EPDM rubber H-645.

Load in Kg	*Deflection in mm*		
	Spring rate of natural rubber	*Spring rate of H-645 compound with HVA 2*	*Spring rate of H-645 compound with SARET 633*
50	42.53	46.11	42.17
100	43.48	45.45	40.00
150	38.76	43.48	42.02
200	38.46	43.48	37.31
250	38.76	45.45	41.67
300	43.48	43.48	43.48
350	45.45	47.62	40.32
400	43.48	43.48	45.45
450	43.48	49.50	43.48
500	50.00	55.56	47.62

Average spring rate: (Kg/mm)	Natural rubber	: 42.61
	Compound with HVA 2	: 46.06
	Compound with SARET 633	: 42.10

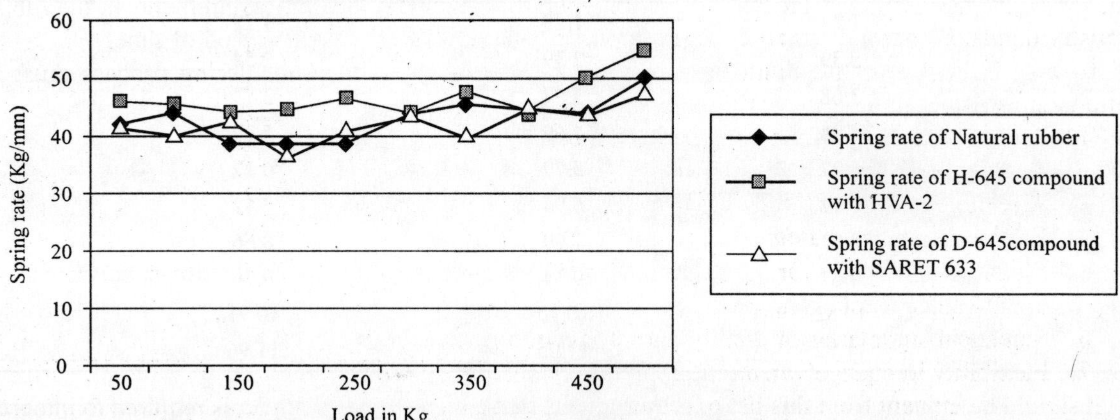

Fig. 21.11. Spring rate of EPDM vs. natural rubber.

Thus formulation has been developed for engine mounting application using ultra high molecular weight Herlene EPDM H-645, which gave comparable results to NR based proprietary compounds in dynamic testings. By using proprietary bonding agents, excellent band strength between EPDM rubber to metal can be achieved in transfer moulded engine mountings.

HVA-2® is N,N′-*m*. Phenylenedimalimide from Dupont SARET-633® is Zinc Di-acrylic Coagent from Sartomer Corporation, USA.

CHAPTER 22

Computer Aided Manufacturing

INTRODUCTION

Experienced rubber calender operators know that it is very difficult to see and control the many kinds of variation that occur in rubber sheet calendering operations. The line operator's job is to produce a uniform sheet of rubber within weight or thickness specifications and within a specified time limit. The difficulty arises when key variables affecting the uniformity of the calendered sheet change. The operator can see large changes in the process over a long period of time, but quality requirements in the rubber industry demand control of small changes in process variables over a short period of time.

To help the operator maintain tight control, key variables of the calendering process must be simultaneously measured and controlled. Here is what is required :

1. An understanding of the calendering process.
2. Accurate sensors to measure key variables.
3. A reliable computer to process sensor signals and information.
4. Effective control strategies to modify machine settings.
5. Meaningful displays for the operator, process engineer, quality and production supervisors, and maintenance technician.
6. Statistical summaries of quality and production parameters.
7. Flexibility to change when requirements change.

It should be evident from this list of requirements that a high speed computer is required to integrate real-time data and to make split-second decisions to stabilise the manufacture of calendered sheets of rubber. The best way to accomplish such integration and decision making is through computer-integrated manufacturing (CIM). The introduction of computers to production planning and design will have the effect of greatly increasing efficiency. Structural analysis using finite element analysis (FEA) allows computer simulation of how a part will respond to loads and forces in its application environment. In other words, it is used to predict the effectiveness of design concepts by indicating, for example, how various designs and materials will work together. Historically, the planning and designing of production facilities for elastomeric materials have been done using the "cut and try" method with in-house empirical data—hence the high cost of prototyping mould products and long development time required. Finite element techniques are a valuable aid in the analysis of elastomeric components and have also attracted extensive research attention; they are useful in evaluating design variations, and they make it possible to get designs into production quickly, as required by the industry.

Nevertheless, among the vast amount of finite element programmes in today's engineering software market, the application of computer codes in the design of elastomeric components is less widespread. This is mainly because elastomeric components exhibit non-linear stress-strain relationships, large deformations during load (sustaining strains of up to 500%), viscoelastic processes such as heat dissipation and time-dependent behaviour, and nearly incompressible characteristics during deformation. Rubbers are used in sophisticated and complex applications because of their unique properties, such as adhesion; damping and energy absorption characteristics; flexibility; resilience; long service life; the ability to seal against moisture, heat, and pressure; non-toxic properties; and mouldability and variable stiffness. Moreover, most analysts recognise that the use of the commercial programmes for elastomer analysis is by no means a simple task.

PROCESS

At the beginning of a production run, the line operator needs to set the calender gaps, start the feed strip from the warm-up mill or extruder, set width and temperature devices, set the line speed, and press the GO button, right? Not really. There are many other factors involved in making high quality rubber sheets. Let's start at the beginning. Warm-up mills or extruders increase the temperature of the rubber compound to re-form the rubber into a flat sheet. As rubber is warmed, it changes dimensions. The dimensional stability of the feed strip to the calender depends on the uniformity of the rubber compound itself and on the uniformity of its raw materials. As the warm-up mill or extruder reaches an equilibrium temperature during start-up after an extended shut-down, the dimensional characteristics of the feed strip also change. The output from the calender changes because the feed to the calender changes. The output also changes because the calender frame actually stretches, which affects the roll gap or opening.

Pre-setting the Calender Gaps

At the beginning of a run, the line operator presets the calender gaps to a weight or thickness specification. Making this preset involves four steps :

1. Linear voltage differential transducers (LVDTs) are accurate only from ± 0.15–0.2 mm because of the capability of the LVDT and because the measurement is indirect. The LVDT measures the movement of the bearing housing relative to the side frame of the calender, rather than taking a direct measurement of the opening between the two rolls. Also, the measurement desired is sheet thickness, not roll gap opening. The difference between the two measurements is usually called die swell, which is the expansion of the rubber sheet as it exits the high pressure squeeze point in the gap.
2. Calenders that have electric screwdown motors usually have sensors to count the number of screw motor revolutions and to translate this number to an indirect gap opening. This setting is adversely affected by backlash of the screw motor and by the bearings and bearing housing. Unfortunately, the number is usually not repeatable to the accuracy required. In addition, electric screw motors are slow (about 1 mm/min). High speed screw motors are available (about 5 mm/min), but the risk of crashing two rolls together increases at high speeds.
3. When calenders have hydraulic gap actuators, pre-set is usually a moot point because the speed of the output is so fast (about 10 mm/min) that a direct thickness measurement on the calender roll can read actual product thickness within a quarter of a roll revolution.
4. Some calenders have electric screwdown motors as course control of the gap settings, and hydraulic actuators for fine control. If LVDTs are used to pre-set the screw motors before the hydraulics take over. the inaccuracies described previously will adversely affect gap positions.

Transforming of Rubber into a Sheet

As the calender starts, the rubber is transformed into a sheet that is dependent on the following factors:

1. The volume uniformity and the rubber temperature in the calender bank have a direct effect on the uniformity of the sheet being produced. The higher the rubber volume in the bank, the higher the risk that some rubber will become stagnant and therefore will cool relative to the rest of the rubber in the bank. When the cooler rubber is squeezed through the forming gap, its dimensional characteristics and die swell will be different, creating a short-term dimensional upset.
2. The sheet uniformity also depends on the individual roll profiles and on the dynamic profile of all four rolls rotating at different speeds. These rolls typically have a crown to compensate for the parting forces that bend the rolls. The theoretical result is a flat roll in dynamic operating conditions. Parameters such as speed, gap position, stock consistency, roll temperature, and cross-axis position are difficult to repeat. Therefore the bending pressure required to flatten the sheet may vary.
3. Reworked or reprocessed material has a detrimental effect on calendered sheet uniformity. Reprocessed material has die swell characteristics different from those of virgin stock. Since the reprocessed material is usually added to the virgin material in inconsistent quantities (it is difficult to meter specific quantities of material on an open warm-up mill), dimensional changes in the calendered sheet will occur on a relative short-term frequency. Or, if reprocessed material is calendered simultaneously, the gap settings probably will not be changed to reflect the change in die swell characteristics.
4. Equilibrium of the calender system is usually reached in about 20–30 minutes, depending on the calender speed, the warm-up characteristics of the stock, the gage of the material, and the temperature and flow rates of the heating and cooling water. During those 20–30 minutes, the side frames of the calender will stretch with the environmental conditions, and with the temperature and speed of the calender. As the operating conditions stabilise, the gap opening will also stabilise, producing a more uniform and flat sheet. Therefore, the start-up period is a very important phase from the standpoint of material uniformity.
5. If equilibrium were maintained after start-up, an operator's job would be much easier. However, there are very often speed changes to the calender train. For example, the feed strip to the calender may break, a wire may jump out of position, or a fabric split may occur. When the equilibrium is disturbed, a new set of gap positions is required to maintain the desired flat sheet. In many cases, each of the four gap actuators and the four roll bending cylinders or four cross-axis actuators must be independently reset, even for a small process upset.
6. On a wire cord calender (Fig. 22.1), the sheet uniformity also depends on the cord diameter, the cord spacing, and the tension of the cord from the creel to the calender. Wire weight may also vary from supplier to supplier, and from the beginning to the end of the creel. Independent tests have shown that the weight of the wire cord can contribute as much as 8–29% of the total variance (standard deviation squared) of the calendered sheet weight variation. Variations of this nature are almost impossible for the operator to control.
7. Shrinkage and cooling are significant factors in the stability and uniformity of the calendered material. As the rubber is formed through the roll gaps, the calender rolls can be a stabilising influence if the surface temperatures are consistent. Large downstream rolls cool the calendered sheet on contact. Within 100–125 m, the material is wound up in rolls at a surface temperature

of about 24°C. If speeds, temperatures, and flows are consistent, the surface temperature will be consistent, therefore the shrinkage characteristics of the rubber will be consistent. If these variables are not consistent, the width and the gage of the sheet will vary, producing an undesirable effect of inconsistent sheet dimensions. In addition, varying levels of textile cord density, denier, and spacing will have an adverse effect on the stability of the sheet's dimensions sheet during cooling.

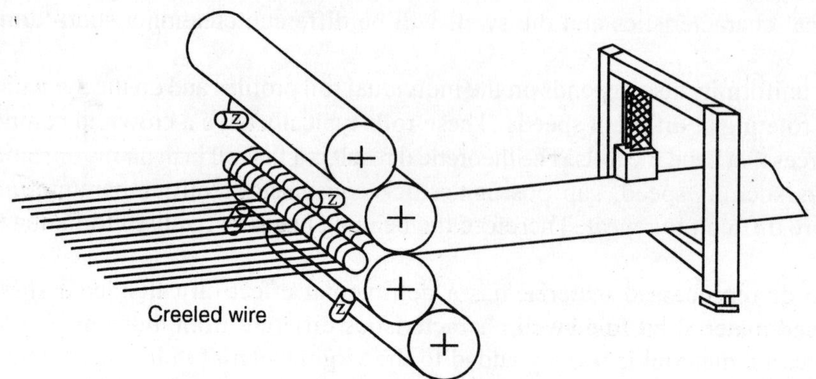

Fig. 22.1. Typical rubber calender (four roll z-wire calender).

Thus the operator has a long list of variables affecting the uniformity of the calendered sheet. If all the process upsets were significant and slow, most operators could control the weight uniformity to a tolerance of 4 or 5% and be relatively satisfied that production and quality standards are being met. However, process upsets are large, small, fast, slow, unpredictable, and predictable. And quality standards are demanding tolerances as low as 1%, depending on the end product.

Sensors

Accurate measurement of these many variables is complex yet necessary for successful process control. Let's review the available sensor technologies.

The most common sensor used to measure the uniformity of a sheet is the beta gage. The beta gage has been used to measure basis weight in the paper industry the rubber industry. The selection of the radioactive source is based on the weight range of the product being measured. For rubber, the source is strontium-90 because the ratio of the transmission to the absorption of the beta source by the rubber is high enough to accurately represent the mass per unit area of the typical rubber sheet. Beta gages with repeatability and static accuracy of measurement of ± 0.25% (2 σ), and dynamic accuracy of 0.50% are available. It should be noted that beta sensors show a very low sensitivity to composition. Specially designed sensor optics can further reduce the sensitivity to heavy elements such as zinc and titanium. The typical beta sensor is mounted on a high reliability scanner travelling at 25 cm/s across the sheet, usually within 5 m downstream of the calender.

When gum sheet measurement on the calender roll is desired, a variety of sensors are available :

1. The X-ray backscatter measures mass because X-rays are actually reflected by rubber. The higher the reflection, the higher the mass.
2. The beta backscatter functions in a fashion similar to X-ray backscatter but is a slower measurement because signal integration times are longer (up to 99 seconds). In addition, the accuracy of beta backscatter is not as good because the flux density of X-rays is much higher than beta sources.

3. Gamma backscatter is also used to measure mass, but it also has inherent accuracy and measurement speed limitations for the same reasons as the beta backscatter.
4. Air gages have long been used for measuring direct thickness. The combination of a rolling knife cutting through the rubber to find the steel roll surface and measurement of air back pressure from a nozzle just above the rubber surface can be substracted to provide a relatively accurate thickness measurement. The problems associated with air gages include high maintenance requirements, the destructive nature of the measurement, and the scarring of the rolls with the knives.
5. A new non-contact thickness sensor has been developed using the combination of laser and inductance measurements. An accurate laser finds the surface of the rubber, while an inductance sensor identifies the distance to the steel roll. A coaxial subtraction of the two signals yields a direct thickness measurement independent of composition and roll runout. The result is that this sensor provides fast measurements (10 readings/s) with a laboratory accuracy of 0.015 mm (2σ) at temperatures up to 104°C.

For as long as mass scanners for rubber calenders have existed, there has been a controversy of measuring weight versus thickness. The operator, of course, has had the benefit of the snap gage for decades. And the design engineer responsible for developing specifications thinks in terms of finished product thickness after curing. The problem is that dimensions of formed rubber sheets change until the sheets are cured. The snap gage is inaccurate. Temperature, handling, ageing, and physical characteristics of the compound and of the reinforcing material (nylon, polyester, wire, fibreglass) affect the sheet's dimensional stability. Studies have shown, however, that mass measurement at the calender is a better indicator of finished product thickness than thickness at the calender because mass at the calender is the same as mass at the curing press.

A digital diode array camera can measure the width of the sheet to an accuracy of 2.5 mm. The most useful location is near the windup of the calender, to indicate the width of material. The beta scanner near the calender also yields an accurate width measurement if the edge detection capability of the sensor is accurate enough.

The rubber sheet surface temperature is a desirable measurement to quantify the effects of temperature on the dimensional stability of the sheet. A radiation or infrared pyrometer mounted just before the windup usually proves to be an acceptably accurate measurement.

Computer

A high speed computer is needed to integrate all these sensor signals and translate them to meaningful control outputs and information. The computer should have enough capacity to process sensor signals, store at least 250 codes or recipes, retrieve control algorithms, and manage information for display and summaries. Advanced, on-line diagnostics will help locate any board problems for quick replacement or repair. The enclosure should be environmentally protected to prevent performance degradation due to temperature or airborne materials. Self-contained power supplies, power transformers to smooth voltage spikes, and battery backup will ensure that the performance of the computer is not interrupted by internal power problems. The computer provides the intelligence, computing power, and integration required to perform all the system's measurement, control, display, reporting, and communications functions. A computer with high reliability will ensure that process control is not interrupted by hardware failures.

Measurements

With our understanding of the calendering process, accurate sensors, and a high speed computer, we are ready to describe the array of measurements available. Product variation is a function of four components: cross-direction (CD), machine-direction (MD), short-term (ST) or uncontrollable, and long-term (LT) or variations controllable by gaging systems (Fig. 22.2). The sensors on the calender train provide a complete statistical history of the process :

1. Total weight.
2. Gum weight.
3. Steel weight.
4. Deviations from specification of each 1 in. slice.
5. Zone averages and their deviation from specification.
6. Bottom coat weight or thickness.
7. Top coat weight or thickness.
8. Sheet width at the calender and at the windup.
9. Surface temperature.
10. Line speed.
11. Length of rolls.
12. Number of wire cords per slice.
13. Total number of wire cords across the sheet.
14. Distance from the edge of the rubber to the first and last wire cords.
15. Number of wire cords per slice whose spacing is out of limits.

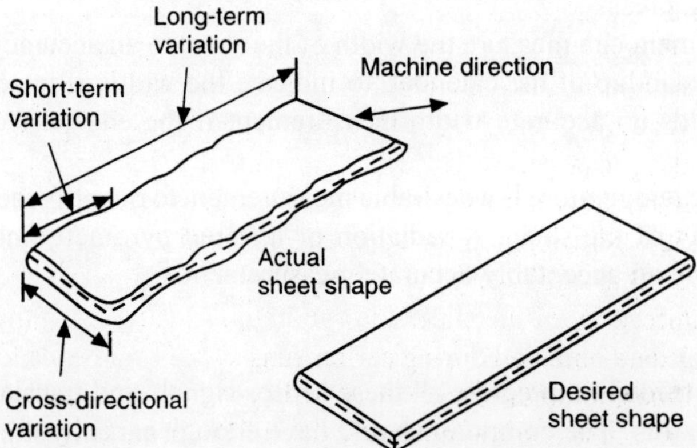

Fig. 22.2. Traditional control objective wire calender system.

Control

The performance of a calender control system depends on measurement accuracy, speed of signal processing, process behavior, and the accuracy with which software algorithms can model the behavior and response of the machine to the process. This section describes some control techniques that have been permitted significant improvements in sheet uniformity and have saved users millions of kilograms in raw materials each year.

The system executive is the real-time software operating system, which includes the input/output drivers, a driver for the video displays, access to setup recipes, programmes for control and data acquisition, and communications links for transmission to remote locations. Two or three zone weight controls use measurement information to feed back to the calender gap, roll bending, or cross-axis actuators to control the average sheet profile to weight targets using cocked sheet and centre zone control. Control tuning is code dependent and should include decoupling of machine and cross-direction variations, as well as compensation for line speed, transport delay, stiction, backlash, and gain of screwdown motors or hydraulics. When thickness of the rubber sheet is measured on the calender roll, gap control uses the thickness measurement to preset and reduce short-term machine direction variations by modifying gap positions through the screwdown motors or hydraulics. When scanner information identifies a deviation from targets, the scanner becomes the outer control loop to override the gap control, thus providing for reduction in long-term machine-direction variations.

Target adaptive control automatically adjusts weight targets based on selectable limits of the specification weight, a lower boundary limit, and a percentage of production that may fall below the lower boundary. This feature minimises the raw material usage by automatically shifting the weight target to a lower limit. Width control utilises feed-forward measurement of width at the scanner and feedback measurement of width at the windup to automatically control the width of the sheet with a width control device, either at the calender or at the windup.

Growth ratio control maintains a desired average weight target at the windup by measuring key variables at the calender and at the windup. When speed, width, and weight at the scanner are compared to speed and width at the windup, a control algorithm corrects for cross-sectional area changes in the sheet by modifying the gap actuators to a new weight target.

Speed gap feed-forward control minimises sheet variations due to speed changes by modelling gap setting versus speed changes. When the side frames of the calender stretch with different dynamic conditions associated with calender speeds, gap openings and throughput of material also change.

Speed ramping allows the gage system to control the line speed rate of change. Line speed can be inhibited until a new code has reached target weight, and it can slow the line when other quality limits are exceeded. This feature maintains quality during planned start-ups, slowdowns, and upsets.

Delta preset control adjusts the forming gap for code changes that occur without stopping the line. If code A is 1500 gsm, and code B is 1600 gsm, the delta or change in target is 100 gsm. The result is minimum disturbance to the uniformity of the sheet and minimum time to reach the new target.

Adaptive preset control uses data obtained during earlier runs of the same code to develop set points for the current run. This technique can be used to present weight or thickness sensors and roll bending with width control devices. Trim knife control automatically adjusts trim knives on wire calenders to obtain a specified gum edge to first-cord measurement. The measurement is obtained from the scanning X-ray transmission sensor in the scanner after the calender.

Display

The video display is the operator's interface to the real-time data that have been transformed into meaningful information. The display frames are also useful to the process engineer, the production and quality supervisors, and the maintenance technician. Typical displays include the following :

1. Code change frames list the information necessary for error-free line setup. Information includes code, compound, fabric or wire, total weight, fabric weight calender speed, widths, gage, top and bottom coat weights, growth ratio, roll temperatures, roll bending settings, crown offset, spread limit, weight deviation limit, and bow bar set point.

2. The production data frame provides a roll average profile of the sheet by slice and by zone, as well as production and quality data.
3. The trend plot frame and data logger allow the operator to monitor selected process variables over time.
4. The distribution plot and Xbar and R charts provide a graphical representation of selectable quality parameters.
5. Alarm displays alert the operator to process variables out of minimum and maximum limits.
6. Profile frames display cross-sheet measurements of selectable variables such as total, gum, top coat, bottom coat, and steel weight; balance of the top and bottom coats; roll average weight; open space; ends per decimeter; and wire cord spacing faults.
7. Displays are also provided for the maintenance technician to monitor sensor status, control tuning constants, calibration data, scanner setup, and system software.
8. A calender system performance frame continuously monitors process uniformity in terms of standard deviations of cross-direction and machine-direction variations, as well as short-term and long-term variations of total, gum, and steel weights.

CIM

On analysing a calender train and explored some of the associated problems, variables, sensors, measurements, controls, and displays, it has been found that CIM is the heart that integrates all these data into meaningful information for effective manufacturing of quality products. The accuracy and reliability of CIM can influence the production and quality efficiencies of the calender line. With CIM, managers and supervisors can use their valuable time for exception management to control the factory rather than monitoring routine functions.

In addition, operators can use CIM to communicate with their supporting departments in the factory to help identify cause-and-effect relationships. CIM helps elevate the role of operators on the factory floor to that of on-line quality control experts—making process corrections before lower quality material can be produced. But the computer is not intelligent. The system depends on accurate sensors, modelling, and interpretation. In addition, the system must be flexible enough to adapt to changing requirements. An important feature of the CIM system, therefore, is the capability to change features without disrupting the process.

The benefits of CIM include reductions in controllable variances from 50% to more than 90% depending on materials, machines, and personnel. Reductions of this magnitude result in shifts in material targets and savings in raw materials that can amount to more than 2000 kg/day depending on production volume. These reductions amount to quite a competitive advantage. Without CIM, the competitive advantage disappears.

Computer Programmes in Design and Planning

Commercially available computer programmes have recently made very significant advances in the development of finite element procedures. It is significant that in several of the programmes as mentioned above, the documentation for the hyperelastic solution recommends procedures for the experimental determination of the parameters of the rubber strains energy (density) functions. It allows the modeling of elastomeric material deformations up to several hundred percent strains and loading changes attributed to large deformation changes.

To achieve more accuracy, more terms are desired in the strain energy function. However, the use of a higher order model means greater complications in determining the material coefficients. The coefficients associated with a higher order model must be chosen such that behaviour in different modes of strain (i.e., tension, compression, shear) is accurately represented. The proper choice of a strain energy function depends on the quality and availability of experimental data, the type of elastomer, and which models have been incorporated in the particular finite element programme being used.

In another instance, contact problems that are very difficult to solve because of the high non-linearities in the physical situation have been handled by automated contact analysis including friction effects. Rubber products and contact problems always seem to involve contact versus non-contact conditions. This very efficient method is indispensable for analysing practical rubber applications. Furthermore, modern non-linear finite element analysis programmes such as the ABAQUS and the MARC account for heat flow and offer the capability to conduct coupled thermomechanical analysis.

That is, the analyst uses the same finite element model for both the thermal and stress analysis, and both thermal and force equilibria are satisfied in each increment before the non-linear analysis proceeds to the next increment. Non-linear finite element analysis of elastomeric products provides a way of testing designs before time and money are spent on building prototypes. With extensive analysis early in the product development cycle, it becomes possible to design parts that go through the prototype phase passing all required physical tests the first time.

This saves money in labour and tooling costs, and material volume. It also allows companies to respond more rapidly to the needs of the market. Computer costs can be justified by factors such as efficiency and better component reliability and performance. Although many more significant and exciting improvements are, of course, still needed to take full advantage of what the finite element method can potentially deliver as an analysis tool, practical engineering applications have been started in several programmes for the automobile and aerospace industries.

Practical Examples

As stated above, the proper choice of a constitutive equation and strain energy function depends on the quality and availability of experimental data, the type of elastomer and its application, and which models have been incorporated in the particular finite element code being used.

A number of analysis using the Mooney–Rivlin model, which is constructed on the assumption that the rubber is isotropic and rate-independent and neglects volume change and hysteresis, has been applied to some engineering applications. The performance of a rubber seal used in an automotive braking disk caliper is a good example of a simulation procedure.

The computer programme used in this analysis is ABAQUS. Inside the brake caliper, a disk brake piston is usually sealed with an O-ring of square cross-section. This seal stretches as the piston moves toward the rotor when the brake is activated. As the braking effort is released, the stretched rubber seal restores its strain energy and retracts the piston to the original position, which eliminates the need for return springs. Moreover, as the lining wears, the piston will automatically slide outward relative to seal, thus maintaining a proper relationship between the lining and the rotor throughout the useful life of the lining. Fig. 22.3 shows a cross-sectional view of an axisymmetric finite element model. The caliper groove (top surface) and the piston (bottom surface) are taken to be rigid bodies. The first step consists of squeezing the seal outside diameter to simulate the as-installed configuration. The contact pressure between the rubber and piston-caliper interfaces developed in this squeezing process prevents the hydraulic fluid from leaking to the outside environment. The second step enforces piston displacement along the piston contact surface to simulate the shearing motion induced by the piston during brake actuation. A seal deformed configuration and the stress distributions are represented in Fig. 22.4. It can be seen that high stresses occur in the corner regions due as a result of the stress concentration effect and, if necessary, this information obtained from the finite element analysis can be used to modify the design.

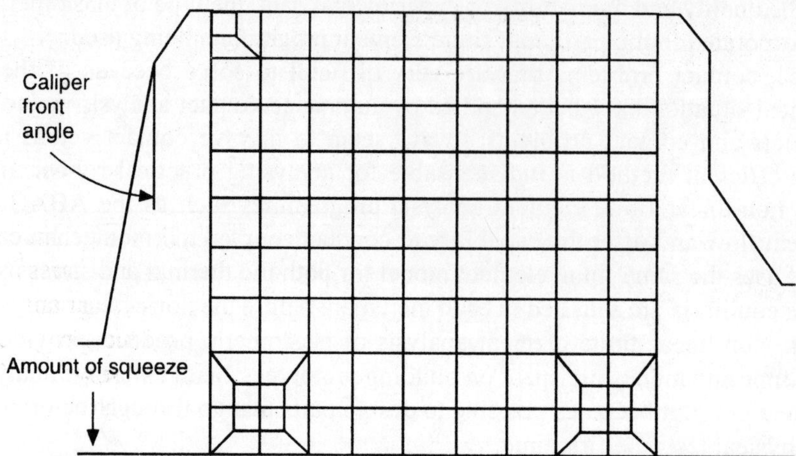

Fig. 22.3. Cross-sectional view of an axisymmetric finite element model (original configuration).

The successful compression and bending analysis, include the contacting problem of constant velocity joint rubber boots. The material composting the rubber elements in this analysis is polychlorprene (Neoprene). In the automobile industry constant velocity joints on drive shafts are generally sealed with rubber boots to keep dirt and moisture out as shown in Fig. 22.5. This elastomer components must accommodate the maximum possible swing angles at the joint and must compensate for changes in the shaft length.

During bending and axial movements, the industrial bellows of the boot must not come into contact with each other or with the joints. Should there be contact, relative movements at these points cause abrasion during the rotation of the shaft and lead to premature failure of the joint due to loss of grease. Above a critical axial elongation, the boot tends to collapse locally. Local bulking can also occur in one of the bellows. This analysis uses the MARC programme and the three-dimensional model. The industrial bellows are loaded in elongation and compression alternately because of rotation of the shaft at the swing angle. To optimise the boot design, it is important to recognise strain distribution.

The deformed shape of a rubber boot at elongation (compression) and bending over wide ranges of swing angle from 0 to 25°. Experimental results also are shown. It can be seen that deformation of the rubber boot advances with an increase of the swing angle; in addition, contact between the individual folds advances as the bellows and the shaft contact. The agreement between the experimental and analytical results is quite good. But design can therefore be modified using this analysis method for more advanced products.

Successful analysis using the hyperelastic material model with five coefficients representing the incompressible behaviour and two coefficients representing the bulk compression are also applied to such engineering applications as modelling contacting surfaces with friction. As a good example, we consider stress and deflection analysis including the contact problem of an engine mount vibration isolator, used in trucks, off-highway vehicles, and agricultural equipment.

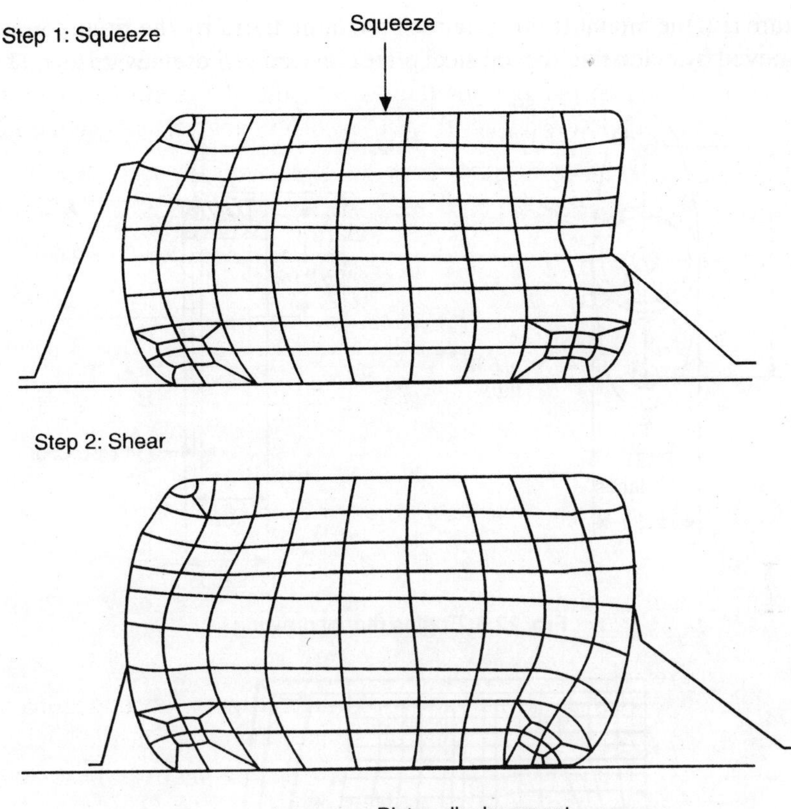

Fig. 22.4. Seal deformed configuration with stress distributions.

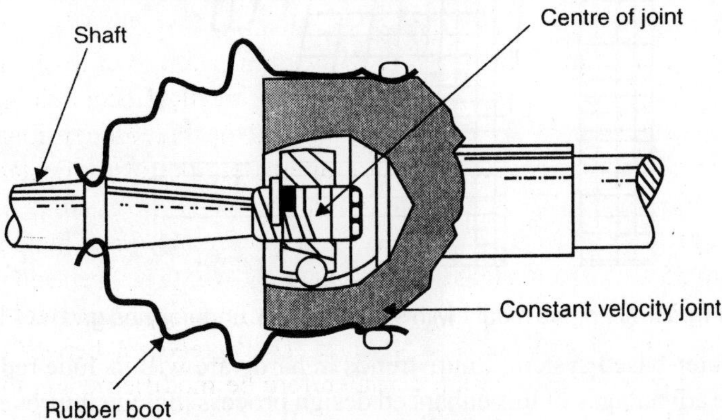

Fig. 22.5. Rubber boot for a constant velocity joint.

The computer programme used is ABAQUS, and the rubber elements are made of polychloroprene (Neoprene). The mount analysed (Fig. 22.6) is also axisymmetric, and the applied load is along the axis of symmetry, so that a two-dimensional axisymmetric analysis was done. Moreover, a snubbing element is usually added to this type of mount to limit the deflection, preventing the engine from impacting with the vehicle structure and the mount from sustaining too large a strain. The finite element model for this mount can be achieved by extending the top steel plate outward and downward as is shown in Fig. 22.7.

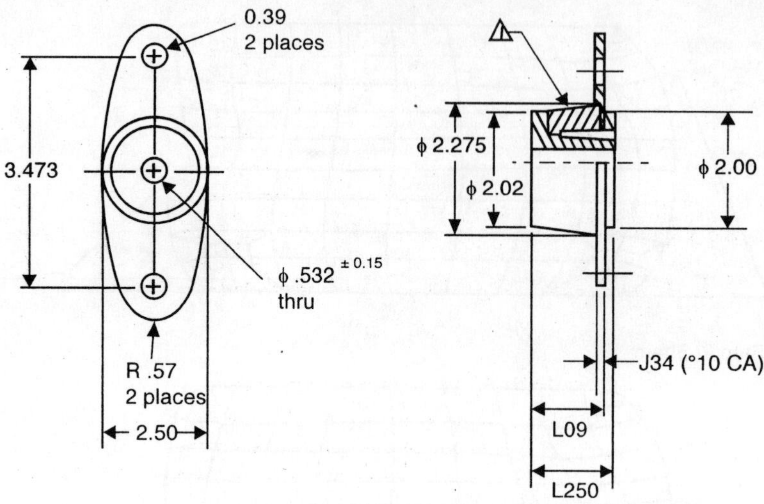

Fig. 22.6. Engine mount drawing.

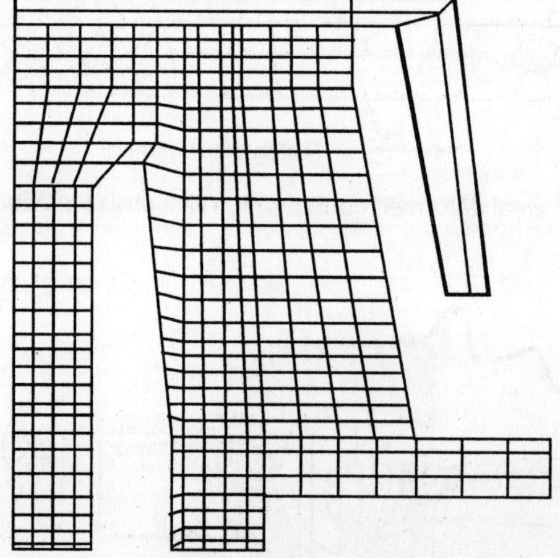

Fig. 22.7. Engine mount with snubbing cup: undeformed grid.

Thus with any computer-based system, future trends in hardware will include reductions in cost and improved reliability. The advantages of this enhanced design process include improved performance and quality of the finished product, faster time to market, optimal use of materials, weight savings, verification of structural integrity before prototyping, and lower production costs, all of which ensure a competitive edge. Furthermore, customising of these commercial finite element analysis programmes for specific applications and the development of in-house programmes are options available to users of computer-aided design. Given the rapid development of this technology during the past few years, the day is rapidly approaching in which finite element analysis is first method used, instead of cut and try.

SECTION IV

Applications/Uses of Rubber Products

CHAPTER 23

Rubber in Automotive Applications

INTRODUCTION

The automobile combines many different materials that must be integrated with the ever-changing design and manufacturing processes. Without rubber components, the automobile in its current state of technology could not exist. Very few people are aware of the role that rubber plays in the total transportation industry. To put everything in proper perspective, consider that elastomers account for approximately 5% of the material content of an automobile. These percentages vary from year to year, especially with increasing government regulations reflecting fuel economy, safety, and environment concerns, but the following breakdown is typical:

Material	*%*
Iron	60
Glass	3
Rubber	5
Plastics	7
Fluids and lubricants	7
Non-ferrous metals	7
Miscellaneous	1

APPLICATIONS OF RUBBER COMPONENTS

Approximately 500–600 applications in the automobile involve the use of elastomers. These uses vary from model to model and from year to year. Some critical uses having safety implications are as follows:

1. Tyres
2. Fuel hoses
3. Brake hoses and cups
4. Various fluid gaskets
5. Engine and TBI gaskets.

Additional uses include weatherstrips, windshield wiper blades, radiator hoses, and electrical wiring insulation. From the application examples, one can see how unique and important elastomers are to the automobile.

Some important automotive areas that need and use elastomers are as follows:

1. Traction for power transmission
2. Reduction or isolation of noise, vibration, and harshness
3. Seals against air, fluids, or other contaminants
4. Electrical insulation
5. Binders for friction materials, adhesives, sealers, and gaskets
6. Energy absorption (safety padding, air bags)
7. Conduits for liquids and gases
8. Flexible body panels
9. Movable impellers or diaphragms for fluid or gas transfer.

No single type of elastomers can fill all the areas or needs by itself. This is why there are so many commercial elastomers, each having a unique property profile. Rubber components need to survive and function under a wide variety of service and environmental conditions. These can include combinations of the following:

1. Extreme hot or cold temperature
2. Exposure to various chemical substances
3. Fuels and lubricants of various types
4. Other fluids, such as hydraulic, brake, and transmission
5. Exposure to ozone
6. Intermittent or continuous mechanical stress
7. Other miscellaneous conditions and environments.

Automotive people are continually faced with material, design, process, and cost decisions in choosing elastomers to meet established performance goals. These goals can include durability, economy, and meeting government regulations, depending on the applications.

Table 23.1, which was assembled to assist in the elastomer selection process, lists the general types of currently available commercial elastomer in order of generally increasing temperature resistance, together with the ASTM D2000/SAE J200 type/class designations the particular elastomer meets. Also listed are qualitative resistances to various conditions and environments important to the automotive industry.

Ethylene Propylene Diene Monomer (EPDM)

The advantages of EPDM include the following :

1. Resistance to weathering, temperature, and ozone.
2. Resistance to break and engine coolant fluids.
3. Cost: compounds can accept high loadings of fillers and plasticisers.

The main disadvantages are fuel and oil resistance and iridescence.

EPDM is used for weatherstrip (primary, secondary, glass run channels, etc.), exterior rubber parts (bumpers), coolant hoses, brake system parts, and ignition cable insulation, and some exhaust hanger isolators.

Butadiene Acrylonitrile (NBR)

The advantages of NBR include the following :

1. Resistance to solvents, fuel, and oil.
2. Mechanical and physical properties: heat ageing, abrasion resistance, compression set, and flexibility.

Table 23.1. Elastomer selection chart.

		Physical properties					Service temperature (°F)			
							Continuous low temperature			
Polymer type	*Material designation: ASTM D2000 SAE J200*	*Gum polymer density range (Mg/m^3)*	*Hardness range (Shore A)*	*Original tensile strength (psi)(max.)*	*Original elongation maximum (%)*	*Compre-sion set resistance*	*1000 H/Air*	*Dynamic*	*Static*	*Ozone and weather resistance*
Polybutadiene (blends)	AA	0.090–0.091	45–80	2500	450	Good	158	–60	–80	Poor
Polynorborene (Norsorex)	AA	–	15–90	2500	600	Fair	158	–50	–60	Poor
Natural and synthetic polyisoprene	AA	0.91–0.94	30–95	4000	700	Very good	158	–50	–70	Poor
Polysulphide (Thiokol)	AK BK	1.34	20–80	1500	400	Poor	158	–40	–60	Very good
Styrene butadiene (SBR)	AA BA	0.93–0.94	40–95	2500	600	Good	158	–40	–70	Poor
Polychloroprene (Neoprene)	BC BE	1.23–1.25	20–95	3500	700	Fair-good	225	–40	–65	Fair-good
Polyurethane (ester, ether) (caprolactone)	BG	0.80–1.21	50–100	5500	600	Poor	235	–50	–70	Good-excellent
Polypropylene oxide (Parel)	BC	1.01	40–90	2500	600	Fair	235	–40	–70	Good
Nitrile (butadiene acrylonitrile)	BF BG BK CH	0.97–1.00	40–95	300	600	Good	235	–30	–65	Poor
Carboxylated nitrile	BF BG BK CH	0.98–1.00	50–95	3500	500	Good	235	–30	–70	Poor-fair
Isobutylene isoprene copolymer (butyl)	AA BA	0.92	30–95	2500	700	Good	250	–50	–70	Very good
Epichlorohydrin ethylene oxide copolymer (ECO)	CH	1.25–1.36	40–90	2500	400	Fair-good	275	–40	–60	Good
Chlorosulphonated polyethylene (Hypalon)	CE	1.10–1.27	50–95	2500	500	Poor-fair	275	–40	–60	Out-standing
Chlorosulphonated polyethylene (Tyrin)	CH	1.16–1.25	40–90	2500	500	Poor-fair	275	–40	–60	Out-standing
Halogenated isobutylene isoprene copolymer	AA BA CA	0.92–0.93	30–90	2500	700	Good	275	–50	–70	Very good

(Contd ...)

		Physical properties					Service temperature (°F)			
							Continuous low temperature			
Polymer type	Material designation: ASTM D2000 SAE J200	Gum polymer density range (Mg/m^3)	Hardness range (Shore A)	Original tensile strength (psi)(max.)	Original elongation maximum (%)	Compre-sion set resistance	1000 H/Air	Dynamic	Static	Ozone and weather resistance
Hydrogenated nitrile	CH	0.97–1.00	40–95	3000	600	Good	275	–35	–60	Good
Ethylene propylene diene monomer (EPDM)	AA BA CA	0.85–0.87	30–90	2500	500	Good	300	–50	–70	Very good
Epichlorohydrin homo-polymer (ECH)	CH	1.25–1.36	40–90	2500	400	Fair-good	300	–35	–55	Very good
Ethylene propylene copolymer (EPR)	BA CA DA	0.86–0.89	30–90	2500	600	Good	325	–50	–70	Very good
Vinyl acetate ethylene copolymer (VAE)	DH	0.98–1.02	50–90	2500	350	Fair	325	–30	–40	Very good
Polyacrylate ethyl, butyl	DF DH	1.10–1.13	40–90	2000	400	Good	350	–20	–40	Very good
Ethylene acrylic (Vamac)	DF DH	1.03	40–90	2000	400	Good	350	–40	–60	Very good
Phosphonitrile fluoro-elastomer (PNF, Eypel)	EF	170–180	35–90	2000	250	Good	350	–65	–80	Very good
Fluorosilicone (fluoro-substituted polysiloxane)	FK	1.3–1.8	30–80	1250	400	Good	450	–65	–80	Very good
Silicone dimethyl, diphenyl, phenylmethyl, and divinyl polysiloxanes	FC FE GE	0.97–0.99	25–90	1500	800	Good	525	–80	–180	Very good
Fluorocarbon fluorinated hydrocarbons, (Aflas, Fluorel, Kel-F, Viton)	HK	1.40–1.95	60–90	2500	350	Very good	525	–20	–50	Very good
Perfluoroelastomer (Kelrez)	HK	2.01–2.02	70–95	2500	200	Good	600	–10	–30	Very good

Its disadvantages are low flammability resistance, low resistance to ozone and weathering, and presence of oxygenated solvents (alcohols and alcohol-containing fuels).

NBR is used for O-rings, diaphragms, gaskets (binders in various gasket materials), oil seals, ar hose liners.

AUTOMOTIVE STANDARDS FOR ELASTOMERS

To have uniform quality rubber components that meet the material, design, process, and other goals, specifications are required. A complete rubber component standard is necessary for the following reasons :

1. To ensure selection of materials that meet requirements.
2. To make certain that the materials are commercially available.
3. To prevent changes in production material(s) without the users' knowledge.
4. To maintain uniform quality in production shipments.
5. To correlate design requirements with material properties.

The requirements of a rubber component specification are as follows :

1. *Material requirements,* such as colour, type of material, finished part quality, and identification markings.
2. *Performance requirements,* such as density, tensile, elongation, and tear.
3. *Description of test methods,* including SAE, ASTM, and other laboratory procedures.
4. *Construction requirements,* for composite materials such as hoses, head gaskets, and weatherstrips.
5. *Construction requirements* (normally shown on the drawing), any information on tolerances or thickness of materials referred to in the standard is required, along with a statement acknowledging the dimensions on the drawings.
6. *Sampling instructions,* including sample size and number of samples to be submitted.
7. *Packaging instructions,* to ensure clean, distortion-free parts after shipment and storage.

The automobile company's own standards for rubber and elastomer are used, along with the classification systems of the Society of Automotive Engineers (SAE) and the American Society for Testing and Materials (ASTM).

THERMOPLASTIC ELASTOMERS (TPE)

Faced with ever-increasing demands for safety and fuel efficiency as a result of government regulation and consumer demands for quality and value, automotive engineers consider the total elastomer spectrum: thermosets and thermoplastics.

Thermoplastic elastomers were developed to bridge the gap between rubber and plastics. Thermoset materials are rubber-based compounds that are capable of becoming permanently rigid when heated or cured. They cannot be re-reprocessed. Thermoplastics on the other hand, are plastics that are capable of softening or fusing when heated and of hardening again when cooled. Thermoplastics can be re-processed. The physical structure of a TPE has two components :

1. Hard blocks—plastic material
2. Soft blocks—elastomer material

Thermoplastics elastomers are formed in two ways: by chemical reaction and by blending. In the first case chemical bonds are formed between the soft and hard blocks during manufacturing process. In blending, fully cured elastomers are dispersed within the hard thermoplastic matrix.

Thermoplastic Elastomer Processing

TPEs are processed on thermoplastic moulding equipment by injection moulding, extrusion, blow moulding, calendering, or thermoforming. The hard segments melt and flow during the injection process. When the material contacts the cool platens, the melt re-solidifies to hard segments. There is no physical change in the soft segment.

Advantages and Disadvantages of Thermoplastic Elastomers

TPEs offer the following advantages:

1. Fast, easy processing on thermoplastic equipment: 45 seconds TPE; 5–10 minutes for the same part in a thermosetting elastomer.
2. No scrap: all parts can be re-cycled if not correct.
3. Greater design freedom.
4. Generally a lower part cost.

The disadvantages are as follows:

1. Higher compression set than some conventional elastomers.
2. Lower chemical resistance than some conventional elastomers.

CHAPTER 24

Thermoplastic Elastomers for Automotive Applications

INTRODUCTION

The 1990's have witnessed the emergence of thermoplastic elastomers (TPEs) as a highly significant segment of the non-tyre rubber product industry, a result of three decades of concerted research and development. In 1996, TPEs enjoyed approximately 12 per cent of the worldwide non-tyre rubber usage, which should grow to 20 to 25 per cent by 2010. Ironically, these materials have found no significant uses in the pneumatic tyre area. However, over the past two decades, non-tyre automotive applications have provided one-fourth to one-third of the usage of TPEs.

THERMOPLASTIC ELASTOMERS (TPEs)

The extensive adoption of thermoplastic elastomers (TPEs) in automotive uses is derived from its advantages and disadvantages relative to thermoset rubber vulcanisates.

Advantages

1. TPEs are fully formulated and require no compounding or mixing. This leads to greater consistency of composition, with batch-to-batch variation eliminated.
2. Processing has fewer steps and is much simpler. Fabrication is more reproducible, with the rejection of fewer rubber parts.
3. Thermoplastic processing enables faster production rates. TPE cycle times are measured in seconds versus minutes for thermoset rubber.
4. Process scrap can be re-cycled. Thus start-up material, off-spec parts, sprues, runners, etc., can be re-used. With thermoset rubber this scrap is discarded.
5. It has been shown that TPE parts which have given a normal service lifetime can be cleaned, re-ground and re-fabricated to give quality use in a new rubber part.
6. Less energy is consumed in fabrication, due to shorter cycle times and no vulcanisation step.
7. Thermoplastic fabrication processes—such as blow moulding, thermo-forming and heat welding–are practical with TPEs, but not with thermoset rubber.
8. TPE parts are better suited for high-speed, automated assembly operations since their dimensional tolerances are tighter than those of thermoset rubber.

9. Quality control of fabricated parts is better, and at reduced cost.
10. Most TPEs have a lower density than that of a competing thermoset rubber. These materials are purchased on a weight basis but used on a volume basis, giving more parts per kilogram.

Offsetting these advantages are certain disadvantages which must be considered:

Disadvantages

1. *TPEs embrace new technology*. Innovating with them requires fabrication skills foreign to rubber people and marketing expertise unfamiliar to plastics people.
2. TPEs generally should be dried before processing. Thermoplastics processors are familiar with this step, whereas thermoset rubber processes are not.
3. They do not function as a rubber above their melting point. This precludes their use near or above that temperature for even a short period of time.
4. Favourable economics generally require volume production, to justify the necessary engineering, tooling and capital costs.

The suitability of a thermoplastic elastomer (TPE) for a given end-use will be determined by the relative importance of these advantages and disadvantages.

THERMOPLASTIC VULCANISATE AND THERMOPLASTIC ELASTOMERS

All commercial TPEs may be classified as either a block copolymer or a rubbery polymer/thermoplastic combination. In the latter category, the rubbery polymer (such as EPDM) will tend to be dispersed in a continuous matrix of thermoplastic (such as polypropylene). Cross-linking of the rubbery phase, usually by a process of dynamic vulcanisation, will give rise to a thermoplastic vulcanisate (TPV) with properties more closely approaching those of thermoset rubber than any other type of TPE. The improved oil resistance, compression and tension set, and properties retention at elevated temperature qualify the TPVs for numerous automotive uses for which the same uncross-linked combination fails to perform.

SPECIFIC AUTOMOTIVE APPLICATION OF TPVs

Rack and Pinion Steering Gear Boots

A highly successful automotive use of these TPEs is in rack and pinion boots to protect the moving parts of the steering system in lubricating grease within the boot. The convoluted design of the blow moulded boot enables the use of the harder (40 Shore D) TPV with sufficient flexibility to adapt to the movements of the steering system. The TPV offers the advantages of the speed and economy of thermoplastics blow moulding, resistance to the environment and lubricating grease, excellent flex fatigue life and puncture resistance. Replacement of a thermoset rubber by a TPV reduced the finished boot cost by a factor of approximately 3, with massive savings resulting.

Air Intake Ducts

A rapidly growing automotive use of TPVs is in the rubber duct which serves as a conduit for the air (purified or unpurified) to the engine. Unfiltered air is directed to the filter from which clean air is conveyed to the engine. The rubber in the duct must perform the following basic functions :

1. Without collapsing at negative pressure, the clean side duct must convey filtered air at 100 to 125°C. A convoluted design with a harder TPV aids this function.
2. Seals must be air tight, thus requiring low compression set and stress relaxation.

3. At temperatures as high as 125°C, the duct must resist fuels, oil, grease, radiator fluid, air oxidation and ozone, and be flexible down to –40°C.
4. The use of convolutes must allow ready installation and provide sufficient fatigue resistance and flexibility to allow for engine roll.
5. The modulus and coefficient of friction must enable a sufficiently high boot pull-off strength.
6. The duct must lessen engine noise.
7. Duct appearance must be acceptable, with embossed lettering feasible. New car buyers expect a clean, nicely-appearing air duct.

A blow moulded EPDM/PP TPV has been found to meet or surpass all of the above requirements. The lower density of a TPV, relative to a carbon black-loaded thermoset rubber, gives a 10 to 20 per cent weight reduction. Further, the blow moulding of the TPV allows design improvement in the convolute system and offers opportunity for generating a composite duct consisting of segments of flexible TPV heat welded or sequentially co-extruded to segments of rigid PP.

Air Bag Cover

TPVs have been found to be a material of choice for air bag cover in automobile safety systems, providing significant cost savings over reinforced polyurethane. The performance requirements of these covers are highly stringent. When called upon to perform, they must never fail since a human life may be at stake. The air bag components must function predictably when the vehicle impacts an object.

The deployment of an air bag is shown schematically in Fig. 24.1, with the sequence of events :

1. Impact activates sensors, sending an electrical impulse to the actuator, igniting a flammable solid in the housing to generate nitrogen gas for inflating air bag.
2. Air bag inflation causes the cover to rupture at the designated area and tear along a specific path through which bag erupts and becomes fully inflated.
3. The person in the vehicle impacts the air bag, which then deflates leaving the empty bag in his/her lap.

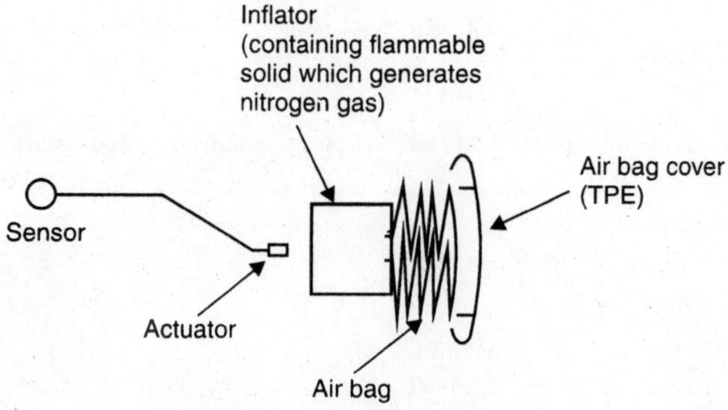

Fig. 24.1. Driver side air bag system.

This entire sequence must take place within one second, in the predicted manner.

The material selection for the air bag is both critical and difficult. At low temperatures the rupture mode of the cover is brittle fracture, whereas at high temperature it is tear. The only reliable way to successfully screen materials for this demanding use is to mould actual parts and deployment-test them at both low and high temperature. Part design is of critical importance.

Other Automotive Applications

Of the approximately 600 non-tyre rubber articles in a new automobile, several of the more demanding ones have been captured by TPVs. Control cables—such as the one connecting the parking brake handle to the rear mechanical brake—give better, longer service if they are protected from corrosion and mechanical abuse by a rubber cover. Successful crosshead extrusion of a TPV over these brake cables have allowed replacement of EPDM thermoset rubber with major improvements in cost, abrasion resistance, durability and service lifetime.

Engine compartment timing belt cover—with a flexible segment of rubber and a rigid segment of Polypropylene (PP) have successfully employed a TPV as the rubber. PP is insert injection moulded over a metal insert (threaded to allow fastening to the engine body), followed by overmoulding of the TPV onto the edges of the cover. The "adhesive-free" application is based on the ability of the TPE to heat weld to the PP and thus perform the sealing function of the cover.

Fuel line covers have also been found to exploit the properties and processing of a TPV. Here, a flame-retardant TPV is co-extruded over the nylon fuel line, alleviating a series of fabrication problems previously encountered with thermoset rubber, as well as giving a cost saving.

CHAPTER 25

High Tension Ignition Cable with Fluoropolymer Insulation for Special Automotive Applications

INTRODUCTION

In a gasoline engine, the fuel-air mixture is ignited with the help of an electric spark provided by a low voltage (12 to 24 volt) battery. This low voltage is converted by the induction coil to a high voltage discharge of 15 to 30 kVolt and in some cases upto 40 kVolt. High tension ignition cable transports this ignition voltage from the induction coil via a distributor to spark plugs in firing order to ignite the fuel.

The frequency of spark depends directly upon the engine speed (r.p.m.), i.e., the number of sparks in a minute are half the multiple of number of cylinders and engine speed. A six cylinder engine, working at a speed of 3000 r.p.m. would undergo 9000 spark ignitions every minute. For a high efficiency engine, it is conventional that 1200 r.p.m. require a suitable ignition cable to transfer 36000 ignition impulses in a minute. Consequently, a high tension ignition cable on one hand has to transfer electrical energy and on the other hand its dielectric behaviour has to be observed since the voltage discharge needed to ignite the fuel-air mixture due to high engine speed is to be taken as high frequency discharge. Ignition voltage peak and the peak height are responsible for the efficiency of the ignition voltage and hence a criterium for a constructive design of a high tension ignition cable.

INSULATION MATERIALS

Fundamental Criteria for the Insulation Materials

Insulsators are available in solid, liquid and gaseous state having very high volume resistivity. Materials with volume resistivity greater than 10^8 ohm.cm at room temperature (RT) are suitable as electrical insulation.

Electrical polarisation takes place in an insulation under the influence of electrical field. The three basic forms of electrical polarisation mechanism, depending upon the molecular and atomic structure of the insulastion material, are as follows :

1. Electron polarisation is generally observed in most of the materials. Ethylene-tetrafluorethylene (ETFE), polyethylene (PE) and polypropylene (PP) are the insulation materials wherein only electron polarisation is observed. These materials are called non-polar.
2. Ion polarisation is observed in ceramics, mica and glass.

3. Orientation polarisation is observed in the polar materials like polyurethane (PUR), polychloroprene (CR) and chlorinated polyethylene (CM).

The typical electrical characteristics of polar and non-polar insulation materials are described in the Table 25.1.

Table 25.1. Electrical characteristics of polar and non-polar materials.

Characteristic	*Polar material*	*Non-polar-material*
Dielectric constant (ε_r) (f = 1 MHz at 20°C)	3 to 7	1 to 3
Loss angle (tan δ) (f = 1MHz at 20°C)	0.5×10^{-3} to 200×10^{-3}	10^{-5} to 10^{-3}
Volume resistivity [ohm.cm]	10^{8} to 10^{16}	10^{14} to 10^{18}

Requirements on Polymers for the Insulation and Sheathing of Ignition Cables

Due to the operational conditions in the engine compartment, ignition cables undergo complexicity of stress and strain. Hence following requirements are necessary to be put on the insulation and sheathing materials.

Mechanical requirements

1. Crush resistance.
2. Tensile strength.
3. Flexibility.
4. Abrasion resistance.

Thermal requirements

1. Long time operational temperatures (3000h) depending upon the application conditions between 90°C to 180°C.
2. Peak temperature upto 250°C.
3. Flexibility at low temperatures (–40°C).
4. Tensile strength at peak temperature.
5. Heat deformation at high temperatures.

Electrical requirements

1. Long time dielectric strength from 20 to 30 kVolt.
2. Break down voltage from 35 to 50 kVolt or more.
3. Resistance to corona.
4. Low capacitance materials.

Resistance to chemicals

Life time resistance to water, moisture, salts, ozone and short time resistance to motor-oil, gas-oil, gasoline, diesel, ethanol, cooler-anti freeze, water anti-freeze and motor detergents.

These requirements are specified in the ISO 738 and the test methods are described. Taking into consideration the individual operational temperatures, various classes for ignition cables are defined in the Table 25.1 of the ISO 738 specifications. The requirements on the respective cables are classified in the Table 25.2 of the ISO 738.

Table 25.2. Test results of the special ignition cable according to ISO 738 E.

No.	*Test method*	*Requirement (7 mm cable)*	*Results (4.5 mm cable)*
4.1	Test for insulation faults	> 30 kVolt	Passed
4.2	Thirty minutes test voltage and breakdown voltage	> 20 kVolt —	Passed > 35 kVolt
4.3	Capacitance	—	< 120 pF/m
4.4	Resistance to corona effect (8 h/15 kvolt)	No cracks on the surface	Passed
4.5	Pressure test at high temperature (+180°C)	Depth of indentation <50%	<10%
4.6	Thermal overloading (+220°C/48h)	No cracks	Passed
4.7	Shrinkage by heat (+220°C/15 min.	=<2%	1%
4.8	Resistance to flame propagation	=<70 sec.	20 sec.
4.9	Flexibility at low temperatures	No cracks	Passed
4.10	Mechanical tensile strength	>250N	Passed
4.12	Resistance to oil (alteration of diameter)	=<10%	6%
4.13	Resistance to fuel (alteration of diameter)	=<15%	10%
4.14	Accelerated life test (test voltage after proceed)	20kVolt	Passed

Requirements on the Ignition Cable Design

1. Lower ohmic resistance of the conductor.
2. Higher ohmic resistance of the insulation material.
3. Good dielectric characteristics of the insulation material.
4. Operational temperature range of the insulation material.

DESIGN OF A COMPLEX IGNITION CABLE

The ignition cables are designed conventionally on the bases of two layers to fulfill the complex requirements. The first layer (insulation) guarantees the super electrical properties where as the second layer (sheath) takes the function of mechanical strength, resistance to chemicals and the flame retardancy. For economical reasons, one layer PVC based cables of class A and B are in some cases still in use.

Insulation and Sheathing Materials for Ignition Cables

Materials classification depending upon the maximum temperature rating from +105°C to +250°C according to ISO 738 are :

Cable class	*A*	*B*	*C*	*D*	*E*	*F*
Max. rated temp. (°C)	105	120	155	180	220	250
Insulation materials	PVC		XLPE		SiR, ETFE, FEP	
Sheating materials	/		CSM, CM		SiR	

PVC: Polyvinylchloride | XLPE: Cross-linked polyethylene | SiR: Silicone rubbr

ETFE, FEP: Fluoropolymers | CM: Chlorinated polyethylene | CSM: Chlorosulphonated PE

SPECIAL HIGH TENSION IGNITION CABLES FOR HIGH TEMPERATURE APPLICATIONS

It is conventional to use ignition cables with 7 mm outer diameter. A new cable for heavy duty application in a very narrow space of a high efficiency engine compartment was designed. The aim was to reduce the diameter by 35% and at the same time increase the break down voltage by 40%.

Design

A two layer design (insulation/sheath) on a 1.0 mm^2 stranded copper conductor (19 × 0.2 mm wires) with outer diameter 4.5 mm was used to implement the desired cable.

Various fluorpolymers (ETFE and FEP) were tried as insulation materials and a special silicone rubber (SiR) compound was used as sheath.

Insulation Material (Fluoropolymer)

Several ETFE and FEP material were tested to obtain a suitable quality of insulation. Finally ETFE from Du Pont (Tefzel 280) was selected to design the high tension ignition cable.

Manufacture

The extrusion process was optimised to extrude ETFE (Du Pont-Tefzel 280) in a single layer. The conventional processing for thin layer extrusion for fluoropolymers is with tube-extrusion and a draw-down-ratio of upto 70 is recommended by the material manufacturer. To extrude a smooth and void free insulation, it became necessary to develope new processing techniques. Special care was taken to design the tools (tip and die).

A semi-pressure extrusion technique with two stage cooling was used to obtain the required quality of insulation. A wall thickness of 0.85 mm of ETFE was possible to produce in a single stage extrusion.

TEST RESULTS

Despite reducing the size of the cable from conventional 7 mm to 4.5 mm, and using a suitable combination of insulation and sheathing materials, successful results were achieved. These are described in the Table 25.2.

Thus a high tension ignition cable, which is lighter in weight and upto 40% thinner than conventional cables, has been developed using ETFE-fluoropolymer insulation and silicon jacket.

This cable offers extended life time through improved corona, fire and chemical resistance. The new series of high tension cables find use in automobile engines and for oil and gas burners operating at an ignition voltage of 30–35 kVolt at between –40°C and +200°C. After many months of reliable operation in Ferrari test car, it is confirmed that the cable is suitable for automotive application in the most demanding and confined underhood conditions.

CHAPTER 26

Uses of Rubber and Composites in Aerospace

INTRODUCTION

The use of non-metallic materials in aerospace applications shows a healthy and continuing growth, with much of the increase in recent years resulting from the very exciting developments in the fields of fibre-reinforced plastics and engineering thermoplastics. Any discussion of recent developments in the aerospace use of rubbers and composites is, therefore, inevitably dominated by a consideration of structural composites, their diversity of forms, and their enormous potential for engineering application.

By comparison with composites, the usage of rubbers is well established in both fixed wing aircraft and helicopters, and consumption can be regarded as being relatively stable in terms of volume per aircraft. Indeed significantly increased application for this class of material is unlikely without the introduction of a rubber with substantially better high temperature properties than can be offered by the best of existing commercial products.

Aerospace requirements certainly exist for such rubbers, and what is needed is an improvement in resistance to both short and long-term exposure to high temperatures, that is, improved strength properties at temperature and improved thermal and thermo-oxidative stability. Notwithstanding this lack of novel and new high temperature elastomers, there remains a considerable impetus for research and development on existing speciality rubbers, and this arises from the persistent pressures from aircraft operators for improved performance and longer service life for existing rubber components on airframes and in aircraft systems.

Without doubt the single most striking influence on the usage of non-metallic materials in aerospace applications over the last fifteen years has been the discovery in the mid-60s of lightweight, strong, and extremely stiff carbon fibres. These materials, which were a product of research at the Royal Aircraft Establishment, Farnborough, quickly became available in continuous filament form, and when used in combination with epoxide resins, they led to products which made it possible, for the first time, to envisage a fibre-reinforced plastic giving serious competition to the light alloys traditionally used as materials of construction in airframes.

In the wake of the carbon-fibre development, there has come a flurry of developments in strong stiff lightweight synthetic fibres (notably the aramids), in speciality resins suitable for use in high performance composites, and in engineering thermoplastics. These developments taken as a whole, represent a substantial challenge to metals in a range of applications going far beyond the aerospace field.

The effective utilisation of carbon fibres in structural composites called for, amongst other things, the ability to produce sheets of accurately aligned fibres pre-impregnated with resin, i.e. uni-directional pre-preg, and the consolidation of stacks of this material into a composite in which the alignment of fibres was retained and the volume fraction of fibre closely controlled.

This 'pre-preg' technology, developed for carbon fibres, has since been transferred to the glass-reinforced-plastics field with excellent results, the upshot being that high quality GRP is also beginning to realise its very considerable potential as a structural material. This is particularly evident in the helicopter field where many of its characteristics are especially attractive.

Composites also feature in another important area of aerospace engineering application, i.e., dry bearings. Something of a quiet revolution has been taking place over the years with the introduction of ever-more-effective plastics composites as liners for flight control bearings in fixed wing aircraft and helicopters.

The obvious advantage of dry bearings is that oil and grease lubrication can be dispensed with, and this in turn should lead to lower costs of maintenance. The life of such dry bearings is governed by the rate of wear of the composite liner in service and it is in this area that substantial research and development has been and still is being applied with notable success.

RUBBER COMPONENTS

Rubbers find wide application in aircraft of all kinds, and although the volume of material used is not particularly impressive, the number and variety of components is. A range of such applications, listed below whilst it is probably not an exhaustive one it does give a feel for the diversity of use.

Tyres, cabin and window seals, seals for fuel, oil, and hydraulic systems, flexible hose, flexible fuel tanks, integral fuel tank sealants, sealants for air-ducting, diaphragms in fuel pumps and pneumatic devices, electrical insulation, and erosion protection.

Apart from the usage in tyres, the requirements are largely for 'speciality' rubbers, and in many of the applications, such as seals, sealants, and hose, crucial properties are resistance to a wide range of aircraft fluids and retention of highly elastic properties over a wide temperature range.

Speciality rubbers finding the most significant usage are the medium and low nitriles, polysulphides, neoprenes, methyl silicones, fluorosilicones, ethylene propylenes, and fluorohydrocarbons, with the nitriles still being the principal work-horse because of their inherent fluid resistance and the ability to formulate to meet an appreciable part of the wide temperature range requirement.

Of the newer speciality rubbers the fluorohydrocarbons, as exemplified by the commercially available Vitons and Fluorels, have probably attracted the most attention, their particular virtues from the standpoint of aerospace application being an exceptional resistance to numerous chemicals and to ageing, and their outstanding high temperature stability.

These properties singled them out for use in aircraft fuel and hydraulic systems (excepting those operating with Skydrol fluids) where they would find application as sealants and static and dynamic seals.

In the latter applications a very influential property in determining in-service life is the resistance of the seal to compression set, and initially fluorohydrocarbon vulcanisates were somewhat disappointing in this respect.

The contrast between the resistance to thermal ageing of the fluorohydrocarbon rubbers as shown by conventional oven ageing and compression set testing was striking (Table 26.1).

Table 26.1. Ageing of fluorohydrocarbon vulcanisates in air at 200°C.

	Conventional oven ageing test		*Compression set test*
Time (h)	*% loss of UTS*	*% loss of EB*	*% set*
24	0	0	47
120	0	0	73
336	0	0	92
5,000	27	7	—
10,000	42	30	—
20,000	58	47	—

Over the time-scale for almost complete set to occur, and by implication the time for a complete loss of sealing force in a static or dynamic seal, there is no apparent loss of mechanical properties in the conventional oven ageing test. Clearly substantial changes are occurring in the vulcanisate structure which are having virtually no effect on the condition of the rubber as judged by the results of the latter tests. It follows that whatever changes are taking place in the rubber network structure are not influencing the overall elastic properties of the material but are leading to a permanent change in dimensions of the compressed rubber with a consequent degradation in sealing force.

Research on the mechanism of thermal ageing in these early fluorohydrocarbon vulcanisates show that the site of instability was the cross-link structure introduced during the vulcanisation process, and that the reaction is thermo-oxidative in nature. Development work subsequently concentrated on introducing modified cross-link structures and stabilising the network by appropriate heat treatments, and the outcome has been a major improvement in resistance to compression set in these types of materials.

COMPOSITE MATERIALS

Composite Dry Bearings

An increasingly important, but relatively unsung area of aerospace application for composites is in dry bearings, and the scale on which such bearings are now being used is indicated in Fig. 26.1. The practical advantage of dry bearings over those lubricated with oil or grease can only be fully realised when composite liner materials with suitably low wear rates become available, and this has proved an extremely fruitful area for research and development on speciality composites.

The composite is used in the form of a thin liner for the bearing and life in service is governed by its rate of wear. As wear proceeds, a gradually increasing backlash develops in the bearing, and in practice this reaches an unacceptable level long before the liner has become worn through. An important area for research has, therefore, been that of studying the effects of variations in materials and forms of construction of the composite liner on the rate of bearing wear. For these thin composite liners the number of possible variations in materials and modes of construction is large, and they include, for example, the weave structure of the reinforcing fabric, the type of fibre and the size applied, the method of incorporating solid lubricant, the type of solid lubricant, and the type of resin binder.

For the last few years, the most widely used aerospace dry bearing material was composed of a thin liner of interwoven PTFE fibre and glass-fibre impregnated with a phenolic resin and adhesively bonded to a metal substrate. A typical interwoven structure is shown in Fig. 26.2. However, a second generation of liner materials began to emerge with a significantly greater degree of wear resistance, particularly in the early stages of sliding. Some of these materials are variants of the original type incorporating different weave structures, different resins, or with the glass fibre replaced by synthetic fibres. Other types are basically synthetic fibre fabrics reinforcing thin layers containing dispersed PTFE particles or flock.

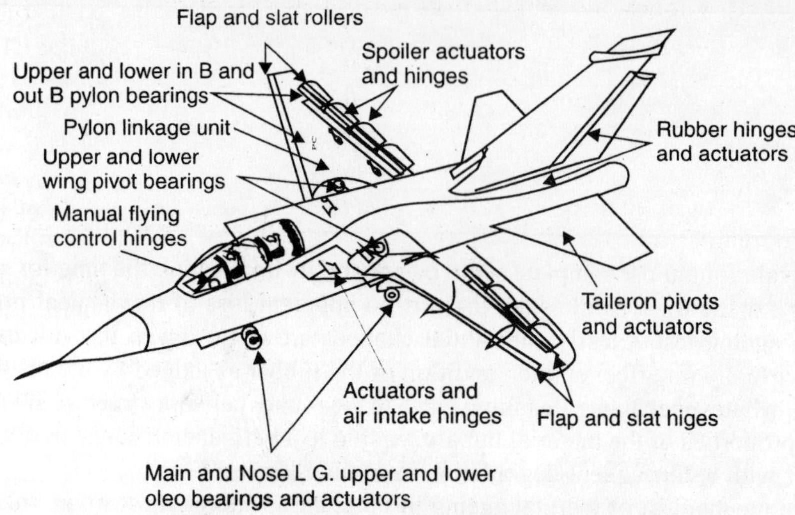

Fig. 26.1. Use of dry bearings in fixed wing aircraft.

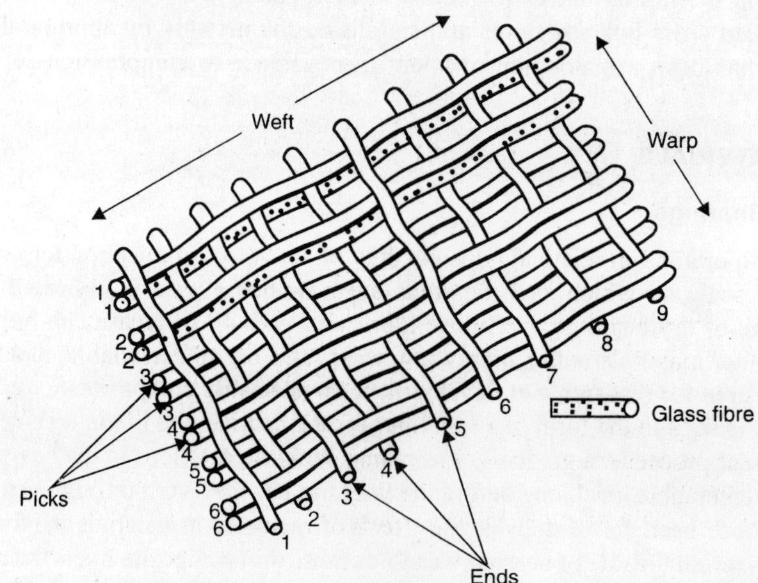

Fig. 26.2. Weave structure for standard dry bearing liner material; PTFE interwoven glass fabric; and PTFE/glass interwoven fabric.

Research has shown that perhaps the most important single factor influencing wear in fabric-based composite bearing liners is the weave structure of the fabric. Some results obtained using different weave structures incorporating both PTFE and glass fibre, are shown in Fig. 26.3, all the liners having been fabricated using the same resin binder. Adopting the commercially available double plain weave material as a yardstick for comparison (Fig. 26.3 curve 1), it can be seen that when the outermost PTFE fibres are replaced by glass it is no longer possible to establish a wear plateau (Fig. 26.3 curve 2). This suggests that too little PTFE is now available to form the desirable 'third bodies' on the rubbing counterface or on the linear surface or on both. This trend is further accentuated when the PTFE content is reduced even further by halving the denier of the PTFE warp fibres (Fig. 26.3 curve 3). By contrast a very slight reduction in the original PTFE content, by replacing 10% of the warp fibres with glass, reduces wear to a small extent (Fig. 26.3 curve 4). With a four-end satin weave (PTFE warp and glass weft) a large amount of wear occurs before the wear plateau is established (Fig. 26.3 curve 5), but with a plain weave the converse is true (Fig. 26.3 curve 6). However, none of the composites containing interwoven PTFE fibres performs as successfully as one containing an all-glass fabric with PTFE particles dispersed throughout the resin matrix (Fig. 26.3 curve 7). This particular configuration has also been found to give low wear when the glass is replaced by other such fabrics, as polyamide, aromatic polyamide, or carbon.

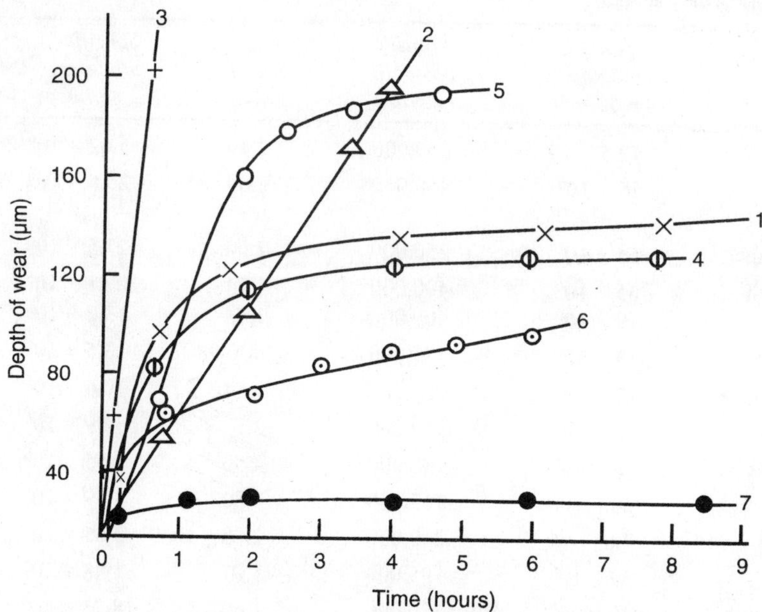

Fig. 26.3. Effect of weave structure on wear rates in PTFE/glass fabric 1. Double plain weave – PTFE warp and top weft; 2. double plain weave – PTFE warp, both wefts glass; 3. as (2) but PTFE warp denier halved; 4. as (1) but 10% of PTFE warp replaced by glass; 5. four-end satin weave – PTFE warp, glass weft; 6. plain weave – PTFE warp, glass weft; 7. all glass plain weave + PTFE flock in resin.

Structural Composites

One of the many attractions of organic polymers and polymer-based composites for aerospace applications is their generally low specific gravity. What has inhibited their wider use has been the inability to match the strength and stiffness of traditionally used metallic materials. However, with the advent of carbon fibres and aramid fibres, and the further extremely exciting prospects for the production of other strong, stiff, and even lighter organic polymer fibres, the whole position is changing. Table 26.2 shows mechanical property levels for a selection of commercially available synthetic fibres and composites based on them, and for some metallic materials favoured for aerospace applications.

The materials which commend themselves most strongly for aerospace structural applications are unquestionably the composites based on carbon fibres, the preferred variant at present being high strength fibre in combination with an epoxy resin matrix. The high levels of strength and stiffness possessed by these materials, coupled with their low specific gravity leads to remarkably high 'specific property' levels. High strength CF/epoxy composites, for example, show a specific modulus 5.6 times greater than that of aluminium alloy (RR58) or titanium alloy (Ti-6Al-4V) and a specific tensile strength 10 times greater than light alloy and 7 times greater than titanium alloy. Apart from carbon fibre, the most commercially advanced of the strong, stiff synthetic organic fibres are the aramids, and in composite form they are also making inroads in the aerospace field, albeit in secondary structure. The other synthetic organic fibres included in Table 26.2 that is ordered polyethylene and ordered polybenzothiazole, are still at the research stage but are nonetheless displaying extraordinary mechanical property levels at low specific gravity. These types of material undoubtedly have potential for further development and could well make an impact in the structural composite field in the not-too-distant future.

Table 26.2. Mechanical property levels.

Materials	*Young's modulus E (lb/in²)*	*Tensile strength (lb/in²)*	*Specific gravity*	*Specific modulus, lb/in²*	*Specific strength, lb/in²*
S-glass fibre	12.5×10^6	650,000	2.49	5.02×10^6	761,000
Boron fibre (tungsten substrate)	35×10^6	450,000	2.69	20.5×10^6	167,000
Carbon fibre, high modulus	55×10^6	250,000	2.00	27.5×10^6	125,000
Carbon fibre, high strength	35×10^6	400,000	1.74	20.11×10^6	230,000
Aramid fibre	19×10^6	400,000	1.44	13.2×10^6	277,777
Ordered polyethylene	13×10^6	435,000	0.95-0.98	13.5×10^6	453,125
Ordered polybenzothiazole	27×10^6	250,000	2.54	10.6×10^6	98,425
Aluminium alloy (RR58)	10×10^6	59,000	2.77	3.59×10^6	21,300
Titanium alloy (Ti-6Al-4V)	16×10^6	150,000	4.50	3.55×10^6	33,333
Steel (FV 520)	28×10^6	155,000	7.80	3.59×10^6	19,870
S-glass/epoxy composite	7.6×10^6	260,000	2.08	3.65×10^6	125,000
Boron/epoxy composite	39×10^6	195,000	1.97	19.8×10^6	99,000
C fibre/epoxy, high modulus	30×10^6	150,000	1.60	18.75×10^6	93,750
C fibre/epoxy, high strength	19×10^6	220,000	1.55	12.26×10^6	141,900
Aramid/epoxy	11×10^6	200,000	1.38	7.97×10^6	144,927

Early exploitation of continuous carbon fibres was principally in the form of highly aligned pre-impregnated tape or sheet material, which could be cut and stacked as required. Pre-impregnation was provided by a partially cured epoxide resin which gave the pre-preg adequate coherence and tackiness for subsequent handling and laminating. Laminates were prepared by the compression moulding of stacks of pre-preg sheet with individual layers arranged to give the fibre orientation desired in the finished component. With regard to this technology one could either be an optimist and see it as providing the designer with a new and considerable freedom to incorporate reinforcement only in those directions where it was required, or be a pessimist and see the requirement to use a highly anisotropic material as being disadvantageous and something giving rise to particular and undesirable difficulties in design.

However, in some areas of application, as for example, in the fabrication of helicopter rotor blades, all were agreed that the use of unidirectional pre-preg technology, offered positive advantages in allowing the mechanical properties of the blade to the optimised in particular and desirable directions by appropriate orientation of the fibre layers.

Carbon fibre is now available commercially in a wide range of tow sizes and a wide range of forms. In high strength fibre, tows of 1000-160,000 filaments are available, and in high modulus fibre, the tows can be obtained in counts 400-40,000. Forms available are unidirectional sheet or tape, woven cloth (plain, square, twill, satin, and bias weaves), random mat (short-fibre sheet or felt), aligned mat, knitted and braided, and the way is, therefore, open for widespread application. Interest is also growing in the use of 'hybrid' composites wherein more than one reinforcing fibre is used, the incentive to usage being either one of cost reduction or of enhancement of specific properties. Commercially available hybrid fabrics, in which the fabric is woven from a mixture of fibres, feature carbon with aramid, E-glass, S-glass, or polyester fibre.

The use of carbon fibre in woven form is technically acceptable in a wide range of aerospace applications, and in practice it can ease difficulties in fabrication and reduce component costs appreciably. A move towards the more widespread use of carbon cloth is already becoming evident in the aerospace industry.

A reduction in weight of an airborne vehicle is welcome from whatever source. The incentive to use lightweight composites in primary and secondary structures is, therefore, particularly strong, and nowhere stronger than in the fields of satellites and other space vehicles and in vertical lift aircraft. Not surprisingly, therefore, the largest pieces of CFRP structure in service to date are to be found on a space vehicle, i.e., the NASA space shuttle, and the most widespread use of composite on a high performance aircraft is at present envisaged for the advanced Harrier Av-8B jump jet. The former vehicle has its payload bay doors, which are 60 ft. long and 1500 sq ft. in area, fabricated in CFRP, with an estimated weight saving of about 23% over the same structure in aluminium alloy, whilst in the latter aircraft, advanced composites are expected to account for about 26% of the airframe structure.

Composite Applications in Helicopters

Lightweight structural composites are naturally attractive for use in helicopters. The most significant area for application so far has been in the main and tail rotor blades where the prospects for improved fatigue resistance over all-metal blades, and the greater freedom in design and fabrication have been quickly recognised and exploited. The use of composites based on unidirectional pre-preg has allowed for a better optimisation of rotor blade mechanical and dynamic characteristics, and compared with metal blades they have demonstrated better static and fatigue strengths. In state-of-the-art composite blades, the principal material of construction is GRP, with discrete areas of usage for CFRP. Substantial numbers of blades have been manufactured using these materials and are fitted to production aircraft. European manufacturers have been in the forefront of composite rotor blade development and the most impressive in-service experience has been gained on helicopters of French and German origin which have seen service world-wide. Many millions of flying hours have been accumulated with high life blades now approaching 7000 hours, and in-service performance is coming fully up to expectations. The blades are demonstrating increased performance and greater reliability than their all-metal counterparts, with the result that significant reductions in direct operating costs are being realised in practice. An attractive feature of GRP for this type of application is the inherently slow rate of crack propagation which is found to occur. This gives the blades a 'fail-safe' quality in that a period of gradual deterioration precedes any major failure.

For the future, consideration is being given by constructors to a much wider use of composites in the airframe structure, and in some quarters it is expected that composites will comprise upto 50% of the empty weight of future helicopters.

Composite Applications in Fixed Wing Aircraft

Following closely on the commercial development of continuous carbon fibre, steps are taken in the UK to design, fabricate, and flight test a limited number of CFRP components of a non-critical nature (i.e., not critical to flight safety). These components are fitted to operational aircraft in order to begin the long process of building confidence in the service usage of such novel materials. The earliest such components were the rudder trim tabs for the Jet Provost trainer aircraft in use with the RAF; six in number, and of thin-skinned honeycomb construction, they are fitted in an unpainted condition. The aircraft fitted with these components spent most of their service life with the RAF Red Pelicans' aerobatic flight, and are subjected to the full range of weather and flying conditions without being given any special treatment. Regular visual and non-destructive inspection confirmed that in-service performance had been entirely satisfactory.

It gives considerable satisfaction that each one of the oldest of the UK flight test components has performed very well under the rigours of in-service use. Of particular note is the absence of corrosion problems on the Vulcan air-brake where CFRP is attached to light alloy, the absence of foreign object impact damage on all the components, and the apparent resistance of the CF/epoxy to aircraft fluids in service.

In parallel with the UK programme, flight tests have also been carried out in the USA on a similar range of non-critical components, but as might be expected, their tests were on a much larger scale. The ability to fit composite components to civil aircraft proved a great boon to the American programmes and has resulted in the accumulation of flying hours at a very high rate, the two most outstanding examples being the Boeing 737 spoilers and the DC-10 upper aft rudder segment. In the former trial 108 spoilers were manufactured and 3 were fitted to each of 27 in-service aircraft. Having this number of components allowed a programme of planned retrievals from flight service and also a limited amount of destructive testing. Six spoilers are selected annually for removal and taken back to the Boeing factory for inspection, using ultrasonic techniques. A further 3 spoilers are selected for destructive testing each year and are tested in the same static test jigs as were used for the initial certification tests on the as-manufactured components. Since flying started in 1973, well over a million flight hours have been accumulated, no significant changes have been observed during non-destructive testing, and the results of destructive tests have been found to fall within the scatter band of strengths of the as-manufactured components. The other component which has been accumulating flight hours at a great rate is the DC-10 upper aft rudder segment, also made in CF/epoxy. Ten of these components were fitted and flying commenced in 1976, the utilisation rate on some of these has been extremely high, sometimes amounting to 400 hours per month. Overall the USA civil aircraft trials involving CFRP components have accumulated an enormous number of hours, and as yet no significant incidents or damage have been reported. The UK and USA experience with these non-flight-critical composite components has been good, and the in-service durability has come fully up to the expectations of those protagonists of composite materials who expected them to show lower maintenance costs than similar standard production metal items. The range of composite items either in use or projected for early usage on production aircraft is indicated in Table 26.3.

Table 26.3. Aerospace applications for composite.

Wing to body fairings	Vertical and horizontal stabilisers
Doors and access panels	Wing skins
Flooring	Engine nacelles and cowlings
Radomes	Helicopter rotor blades
Air brakes	Dry bearings
Air ducting	Drive shafts
Control panels	Leading and trailing wing surfaces

The major step of introducing advanced composites into primary flight critical structure is understandably proceeding more slowly, but significant advances are nevertheless being made. There have been, and continue to be, a number of major structural demonstrator programmes in both the USA and UK wherein design and fabrication processes are validated. Major ground tests of static and fatigue strength and environmental resistance have been carried out successfully and this has led, more particularly in the USA, to the manufacture of some large aircraft components. The Av-8B wing structure has already been mentioned, and another heavy user of composite structure in production is the McDonnell Douglas F-18 Hornet.

Thus, it is fitting to mention what is by far the most ambitious use of structural composites (carbon fibre/epoxy resin) in aircraft construction to date, that is the Lear Fan 2100 seven-seater light aircraft. This aircraft has an all carbon fibre/epoxy airframe which gives it an empty weight of only 3850 lbs, and it made its maiden test flight successfully earlier 1997.

CHAPTER 27

Some Novel Applications of Rubber in Railway Engineering

INTRODUCTION

The long range elasticity and resiliency of rubber and rubber-like materials, offer advantages in eliminating undesirable shocks, vibrations and noise associated with vehicles moving at high speed. Shocks and vibrations in rail vehicles are much more severe due to rail-wheel inter-sections, track irregularities, crossings and rail joints, firing impulses of engines, etc. Railway engineers too have taken advantages of rubbery state of matters in designing and construction of all types of rolling stock and track. The Indian railways utilises a large number of functionally different rubber components of which one of the oldest use of rubber which immensely interested the railway men is buffer spring against the longitudinal shocks.

Indian railways recently adopted ballastless track structure especially for Metropolitan Transport System. Function of the rubber pad is critical, providing entire elasticity to the track and minimising undesirable vertical and lateral vibrations. Different design features and physical characteristics which have been optimised for the purpose have also been discussed here.

The ability of rubber to change shape under load, to resist this change in shape and subsequently to recover from the change in shape left no competitors for the purpose of elimination of shocks. In Indian railways, helical metal springs inside buffer of wagons have been replaced by metal bonded rubber springs with improved serviceability and durability.

The design has undergone some modifications and high capacity buffer spring of more than 1280 kg-m has come to stay. These rubber spring elements are assembled in series with varying numbers depending on the type of rolling stock, to form a pack as shown in Fig. 27.1. The nominal thickness of each spring is 32 mm and the free height of the assembly of 16 springs is 512 mm. The stroke of assembly is limited to 127 mm and pre-compression 45 mm nominal.

RUBBER BUFFER SPRING

This design is considered to be unique having 'U' section circular channel and optimum 'shape factor' for improved performance and durability. Load-deflection characteristics as shown in Fig. 27.2 nearly follow the pattern of a metal spring upto a load of 12-15 tonne which is within the normal operating zone.

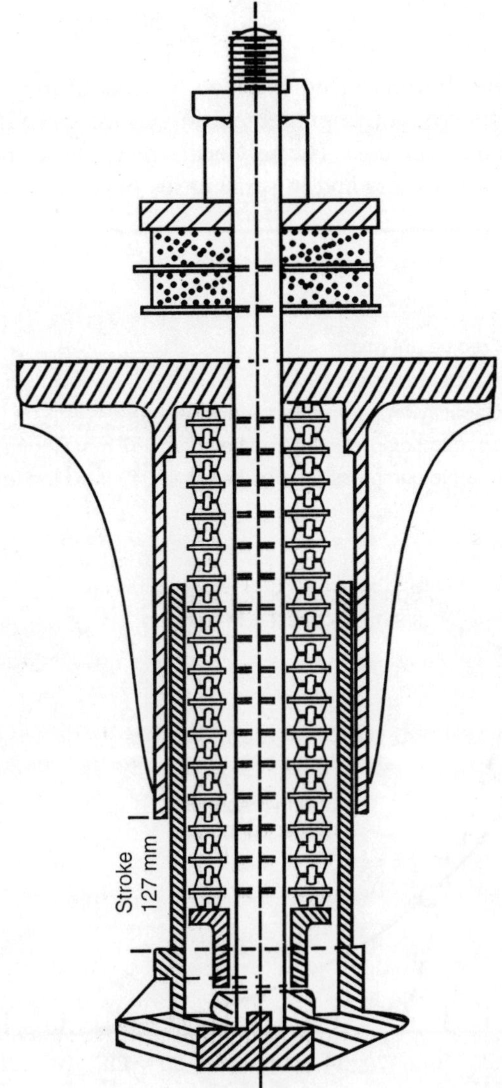

Fig. 27.1. Rubber springs in buffer assembly

This type of load deflection characteristics in rubber is attributable to the combination of air cushioning of the entrapped air and the 'shape factor'. These combinations of design features also enhanced the capacity more than 1280 kg-m. The impact endurance test on the assembly indicates retention in capacity more than 85%. The various parameters stipulated for the assembly of 16 springs are given as under :

1. Capacity for a pack of 16 springs, minimum — 1030 kg-m
2. Load to compress the free height to 467 mm, maximum — 2 T
3. Load to compress the free height to 340 mm, maximum — 40 T

The typical physical properties of the rubber are as follows:

1. Hardness (IRHD) — 70 – 80
2. Tensile strength (kg/sq cm) — 150

3. Elongation at break (%) 250
4. Compression set (%) at 70 ± 1°C for 24 hrs. 30

The introduction of the rubber buffer spring in place of steel spring has resulted in substantial economic gain to the railways in terms of improvement in serviceability, operational efficiency and reduction in maintenance and on material cost. The serviceability of these springs under Indian climatic conditions is observed to be around 8 years and in some cases beyond 12 years.

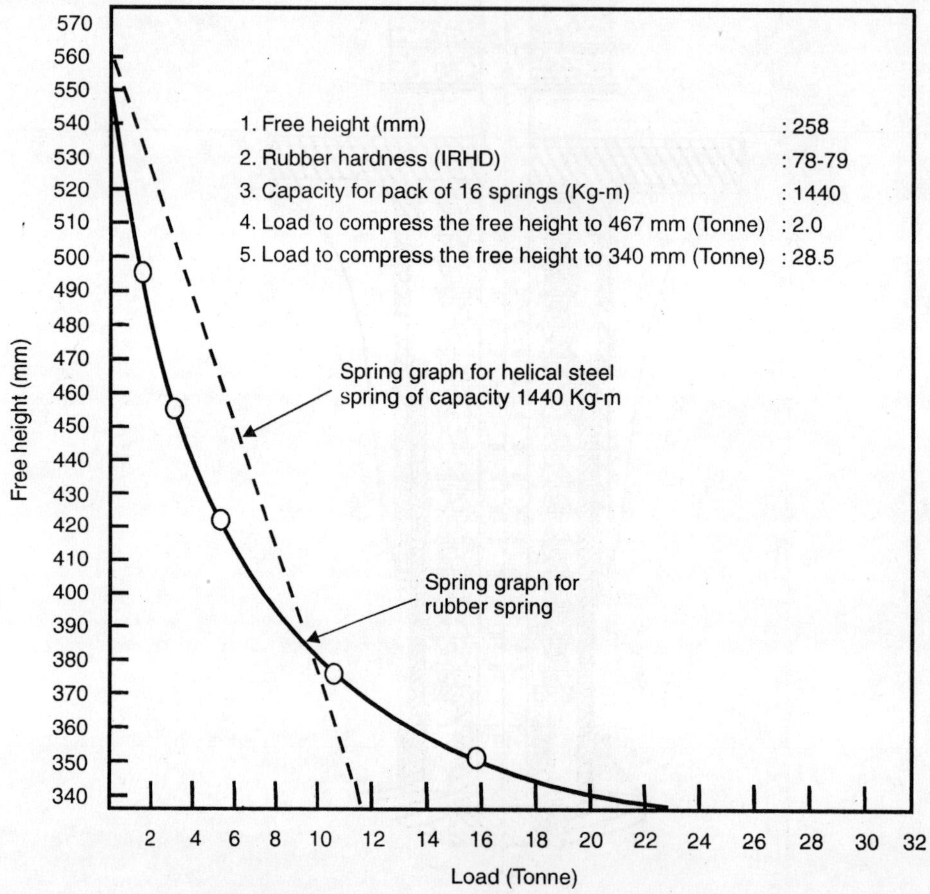

Fig. 27.2. Load and deflection characteristics.

RUBBER PADS FOR BALLASTLESS TRACK

In ballastless track assembly, rubber pad provides entire elasticity to the track. The rubber pad is placed between the bearing plate and the concrete bed. From assembly design consideration, thickness of the pad has been kept 12 mm nominal. Semi-circular grooves in the transverse direction have been provided on the pad to obtain 'shape factor' and serviceability. With this typical design, the fatigue is expected to be minimum.

The static spring graph is shown in Fig. 27.3. The spring characteristic is nearly linear except in initial range of loading. This non-linear portion of the curve is compensated by the pre-compression of the assembly and therefore, spring characteristic of the pad is linear under actual service conditions. This unique spring graph is the result of combination of 'shape factor' and composition of rubber. Various physical properties of rubber are given in Table 27.1.

Table 27.1. Physical properties of rubber.

	Properties	*Observed values (range)*
i.	Hardness (IRHD)	55–65
ii.	a. Tensile strength (kg/sq.cm)	150–200
	b. Per cent retention after ageing*	80–90
iii.	a. Elongation at break (%)	450–500
	b. Per cent retention after ageing	75–85
iv.	a. Stress at 100% elongation (kg/sq.cm)	20–35
	b. Change after ageing (%)	± 25
v.	Set (%)	
	a. Subjected to 50% strengthening	15–20
	b. Subjected to 50% compression	15–25

* Air ageing at 100 ± 1°C for 96 hrs.

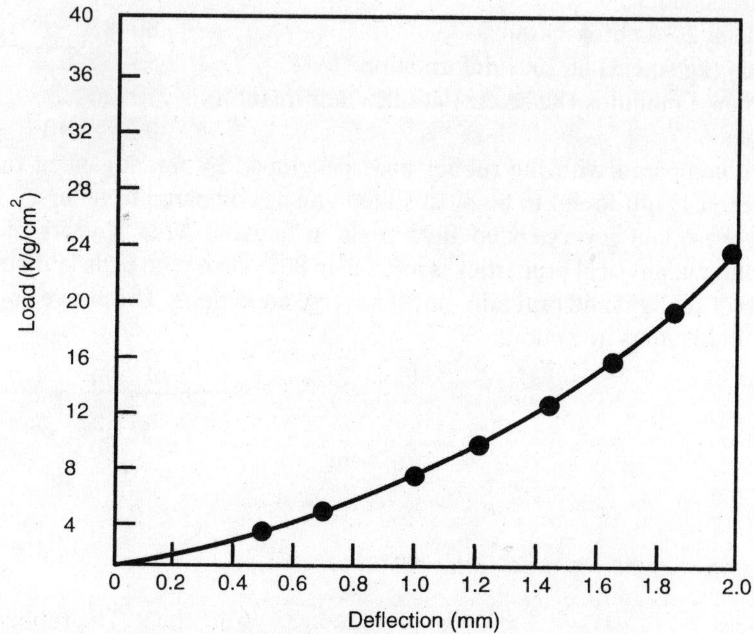

Fig. 27.3. Spring graph for 12 mm thick of rubber pad.

In consideration of durability in service, the stipulations of the ageing characteristics have been kept stringent and achieved after long strive.

With preliminary assessment of the spring characteristics and the physical properties of rubber, the pads are subjected to various simulated service tests in Vibrogir Pulsator and limited service trial on experimental track. Test parameters for Pulsator and Vibrogir are given in Table 27.2.

Table 27.2. Test parameters for Pulsators and Vibrogir.

Pulsator	
a. Vertical load	Varying between 7.5 to 2 T
b. Lateral load	Varying between 4.6 to 1.2 T
c. Frequency	300 cycles/minute
d. No. of cycles	2.5 million
Vibrogir	
a. Pulsating load	± 4.3 T
b. Induced cyclic load	+ 8.3 to –0.3 T
c. Frequency	50 cycles/sec.
d. Acceleration	100 'g'
e. Test duration	50 hours

Evaluation of the Pads

For evaluation of the pads during procurement besides physical properties and spring characteristics it is intended to incorporate certain dynamic properties based on Yerzley oscillograph studies. The parameters obtained in this test are as under :

1.	Resilience (%) at 20% deformation	80–85
2.	Static modulus (kg/sq.cm) at 20% deformation	55–65
3.	Effective dynamic modulus (kg/sq.cm) at 20% deformation	70–90
4.	Oscillating decay	0.35–0.40

The test results are compared with the rubber pads developed by the advanced railway systems. Pads developed indigenously are found to be most suited one as compared to others.

In test track, pads were laid for extended field trials in tunnels. After 8 years of service it was observed that the retention in physical properties is more than 80% even with pads laid outside the tunnel having exposure to direct sunlight and rain fall—most adverse conditions. The above application is also considered as a major indigenous invention.

CHAPTER 28

Special Applications of Polymers and Rubber

INTRODUCTION

In some areas of applications, plastics materials have been long established. A well-known example is in the electrical industries where the excellent combination of properties such as insulation characteristics, toughness, durability, and, where desired , flame-retardation capacity have led to increasing acceptance of plastics for plugs, sockets, wire and cable insulation, and for housing electrical and electronic equipment. This has lead to the increased use of the more general purpose plastics.

Plastics have been widely accepted in industrial equipment development as materials for many critical components and for accessories where such features as toughness, abrasion resistance, corrosion resistance, electrical insulation capability, nonstick properties, and transparency are of importance. For example, PVC, polyacetal, and PTFE are used for pipes, pumps, and valves because of their excellent corrosion resistance. Nylons are used for such diverse applications as mine conveyor belts and main drive gears for knitting machines and paper making equipment because of their excellent abrasion resistance, toughness, and low coefficient of friction.

In agriculture and horticulture, the most widely known use of plastics is in film form. But plastics also find use for automotive watering equipment, water piping containers, and potato chitting trays, to name but a few uses.

In the medical sector, plastics uses range from medical packaging and disposables (e.g., bandages and injection syringes) to nontoxic sterilisable items (such as tubing and catheters) to spare-part surgery (such as hip joints and heart valves). A wide-range of polymeric materials are found in these applications though medical grades of PVC and polystyrene are the most commonly used. Engineering resins, however, are finding an increasing number of uses in this area.

While natural and synthetic fibres have remained the major materials for clothing, plastics have also found increasing use in this area. Footwear has provided a major outlet for plastics as they are used not only in soles and uppers, but increasingly in an all-plastics moulded shoe that is in high demand in underdeveloped regions of the world. In rainwear, plastics and rubbers continue to be used for waterproof lining and in the manufacture of all-plastics packable raincoat. In cold-weather apparel, polyurethane foam has found increasing use as an insulation layer and for giving “body” to clothing. While the above paragraphs indicate some of the major uses of plastics in a number of areas, the more recent trends in plastics applications in these areas form the subject of subsequent discussion in the chapter.

POLYMERS IN ELECTRICAL AND ELECTRONIC APPLICATIONS

Plastics have played critical roles in every aspect of the ongoing revolution in electronics. They are now found in everything from computer housings and floppy discs to encapsulants and connectors. The wide range of electrical and electronic applications of polymers includes wire and cable coverings, housing adhesives, printed circuit boards and insulators.

One of the key trends in electronic components today is miniaturization,. This development is bringing new opportunities for plastics in the thin-walled and downsized products now reaching consumers. In consumer and office electronics, portability is a key design and marketing criterion. Portable computers, printers, copiers, cellular phones, and measuring and monitoring devices are highly plastics intensive. Today's portable (laptop and notebook) computer, for example, has 50–75% of its weight (and even more of the volume) represented by plastics.

Wire and Cable Insulation

Neoprene synthetic rubber has been used for many years in the cable industry as it provides good low voltage insulation and mechanical toughness as well as good resistance to oil, high temperature, and oxidation. Alcryn, (from Du Pont), is a melt-processable rubber cable sheathing for signal and power transmission. It is claimed to have excellent oil resistance even at high temperatures, and good resistance to severe mechanical treatment. Another key product (from Du Pont) is Hypalon, a chlorosulphonated polyethylene that exhibits high resistance to oils and a host of chemicals. One application of this product is the jacketing of cables in chemical plants.

The largest single outlet for plastics in electrical and electronic applications—roughly half of the volume consumed—is the wire and cable, where PVC dominates on the basis of its formulating flexibility, ease of processing, and low cost. However, the halogen-containing polymer systems tend to be more corrosive to copper than the non-halogen-based systems. Chlorinated compounds can also result in unacceptable emissions from conventional incinerators. Because of industry and regulatory concern over combustion, corrosively, toxicity and environmental impact at the points of manufacture and disposal, extensive use of PVC in future thus appears uncertain.

Fluoro-polymers such as Teflon and Tefzel that have high mechanical strength and dielectric properties with heat resistance enable thin wall insulation for high current densities. Du Pont's fluoro-elastomer Viton is used for insulation and heat shrinkable sleeving. In addition to excellent heat and chemical resistance it has high resistance to fuels, oils, and lubricants. Fluon, a polytetrafluorethylene (PTFE) product from Fothergill Cables, which specialises in electronics cables for aerospace and defense industries, is applied to thin wall insulation for equipment wires, multicore cables, thermocouple wires, and radio-frequency coaxial cables. Fluon has a service temperature range of –75 to +260°C, non-flammability, high chemical resistance, good dielectric properties, and resistance of damage by soldering.

Polyesters in wire and cable applications include Hytrel (Du Pont), a thermoplastic polyester elastomer for the offshore industry; Kapton (Du Pont), a polyimide film offering weight saving and chemical resistance; and Mylar (Du Pont), used in primary insulation, shield isolation, chemical barrier, and mechanical protection.

Large number of products to suit different applications in the wire and cable industry have been developed. While for general purpose insulation and sheathing, poly (vinyl chloride) (PVC) is used with chlorinated plasticisers such as polyether ether ketone (PEEK) wire and cable covering is designed to operate in high-demand environments. PEEK has high-performance electrical properties that permit the use of smaller diameter cable covering without loss of electrical power or reduction in insulation.

Because of the product's good inherent flame retardancy, low smoke emission, and low toxic and corrosive gas emissions it is used in underground mass transit wiring and cable insulation. PEEK has also been used for insulation a coaxial control cable operating in a high radiation environment at a nuclear installation in the United Kingdom.

A new technology has been introduced for silane cross-linking in low-voltage cable insulations to produce more uniform cross-linking and greater shelf-life. The new silane-linked polyethylene compounds can be run on conventional extrusion equipment. A strippable double layer cable shield in cross-linkable polyethylene for medium voltage cables has also been produced. The shield has a strippable outer layer formed by a coextrusion of 20% strip material and 50% standard semi-conductive material.

Polymer Insulators

Transmission insulators provide mechanical support and electrical isolation for overhead, open wire, power transmission lines. The primary insulation on these lines is air. All insulators use an innovative and unique silicone compound to provide an unbonded interface between insulating rubber housing and load-bearing fiberglass core. This interface has proven to be more reliable than the chemical bonded type of interface.

Printed Circuit Boards

Developments in the electronics industry are focused on two main areas—miniaturization and improved cost/performance ratios—creating the need for polymers that can tolerate higher temperatures for longer periods with good dimensional stability and mechanical strength.

Though the majority of printed circuit boards (PCBs) at present are two-dimensional and made of glass-filled epoxy resins, increasing miniaturization is leading to more PCBs being multi-layered with several copper circuits sandwiched between insulating layers. In these applications, because of the added stress due to thermal expansion, speciality materials such as polyamides and fluoro-polymers are being increasingly used.

The present trend is towards injection-moulded thermoplastic boards in place of thermoset laminate PCBs. The two main reasons for this shift are the difficulty of etching on thermoset laminates, and the greater design flexibility and potential cost saving in high-volume production by injection moulding. A further advantage of injection moulding is the ability to produce circuit board with three-dimensional features such as ribs, bosses, heat sinks, and holding devices for snap-parts. Engineering polymers that can be used for moulded circuit boards include polyethylene terephthalate (PET) and glass-reinforced grades of polyethersulphone (PES).

In the electronic industry, one of the most successful uses of advanced composites has been in PCBs. Polyetherimide is being used in transport PCBs, especially aerospace and missile guidance systems. Polysulphones and other high-performance thermoplastics are being used in multilayer PCBs for mainframes and supercomputers operating at gigahertz frequencies.

Process developed for the injection moulding of the circuit boards include PCK Technology's mould-n-plate process and an in-mould metallisation process. The mould-n-plate process uses two shots, the first being one of catalytic resins, to form paths on plastics parts that are then processed in a copper bath to form circuitry. Special grades of polysulphone, polyethersulphone, and polyarylsulphone are used for the process. In the in-moulding metallisation process, various conductive, resistive, capacitive, or inductive components are embedded directly into the plastic part in the tool.

Connectors

The miniaturization of PCBs has created the need for connectors to be made from special materials suitable for shaping into thinner parts that can be packed into higher-density spaces and that can withstand higher soldering temperatures. Such materials available on the market include tailored grades of polyphenylene sulphide (PPS), PEEK, PET, polyetherimide, sulphone-based alloys, and liquid crystal polymers.

Enclosures

Plastics are increasingly being used for electrical/electronic enclosures. Those used for enclosures of circuit chips should, however, be of high purity and high flow. Since miniaturization reduces the surface area of heat dissipation, the materials used are required to have high heat resistance in addition to high purity. While epoxy resins dominate the market, other materials used include polyurethanes, polyimides, and silicones.

Plastics used in electrical housings are now required to provide protection against electromagnetic interference (EMI), radio frequencies, and electrostatic charges that could damage sensitive equipment circuits. To impart EMI shielding properties to a plastic component, a coating of conductive layer may be provided by post-moulding or in-moulding processes or polymers may be used that have been metal or carbon black filled to make them conductive.

The methods mostly used to produce EMI shielding on plastic moulding are conductive paints, zinc arc spraying, vacuum metallising, electroplating, and foil application. There are also a range of filled plastic compound on the market that are based on metal fillers such as aluminium flake, brass fibres, stainless steel fibres, graphite-coated fibres, and metal-coated graphite fibres. The most cost effective conductive compounds are, however, based on carbon black.

One of the key determinants of future material selection for enclosures will be the growing global demand for recycling of these units. The recycling of electronics equipment enclosures and other moulded electronic parts, however, presents formidable obstacles, since there are many different resins in use, as well as a wide range of performance-enhancing additives and functional coatings for EMI shielding and ESD protection. To overcome these problems, industry is likely to narrow the enclosures resin menu to the usable plastics stream and focus on a well-defined groups of additives, coatings, adhesives, and other ingredients.

Optical Fibres

A major advance in communications has been the introduction of optical fibre transmission. In optical fibre transmission, voice and data signals are converted into beams of light that then travel along extremely thin glass fibres. Major advantages of this transmission over the conventional current carrying metal wire are higher message speed and density, lighter weight, and immunity to electromagnetic interference. The main growth in the use of optical fibres is expected to be in telephone networks.

Plastics find extensive use in several areas of fibre optic cable. Buffer tubes, usually extruded from high-performance plastics such as fluoropolymers, nylon, acetal resins, or polybutylene terephthalate (PET) are used for sheathing optical fibres. A blend of PVC and ethylene vinyl acetate (EVA) polymer, does not require a plasticiser, which helps the material maintain stability when in contact with water-proofing materials. PVC and elastomers blends are also used for fibre optic applications. (Stiffening rods for fibre optics are either pultruded epoxy and glass or steel. Around these is the outer jacketing, which is similar to conventional cable).

Japan's Mitsubishi Rayon Co. Ltd. developed a plastic optical fibre cable "Eska Premium" with a wide range of freedom, excellent durability and which can be used for wiring in narrow spaces. It is claimed that the cable can be bent or wound without deteriorating the characteristics of the fibre cable. The cable is lighter and easier to work with than glass-based cables, and is ideal for short communications and transmission wiring between the controllers of machine tools, personal computers, and peripheral equipments.

Information Storage Discs

In recent years, information has begun to be stored magnetically, on tape and on disc. Today their position is being challenged by optical discs, because an optical disc can store 1,000 times more information than its magnetic equivalent, has a longer guaranteed lifetime (usually 10 years), and more easily accessible data.

Although magnetic storage discs are beginning to be regarded as archival as compared to optical discs, the advantage they have over most optical systems is that information can be added and erased.

The compact audio disc (CD) based on digital recording and playback technology is 4.75 inches in diameter and can store 16 million bits of information in the form of minute pits in the substrate. It is the presence or absence of the pits that is read by the laser. The pits are 0.1 microns deep, 0.5 microns wide, and between 0.833 and 3.56 microns long. Each "track" comprises a spiral of these pits. The track is laid in polycarbonate that is backed with reflective aluminium and coated with a protective acrylic lacquers. The CD can thus be played only on one side.

The CDs are manufactured either by injection compression moulding or injection moulding. Injection compression moulding minimises moulded-in stress but requires longer cycle times than injection moulding and produces flash. For both processes very clean room conditions, up to 1000 times cleaner than medical moulding, are essential.

The processability of the polycarbonate, or any other material used as the substrate, is crucial in the manufacture of all optical discs. The material must have low melt viscosity and high melt flow to minimise internal stresses caused by molecular orientation of the polymer during filling. Also important are high long-term dimensional stability, resistance to moisture and gas permeability, low birefringence, and superior optical properties. The molecular weight of the polycarbonate has been reduced to give an exceptionally high melt flow rate compared to a standard polycarbonate. In addition to polycarbonate, a number of other materials including acrylic resins, polyacrylate, thermoset epoxies, and blends on co-polymers have been tested for their suitability in optical discs.

Polymeric Field Effect Transistor (FET) and Light Emitled Diodes (LED)

The fact that many conducting polymers are semiconductors in the undoped or slightly doped state has prompted attempts by both industrial and academic groups to use them in microelectronics such as semiconductor devices and especially field effect transistors (FETs). For the semiconducting layer between the source and drain electrodes of a FET, micrometer-thick films are required. Film as thin as this has been prepared by spin coating a solution of precursor polymer on a substrate having the required electrode pattern and subsequent heat treatment in a stream of gaseous dopant to convert it into a conducting polymer. Improvement of technique has led to carrier mobilities as high as 10^{-1} cm^2/V-s in polymeric FETs. This has opened the way to uses such as flat-panel colour displays for computers and possibly flat colour televisions when tied to a liquid crystal matrix.

Light emitting diodes (LEDs) have also been fabricated from the conjugated polymers such as poly(p-phenylene-vinylene) as the emissive layer. The advances in the chemistry of processible conjugated polymers and focused work on the physics of electroluminescence in these materials have led to the development of flexible, almost entirely metal-free LEDs,. These polymers-based LEDs could be competitive in display applications because of the potential ease, low cost of fabrication, and large surface area of devices based upon processible polymers.

POLYMERS IN AGRICULTURE AND HORTICULTURE

The application of polymeric materials in agriculture and horticulture has increased considerably in recent years, not only as replacement for traditional materials but also as a means of effecting improvement. For example, the application of plastics has led to a significant improvement in technological processes in the growing and storing of agricultural crops, construction of storages for fertilisers, vegetables and animals, and in agricultural equipment and drainage technology.

Plastic Film

The main uses of plastic films in agricultural and horticultural applications are for covering fodder and cultivation units such as seed-beds, for mulching and warming soil, for the building of canals and reservoirs, and as packaging materials for storing and transporting food.

The most widely used plastic film is polyethylene (PE). PE film tunnels and perforated flat PE film allow better use of natural resources such as solar energy, water, and soil. Crops may be made to ripen early making cultivation possible all year round. In addition to covering produce, PE film has a number of other uses. Livestock feeding stuff such as green foliage grass and maize can be preserved by covering the forage in a silo with large gastight black or white PE films. Shrinkable PE films are used for sheet steaming in horticulture.

Polyethylene films for agricultural applications needs to have high strength and elasticity, resistance to wind forces, and a long service life. They are mostly produced from low density PE combined with linear, low density PE and ethylene-vinyl acetate copolymer.

As PE films, like all synthetic polymers, suffer from decomposition by environmental influences such as light and atmospheric oxygen, special photoprotective systems are added to delay these effects. Systems that are commonly used are ultraviolet (UV) light absorbers, quenchers, radical scavengers, and hydroperoxide decomposing agents.

Photodegradable PE film is used for mulching the soil in vegetable growing. It improves soil drainage and discourages weed growth. The film breaks down within two weeks to three months as a result of solar radiation, combines with the soil, and is broken down by micro-organisms.

Pipes and Hoses

Plastic pipes are being increasingly used in the agricultural industry for drainage and irrigation. The improved strength and greater crack resistance of new grades of polyethylene have allowed the wall thickness of such pipes to be reduced and their diameters increased.

Polyethylene has been irradiated to produce cross-links thus giving it improved strength. Drip irrigation for raw crops uses such irradiated LDPE tubing. For vine and tree crops the drop emitter is made of PE moulded in large multicavity moulds.

Plastics are now being used in a variety of ways to replace metal components in different irrigation systems. Interestingly, one of the first uses of PVC pipe in the 1960s was in sprinkler irrigation system as a replacement of brass, gun-metal, or zinc alloys.

Hoses used in irrigation are invariably made of PE. Crimped PE pipes are being more frequently used for drainage and irrigation of soils. Flexible hoses that can be rolled up are used in spraying and sprinkling machines. Hoses reinforced with a rigid spiral are being used with machines to introduce chemicals into the soil for the protection of plants.

Greenhouses

The use of plastics in horticultural industry has grown substantially in recent years. Several plastics, and PVC in particular, are used in applications such as hosepipes, glazing strips, paint, and wiring. However, it has come to light that as greenhouses become airtight, some chemical vapours released by plastics may build up to levels toxic to plants. The plasticiser dibutyl phthalate (DBP) used in PVC has thus been found to the phytotoxic in a greenhouse environment. Research has shown that a fire retardant used in polystyrene trays and granules reduced the growth of some seedlings. Similarly, a silicone joining compound used in green houses and fumes from a paint used on an electric boiler were also found to be harmful to test plants.

Responding to the toxicity problem of plastics in greenhouse applications, the Dutch horticultural industry developed a test for screening. Plastics that pass the test qualify as safe horticultural plastics can be used in a greenhouse environment.

POLYMERS IN DOMESTIC APPLIANCES AND BUSINESS MACHINES

As plastics are more lightweight than glass or metal, are generally impact resistance, lend themselves well to manufacturing techniques, and allow designers greater freedom in design, they are continuing to replace metals in domestic appliances and equipment housings. The use of plastics has been well exploited in both these areas.

Large Appliances

While the use of plastics in refrigerators is well established, product development has led to replacement of some of the conventional plastics with newer ones of improved properties. For insulation rigid polyurethane foam has, on the whole, replaced fiberglass in cabinets and doors. In this application, diphenyl-methane diisocyanate (MDI), which has lower vapour pressure than toluene diisocyanate (TDI), is now preferred.

For refrigerators lines low-coast polystyrene (PS) was the first choice, but this was soon replaced by ABS, which has higher stress crack resistance to trichloromethane, a foaming agent that is used to blow polyurethane insulation behind the liners. However, because of the higher cost of ABS there has been a swing back to PS with special grades of PS now being available.

Plastics are used to make a number of dishwasher components, such as impeller and other pump parts, inner door, tubs, and consoles. Polyphenylene sulphide is used to make dishwasher impellers. The material shows processing ease, good durability, and resistance to high heat and detergents. For dishwasher pump parts, tubs, and door liners, glass-filled polypropylene has all but replaced diecast zinc. Noryl-modified polyphenylene oxide (PPO) is replacing stamped metal for the dryer for blower wheels. Both rigid PVC and PC are used for consoles. Rigid PVC is used because of its inherent flame retardancy, rigidity, and high heat distortion temperature. Polycarbonate (PC) is used for its high heat resistance, dimensional stability, and good electrical insulating properties. Plastics are replacing metals in washing machines.

Small Appliances

The market for plastic cookware that can be used in the freezer and microwave is set for a big expansion. Most cookware is made up of fiberglass reinforced thermoset resin. High-barrier multilayer containers that have gained popularity in the market are heat retortable, microwaveable, and provide a shelf life of up to two years without refrigeration for a variety of convenience foods. Liquid crystal polymers have also been used to make cookware as they are transparent to microwave, offer high heat resistance, and give an excellent finish. The most notable development in small appliances has been that of the electric kettle. Nearly all new kettle bodies, lids, and spouts are moulded from plastics.

Business Equipment

Business machine and equipment housings represent a fast growing area for ABS polymers that are commonly described as tough, hard, and rigid—a combination that is unusual for thermoplastics. It's light weight and the ability to economically achieve a one-step finished-appearance part have contributed to large volume applications of ABS. A variety of special ABS grades are used for equipment and housings. These include high-temperature resistant grades (for instrument panels, power tool housings), fire-retardant grades (for appliance housing, business machines, television cabinets), electroplating grades (for exterior decorative trim), high-gloss, low-gloss, and matte-finish grades (for moulding and extrusion applications), and structural foam grades (for moulded parts with high strength-to-weight ratio).

Air Filters

High-efficiency particulate air (HEPA) filters almost entirely use extrafine glass fibres and generally feature excellent functions, but if the fibre diameters are made finer to improve particle collection, the power consumption rate is increased considerably, clogging occurs, the filter becomes vulnerable to chemicals, and combustion treatment becomes difficult. Japan's Toyobo Company Ltd. has developed for air conditioning units a super-high-performance filter, the Eliton Super, made of a polypropylene polymer fibre. It is to be used as HEPA filter in clean rooms for removing over 99.97% of dust particles of 0.3 μm and as an air-cleaning system. The fibres in this new filter are charged with static electricity to improve the particle-collection efficiency. Compared with glass fibre, it consumes less power, has greater resistance to chemicals, and has a substantially extended service life. The filter can be combusted with ease and reduced into a small volume. The company also supplies filters for industrial clean rooms, primarily to semiconductor and air-conditioning systems manufacturers.

Solar systems

Polymers have a share in the development of each major solar technology because they offer not only potentially lower costs, easier processing, and light weight, but also a greater design flexibility. Polymers are used in most of the solar systems and equipment such as covers, thin-film honeycombs and housing for flat-plate (non-concentrating) collectors, reflecting surfaces and optical lenses for concentrating collectors, reflector shells, structural and support members, insulation, piping, moisture barriers, adhesives, and sealants.

POLYMERS IN MEDICAL AND BIOMEDICAL APPLICATIONS

A wide variety of polymeric materials are used in various medical and biomedical applications, which range from medical packaging and products used in medical treatment and diagnosis equipment to items that are actually implanted in the body, e.g., pacemakers, hip joints, and artificial heart valves. These products have unique requirements from absolute non-toxicity and compatibility with human tissue to mechanical strength, resistance to sterilisation techniques, and, in the case of disposable end products, mass production on a large scale. Due to differing requirements a wide variety of materials are used. While medical grades of poly (vinyl chloride) (PVC) and polystyrene (PS) are the most commonly used polymers in the medical sector, high-performance engineering resins are now finding an increasing number of uses in this field. The areas in which plastics find use can be divided broadly into the following: (i) medical packaging; (ii) nontoxic sterilisable items; (iii) appliances; and (iv) disposables.

Medical Packaging

Significant advances have been made in recent years in the packaging of medical products requiring medium and high barrier protection. The demand for sterilizable packing for hospitals and industry is being met by special paper, plastic films, and other non-woven materials.

The chlorinated tetrafluoroethylene (CTFE) fluoro-polymer is one of the highest barrier materials against moisture transmission. It is used in an antibiotic test kit where even trace amounts of moisture would violate the accuracy of the test.

A large number of small medical containers are now moulded in poly (ethylene terephthalate) (PET). Its advantage over glass include no breakage of filling liners and higher output. Tamper-evident closures, droppers, and dispensers are usually moulded in polypropylene and high-density polyethylene.

Nontoxic Sterlisable Items

Polymers used in nontoxic sterilisable items, such as tubing, artificial organs, and wound coverings, must be able to withstand sterilisation by ethylene oxide, steam autoclave, or gamma radiation. For medical products sterilised by gamma irradiation, only plastics that do not degrade or discolour on exposure to radiation, such as polyester and polycarbonates, can be used.

The biological compatibility of materials used in the nontoxic category is vital and is the subject of continuing research. Since no material is completely inert, it is the level of interaction between an implant and the surrounding tissue that determines the acceptability of the material. Several other facts that are important in determining acceptability include mechanical properties of the polymer, e.g., wear resistance and fatigue, and bulk chemical properties such as resistance to degradation by hydrolysis, sensitivity to enzymes, and the way it reacts to the deposition of protein.

PVC is used for nearly all surgical tubing. Recent developments include flexible PVC compounds with greater resistance to body fluids, and X-ray-opaque PVC compounds that enable PVC catheters and tubing to be traced after insertion into the body.

The development of artificial wound coverings has received much attention. The major requirements for such coverings are protection from microbial attack from the environment, optimal water permeability, capability of adhering well to the wound, and ready removability without causing tissue damage. Other important factors are prevention of excessive formation of granulation tissue and optimal elasticity to facilitate an intimate cover of the wound.

Sustained-release wound dressing capable of delivering antibiotics or other biomedical agents in small concentrations have been obtained by micro-capsulating a variety of drugs into a UV curable urethane elastomer. Special UV-curable polyurethane elastomers has been developed for such applications.

Development related to medical implants, artificial organs, and prosthesis have improved the quality of life and increased the life expectancy of many individuals.

Silicone rubber is a highly biocompatible thermosetting elastomer that has found applications in prosthesis for opthalmology, neurology, facial reconstruction, replacement of finger, toe and wrist joints, cardiovascular applications, such as pacemaker coatings and lead wires, and tendon replacements. It is also used in drug delivery systems and tubes for carrying blood, drugs, and nutrients.

The artificial heart, which has been successful in keeping a recipient alive for more than a year, has valves made from modified polypropylene and the two ventricles made of polyurethane supported on an aluminium base. For arterial replacements or by-pass of clogged blood vessels, polyester fibre remains the preferred material.

Ultrahigh molecular weight polyethylene (UHMW PE) is most commonly used for articulation surfaces in joints. It is also used for prosthesis components in total hip replacement. In the latter application, it may be reinforced with carbon fibres to increase wear properties.

Expanded polytetrafluoroethylene (PTFE) grafts have gained increasing popularity as synthetic or non-textile grafts for reconstructive procedures, such as, above or below the knee bypasses for limb salvage.

Appliances

Plastics components are now used in a number of appliances such as crutches, walking sticks, and wheelchairs. The comfort of the product is an important consideration in all of these applications.

Thermo-formed high density polyethylene (PE) and low density PE foams that can be moulded into a variety of complex shapes are being increasingly used in supporting devices in the treatment of orthopaedic and disabling conditions. Some examples are wheelchair seats, spinal jackets, medical shoe inserts, and splint supports. Specialised high-temperature plastics such as polyether sulphone and polyetherimide are used for components for medical analysis and diagnosis equipment.

Disposables

Disposable medical products such as syringes, test tubes, petridishes, dialysis products, and drainage devices are proving to be high-volume markets for transparent plastics such as rigid PVC, polystyrene, styrene acrylonitrile (SAN) co-polymer acrylic resins, and polycarbonate. The possibility of mass production is an important consideration in these applications.

POLYMERS IN MARINE AND OFFSHORE APPLICATIONS

Polymers find a wide range of applications in offshore and marine industry. The properties of materials used in these applications invariably include resistance to marine fouling, sea water, and crude oil. The main applications are in marine cables, coatings, and adhesives.

Cables

Due to the damage caused by smoke-filled or halogen-based fumes from cables in the event of a fire, the offshore industry is showing increased interest in low smoke, zero toxic gas emission cable sheath. Research has centred on obtaining halogen-free polymers having fire retardance as well as oil, chemical, and weathering resistance. One example is a series of formulations called NONHAL-E (elastomeric) and NONHAL-T (thermo-plastic) developed for the offshore industry. The elastomeric formulations are based on nitrile rubber and are resistant to oil and other commonly encountered fluids. Cables insulated with ethylene propylene diene terpolymer (EPDM) and sheathed with NONHAL-E are suitable for very tough applications. NONHAL-T was developed to replace PVC in cable sheathing. Halogen-free cables produced with a cross-linked polyethylene and bedded and sheathed with NONHAL-T gives excellent performance in fire conditions.

Other Applications

A wide range of coating materials based on polyurethanes and epoxy resins is available to protect drilling rigs from corrosion.

Hoses for offshore applications are in most cases made of abrasion-resistance materials such as poly (vinyl chloride) or polyurethane. Polytetrafluoroethylene hoses have been developed for use as interconnectors to oil or gas combustion chamber burners or turbines. The inert qualities, ability to withstand the hostile environment, and high-pressure capability suit them to this application.

Special polymeric material is needed to make heavy, finned pistons used in through the flow line well servicing operations.

Several methods are available for the treatment of oil spillages. The two components, a polymer and a cross-linking agent have been developed. The two components are sprayed simultaneously onto the spill where they mix and rapidly cross-link to produce a rubbery solid in which the oil becomes physically entrapped. The solid assembly is removed easily.

POLYMERS IN SPORTS

The properties of polymers—resilience, toughness, and good strength-to-weight ratio—ensured the quick entry of polymers into the sports industry. In most cases, the polymer has been used as a replacement for another material and, being a competitive business, the sports industry has also been very receptive to new developments in polymer formulations and advances in design that contribute to improved sporting performance. Uses of polymers in the sports industry fall into three main categories: synthetic surfaces, footwear, and equipment.

Synthetic Surfaces

In sports such as athletics, traditional materials have been almost completely replaced by synthetic materials. Sports surfaces in modern indoor stadiums are usually made from a porous structure of rubber crumb and a binder such as polyurethane. Other polymeric materials used include flexible PVC or rubber sheet, polyolefins, polypropylene (PP) grasses, and foam laminates.

Footwear

There has been significant development in sports footwear in recent years, greatly aided by polymeric materials that allow for better shape, manufacturing techniques, and material performance.

In the design of running shoes, much emphasis has been placed on anatomical comfort leading to the development of toe bars and heel canters, wedges, and midsoles, usually moulded in polyethylene foams. Greater attention has been focussed on the outsole of sport shoes in order to cope with the increased use of indoor facilities and artificial surfaces. Ski boots use a wide variety of plastics adapted for low temperatures. Thermoplastic polyurethane/ABS blends and polyamides are used for the outer shell of ski boots and modified polyethylene terephthalate for the binding. Polyurethane (PU) foam is often used to line the boots. Micro-sphere fillers may be incorporated into foam-lined boots to add further thermal insulation to the wax binder. The antivibrational characteristics of PU foams have also led to their use in ski fittings. A sandwich construction of polyurethane elastomer and aluminium alloy has been fitted between the ski and the binding to reduce shock and vibration.

Equipment

Developments in plastics sports equipment are mainly aimed at improving player performance and increasing player safety. Items that fall into the first category include balls, bicycles, tennis rackets, and windsurfing boards. Crash helmets belong in the second category.

Since the maximum performance of sports balls is governed by strict specifications, the scope of using plastics to improve them is rather limited. Improvements can however still be made. For example, micro-sphere fillers have been used in golf balls and bowling balls. The travel distance of golf balls is decreased by up to 30% by the incorporation of micro-spheres. In bowling balls, incorporating of micro-spheres reduces the density of the ball's core enabling lightweight balls to be made for children and women players. Plastics are being used to make a number of bicycle parts replacing metals. Examples include polyamides for structural components such as the bike frames and wheels, and polyacetals for tyre valves, brake levers, and gear change and power switches. Sailboard accessories use thermoplastic blends; for example, boom fittings and mast steps that have to withstand very high dynamic loads are made of glass-fibre reinforced polyamide blends. Polycarbonate/ABS blends are commonly used for daggerboard cases. The most popular type of sails are film sails, which are laminates of a plastic film, usually a polyester such as bonded to woven fabric. They are claimed to offer superior tear resistance.

The tennis racket is another example of sports equipment that has been improved significantly due to the availability of high-performance plastics. The carbon fibre-reinforced nylon tennis racket frame first introduced by Dunlop has become the industry's standard for this equipment. An improved design incorporates Atochem's Pebax nylon ether block copolymer into the racket frame where maximum vibrations are concentrated, i.e., between the two profiles of the frame in the handle and across the throat of the frame.

An item of sports equipment that has been improved greatly with regard to player safety is the crash helmet. American football helmets are polycarbonate or ABS shelled, while impact nylons are often used on motorbike crash helmets. Another item of sports equipment in the United States where safety has been improved is a Reduced Injury Factor baseball that has a novel polyurethane core material.

EXANE (POLYOLEFIN) IRRADIATION CROSS-LINKED MATERIAL FOR SPECIALITY APPLICATIONS

Exane (Polyolefin) is a trade name for a polyolefin based compound which when processed under tightly controlled mixing, extrusion and irradiation cross-linking conditions, exhibit a unique blend of electrical, mechanical and environment resistance characteristics. This has allowed the material to be successfully used as a combined electrical insulation/protective sheath application in power, control and instrumentation cables used in harsh environmental conditions like off-shore oil rigs, high powered locomotives and inside wiring for transit cars.

Unique Performance Characteristics of Exane

Exane material when processed under controlled mixing, extrusion and irradiation cross-linking conditions exhibits the following unique combination of performance characteristics. (i) wide operating temperature range (–55°C to 125°C); (ii) excellent resistance to oil, solvents and chemicals; (iii) excellent moisture resistance; (iv) excellent electrical properties; (v) good flexibility; (vi) superior resistance to: abrasion, crush and cut-through; (vii) excellent flame resistance; (viii) vertical flame test: flame extinguishes within 3 seconds); (ix) low acid gas generation; and (x) excellent thermal overload withstand.

Some of the important properties of exane (polyolefin) are given in Table 28.1.

Table 28.1. Properties of Exane.

Physical properties	
Original	
Tensile strength min. PSI	2000
Elongation at rupture, min., %	250
Ageing requirements	
After air over 7 day @ 158°C ±2°C for 168 hours	
Tensile strength min. % of unaged value	90%
Elongation, min. % of unaged value	50%
Ozone (Test method ICEA S-19-81)	
After 24 hours exposure to an ozone concentration of 0.03% by volume at 90°C ±20°C, there shall be no insulation cracks	
Flammability	
Vertical flame test in accordance with ICEA S–19–81 except that flame must extinguish before next application of flame applied	3 Seconds (max.)
Electrical test	
Insulation resistance constant @ 15.6°C	20,000
Electrical method (ICEA S-66-524)	
Increase in capacitance, max. %	
1–14 days	3.0
7–14 days	1.5
Stability factor after 14 days	1.0
Alternate to stability factor stability factory difference, 1–14 days maximum	0.5
Hot oil resistance	
After 96 hours immersion in ASTM oil # 2@ 150°C ± 2°C	30% swell (max.) no cracks, splits

Benefits of the Unique Characteristics

The key basis for the unique blend of characteristics are:

1. Innovative formulation.
2. Controlled cross-links process utilising irradiation technology.

The resulting benefits are:

1. Extrusion thicknesses as low as 0.125mm are possible. This results into a smaller and lighter insulated wire.
2. Material has the ability to perform as an electrical insulation as well as protective sheath at the same time. This again allows a smaller wire as compared to the conventional two layer designs where the first layer gives the electrical performance and the second layer protects the electrical layer from the environment.

Thus, exane, a polyolefin based, irradiation cross-linkable material under controlled compounding, extrusion and irradiation cross-linking processes exhibits a unique combination of properties. The unique combination of properties, allows smaller/lighter and cost competitive cable designs to be produced.

These cables have developed an outstanding performance history in harsh and demanding environments like, high power locomotives, off-shore oil rigs and train car wiring applications all over the world.

SPECIALITY POLYMERS IN ROAD APPLICATIONS—SOME NOVAL TECHNOLOGICAL ADVANCES

Laboratory and field studies conducted at Central Road Research Institute confirm that ethylene vinyl acetate copolymer with vinyl acetate content ranging 18 to 28%, when added to bituminous binders to the extent of 2.5% to 5.0% improve flexibility at low temperature, prevent permanent deformation of road surfacing at high temperature and improve fatigue life leading to safer, durable, longer lasting pavement with low maintenance cost. Performance of road surfacings modified with ethylene vinyl acetate copolymer as an ingredient have been recorded superior under heavy loading.

Increased number of traffic loads with overloading of axles in excess of permissible limits and higher tyre pressure have caused widespread problems with bituminous roads on National Highways and available 60/70 grade bitumen is not suitable for heavy duty roads in industrial complex, ports and airfield pavement.

The accelerated deterioration of bituminous surfacing prematurely burdens the maintenance budget and pose funds constraints on new development. It is a global view that the quality and longevity of pavement must therefore be restored in order to reduce maintenance budget and to save road user's cost. Polymer modified binders are premium binders, which allows an engineer to design and build a stronger and durable pavement. Scientists have conducted extensive laboratory and field studies in recent years using a variety of polymeric additives for their use in heavy duty road and airfield surfacings. Ethylene vinyl acetate copolymer having wide range of properties, which depending upon the vinyl acetate content, extends from plastic to rubber-like products. At a very low vinyl acetate content, the co-polymer behave like LDPE and with increasing vinyl acetate content amorphous character become predominant resulting in flexible rubbery material. Hence, such wide range properties of ethyl vinyl acetate (EVA) make it, a polymer with diverse applications. Flexibility, polarity, low temperature toughness, shock resistance and good weathering has made it a material suitable for some potential applications in highway and transportation engineering viz., bitumen modification, bitumen packing, traffic cones and traffic lane dividers.

The physical and engineering properties of modified bitumen are better than conventional 80/100 or 60/70 bitumen. Higher value of penetration ratio indicates improved temperature susceptibility of modified bitumen. Improved value of low temperature ductility further indicate resistance of modified binders to abrasion and thermal cracking. Higher values of softening point for modified binders indicate reduced chance of bleeding and fatting up at higher temperature besides resistance to deformation.

CHAPTER 29

Anti-static and Conductive Uses of Rubber

INTRODUCTION

We all are aware of the insulating properties of rubber—most of our domestic and industrial electrical cables are made using rubber as insulant. Rubbers can be modified to make them conduct electric charges or currents, with the usage of certain additives and modifiers. The major use of conductive or anti-static rubber is the dissipation of electrostatic charges, usually known as static charges. This is the reason why rubbers for this type of application are termed 'anti-static'.

Rubbers can be rendered conductive by incorporating adequate loadings of carbon black, i.e., 20 phr to 100 phr depending upon carbon grade. One other method, though not common, is by usage of fine metal powders like that of nickel and the resistivities by this method may be brought down substantially. With metal powders, especially when used in a liquid polymer, the conductivity may be increased in one direction by applying a magnetic field during hardening.

TERMINOLOGIES

The terms 'anti-static' and 'conductive' must be defined with respect to the resistance between certain points on a product. When static electricity is merely a nuisance, resistivity values of 10^{10} to 10^{12} ohms may be tolerable, but in case it poses a hazard, the value must be below 10^8 ohms. The conductive products which are also used to prevent static electrification would not have any lower limit of resistance. However, these should not exceed a resistivity limit of maximum 10^5 ohms, though this value is somewhat arbitrary. British standard 2050 defines the dividing line between conductive and anti-static products at a resistance of 5×10^4 ohms. These conductive rubbers, in some cases, exhibit a thermo-emf and also Hall effect. Hall effect is the generation of an emf in a body when this is subjected both to a magnetic field and to a perpendicular current flow.

GENERATION OF ANTI-STATIC CHARGES

Whenever, two dissimilar solids or a solid and a liquid are placed in contact, there may be re-distribution of charges between them owing to transfer of electrons or ions, so that when they are separated, one surface may carry an excessive positive charge, while the other an excess of negative charge. It is often thought that rubbing of surfaces is necessary and hence the term tribo electricity has come, but this is not the case. Charges may be separated when the surfaces are separated, even if no sliding has taken place. Owing to the sliding, contact area of two surfaces increases, thereby, enhancing charge generation. Besides the above, due to sliding a slight transfer of material may take place as well as localised heat generation which also plays a role in charge re-distribution.

When an article on which a charge has developed, is discharged rapidly, voltages less than the original voltage and of the opposite signs are developed by electrostatic induction on neighbouring objects. This may result in an incendiary spark discharge to the earth from one of these neighbouring objects even if the resistance to earth of the latter is sufficiently low to prevent the energies of directly generated static charges from reaching incendiary levels.

APPLICATIONS OF ANTI-STATIC

Electrostatic charges of the sparks resulting, as mentioned above, are normally quite harmless and unimportant. However, these become considerably important under any of the following conditions.

1. Where inflammable liquids or gases are being handled.
2. Where explosive dust, e.g., coal dust or flour are being handled.
3. Where detonators are being handled.
4. Where the continuity of production of a light material depends on relatively accurate registration between the material and machine, e.g., textile fibres, which are required to pass evenly through spinning machinery and which bow apart when they are charged.
5. Where an insulating material becomes charged and attracts dust.
6. Where highly insulating covers or parts are used for mechanical or electrical equipments which must respond to very slight forces and these forces may be effected (countered or simulated) by electrostatic forces e.g., in very sensitive meter movements.
7. Where sensitive electronic equipment, such as communication equipment, is disturbed by sparking.
8. Where a slight shock or appearance of sparks is disturbing to a person.

Usage of anti-static paints, ionising the ambient air, usage of surface treatments which reduce transfer of charges, are some of the other methods of reduction of static electrification.

PRODUCTS WITH ANTI-STATIC USAGE

Anti-static Vehicle Tyres

In earlier years, it was common for motor vehicles to become charged due to separation of the tyre tread and the road by rotation of the tyres. Though the problem is normally mildly unpleasant, it is aggravated if the climate is dry. These complaints led to the introduction of anti-static tyres, based on special carbon blacks, but the use of furnace black in place of channel blacks led to all tyres having sufficient conductivity to prevent shock from vehicles.

Static charge interferes with radio communications in vehicles. A further problem was the occurrence of ozone punctures or spark-bite of the inner tube, where the voltage gradient along the internal surface of the tyre was locally sufficient to cause ionisation and the production of ozone, which caused cracking of the strained rubber of the tube.

When an aircraft lands, it may carry a considerable charge (acquired during flight). This could be dangerous to the first person bridging a gap between the aircraft and an earthed object and, therefore, an anti-static tyre is used normally on the tail or nose-wheel of the craft. Vehicles handling explosives should have tyres with resistance below 5×10^4 ohms.

Belts

Voltage builds up rapidly on conveyer and transmission belts and may be vary high in case of V-Belts due to their high rate of separation of surfaces. It could be dangerous in usages like coal mines and places where explosive vapours and dusts are handled, if anti-static compounds are not used.

Hoses

Hoses carrying inflammable powders or liquids or which pass through an inflammable atmosphere are big static hazards. For such cases both hose and container are required to be anti-static. It is required to use single or multi-static conductive wires and fibres in the hoses and these should protrude out at the ends so as to make contact with the hose nozzles. In such applications it is required that the resistivity should not be more than 10^4 ohms at a time. The tyres of the vehicles for this type of applications should also be anti-static.

Material Processing Machinery

Anti-static rubbers have been used for roll coverings and other drafting items in textile and other light material processing machinery as static charges would tend to bow the materials apart.

Hospital Equipment

In earlier years, a number of cases of explosions were reported in hospital operating theatres, resulting from electrostatic discharges. This is critical in the case of some anaesthetics which are highly explosive and have very low ignition energy, especially when mixed with oxygen. With anaesthetic equipment it is necessary to use considerable lengths of rubber tubings in intimate contact with unanaesthetic gases. Some items such as warm blankets, when pulled on an insulated trolley, leave considerable static charges on the trolley, giving rise to the need for the trolley wheels to be conductive.

The floor of an operating theatre should also be anti-static. Other articles of hospital operating theatre equipment which should be anti-static include footwears, articles made from sheeting including aprons, cellular rubber pads and mattress covers, tyres of all trollies and beds, furniture feet, corrugated tubings, anaesthetic and other tubings, anaesthetic masks, bellows, buffers and non-rubber articles also.

An important factor is the increase in resistivity during service and care should be taken for this. For instance, anti-static rubber footwear may be within specified limits when new, but they may leave this range during service. It is recommended for anti-static floorings, that resistance should be so low that use or ageing does not result in resistance rising above 10^6 ohms. However, any change in resistance of flooring are likely to be smaller than those items which are deformed considerably in usage.

Cables

Conductive rubbers are used in cables for various purposes. Where a cable is constructed of a central conductor, an insulant and an outer conductive sheath, minute voids occur at the metal/insulant interfaces. Under the influence of high voltages, discharges take place, causing deterioration of the insulation. If, however, a layer of conductive rubber or plastic is interposed between each of the metallic members and the insulant and is firmly bonded to the insulant, the conductive material takes up substantially the same voltage as the adjacent conductor and hence the air in the voids against metal is not stressed electrically and the voids are innocuous. At the new interface between the conductive and insulating polymers no voids need occur to avoid corona effects in the voids. This makes possible the satisfactory use for power cables up to at least 30 KV and for X-ray cables up to 200 KV. In case of high voltage rubber insulated submarine cables, dry cotton tapes frictioned with an unvulcanised neoprene compound containing acetylene black, are commonly used around metal strand conductors.

Conductive rubber usage has been proposed in microphone cables also. Owing to the local separation of the dielectric from the core and sheath which takes place when an instrument cable is flexed, spurious signals known as 'noise' appear at the end of the cable. By using conductive rubber at each interface, this noise is eliminated.

Usage of conductive rubber in a highly flexible conductor and high voltage cables is common in other applications also. Such materials may be used as sheathings, which can carry sufficient current to operate earth leakage trips.

Conductive rubbers have been used as heating elements (especially silicone rubbers), and have been proposed as solders, conductive adhesives, radio frequency screening, battery components and optically transparent conductive layers, but the extent of such uses is not known.

COMPOUNDING

Carbon blacks with small particle-size, a high specific surface, permanent chain structures and low volatile content, give most conductive rubber compounds. Graphite, though giving good conductivity, is rarely used because of the poor reinforcement provided.

For many years acetylene black held the field as the only carbon black conferring a high degree of conductivity at moderate loadings and was, therefore, termed a conductive black. With the introduction of furnace blacks in place of channel blacks, it is now rather difficult to produce insulating black mixes, except when carbon blacks of large particle-size or very low concentrations of other carbon blacks used. Fig. 29.1 illustrates resistivities vs. concentration values for channel blacks while those in Fig. 29.2 are for furnace blacks.

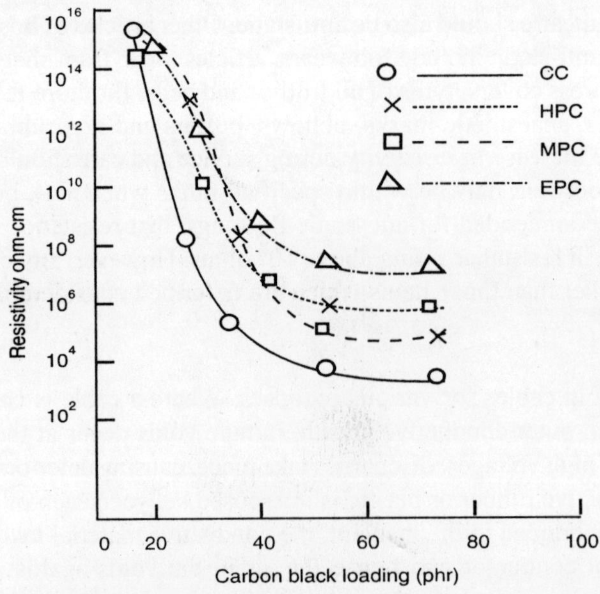

Fig. 29.1. Effect of channel blacks on resistivity.

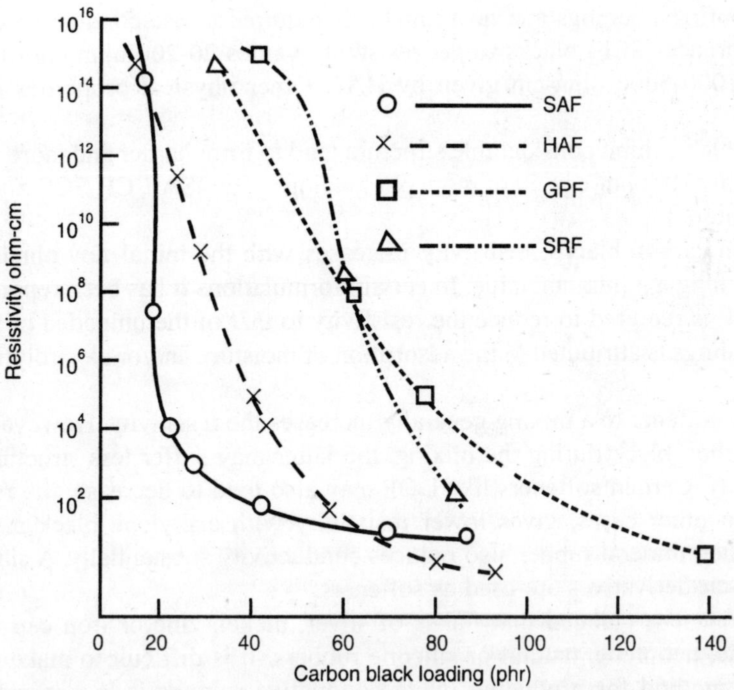

Fig. 29.2. Effect of furnace blacks on resistivity

Resistivity change with progressive replacement of a HAF by a thermal black is given below in Fig. 29.3.

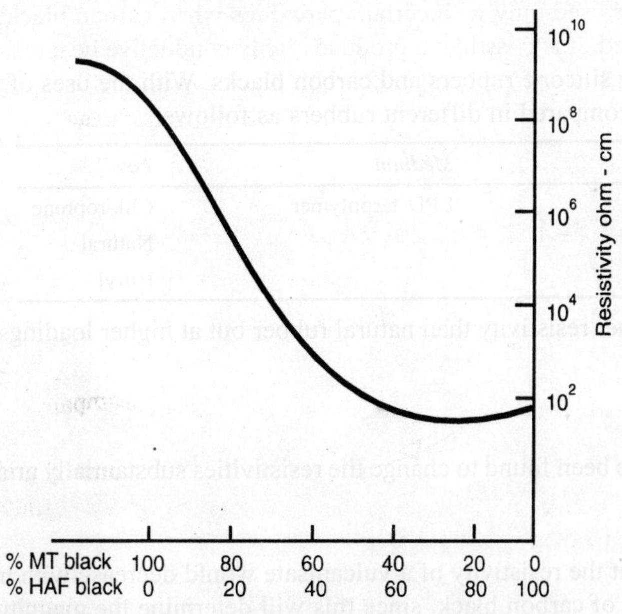

Fig. 29.3. Effect of progressive replacement of HAF by MT black.

It may be noted that anti-static protection is not sufficient with HAF or ISAF type blacks for aircraft tyres, footwears, floorings, beltings and hoses and it is required to use conductive furnace (CF) and a super-conductive Furnace (SCF) blacks to get resistivity values 20-200 ohm-cm in tread type stocks, as compared with 1000-5000 ohm-cm given by HAF. Other physical properties of compounds are similar.

Carbon blacks which exhibit considerable structure tend to form harder and more conductive rubber. To get resistivity below 10^6 ohm-cm, 50 or more phr loadings of an ISAF, CF, SCF, and extra-conductive furnace (ECF) is required.

In case of certain carbon blacks, resistivity increases with the initial few phr loadings and starts decreasing after reaching a minimum value. In certain formulations it has been reported that around 25 phr of a FEF or HAF is required to reduce the resistivity to that of the unloaded mix. This increase in resistivity at low loadings is attributed to the adsorption of moisture and/or electrolytic impurities in the carbon blacks.

The addition of a softener to a mixing generally increases the resistivity. However, if the softener is added before the carbon black during the mixing, the latter may suffer less structure breakdown and give high conductivity. Certain softeners like DOP may also tend to decrease, the resistivity. Pine-tar, though increasing in other cases, gives lower resistivity with acetylene black at 15 or higher phr concentrations. Further, mineral rubber also reduces conductivity substantially. A similar effect is seen if certain other coal-tar derivatives are used as softeners.

Apart from carbon, it is claimed that flakes of silver, nickel, zinc or iron can be used to impart conductivity to nitrile, neoprene, natural or silicone rubbers. It is difficult to make highly conductive white rubbers. One method for producing white conductive rubbers is incorporating glass powder treated with conductive metallic salts. The addition of some grades of aluminium silicate may make it anti-static, if the conditions are not unduly dry. Moreover, the addition of glycerol lowers the resistivity even when humidity is low.

Silicone rubbers can be used only with certain peroxides when carbon blacks are incorporated and acidic blacks cannot be used. It is possible to produce highly conductive heat resistant compounds from the vinyl group containing silicone rubbers and carbon blacks. With the uses of furnace and acetylene blacks, resistivity can be compared in different rubbers as follows:

High	*Medium*	*Low*
Nitrile	EPD terpolymer	Chloroprene
SBR		Natural
		Butyl

Chloroprene gives higher resistivity than natural rubber but at higher loadings of carbon blacks, the situation reverses.

PROCESSING

Processing conditions have been found to change the resistivities substantially and sometimes manifold.

Mastication

It would be anticipated that the resistivity of a vulcanisate would decrease with increasing plasticity of the rubber before addition of carbon black, since this will determine the magnitude of the mechanical forces to which the carbon black structure is subjected during the mixing process.

Mixing Time and Rotor Speed

Both higher mixing time and higher rotor speed result in manifold high resistivity values and the effect is, in some cases, a few hundred-folds also. The effect of mixing time is represented in Fig. 29.4.

Re-milling

This results in increases in resistivity though the effect is only marginal in case of high black loaded compounds and the increase is noticed with each re-milling. Increasing the mixing temperature also results in an increased resistivity. In general, any process which results in breaking of carbon structures, results in an increase in resistivity.

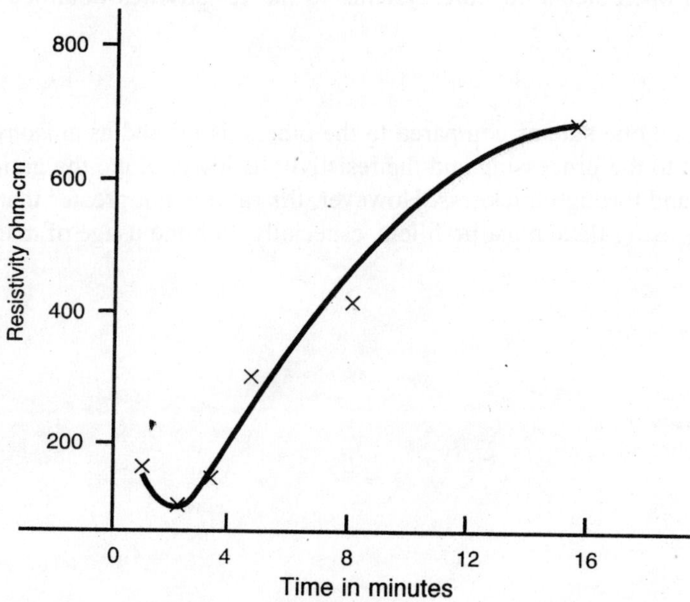

Fig. 29.4. Effect of mixing time on resistivity.

Heat-treatment

Heat treatment is desired for technical reasons in many mixes, especially butyl mixes. It has been found that the resistivity values increase invariably with this. In the Table 29.1 below are shown resistivity of different black field normal and heat-treated stocks.

Table 29.1. Resistivity (ohm-cm) of vulcanisates from normal and heat-treated N.R. stocks.

Carbon blacks	*Normal*	*Heat treated*
HMF	1×10^7	2×10^3
HAF	4×10^4	7×10^7
Acetylene	1×10^3	5×10^3
C.C.	9×10^2	1×10^4

Extrusion

Drawing down the material after it leaves the die reduces the longitudinal resistivity, thus, partly off-sets the increase in resistivity due to the extrusion.

Moulding and Vulcanising Conditions

The resistivity increases with the increasing period of standing before vulcanisation. However, the resistivity of press-cured articles generally increases with the increasing moulding pressures. The effects quoted are relatively slight over the range of pressures generally employed. Mould flow also increases the resistivity greatly. It is preferred that the vulcanisates used for anti-static purposes should be press-moulded as compared to open steam/air cure systems as the resistivities obtained are lower in the former case.

Anisotropy

To show more resistance on one side as compared to the others, is termed as anisotropy in a product. This generally results due to the processing and the resistivity is lower along the grain as compared to the side across the grain and through thickness. However, the ratio is not greater than 1:6. Anisotropy can be made very high by using flake metallic fillers, especially with the usage of magnetic fields.

CHAPTER 30

Uses of Rubber in Bearings Piers and Fenders

INTRODUCTION

There are a number of polymeric materials that distinguish themselves from others by virtue of their limited use, high prices, or very specific applications or properties. In this chapter the applications of speciality rubbers which are high pressure, high temperature and weather resistant are discussed. These rubbers find wide applications in: bearing and bridges; piers and fenders; structures; aircrafts, ships and machines; buildings and packaging etc.

RUBBER IN BEARINGS FOR BRIDGES

Bearings for bridges must be so designed that they permit expansion and contraction of the bridge with changing temperature in summer and winter, they must be able to rotate slightly from deflection under load, and they must transfer thrusts from heavy traffic smoothly to the sub-structure of the bridge.

Modern vehicles travel at high speeds and have improved braking and acceleration. These factors, together with increasingly heavy loads, are increasing the problems to be overcome. In the past bridge bearings have been mainly in the form of mechanical devices designed to roll or slide, in order to permit movements of the ends of bridge. Such bearings are difficult and expensive to maintain, and need frequent lubrication and attention to keep them free from corrosion. They suffer from the very small movements which they are required to perform, which result in a high rate of wear over small areas. Concrete bearings have been used also, but these take up a large amount of space and are expensive. Bearing pads consisting of layers of canvas and red lead have been used as well. These are able to compensate to some extent for minor irregularities in alignment, and to prevent moisture, grit, and dirt from getting between the bearing surfaces and causing abrasion and rusting. They have, however, little or no cushioning effect, and as they are built up on the spot, considerable time and labour is involved. They are also susceptible to weather conditions during installation, and should only be built up during good weather to obtain the best results.

Advantages of Rubber Bearings

The comparatively recent introduction of rubber bridge bearings has enabled engineers to obtain the necessary flexibility in bearings without the attendant disadvantages of frequent attention, and/or high cost, cited above. There are four main types of bridge bearings which make use of rubber.

1. Fabric-laminated pads designed primarily for load distribution.
2. Moulded pads of solid rubber which are designed to support moderate loads on large bearing areas and to provide small expansion movements.
3. Pads to improve bearing surfaces and to prevent electrolytic action (e.g., between a rail and a bridge structure).
4. Steel-reinforced pads designed for heavy loads on limited bearing areas.

Types of Bridge Bearing

(i) Steel-reinforced neoprene; (ii) steel-reinforced natural rubber; (iii) butyl rubber bearing pads; (iv) fabric-laminated pads; (v) butyl bearings; and (vi) neoprene bearings.

RUBBER IN PIERS AND FENDERS

For energy absorption of either impact or vibration rubber is pre-eminent. It provides a variable spring rate which, whilst yielding to a light blow, stiffens up as the shock pressure increases. It is largely self damping, its reaction to a blow dying out quickly, and it needs no lubrication, painting or other attention. Rubber correctly compounded for the job has stood up for years, even decades, to tropical sunshine or arctic cold, to salt water, oil scum, Toredo worm and other agents to be found in enclosed waters, with the result that there is now a diversity of fender designs.

Simple Fenders

Simple fenders of the type consisting of baulks of timber, often further protected by rows of old tyres or other scrap rubber, or by rope fenders are very common. Other non-rubber fenders which have been used for a very long time are floating wooden rafts, which are moored to the jetty and keep the ship away from the main structure, and wooden piles which are driven into the harbour bed in front of the jetty, and slightly away from it. A more modern development is the gravity fender, consisting of a heavy block of material such as concrete, suspended from the jetty by chains, with its outer face projecting beyond the face of the jetty. The weight of the block exerts an outward force on the side of the ship if the ship moves towards the jetty and pushes against the block. Simple fenders using rubber are usually heavy rubber tubes, rubber balls, fender blocks, or devices using old tyres.

More Complicated Systems

While the use of rubber in the simple fenders noted above leads to considerably increased protection of both ships and jetties, more complicated systems have been specially devised to give maximum absorption of impact.

Tweddell fender

A more sophisticated fender based on the use of tyres, in combination with rubber blocks and baulks of timber is a floating fender of the type known as dummies or catamarans, and constructed of baulks of timber.

Compression cylinder type

In this fender the fender pile is recessed into the jetty and the shock of a vessel coming into dock is taken by the main rubber cylinder, while secondary cylinders control and restrain sideways movement. The rubber acts purely in compression or rolling compression and there is no mechanical connection with either the jetty or the fender pile.

Rubber buffer systems

Notable features of this design of fender are that the fender face can move in any plane to suit the varying shape of the ship's sides and its angle of approach, and that it provides smooth operation and a complete absence of moving parts requiring maintenance.

Buckling column fenders

Buckling column type rubber fenders have been developed in the United States, and possess certain advantageous characteristics. When used in fenders, these units will :

1. Build up relatively high load for small initial deflection
2. Collapse at relatively small initial deflection
3. Maintain a constant force over the range of buckling defection
4. Buckle in a pre-selected direction.

Raykin fender

This fender unit, which is manufactured in the United Kingdom, consists of a series of sandwiches of rubber bonded between steel plates and assembled together to form a V-type assembly. The apex of the V faces seaward and is thrust inwards by berthing ships. The rubber components are stressed in both shear and compression and it can be readily seen that the flexibility of the unit is such that it will yield to either a direct or glancing impact.

Andre-neidhart system

The unit consists of square inner and outer metal parts with cylinders of rubber forced into position between them but not bonded. The compression and torsional rates are dependent on the ratios between the dimensions of the metal parts and the diameter of the rubber parts.

Shear type fenders

These units consist of a rubber sandwich bonded between two metal plates. One metal plate of each unit is fixed to the wharf or jetty and the other to either a pile or, when using two units, to a suspended timber or metal fender.

RUBBER IN DAMPING PANELS AND STRUCTURAL MEMBERS

The capacity of a homogeneous engineering structure to dissipate damping energy is often seriously limited unless high stress amplitudes are involved. Material composites, particularly those having a high damping visco-elastic adhesive in the composite, provide new possibilities and are becoming increasingly important.

Types of Visco-elastic Damping Treatment

There are two main types of damping treatment, the type in which extensional deformation of the damping layer is dominant, and the type in which mainly shear movements occur. Extensional deformation accounts for the losses when a single layer of damping material is applied to a plate (e.g. mastic on a car door) while shear damping is likely to dominate whenever a viscoelastic material is used between the plate to be damped and a relatively stiff cover plate or septum.

ISOLATORS IN AIRCRAFT, SHIPS AND MACHINES

Manned Aircraft and Guided Vehicles

Some general factors are common to the majority of requirements, which can be extremely wide and varied.

1. Any type of power plant will produce excitations extending to relatively high frequencies, although reciprocating engines and propellers produce a more pronounced periodic vibration at significantly lower frequencies than the more random characteristics of rocket type power plants. The disturbances found in guided missiles may be mechanically or aerodynamically generated and are generally anharmonic and aperiodic. The frequency spectra involved usually cover a range from 5 to 2000 cycles per second. Other vibratory excitation comes from aerodynamic conditions such as buffeting and boundary layer noise.
2. The shock occurring in airborne vehicles is not usually severe, although in manned aircraft landing or the firing of armaments may produce shocks which have to be taken care of in design. In guided missiles shock may result from action of the guidance system, separation of booster sections inflight, and handling hazards prior to launching.
3. Guided missiles and several types of manned aircraft can operate at any altitude, and the forces on the equipment may become several times that of gravity.
4. Of necessity aircraft structures are light, so in consequence there is probably a relatively low stiffness in the members to which the isolators are attached. The lower stiffness must be considered in isolator design.
5. In many situations the isolators may be required to withstand temperatures down to –54°C (–65°F) especially for components outside the pressure cabins, etc. The effect of aerodynamic heating of guided weapons, space vehicles or high speed aircraft may require the rubber to work at high temperatures. High temperature performance is often required for isolators situated near the propulsion units.

The installation and servicing of equipment in aircraft and missiles raises several problems of isolator design, both because space is at a premium and because the vehicle is meanwhile unavailable for use. Therefore, equipment must be mounted with the minimum number of thumb-screws or other quick release devices. Often it is simpler to regard the isolators as a permanent part of the structure and to release the equipment itself rather than try to release the whole assembly.

Ships

Typical shipboard vibration factors to be considered in isolator design are as follows :

Firstly, the predominant vibration of the ship is that induced by the propeller shaft and depends on the frequency and the number of blades involved. The transmitted frequency will vary from a minimum upto approximately 20 cycles per second.

Secondly, the predominant shocks will probably arise from underwater explosions or from the ship's own armament, and the severity of these shocks will obviously depend on the type of ship involved.

Thirdly, all the machinery operating on a ship creates an ambient vibration which is not usually objectionable. However, in some types of naval shipping, such as submarines, this ambient vibration may be capable of radiating acoustic power into the water, and this may constitute a hazard to the ship as it can be detected by underwater listening devices.

Grinding and Shaping Machines

The grinding machine operates in a steel works, where it can maintain the necessary standard of accuracy and finish only if it is insulated from vibration due to a rolling mill. This installation could be mounted on a small number of self-contained isolators. Shaping machines are sometimes mounted as being treated as systems with two degrees of freedom.

Reciprocating Machines

Machines involving reciprocating motion include internal combustion engines, air compressors, vacuum pumps, etc. The vibrations in a reciprocating machine result from two unrelated sources, namely torque impulses and mass unabalance.

Lathes

When lathes are mounted upon isolators there is the problem of a possibly high out of balance force at the highest part of the machine (i.e., on the face plate). Even though it is bad practice to have an out of balance force on the face plate, it is seldom that complete balance is obtained. This force tends to excite the rocking modes of the machine and mounts must be selected that are stiff enough to limit the rocking and flexibility of the installation whilst ensuring adequate isolation from outside sources.

Electric Motors

The problem of isolation of vibration of electric motors results from two sources -

1. Fluctuation in the torque characterised as an angular reaction upon the stator and thus upon the supports for the motor.
2. Unbalance of the rotating parts, characterised as a rotating radially directed force vector at the rotational frequency of the shaft.

RUBBER IN BUILDINGS

Acoustics

The term "acoustics" in building practice refers to the study of listening conditions in auditoria. It is concerned with the perfection of desired sounds and the elimination of echo. Rubber possesses properties which command its use for such purposes in wall and ceiling linings, upholstery and floors.

Where interiors of large buildings are finished with hard surfacing materials such as glazed tiles, hard plaster, stone and glass, they have a long reverberation period. This is due to the sound waves impinging on the non-absorbent surfaces and being reflected from surface to surface for some seconds before gradually fading away to inaudibility. In such auditoria normal rate speech and rapid passages in music are not clearly defined due to the confusion of initial and reflected sound waves. To correct this condition the reverberation period must be reduced, and this is achieved by the careful use of sound absorbing surfaces. Porous or cellular materials are ideal for this purpose; when sound waves impinge on them they penetrate the surface, causing vibration of the air in the pores of the material. The sound is weakened by frictional losses, sound energy becoming heat energy.

Sound Insulation

Sound insulation deals with the insulation of parts of a building either from internal or external noises. It is concerned with the elimination and reduction of unwanted sound. Sound can be transmitted through a structure by wave transmission, by transmission due to resonance of the structure or structural element, or by a combination of these two.

In the case of airborne sounds generated in a room and having frequency components remote from the resonant frequency of the structure, most of the transmission is wave motion. If an airborne sound has a frequency identical to, or very close to, the resonant frequency of the structure, then most of the transmission is due to the movement of the structure as a whole. Unwanted sounds from sources either inside or outside a structure to be insulated are dealt with as follows.

1. By isolating the source of the unwanted sound.
2. By constructing all floors, ceilings, partitions and walls to prevent the transmission of noise from one part of the building to another. Vibration will travel long distances through homogeneous structures with very little attenuation, consequently the only way to reduce structure borne or airborne transmission is to provide discontinuous construction as far as possible.
3. By making all openings airtight and, therefore, soundproof. In sound insulation it is obvious that the final result is only as good as its weakest link. Noise is conveyed about buildings by the air and it is important that free air paths should not be overlooked. Doors and windows should be made to fit closely and shut tightly, and holes for the passage of services through the structure should be adequately stopped.

Antivibration pipe fastenings

Vibration and noise carried along piping for water supplies, radiators and air conditioning systems are an everyday nuisance in buildings. To prevent the transmission of such vibrations to the pipe supporting structure, anti-vibration pipe fastenings have been devised.

Mounting of a complete building on rubber

One of the most difficult problems in building technology has been the design of bearing support pads which could withstand the normal expansion and contraction of metal structures, as well as being able to stand up to the stresses and strains arising from vibrations, shocks and tremors. Recently, however, some bearings have been produced which are claimed as suitable for general building construction.

RUBBER IN PACKAGING

The control of shock impulses on packaged goods has traditionally been a matter of incorporating into the pack shock absorbing or cushioning material, usually chosen by type and quantity and arranged according to the packer's knowledge of existing practice and of the handling and storage hazards expected. Particularly, in the industrial and the military field, packs are becoming increasingly complicated and costly products, to which purchasers such as government departments wish to apply strict tests of performance.

An important problem is that of discovering what material and what thickness of pad, or what type of solid mounting, will give protection against damage to an article when its container is dropped. The complete problem is complex because the result depends, among other things, on the shape and rigidity of the container, how it drops, whether flat, edgewise or cornerwise, and so on.

Typical Shock and Vibration Environment

(i) Transportation by rail; (ii) transportation by road; (iii) transportation by sea; (iv) transportation by air; and (v) climatic conditions.

Bulk Cushioning Materials

(i) Latex foam; (ii) cellular rubber (open cell); (iii) cellular neoprene; (iv) foamed polyurethane (polyester); (v) closed cell rubbers; and (vi) rubberised hair.

Solid Rubber Mats

Ii) Low cost; (ii) high shock absorption efficiency; (iii) low crepe; (iv) characteristics which do not change very much over a wide temperature range; (v) availability in a mat form; (vi) non-susceptibility to varying humidities.

Additional Air Damping

Additional air damping can be introduced with bulk cushioning and mats by enclosing the cushion in an envelope of PVC sheet, or similar material, with suitably sized holes in the side surfaces of the sheet. The ratio of the area of these holes to the volume of cushioning is critical and depends on the type and nature of the cushion used.

Buckled Level Systems

Buckled lever systems, which have been developed for packaging and vibration absorption applications, are systems in which a suitable material or spring is compressed (or tensioned) between jointed levers that buckle under load.

Buckled Columns

The elastic instability of columns under compressive loads is also used for packaging purposes. A column becomes unstable and buckles if the applied compression load exceeds a certain critical value.

Solid Rubber Mountings

The biggest difficulty encountered in designing isolators for portable equipment is allowing for effective isolation in a wide range of possible situations. For instance a shock isolator employing rubber in compression and having a natural frequency of approximately 25 cycles per second affords satisfactory protection against shocks of the type met in naval service, but if this same isolator is subjected to the vibration experienced in aircraft, which includes the frequency range of 10-50 cycles per second, any mounted equipment may be subjected to excessive vibration unless the isolator contains sufficient damping. Unfortunately to obtain sufficient damping usually requires added complex equipment, and this is often not possible because of space and cost.

Shear Mountings

The simplest type of unit used in packaging applications is the straight-forward shear mounting, consisting of metal plates with rubber sandwiched in between. Suitable lugs or holding bolts are provided for fixing to the supported equipment, and to the outside container.

Shear mountings are used for a variety of shipping container applications protecting missile and aircraft components, delicately balanced jet and reciprocating aircraft engines, and extremely fragile electronic equipment and parts. (i) tube form mounting; (ii) metalastic buckling type shock mounting; (iii) silentbloc delta mountings; (iv) lord "dynashock" mounting; (v) stud ring mountings.

CHAPTER 31

Nitrile Rubber and its Application in Construction Industry

INTRODUCTION

Construction activities in the country are seeing its boom time at present. The stress by the government is on road infrastructure development. There are a number of projects such as Golden Quadrangle (Joining India across its length and breadth by expressways) or flyover projects. These projects are using state of art technologies involving more sophisticated machineries, improved concrete mix or moulding and forming patterns. The essence is also on aesthetics, fine finish and strength of concrete design. With these objectives, newer products and technologies are being used in individual construction projects. In one of the flyover project site, appropriate shuttering fluid along with rubber as liner (in particular nitrile rubber liner) in the concrete moulds was used.

It was established that whenever quality shuttering fluids along with nitrile rubber liners were used, one could achieve superior surface finish , coupled with its antistatic properties, low compression set etc., as other benefits for concrete moulding operations. It was also found that there stands a wide gap in technology upgradation in the normal construction patterns presently followed and the present day knowledge available. It was established an ease in mould release, superior surface finish ; but it was also established that the nitrile rubber liners had more than doubled the life and provided a clean, shining surface as compared to the natural rubber liners in use resulting in overall significant savings.

Construction methods and construction activities are undergoing sea change, with new, modern, faster and more efficient techniques of construction being adopted and applied every day. This has led not only to the development of more efficient techniques of construction but has also led to equally high performance product requirement as shuttering liners, fluids etc.

Shuttering liners are desired in the manufacture of pre-cast concrete products and are used in building and construction applications for various reasons. Among these, the most prominent is the easy release of concrete block from the mould and also the surface finish of such mouldings. In the past, various types of shutter liners and release agents were used, with major use being wood and had dominated this industry for quite some time until the polymer liners made an entry. Today, the vast majority of liners are made out of polymers such as rubbers, epoxy, polyolefins, nylons etc. In India, the first rubber lined concrete moulding concept was introduced by the construction major Larsen and Toubro Limited in the construction of flyover at Bangalore in the late 90's.

With this background, detailed and exhaustive studies were undertaken at a site provided by a Construction major using both Power shuttering fluids and nitrile rubber based liners.

Oils and Lubricants are in use in large quantities in the construction industry. These are used either by way of lubricants for plant and equipment or as mould release agent. Except few major players and considering the unstructured and un-organised that this industry is, oils in particular for use in mould have received little attention or had any significance.

People use diesel, kerosene, spent engine oils etc for the same purpose. However, these products have their own drawbacks and deficiencies, which over the years have been given little importance; however, presently, there is a growing realisation, that with upgradation of construction techniques, the need for use of quality construction materials including mould liners, shutter release oils and process techniques are imperative.

RUBBER IN CONCRETE

Rubber in Concrete can be classified into following major groups.

1. As a quality improver to concrete such as impact modifier.
2. As an 'aid' in the processing and application of concrete etc.

The use of rubber lattices, both synthetic and natural and many block polymers find their way in improving the impact properties besides other qualities such as adhesion, water proofing etc. There are special products available for each of the above specific requirement. Presently, in-order to improve the aesthetics of the concrete surface, use of polymer lined moulds have become increasingly popular compared to the conventional methods of using wood or metal. Though, the use of natural rubber liners predominated the market, it became mandatory to improve the quality and cost performance of these liners. Today, use of synthetic rubbers such as nitrile rubbers and their blends, polychloroprene etc., were tried and have became popular.

Nitrile rubbers can be successfully used as a shuttering-aid in the construction of high quality finish bridges, where the concrete moulds are provided with rubber. The rubber liners are stuck to the mould surface using special adhesives , the gaps and joints are filled with good caulking compounds Before the concrete is filled inside the mould, the liners inside the mould are cleaned from past residues and a fresh coat of release agent is applied. These release agents are normally based on synthetic oils. Several types of packages based on synthetic/mineral, oils or greases are available today. Typical detail is available in Table 31.1.

Table 31.1. Properties of various release agents.

Release Agent	*Conventional Shuttering Fluid – 1*	*Conventional Shuttering Fluid – 2*	*Shuttering Fluid (Spl.)*
Base material	Kerosene-silicone combination	Mineral oil Base	Power oil
Viscosity @ 40°C, c.St	Irregular	8–9	24–26
Stability	Non uniform showing separation	Homogenous solvent Based product	Homogenous uniform product

Selection Criteria for a Good Liner Material

The selection criteria for a good liner material are based on following characteristics:

1. It should provide excellent abrasion resistance during moulding cycles.
2. Should possess good weather and age resistance properties.
3. Should have good oil and solvent swell resistance properties.
4. Excellent tensile and flex properties.
5. Excellent compression properties.
6. Good and easy adhesion to metal surfaces of mould.
7. Should be economical and easily Available.

As nitrile rubbers meet all the above requirements, they were ideally selected for this work. Some of these properties are evident from the results obtained and are discussed later in this section.

EXPERIMENTAL

Release Agents: The following options were used in this work.

1. Conventional Shuttering Fluid 1 -Kerosene and Silicone combination.
2. Conventional Shuttering Fluid 2 -Shuttering Fluid in use.
3. Power Shuttering Fluid (Spl).

All the release agents were tested for following and the results are tabulated below in Table 31.1.

Rubber liners: The following options can be used:

1. Natural rubber liners.
2. Nitrile rubber liners

Both the rubber liners were compression moulded in standard contoured surface type and are approximately 6-8 mm thick. The typical geometry of rubber liner is given in Fig. 31.1.

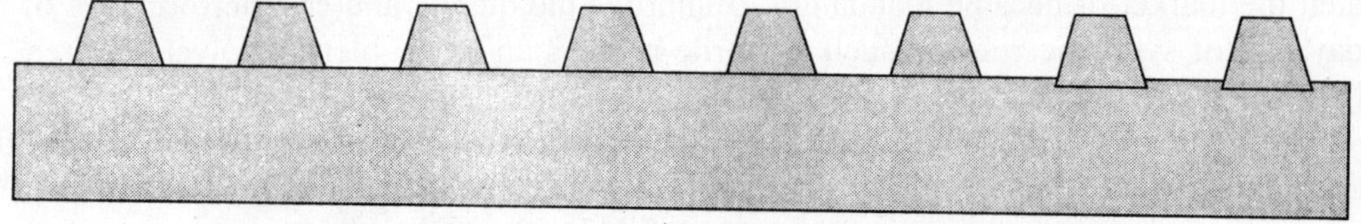

Fig. 31.1. Typical cross section view of a rubber liner.

It was observed that Natural Rubber, which was used as lining material for concrete moulds was found to absorb and swell with petroleum products. Hence, the best answer was to use a more viable option – Oil Resistant Synthetic Rubber -Aparene-NBR.

As discussed earlier, the rubber compound used in the manufacture of liners shall not only provide good ageing and oil resistance properties but shall also provide excellent durability. Normally, good compression set properties give better ageing resistance properties to the compound; however, this is not reliable. Hence, besides, compression set, stress relaxation can be used as a tool in predicting the life of rubber compounds in various applications and in particular for seals.

Extensive studies were carried out on stress relaxation properties on many rubber compounds. From these studies, it can be revealed that many synthetic rubbers such as EPDM, CR, NBR perform much better in service than their counterparts. Nitrile rubbers are also found to be far superior than natural rubbers. As a rule of thumb, the relaxation properties of nitrile rubbers can be two to four folds better than those of natural rubber based compounds. This should improve the life cycle of the product in use. It also provides better dimensional stability to the product which will enable to provide consistent concrete surfaces in repeated cycles. A typical recipe for the manufacture of rubber liner is given in Table 31.2.

Table 31.2. Recipe for rubber liner for concrete moulds.

Aparene NBR	100.00	Hardness, Shore A	75
Zinc oxide	3.00	100% Modulus, kg/sq.cm	28
Stearic acid	1.00	300% Modulus, kg/sq.cm	53
Carbon black (HAF)	20.00	Tensile strength, kg/sq.cm	64
China clay	100.00	Elongation at break, %	401
Calcium carbonate	30.00	Specific gravity	1.52
Di-Octyl phthalate	10.00	Ash content, %	45.02
SP	1.00	*Oil swell properties @ 100°C, 70 h*	
Sulphur	1.00	Volume swell in P-50* (%)	+ 2
CBS	2.00	* Paraffinic oil of 50 sus viscosity.	
TMTD	0.35		
Total	268.35		

OPERATIONS

The operations involved casting of parapets and splines for a major flyover are given below:

1. Mould preparation : wherein the moulds were cleaned and liners were stuck on it. These liners were approximately 120 cm in length and 30 cm in width.
2. Application of release oil : Mould release oil was applied on the rubber lined surface to achieve effective and easy release during the de-moulding of concrete. It was found that approximately 3 litre of oil was required for a spline and 1.5 litre for a parapet mould assembly.
3. Mould assembly: Here various parts of the mould were assembled to provide it the proper shape as per the requirement.
4. Moulding: In this step, the concrete was poured into the mould assembly. Thereafter it was left for 48-168 hours depending on the product.
5. Demoulding and lifting: It was subsequently demoulded, lifted, kept in the stock yard where remaining curing was effected or sent for rework etc.

The mould to concrete bond/grip after curing is very strong , resulting in difficulties in demoulding; besides, the liner channels were very narrow which aggravated the problem further, So much so that in case of parapet, if an incorrect shuttering fluid was used, the mould release was difficult and it shook the complete mould from its base apart from pulling out the stuck rubber thereby spoiling the rubber liner assembly. This not only added to delay in operations but further re-work in mould preparation for subsequent cycles. Thus, this operation can be an extreme test for a shuttering oil to perform and deliver.

EXPERIMENTAL TRIALS

Exhaustive trials were undertaken involving shuttering fluids and nitrile rubber liners.

The following steps were involved.

1. Application of shuttering fluid and analyses of the result.
2. Analyses of the properties of rubber liner as well as shuttering fluid being used along with their compatibility.
3. Developing appropriate nitrile rubber liner and compatible shuttering fluid for use in this operation.

4. Undertaking trials in actual condition to study and establish the product both technically and commercially for its use in the concrete moulding operation.

To improve performance properties and ease in shutter release, quality of the rubber liner were examined and subsequently use of aparene nitrile rubber liner along with power shuttering fluid were recommended. The detailed analyses of results are discussed in the following section.

RESULTS AND DISCUSSIONS

Prior to initiating the trials exhaustive studies on the type of rubber being used, the shuttering oil in use, their demoulding characteristics etc were conducted. studies are also undertaken to check the compatibility of the rubber in use with the shuttering fluid.

These studies helped us to develop an appropriate and cost effective Nitrile Rubber formulation along with shuttering fluid. The detailed studies of various physical and ageing properties on these liners in the laboratory is provided in Table 31.3.

Table 31.3. Physical and ageing properties of different liners.

Aparene NBR based shuttering liners		
Hardness (Shore° A)	75	76
Tensile strength (kg / cm^2)	64	134
Elongation at Break (%)	401	392
Ash content (%)	37	20.8
Change in hardness (Pts)	–29	–44
Change in volume (%)	+25	+111
Observation: conventional shuttering oil in use effected natural rubber much more than nitrile rubber		
Change in hardness (Pts)	–7	–29
Change in volume (%)	+2	+102
Observation: NBR liner had better oil compatability than natural rubber		

These results indicate that the solvent used in the conventional oil was swelling the rubber more and bringing down the hardness significantly, thereby affecting it's overall consistency and the life. It was decided to overcome this problem by providing nitrile rubber liner and shuttering fluid that would provide quicker release, better surface finish ease of operation with higher consistency levels etc. These were evident from our observation and are listed below.

Observations from the experiments conducted on NRL and NBR

Some of the observations from the experiments conducted on Natural rubber liners and NBR using the facilities of construction major are given below. The construction company were moulding concrete cantilevers, spline, and parapets at their construction yard site.

1. It was observed that natural rubber has more surface adsorption and gripping tendency thereby creating difficulties in shutter release. This effect was much less pronounced in nitrile rubber liners.
2. The currently used shuttering fluid mixture as well as shuttering oil provided by other oil company had tendency to swell the rubber liner and affect the rubber consistency, thereby affecting its overall performance in shutter release.

3. Both Nitrile rubber liner and shuttering fluid, performed effectively and satisfactorily by providing better surface finish and quicker mould release.

Because, the size of the rubber liners are much smaller (120cm × 30cm) and constrained as per the mouldings and maufacturing capabilities, number of liners are required in the specific type of mould depending on its size.

Assumptions

1. Life of NBR Liners are at least twice that of conventional system, though in practice it can be much higher.
2. The cost of NBR compound is approx. 20 per cent more than the natural rubber.

The above results indicate that in shuttering operations, it is essential that we use rubber liner technology with effective and efficient shutter fluid as mould release. This will help to obtain good surface finish and effective shutter release. The following conclusions were arrived at:

1. Nitrile rubber liner alongwith appropriate shuttering fluid are economical both in short and long runs.
2. Nitrile rubber can provide better quality concrete surface in repeated cycles of moulding in conjuction with compatible shuttering fluid.
3. Nitrile rubber liners and shuttering fluids can be successfully employed in the manufacture of concrete shuttering operation.

SECTION V

Special Topics

CHAPTER 32

Chemical Modification of Polymers

INTRODUCTION

The terms rubber and elastomer embrace those polymers which have useful rubberlike, highly elastic properties at ambient temperatures; however, many polymers which are non-rubbers by themselves can be chemically modified to a relatively small extent to give products with very useful viscoelastic properties. For example, the introduction of a few chlorine atoms and sulphoxide groups into polyethylene changes the macromolecules so that they no longer tend to form crystalline regions, but become elastomers over a wide temperature range. These groups also allow vulcanisation, so that reversibly deforming, solvent-resistant products can be formulated. On the other hand, starting with a conventional rubber, it is possible, by means of chemical reactions, to convert the macromolecular chains so that they lose their viscoelasticity and become thermoplastic at ambient temperatures (e.g., the cyclisation of polyisoprenes). Quite apart from the question of the temperature at which observations are carried out, the borderline between the viscoelastic and the plastic states is a relatively ill-defined one. The common factor of rubbers and plastics is, of course, their macromolecular nature. Now that we have a better understanding of their structure at a molecular level (not forgetting rubber-modified plastics and 'plasticised' rubbers), we are able to say that, with some exceptions, a modification that can be carried out on one polymer species may, under suitable conditions, be carried out on a polymer of a different species. Whether the same type of chemical modification will give us the properties we are seeking without the loss of properties we would wish to retain is a matter, first, for conjecture and, subsequently, for experimental verification.

In the early days of polymer chemistry and technology, a new chemical modification successfully applied to one polymer was quickly evaluated with a whole range of chemically similar reagents (e.g., the esterification of cellulose), and the same type of reaction was attempted with the available range of similar material. Today, the number of homopolymers available runs into the hundreds, and the number of random co-polymers runs into the thousands; therefore, the number of possible chemical modifications, including block and graft co-polymers, runs into astronomical figures. Only a fraction of these potential systems has been evaluated (and then frequently only in a superficial way to obtain patent coverage), so that the field is wide open for research and development. Because of its breadth and depth, the field of the chemical modification of polymers can be treated only in outline in a single chapter and only the more important reactions can be described. To serve the interest of the reader, however, this broad survey is punctuated by discussions in greater depth of areas that are of interest to *rubber chemists* and have considerable potential for further development. The emphasis is on the principles underlying the chemical modification of polymers; specific details of reactions conditions are deliberately omitted.

ESTERIFICATION, ETHERIFICATION, AND HYDROLYSIS OF POLYMERS

The chemical modifications discussed in this section are historically and scientifically so closely linked to one polymer, cellulose, that although the latter occurs primarily as a fibre and not an elastomer, a discussion of this group of cellulose modifications seems appropriate. Apart from the fact that some cellulose derivatives, like ethyl cellulose, when plasticised, can be quite elastomeric, the effects of modification of a basic polymer are particularly well demonstrable on a substance as stiff and highly crystalline as cellulose. Moreover, in view of the expected hydrocarbon shortage, cellulose may soon gain a new role as a polymeric starting commodity.

Cellulose, identified chemically as β-1,4-glucan, is the most widely distributed natural polymer, constituting the permanent structure of plant cell walls.

Much of the early history of the chemical modification of cellulose is related to the attempts to find a solvent for it, as its macromolecular structure was not understood at that time. *Mercer* discovered and commercialised the interaction of alkali with cellulose fibres, a process still in use under the name mercerisation. The initial product of the reaction, alkali cellulose, is not a chemical modification but a physical form in which water and sodium ions penetrate the macromolecular structure and reduce the hydrogen bonding, with consequent swelling of the fibres. The initial product, cellulose I, is converted to cellulose II, a complex physicochemical modification, in the final washing stage. The degree and rate of swelling in this process are dependent on the source of the cellulose, and if the fibres are stretched prior to and during the reaction, optimum interaction is achieved. Many other inorganic salt solutions swell cellulose and of these, zinc chloride has found the widest application. Aqueous solutions of thiourea, resorcinol, chloral hydrate, and benzene sulphonates also lead to limited swelling of cellulose. In all cases the reduction in physical cross-linking can be followed by a study of the X-ray diffraction diagrams of the crystalline content.

The complete solubility of cellulose in cuprammonium solutions, led to the development of the rayon industry but, as in the case of alkali cellulose, the regenerated polymer is chemically the same as the precursor. Regeneration via cellulose xanthate solutions, invented by *Cross* and *Bevan*, is another process still in use; it forms the basis for the manufacture of Cellophane. The first 'chemical' modification of cellulose was achieved by *Braconnot* with the production of cellulose nitrate from a wide range of cellulosic materials. The products were highly inflammable powders which could be dissolved in concentrated acetic acid to give clear tough varnishes. (Note the conversion of a fibre to a film by chemical modification). Later highly nitrated cellulose, guncotton, was produced by acetylated cellulose using hot acetic anhydride as the reaction medium. These reactions have a common mechanism, namely, the esterification of the hydroxyl groups in the basic cellulose moiety:

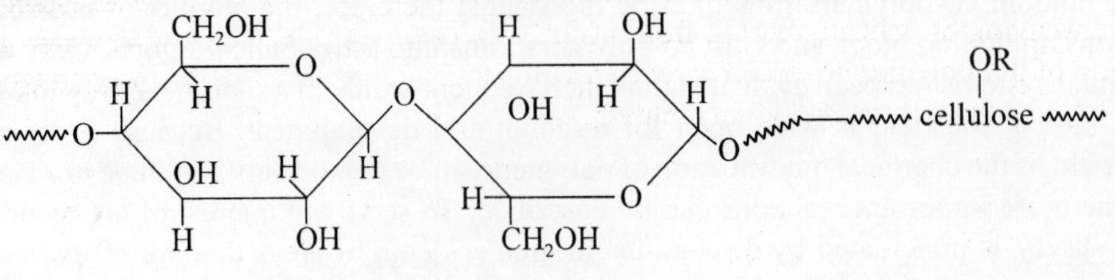

... (32.1)

Since this early work, a very large range of organic acids have been used to prepare cellulose esters, mixed esters, and ethers esters. A typical example of considerable commercial importance is the acetylation of cellulose. As in all esterifications of macromolecular materials, the accessibility of the hydroxyl groups to the esterifying acid is of prime importance. Reaction (32.1) represents complete esterification, a process which is probably never fully achieved. The identification of the esterified product is, therefore, dependent not only on the content of acetyl groups but also on the location of these groups on the macromolecular backbone. Both factors are affected by the method of preparation and the esterification conditions.

Although many esterification reactions are based on inorganic acids, for insoluble hydroxyl compounds like cellulose, xanthation is more important. Sodium hydroxide is normally used to produce the swollen alkali cellulose, which (after ageing) is reacted with carbon disulphide to form the sodium salt of cellulose xanthate:

$$\sim\sim \underset{\text{cellulose}}{\overset{O'}{|}} \sim\sim + CS_2 \longrightarrow \sim\sim \underset{\text{cellulose}}{\overset{O-C(=S)S'}{|}} \sim\sim \qquad \text{... (32.2)}$$

The cellulose can be regenerated by spinning (or extruding as a film) into an acid bath containing salts such as sodium and zinc sulphate. During spinning or extrusion, the macromolecules are oriented in the direction of flow to give high strength to the viscose fibre or the Cellophane film. The occurrence of macromolecular orientation during spinning is very important and it is used in the chemical modification of many polymers.

The cellulose ethers constitute another important group of cellulose derivatives prepared from alkali cellulose by standard etherification reactions between the hydroxyl groups and an alkyl halide. The properties of the ethers depend on the extent of the reaction, i.e., the degree of etherification. In general the ethyl celluloses are water-insoluble thermoplastic materials, whereas methyl ether, ethyl hydroxyethyl cellulose, and carboxymethyl cellulose are soluble in cold water and are used as viscoelastic thickeners and adhesives.

For the preparation of synthetic hydroxy polymers, hydroxyl groups can be introduced by co-polymerisation of the base monomer with a hydroxy monomer. These groups can then be used for esterification or etherification, but the relatively high cost of hydroxy monomers detracts from the wide spread use of direct co-polymerisation. Instead, one introduces the groups required by the complete or partial hydrolysis of the ester groups in an appropriately hydrolysable polymer, such as poly(vinyl acetate). Complete hydrolysis yields poly(vinyl alcohol) (PVA), a water-soluble polymer with considerable utility as a stabiliser and viscosity modifier for aqueous systems:

$$\sim\sim CH_2-\underset{OH}{\underset{|}{CH}}-CH_2-\underset{OH}{\underset{|}{CH}}\sim\sim \qquad \text{... (32.3)}$$

PVA has a unique use as a strengthening fibre in conjunction with weaker materials such as merino wools in the weaving of delicate fabrics, from which it can afterward be removed by water washing. A major portion of the polymer produced is reacted with aldehydes to form the corresponding poly(vinyl formal), poly(vinyl acetal), and poly(vinyl butyral):

$$\overset{OH}{\overset{|}{\sim\sim}} + RCHO \longrightarrow \overset{R\diagdown C \diagup O}{\overset{|}{\sim\sim}} + H_2O \qquad \text{... (32.4)}$$

There are different grades of each of these materials according to the overall molecular weight and the degree of substitution. These polymers are used as components of systems with unique adhesive properties, e.g., in the manufacture of safety glass laminates [poly(vinyl butyral) and mixed derivatives] and of metal-to-metal adhesive [poly(vinyl formal) cured with phenolics and other resins]. Reactions of poly(vinyl alcohol) with acids or anhydrides occur as normal esterifications, a route used to synthesise polymers and co-polymers which cannot be readily formed by conventional polymerisation (e.g., when the reactivity ratios of the monomers are not suitable).

Natural rubber and synthetic rubbers in general do not have hydroxyl groups in sufficient numbers for them to be used for esterification reactions. Terminal hydroxyl groups may be introduced into synthetic rubbers as terminal catalyst or initiator fragments and used for coupling or extension reactions.

HYDROGENATION OF POLYMERS

Any polymer with unsaturated hydrocarbon groups, either in the main chain or as side groups, can be hydrogenated. The most recent work in hydrogenation has produced excellent products, such as linear polyethylene from the hydrogenation of poly(1,4-butadiene) and poly(ethylene-co-propylene) rubber from the hydrogenation of polyisoprene.

$$-\!\!\left(CH_2-C=C-CH_2\right)\!\!- \xrightarrow{H_2} -\!\!\left(CH_2-CH_2CH_2-CH_2\right)\!\!-_n \quad ...(32.5)$$

$$\left(-CH_2-\underset{\displaystyle CH_3}{\underset{|}{C}}=C-CH_2-\right) \xrightarrow{H_2} \left(-CH_2-\underset{\displaystyle CH_3}{\underset{|}{C}}-CH_2-CH_2-\right)_n \quad ... (32.6)$$

These reactions were carried out using a heterogeneous catalyst.

Homogeneous soluble transition metal catalysts for hydrogenation have been used to create novel polymers. Homogeneous hydrogenation catalysts are usually generated from Ziegler-type catalysts based on nickel or cobalt organic salts reduced in the presence of organo-aluminium or organo-lithium compounds. These catalysts are used to form saturated elastomers by hydrogenating unsatured elastomers. The resulting polymers have vastly different viscoelastic properties than their unsaturated parent polymers. For example, the hydrogenation of a 99% poly(1,2-butadiene) has resulted in the formation of polybutene which has a lower glass transition temperature than its parent elastomer. It is interesting to note that hydrogenation does not affect the polymer molecular weight or backbone architecture.

The ease of hydrogenation and the resultant degree of saturation achieved reflect the microstructure of the polymer. Hydrogenation of unsatured elastomers usually proceeds in a blocky way. This is due to the different reactivities of the various double bonds. In general, double bonds of 1,2 structure are four times more reactive than double bonds of 1,4 structure. *cis*-1,4 units are more reactive than *trans*-1,4 units. *Chamberline* have shown that hydrogenation of 1,2 units is statistically random, whereas the hydrogenation of 1,4 units is not.

The complete hydrogenation of poly(1,4-butadiene), *cis* or *trans* structure, forms a polyethylene with a low melting point of about 115°C. It is believed that this linear polyethylene is a low-density material. It has been found that on the partial hydrogenation of a 99% *cis* poly(1,4-butadiene) showed that the hydrogenation proceeded in a blocky fashion.

The 40–50% hydrogenation of poly(1,4-butadiene) (*cis* content 98%) made by nickel catalyst gives a polymer with a melt point of +98°C and a crystallisation temperature of –13°C as measured by differential scanning calorimetry (DSC). This confirms the fact that hydrogenation of *cis*-poly-1,4-

butadiene proceeds in a blocky fashion to produce a block of polyethylene and a block of poly(*cis*-1,4-butadiene).

Polybutadienes made by anionic catalysts in the presence of polar modifiers contain a mixed microstructure of *cis*-1,4, *trans*-1,4, and 1,2 units. Hydrogenation of these polymers leads to interesting products. As mentioned previously, hydrogenation favours the 1,2 units over the 1,4 units by a 3 (or 4)-to-1 ratio. Because of this mismatch in reactivity, hydrogenation of a polybutadiene containing 40–50%, 1,2 units produces a polymer containing a polyethylene portion with a T_m of 85–95°C and a rubbery portion with a T_g of –62°C.

Block co-polymers can also be hydrogenated to produce unique products. Hydrogenated triblock co-polymers of poly(styrene-co-butadiene-co-styrene) (SBS) are commercially available from various companies. The middle block is usually a mixed microstructure of poly(1,2-butadiene) and poly(1,4-butadiene) units. The resulting product is a hydrogenated unsaturated polymer which exhibits greater thermal and oxidative properties than the parent SBS triblock.

Similar procedures have been used to hydrogenate poly(1,4-butadiene–co-1,2-butadiene) diblocks and poly(1,4-butadiene–co-1,4-isoprene–co-1,4-butadiene) triblocks. Hydrogenation of these diblock and triblock co-polymers forms thermoplastic elastomers with crystalline and amorphous segments. All these materials exhibit crystallinity, glass transition, solubility, and dynamic mechanical loss spectra different from those of their unsaturated counterparts.

Another method of preparing saturated rubbers was developed using the diimide reduction. This method can be used to produce a high degree of saturation dependent on the type of reagent used; however, side reactions can occur in this method. Generation of the diimide from *p*-toluenesulphonyl hydrides leaves an acidic fragment which may cause cyclisation in some unsaturated elastomers.

DEHALOGENATION, ELIMINATION, AND HALOGENATION REACTIONS IN POLYMERS

Dehydrochlorination of Poly(vinyl Chloride)

The dehydrochlorination of poly(vinyl chloride) has been the subject of much investigation, particularly with the view of developing greater stability in PVC polymers and co-polymers. Like many polymeric reactions, dehydrochlorination is a complex process. The vinylene groups, created by the elimination of HCl from adjacent carbon atoms in the chain, may be the result of free radical ionic, or ion-radical steps.

$$\sim\sim CH_2-CHCl-CH_2-CHCl\sim\sim \longrightarrow HCl + \sim\sim CH{=}CH-CH_2CHCl\sim\sim \quad \text{... (32.7)}$$

The presence of a small proportion of head-to-head, tail-to-tail, and other configurational irregularities in the backbone structure of poly(vinyl chloride) leads to more complex elimination steps by thermal degradation (alone or in the presence of catalysts such as aluminium chloride). The introduction of ring structures, a major process during dehydrochlorination, is likewise affected by the distribution of the chlorine atoms along the polymeric backbone. Hydrogen bromide can be effectively eliminated thermally from poly(vinyl bromide), as dehydrohalogenation is a universal thermal reaction the complexities of which increase from chloride to iodide.

Thermal Elimination

The thermal elimination process can be applied to most 'substituted' groups in vinyl polymers by controlled pyrolysis at 600–700°C, producing polyvinylene compounds, e.g., by the splitting off of acetic acid from poly(vinyl acetate). By careful temperature control one can achieve bifunctional reactions and/or intramolecular cyclisations. This has been developed commercially at relatively high temperatures, in the case of the polymerisation of methacrylamide above 65°C, to yield a polymer with a substantial proportion of imide groups:

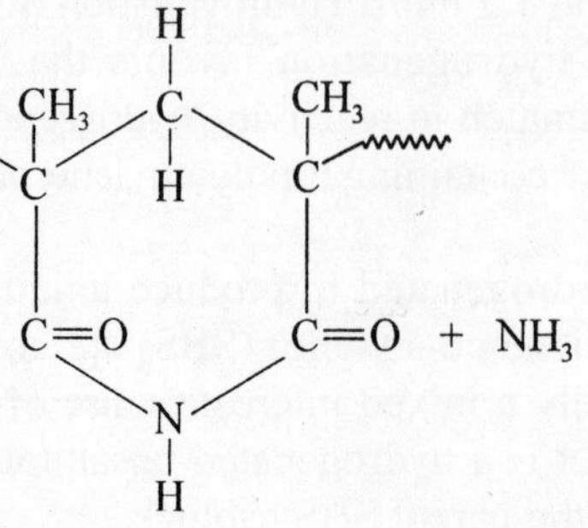

$$\sim\!\!\sim C(CH_3)-CH_2-C(CH_3)\!\sim\!\!\sim \quad (\text{ring: } C(=O)-NH-C(=O)) \; + \; NH_3 \qquad \ldots (32.8)$$

Polymers and co-polymers of multifunctional vinyl monomers (using the term to cover the presence of a halogen or other reactive group in addition to the vinyl group, rather than in the sense of more than one polymerisable group), such as α-chloroacrylic acid, often undergo partial lactonisation and hydrolysis during polymerisation. Heating in alcohol solution or electrolysing alcoholic solutions, one obtains, e.g., the introduction of double lactam rings during the acid hydrolysis of poly(α-acetamineoacrylic acid)

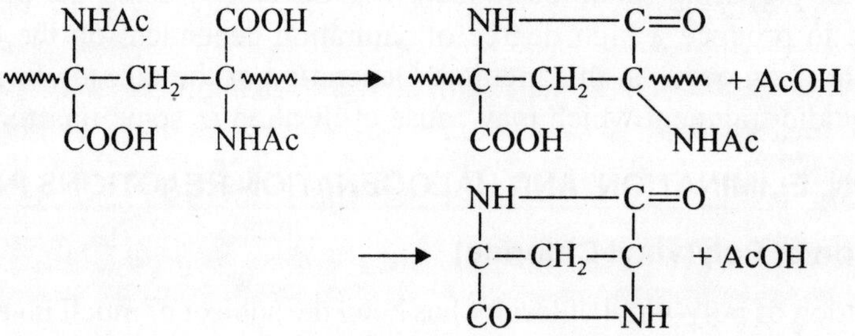

$$\sim\!\!\sim C(NHAc)(COOH)-CH_2-C(COOH)(NHAc)\sim\!\!\sim \;\longrightarrow\; \sim\!\!\sim C(COOH)(NH-)-CH_2-C(-C{=}O)(NHAc)\sim\!\!\sim \; + AcOH$$

$$\longrightarrow\; C(NH-)(CO-)-CH_2-C(-C{=}O)(-NH) \; + AcOH \qquad \ldots (32.9)$$

The discolouaration of polyacrylonitrile is due to a similar type of elimination reaction, which in this case occurs intra- as well as intermolecularly to give cross-linked insoluble ring products:

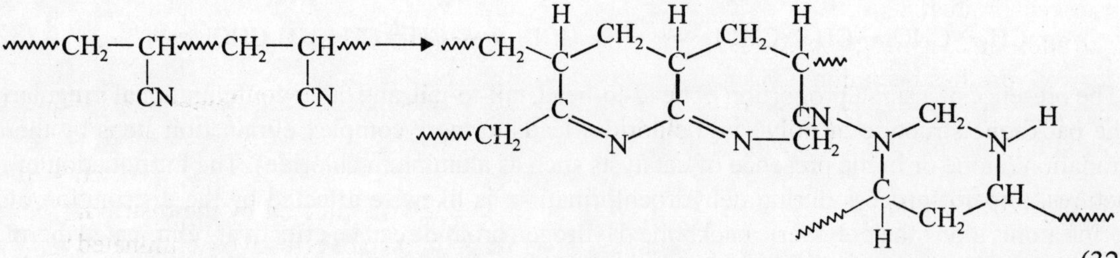

$$\sim\!\!\sim CH_2-CH(CN)\sim\!\!\sim CH_2-CH(CN)\sim\!\!\sim \;\longrightarrow\; \sim\!\!\sim CH_2-CH-CH_2-CH-CH_2-CH\sim\!\!\sim \;(\text{rings: } C{=}N-C{=}N-CH_2-N(CH_2)-N(H)-CH(CH_2)\ldots;\ CN) \qquad \ldots (32.10)$$

The controlled heating of polyacrylonitrile fibres under tension also causes an elimination of nitrogeneous products to leave a 'carbon fibre' of high tensile strength which can be considered as the end product of the line of chemical elimination reactions. Carbon fibres from cellulosic materials, lignin, and various interpolymers and blends have been developed. The structures of these products consist largely of three-dimensional carbon networks, partially crystalline and partially graphitic or amorphous.

Halogenation of Polymers

In case of halogenation and hydrohalogenation the main problems with these reactions are the cyclisation and chain scission that occur parallel to the halogenation reaction. These introduce difficult problems in the characterisation of the resulting products. Despite these problems, several products have been prepared and commercialised. Chlorination of poly(1,4-butadiene) to prepare a product similar to poly(vinyl chloride) has been reported and this process had extensive side reactions and chain degradation. The chlorination of butyl rubber and conjugated diene-butyl rubbers gives end products that are used in the tyre industry as inner liners for air retention.

Ethylene-propylene co-polymers (EPDM) are, by their random co-polymerisation, amorphous in structure and therefore easily halogenated. EPDM has been chlorinated to improve its properties and co-curability with other rubbers. The chlorination was directed toward the termonomer dicyclopentadiene to form the allylic chloride. In this manner, EPDM was chlorinated and the resulting products had improved properties.

EPDM rubbers are modified by 1,2-addition of *N*-chlorothiosulphonamides to their olefinic sites. Such additions may be carried out in solution or without solvent in an internal mixer or extruder. The solventless reactions are facilitated by added carboxylic acids or by certain metal salts of weak acids. The modified EPDMs are of interest because of their ability to co-vulcanise in ozone-resistant blends with polydiene rubbers. Although less fully explored, *N*-chlorothiocarboxamides and imides also react with EPDM to produce modified products which co-vulcanise in blends with polydienes.

In the absence of oxygen, the chlorination of polyethylene, with or without a catalyst, can be controlled to provide products with varying chlorine content. The chlorination process is statistically random so that chlorination of polyethylene to the same chlorine content as poly(vinyl chloride) (60 per cent) gives a product which is chemically different from PVC yet fully compatible with it. This random chlorination of polyethylene destroys its crystallinity. At a degree of chlorination corresponding to the loss of all its crystallinity, the chlorinated product becomes soluble at room temperature. The *p*-bromination of polyethylene follows a similar course to yield a rubberlike polymer at 55 per cent bromine content.

Both chlorination and bromination of polypropylene and isobutylene lead to degradation of the main chain, with the loss of many useful properties. Degradation during chlorination can, however, be avoided at low temperatures by limiting the reaction to a maximum of about 2%. This procedure forms a useful commercial product.

The addition of hydrogen chloride to unsaturated elastomers has also received considerable attention. Extensive work has been done on the hydrochlorination of *Hevea* [poly(*cis*-1,4-isoprene)] and *Balata* [(poly(*trans*-1,4-isoprene)] rubbers. Both *cis*-1,4 and *trans*-1,4-polyisoprenes readily add hydrogen chloride following *Markovnikov's* rules with only a small amount of cyclisation.

Cyclisation of Polymers

Cationic cyclisation of unsaturated elastomers such as poly(*cis*-1,4-isoprene), poly(3,4-isoprene), poly(1,2-butadiene), and poly(1,4-butadiene) usually leads to the formation of cyclised resinous products of no commercial value.

Cyclisation of unsaturated elastomers, such as polyisoprene can be carried out in the solid state, in solution, or even in the latex. The process involves the transformation of linear macromolecular chains into much shorter ones consisting of mono-, di-, tri-, and tetrapolycyclic groups distributed randomly along the chain. Cyclisation of polyisoprene increases its glass transition temperature by 20–30°C. The mechanism of cyclisation of elastomers depends on the catalyst employed in the process.

OTHER ADDITION REACTIONS TO DOUBLE BONDS

Ethylene Derivatives

Besides the addition of halogens and hydrohalogens 'across the double bond' there are many other reagents which will react similarly with unsaturated polymers by free radical, ionic, or radical-ion mechanisms. Of prime importance is the addition of ethylene derivatives to polydienes. One of the earliest reactions of natural rubber to be studied in detail was the combination with maleic anhydride. Depending on the reaction conditions and the presence or absence of free radical initiators, one or more of four basic reactions may take place, with the products shown (the arrows indicate where the addition has taken place and the new bonds formed).

1. Intramolecular addition to the double bond within polyisoprene chains:

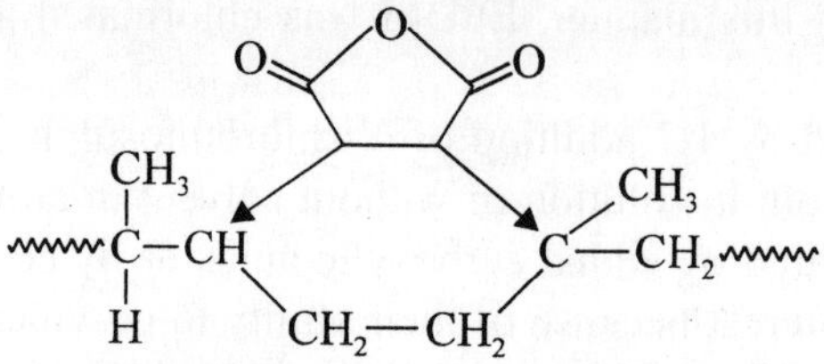

... (32.11)

2. Intermolecular addition to double bonds in different polymer chains in this group should be included, the statistically possible reaction between widely separated double bonds within the same molecule:

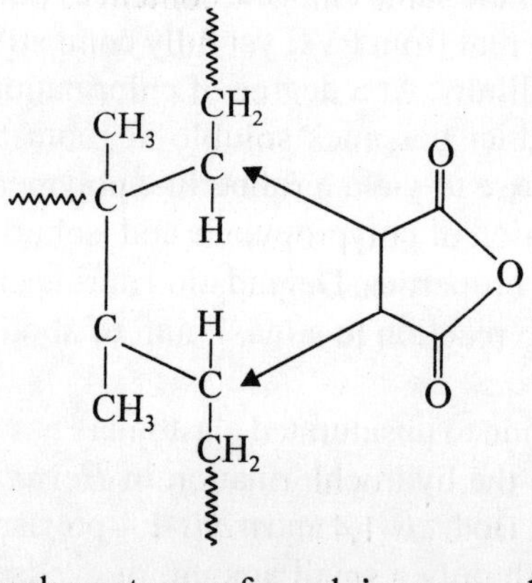

... (32.12)

3. Addition to α-methylenic carbon atoms of a polyisoprene chain:

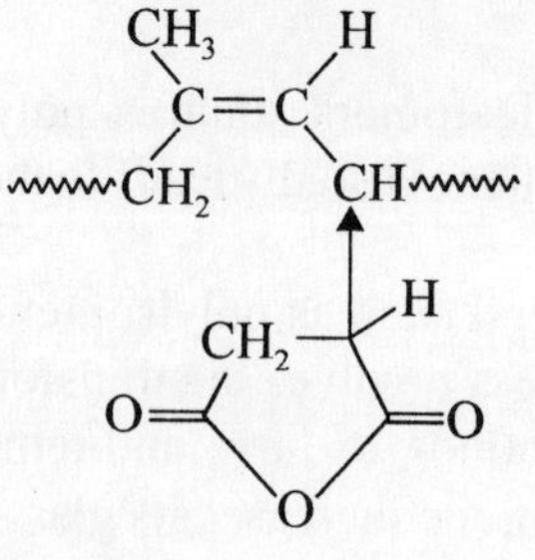

... (32.13)

4. Intermolecular addition to α-methylenic carbon atoms in adjacent chains (or widely spaced α-methylenic carbon atoms in the same macromolecule):

$$\begin{array}{c} \sim\!\sim CH_2(CH_3)C{=}CH{-}CH\!\sim\!\sim \\ \uparrow \\ \text{maleic anhydride } (O{=}C{-}O{-}C{=}O \text{ ring}, \; H{-}C{-}C{-}H) \\ \downarrow \\ \sim\!\sim CH_2(CH_3)C{=}CH{-}CH\!\sim\!\sim \end{array} \quad \text{... (32.14)}$$

In general, the overall reaction rates increase with rising temperature and in the presence of oxygen or free radical initiators, but these same conditions promote intermolecular reactions leading to gel formation. Similar reactions take place with *gutta-percha*, synthetic poly(*cis*-1,4isoprene), and poly(*cis*-1,4-butadiene). Many scientists used two-roll mills and other mastication techniques as convenient ways of blending the maleic anhydride with rubbers at elevated temperatures, but where these techniques have been used mechanochemical reactions have complicated the overall process. Reaction products of natural rubber containing 5–10% combined maleic anhydride can be vulcanised by conventional sulphur cures: of greater interest is the possibility of creating cross-linking by the use of oxides of calcium, magnesium, and zinc.

Other compounds reacting similarly via activated double bonds (excluding here block or graft co-polymerisation) include maleic acid, *N*-methyl-maleimide, chloromaleic anhydride, fumaric acid, γ-crotonolactone, *p*-benzoquinone, and acrylonitrile. Other polymers with unsaturated backbones, such as polybutadiene, co-polymers of butadiene with styrene and with acrylonitrile, and butyl rubber, react in similar ways, but the recorded reaction with poly(vinyl chloride) is largely mechanochemical in nature.

Prins Reaction

Another addition to polymers with main-chain unsaturation is the Prins reaction between ethylenic hydrocarbons and compounds containing aldehydic carbonyl groups. *Kirchof*, described the reaction of natural rubber in benzene solution with aqueous formaldehyde in the presence of concentrated sulphuric acid. The general reaction of an aldehyde, RCHO, with a polyisoprene in the presence of an inorganic or organic acid or an anhydrous metal salt is represented by

$$\sim\!\sim C(CH_3){=}C(H)\!\sim\!\sim \; + \; H{-}C(R){=}O \xrightarrow[X_c]{H_c} \sim\!\sim C(X)(CH_3){-}C(H)(CHOH\text{-}R)\!\sim\!\sim \quad \text{... (32.15)}$$

In the absence of such catalysts, the reaction leads to a shift in the double bond rather than its elimination:

$$\sim\!\!\sim C(CH_3){=}CH\!\sim\!\!\sim \; + \; R{-}C({=}O)H \longrightarrow \sim\!\!\sim R{-}CHOH{-}C({=}CH_2){-}CH\!\sim\!\!\sim \qquad \text{... (32.16)}$$

These reactions can be carried out in solution or in dispersion or by reaction in the solid phase; in the last case it is again difficult to differentiate the Prins reaction from mechanochemical reactions initiated by chain rupture during mastication.

Other aldehydic compounds, such as glyoxal and chloral, also react in a similar way with polyisoprenes and unsaturated rubbers [e.g., poly(*cis*-1,4-isoprenes), poly(*cis*-1,4-butadiene) and co-polymers of isobutylene and isoprene]. The use of strong acids, or Lewis acids, causes complications, as the acids themselves, under suitable conditions, catalyse cyclisation and *cis-trans* isomerisation, and these reactions may occur simultaneously with the addition reactions.

OXIDATION REACTIONS OF POLYMERS

Uncontrolled oxidation of rubber is detrimental to its physical properties. Oxidation reactions takes place readily at unsaturated groups in polymers and are often referred to collectively as epoxidation; however, oxidation under controlled conditions can lead to useful products such as the epoxidised natural rubber introduced by the Malaysian Rubber Producers Association. Natural rubber in the latex form is treated with hydrogen peroxide dissolved in acetic acid. This gives 50 per cent epoxidised natural rubber. This rubber shows very interesting physical properties and excellent carbon black dispersion. Similarly, non-aqueous epoxidisations of synthetic polyisoprene can be achieved using either hydrogen peroxide or hypochloride in *t*-butanol. The controlled degree of epoxidation usually leads to some interesting products. For example, the 25 per cent epoxidised synthetic polyisoprene is an elastomer with viscoelastic properties similar to those of the unepoxidised material, but has better carbon black dispersion. This gives high modulus and tensile strength, however, higher degrees of epoxidation (60–75 per cent) produce a resinous material which is not rubbery.

FUNCTIONALISATION OF POLYMERS

Polymers with stable backbones such as polystyrene, polyethylene, and polypropylene can be functionalised. Functionalisation of polystyrene has received considerable attention, because it is a unique polymer with aromatic rings which are capable of undergoing many nucleophilic as well as electrophilic reactions. Resin based on sulphonated polystyrene are used in ion-exchange material and catalyst binding materials.

Electrophilic substitution on polystyrene through a chlorometallation reaction yields chlorine functionality. This has opened up the possibilities of making many derivatives of polystyrene. Starting with chlorometallated polystyrene, derivatives such as quaternary, ammonium, or phosphonium salts have been made. Similarly, ethers, esters, sulphonamides, silanes, and ketone derivatives have been made by replacing the chlorine atom on chlorometallated polystyrene. In the case of polystyrene, however, it was discovered that chain end functionalisation can be realised if the chain ends were terminated by group I metals such as lithium and potassium.

MISCELLANEOUS CHEMICAL REACTIONS OF POLYMERS

Direct replacement of the hydrogen atoms of aromatic rings such as styrene or the allylic hydrogen in poly(1,4-butadiene) or poly(1,4-isoprene) can be carried out via metallation with organometallic compounds of group I such as lithium, sodium, and potassium. Usually, the yield tends to be low and the product is insoluble; however, the use of chelating diamines with organolithium compounds has increased the yield and the products are soluble in cyclohexane. For example, polystyrene has been metallated in high yields to give polylithiated polystyrene in which several functional groups have been successfully introduced. Similarly, polyisoprene and polybutadiene have been successfully metallated with either *s*-butyllithium or *t*-butyllithium in the presence of tetramethylethylenediamine (TMEDA) at 50°C. In the case of the polyisoprene, chain scission and reduction in molecular weight resulted at longer metallation temperatures and times. In many cases, these lithiated polymers have been used to prepare graft and block co-polymers.

BLOCK AND GRAFT CO-POLYMERISATION

Effects on Structure and Properties of Polymers

Some of the most significant changes in structure and properties of polymers can be brought about by either block or graft co-polymerisation. The term block co-polymer is applied to macromolecules made up of sequences with different chemical (or physical, i.e., tactic) structures, usually represented by A, B, and C. The sequences are of a molecular weight which would give them polymeric features even if separated. The manner in which these sequences are arranged defines the type of block co-polymer prepared. A diblock co-polymer is represented by AB, indicating that a segment with chemical composition A is connected to a segment with composition B. Other possible types of block co-polymers include triblocks ABA and ABC, for example. They may be linear or branched; the linear structures are called block co-polymers,

$$\text{AAAAAA}\cdot\text{BBBB}\cdots\text{BBBBB}\cdot\text{AAAAA} \quad \ldots (32.17)$$

and the branched structures graft co-polymers.

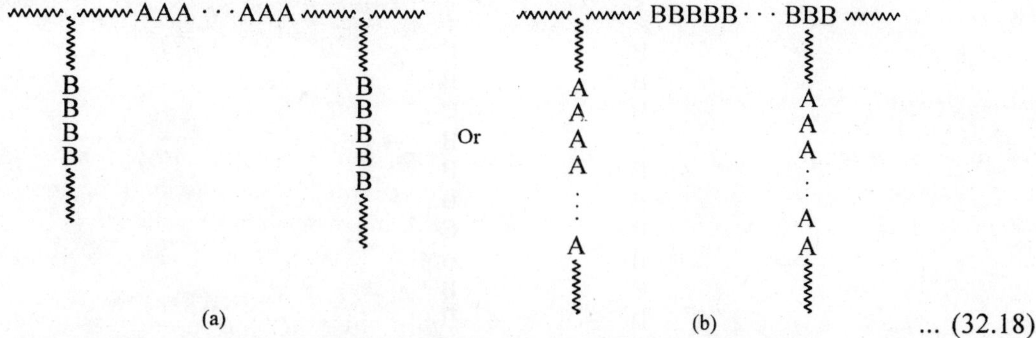

... (32.18)

Thus the same polymeric sequences may be put together as block or as graft copolymers, with differing properties, though the major differences between the properties of block and graft copolymers of the same constituent polymers are pronounced only in solution or in the melt. For example, natural rubber may be block co-polymerised with poly(methyl methacrylate), or methyl methacrylate monomer may be grafted onto natural rubber. In an attempt to distinguish by nomenclature one structure from the other, insertion of the letters *b* and *g*, for block and graft, respectively, between the names of the specific sequences was introduced, e.g., natural rubber-*b*-poly(methyl methacrylate) and natural rubber-*g*-poly(methyl methacrylate) in the case of the examples cited. The structures of these two macromolecules represented in equations (32.17) and (32.18a), where the ··· AAAA ··· sequences represent natural rubber and the ··· BBBBB ··· sequences represent poly(methyl methacrylate).

The number and order of sequences may be more complicated. Block co-polymers are usually made by free radical or living polymerisations. These processes can produce polymers that consist of a pure A block connected to a pure B block, with no interphase zone of mixed A and B structure. The preparation of block co-polymers is not limited to monomers A and B, but can also encompass segments of random co-polymers. For example, a block of a random co-polymer AB can be connected to a block of polymer A or B. Moreover, the point of attachment of the blocks can be either at the end or the middle of the polymer chain. Several examples of the various types of block co-polymers possible follow:

```
~ AAAAAA BBBB AAAAA ~     or     ~ AAAAAAAAAAAA ~
                                     B      B
                                     B      B
                                     B      B
                                     B      B          ... (32.19)
```

```
~ ABABAB  AAAAAA ~     or     ~ ABABBABB  AAAAAAAA ~
                                   A            B
                                   A            B
~ ABABA   BBBBBBB ~                A            B
                                   A            B      ... (32.20)
```

When the sequences making up the segments are random co-polymers, the prefix *co* may be introduced, with the major component monomer preceding the minor constituent. A backbone polymer of butadiene-styrene rubber grafted with styrene containing a small percentage of acrylic acid would be described as poly[(butadiene–co-styrene)–(styrene–co-acrylic acid)] and could be schematically represented as:

```
~~~~AABABAAABBABAAAABBA~~~~
    ≀              ≀
    B              B
    B              B
    B              C
    C              B
    B              B
    B              B
    B              C
    C              B
    B              B
    ≀              B
                   B
                   ≀                 ... (32.21)
```

where A represents butadiene, B styrene, and C acrylic acid.

Block Co-polymer Synthesis

Several methods can be used to synthesise block co-polymers. Using living polymerisation, monomer A is homopolymerised to form a block of A; then monomer B is added and reacts with the active chain end of segment A to form a block of B. With careful control of the reaction conditions, this technique can produce a variety of well-defined block co-polymers. Mechanico-chemical degradation provides a very useful and simple way to produce polymeric free radicals. When a rubber is mechanically sheared, as during mastication, a reduction in molecular weight occurs as a result of the physical pulling apart of macromolecules. This chain rupture forms radicals of A and B which then re-combine to form a block co-polymer. This is not a preferred method because it usually leads to a mixture of poorly defined block co-polymers.

These block co-polymers have found many uses in the shoe sole and adhesive industries. The physical properties were dependent on the macrostructure and microstructure of these block co-polymers.

Examples

As major examples, let us consider the three monomers butadiene, styrene, and acrylonitrile, and see how they can be block co-polymerised together by mechano-chemical means. From the larger number of theoretical possibilities, eleven have been selected for discussion; these may be prepared by mastication of the following:

1. A butadiene-styrene co-polymer rubber with acrylonitrile monomer.
2. Polyacrylonitrile (plasticised) with a mixture of butadiene and styrene monomers.
3. A butadiene-acrylonitrile co-polymer rubber with styrene monomer.
4. Polystyrene with a mixture of butadiene and acrylonitrile monomers.
5. A styrene-acrylonitrile resin with a mixture of styrene and butadiene monomers.
6. Polybutadiene with a mixture of styrene and acrylonitrile monomers.
7. A butadiene-styrene rubber with polyacrylonitrile (best plasticised).
8. A styrene-acrylonitrile resin with polybutadiene.
9. A butadiene-acrylonitrile rubber with polystyrene.
10. A high styrene-butadiene resin with acrylonitrile monomer.
11. A high styrene-butadiene resin with polyacrylonitrile (plasticised).

In each example, the products would be chemically and physically different in terms of the makeup of the structural sequences, and all properties would also depend on the relative proportions of the initial components. In all mechano-chemical reactions, some of the starting polymer or co-polymer remains unchanged, mainly the low-molecular-weight fraction, which is not effectively sheared, and some homopolymer may be formed from the polymerising monomers by chain transfer reactions. Varying the mastication conditions greatly influences the yield and rate of reaction; the chemical nature of the products is less affected, except that the presence of butadiene as one of the constituents (either polymer or monomer) will cause increasing gel contents with continued mastication. Processes 3, 5, 6, 8, and 9 are known to give products in which a rubber phase is dispersed in a resinous matrix; i.e., they are alternative methods for producing an A–B–S-type co-polymer (It has been found in practice that a number of monomers that normally do not polymerise by free radical processes in the temperature range 10–50°C can be block co-polymerised by cold mastication techniques, indicating ionic initiation via heterolytic scission.) The presence of a proportion of block or graft co-polymer in the system assists in stabilising the dispersion of the rubber phase in the resin matrix by acting at the phase boundary as a 'soap', i.e., a compatibilising agent at the phase boundary.

Other Methods of Effecting Mechanico-chemical Reactions

Mechanico-chemical degradation is the term used in describing chain scission of polymer backbones through the application of shear during a processing operation. It was previously believed that this type of process led to carbon-carbon chain scission, which usually causes a dramatic change in rheological properties. It has been found by *Watson* that, in the absence of oxygen, radicals produced by mechanical shear can be used to initiate the polymerisation of vinyl monomers to form block co-polymers. For example, vibromilling of natural rubber below its glass transition temperature has enabled block copolymerisation of natural rubber with methyl methacrylate to be carried out on a small scale, with conversions as high as 86%. Similar results were achieved with styrene and with acrylonitrile.

This type of approach has also been used to attach antioxidants to unsaturated polymers. The novel approach of *Scott* using the technique employed by *Watson* enabled the attack of substituted allyl mercaptans and disulphides to olefinic double bonds employing the *Kharasch* reactions.

Mechanico-chemical reactions can occur during processing, when the polymer is converted to a finished product. The chain scission can occur in both saturated and unsaturated polymer backbones. For example, during processing in a screw extruder, backbone scission in polypropylene produces long-tail free radicals which can form macroalkyl radical peroxides. These peroxides are responsible for the observed decrease in melt viscosity. In the absence of oxygen, these monoallyl macroradicals can be used to graft new monomers or such polymers as polypropylene and polyethylene to the backbone. In this manner, maleic adducts of polypropylene have been prepared, giving improved dyeability, hydrophilicity, and adhesion.

Ionic Mechanisms

Ionic mechanisms for the preparation of block co-polymers are a very important tool of the synthetic polymer chemist. A feature of many homogeneous anionic polymerisations in solution is that termination can be avoided by careful control of experimental conditions. In fact, an infinite life of the active chain end is theoretically possible, and this has led to the term living polymers. Polymer carbanions can resume growth after the further addition of monomer.

By changing the monomer composition, block co-polymerisation is readily initiated, and this process can be repeated. A major advantage of this type of synthesis over most free radical processes is the ability to control the chain length of the sequences by adjusting the concentrations of initiating sites and of monomer at each stage of the block co-polymerisation.

Anionic block co-polymerisation employs organolithium initiators, which have wider use because of their extended range of solubility, which includes hydrocarbons. Organolithium compounds act as initiators by direct attack of the organic anion on the monomer species, again a fast reaction and, in the absence of compound with active hydrogen atoms, without transfer or termination steps. If carefully executed, the reaction permits one to have precise control over molecular weight and (the narrow) molecular weight distribution.

The convenience of this technique has led to the development of many commercial products, including thermoplastic elastomers based on triblocks of styrene, butadiene, and isoprene. The initiator used in these systems is based on hydrocarbon-soluble organolithium initiators. In some cases, a hydrocarbon-soluble dilithio initiator has been employed in the preparation of multiblock co-polymers. Several techniques are used to prepare thermoplastic elastomers of the ABA type. A short summary of these techniques is given here.

Three-stage process with monofunctional initiators

In this technique, used, e.g., for the synthesis of block co-polymers of poly(styrene–*b*-butadiene–*b*-styrene) (SBS), a polystyrene block is formed by employing *n*-butyllithium as the initiator in an aromatic solvent. Butadiene monomer is then added to react with the polystyrene-lithium chain end to form the poly(butadiene) block. If the reaction was terminated at this stage, a poly(styrene–*b*-butadiene) co-polymer would result, which has no thermoplastic properties. Therefore, styrene monomer is added to produce the triblock SBS. The process for the preparation of SBS is very carefully controlled to avoid the formation of a diblock, as the presence of any appreciable amount of SB dramatically reduces the thermoplastic properties of SBS.

Two-stage process with difunctional initiators

Several commercial processes using difunctional initiators based on soluble organolithium compounds have been developed. These compounds can polymerise at both ends. Difunctional initiators are useful in the cases of ABA block co-polymers where B can initiate A but A cannot initiate B. These difunctional initiators are useful in the preparation of SBS. The elastomeric butadiene block is polymerised with hexane as a solvent. The added styrene monomer is also soluble in hexane. This method is also useful in preparing triblocks with hydrocarbon middle blocks and polar end blocks such as poly(methacrylonitrile–*b*-isoprene–*b*-methacrylonitrile).

Monofunctional initiation and coupling

In this two-stage process B is sequentially polymerised onto A, and then the two chains are coupled to yield an ABBA block co-polymer. Triblocks of SBS have been prepared using this method, with methylene dichloride as the coupling agent. The disadvantage is the formation of radical anions which can lead to contamination of the triblock with multiblock species.

Tapered block co-polymers

This method is used to form a block co-polymer which consists of two segments of essentially homopolymeric structure separated by a block of a "tapered" segment of random co-polymer composition. These are usually prepared by taking advantage of the differences in reaction rates of the component monomers. When polymerised individually in hexane, butadiene reacts six times more slowly than styrene; however, when styrene and butadiene are co-polymerised in a hydrocarbon solvent such as hexane, the reaction rates reverse and the butadiene becomes six times faster than the styrene. This leads to a tapering of the styrene in a co-polymerisation reaction.

Graft Co-polymer Synthesis

The synthesis of graft co-polymers is much more diverse, but can nevertheless be divided into groups of related processes; (i) polymer transfer; (ii) co-polymerisation via unsaturated groups; (iii) redox polymerisation; (iv) high-energy radiation techniques; (v) photochemical synthesis; and most importantly; (vi) metallation using activated organolithium with chelating diamines.

Polymer transfer

In a free radical polymerisation, chain transfer, is an important reaction. Chain transfer to a monomer, solvent, mercaptan, or other growing chain can take place. When a chain transfer reaction to another chain takes place, it creates a radical which acts as a site for further chain growth and grafting.

$$\begin{aligned} P + P &\rightarrow PH + P\cdot \\ R\cdot + P &\rightarrow RH + P\cdot \\ P\cdot + P\cdot &\rightarrow PH + P\cdot \end{aligned} \qquad \text{...(32.22)}$$

The reaction proceeds by the transfer of a hydrogen or halogen (in the case of halogenated polymers) atom from a macromolecule P to the growing chains P· (or to an excess initiator free radical R·, thereby 'terminating' them). The reactivity is now located on the transfer molecule, which in turn initiates co-polymerisation, i.e., the growth of a grafted side chain of a newly introduced second monomer. A measure of grafting occurs with most monomer-polymer systems, especially those initiated by benzoyl peroxide, if the concentrations of polymer and initiator are high.

The simplest technique is to dissolve the polymer in the appropriate solvent; add the peroxide initiator, which abstracts a hydrogen radical and generates a radical on the polymer chain; and then add fresh monomer fcr grafting onto this site. This technique has been employed in grafting methylacrylate onto natural rubber and synthetic polyisoprene. In this manner, several commercially useful products such as ABS resins have been prepared; however, tyre elastomers are not made in this manner because of the generation of micro and macro gel particles, which are detrimental to physical properties. In many cases when latex grafting has been used, the product has usually been targeted toward thermoplastic applications rather than rubber applications.

Co-polymerisation via unsaturated groups

Other methods (e.g., most car body repairs) are based on the polyester-styrene co-polymerisation process (reinforced with various types of inert mesh or glass fabric), a graft co-polymerisation of styrene onto backbone unsaturated polymer of relatively low molecular weight. In general, for high grafting yields, a reasonably high concentration of pendant vinyl groups is required on the backbone polymer. For glass-reinforced plastics, the polyester resins are selected with this in view. In natural rubber, a few such groups per molecule are always present and these undoubtedly participate during normal grafting. The content of pendant vinyl groups can be increased by mastication of unsaturated rubbers under nitrogen, because the resonance structures re-combine as

$$\begin{array}{c} \sim\sim C(CH_3){=}CH{-}CH_2\cdot \\ \Updownarrow \\ \sim\sim \dot{C}(CH_3){-}CH{=}CH_2 \end{array} \quad + \quad \begin{array}{c} \cdot CH_2{-}C(CH_3){=}CH\sim\sim \\ \Updownarrow \\ CH_2{=}C(CH_3){-}\dot{C}H\sim\sim \end{array}$$

$$\downarrow$$

$$\sim\sim C(CH_3)(CH{=}CH_2){=}CH_2{-}C(CH_3){=}CH\sim\sim \quad \text{and} \quad \sim\sim C(CH_3){=}CH{-}CH_2{-}CH(C(CH_3){=}CH_2)\sim\sim \text{ etc.} \qquad \text{... (32.23)}$$

The direct introduction of peroxide groups into the backbone of polymers, such as poly(methyl methacrylate), has been used to produce macromolecular initiators for the synthesis of block co-polymers, e.g., poly(methyl methacrylate–*b*-acrylonitrile) and poly(methyl methacrylate-*b*-styrene). Ozonisation can also be used, with careful control of the degree of ozonolysis, to introduce epoxy ring structures into natural rubber:

$$\sim\!\sim CH_2-\overset{\overset{CH_3}{|}}{C}=CH-CH_2\sim\!\sim \xrightarrow{O_3} \text{(ring: } (CH_3)(\sim\!\sim CH_2)C-O-CH(CH_2\sim\!\sim)-O-O \text{)} \quad \text{... (32.24)}$$

By carrying out the reaction to about 4% of the available double bonds in a solvent such as toluene at a low temperature followed by a nitrogen purge, grafting can be effected by addition under nitrogen of methyl methacrylate (MMA) monomer (reacting at 80°C in sealed ampules) and formation of two MMA chains attached to the oxygens of the opened —O—O— bridge. This technique should be applicable to isoprene and butadiene co-polymers.

Redox polymerisation

Redox polymerisations are among the most popular techniques for grafting reactions, and of the possible initiator systems, ferrous ion oxidation and those based on ceric ion reduction are widely used. In a redox polymerisation, a hydroperoxide or similar group is reduced to a free radical plus an anion, while the metal ion is oxidised to a higher valency state, and at the same time a monomer is added. When the reducible group is attached to a polymeric chain, the free radical grafting sites thus formed on the macromolecular backbone act as initiators for graft co-polymerisation

$$\sim\!\sim CH_2-\underset{\underset{O-OH}{|}}{\dot{C}H}\sim\!\sim + Fe^{2-} \xrightarrow[\text{aldehyde}]{\text{amine of}} \sim\!\sim CH_2-\underset{\underset{\dot{O}}{|}}{CH}\sim\!\sim + Fe^{3+} + OH^- \quad \text{... (32.25)}$$

This method has been used to graft methyl methacrylate to natural rubber latex. (Actually, fresh latex contains a few hydroperoxide groups per macromolecule, which can take part in grafting reactions). Re-centrifuged latex concentrate is mixed with methyl methacrylate and a solution of tetraethylene pentamine is added, followed by a small quantity of ferrous sulphate solution. The homogenised blend is allowed to stand, often overnight. The graft co-polymer is isolated by coagulation. As practically all free radical sites are formed on the rubber backbone, there is very little free poly(methyl methacrylate) in the grafted system; on the other hand, some rubber chains are without grafts, as not all chains have hydroperoxy groups. Higher yields of graft co-polymer are obtained by allowing the monomer to dissolve in, and equilibrate with, the latex particles before adding the amine and ferrous ion initiator. It has been claimed that passing oxygen (air) through the latex for several hours reduces the free rubber content of the polymerisation product, but nitrogen purging is then necessary to prevent dissolved oxygen from acting as a polymerisation inhibitor.

Hydroxy polymers can be grafted by redox polymerisation by using a water-soluble peroxide, such as hydrogen peroxide in conjunction with ferrous ions. The OH radicals thus produced abstract H atoms from the hydroxy groups in the polymer, giving free radical grafting sites on the backbone. This method has been used with starch and cellulose derivatives, but considerable quantities of homopolymer are formed from the initial hydroxyl radicals in parallel with the H abstraction. By introducing a few hydroxyl groups into a co-polymeric synthetic rubber, grafting can be effected, provided the presence of homopolymer can be tolerated. Mixtures of ferrous ammonium sulphate and ascorbic acid are suitable redox initiation systems. Many patents claim the preferred use of ceric ions, which easily oxidise hydroxyl groups by a radical-ion reaction:

$$\begin{aligned} R - OH + Ce^{3+} &\longrightarrow R - \overset{+}{\underset{\bullet}{O}}H + Ce^{2\bullet} \\ R - \overset{+}{\underset{\bullet}{O}}H + OH &\longrightarrow R - \dot{O} + H_2O \end{aligned} \quad \text{... (32.26)}$$

The advantage of this reaction lies in the fact that only hydroxyls on the polymer are converted into R — O free radicals, so that no homopolymer can be produced and pure graft is obtained.

High-energy radiation techniques

During high-energy irradiation *in vacuo*, e.g., from a Co source, some main-chain degradation of natural rubber and other polyisoprenes occurs:

$$\sim\!\sim CH_2-\overset{CH_3}{\overset{|}{C}}=CH-CH_2-CH_2-\overset{CH_3}{\overset{|}{C}}=CH-CH_2 \sim\!\sim \longrightarrow \quad \text{... (32.27)}$$

Much of the irradiation energy is also adsorbed by the removal of hydrogen atoms:

$$\sim\!\sim CH_2-\overset{CH_3}{\overset{|}{C}}=CH-CH_2\cdot \; + \; \cdot CH_2-\overset{CH_3}{\overset{|}{C}}=CH-CH_2 \sim\!\sim$$

$$\sim\!\sim CH_2-\overset{CH_3}{\overset{|}{C}}=CH-CH_2 \sim\!\sim \xrightarrow{\text{Irradiation}} \sim\!\sim CH_2-\overset{CH_3}{\overset{|}{C}}=CH-\dot{C}H \sim\!\sim \quad \text{... (32.28)}$$

The irradiation of natural rubber in the presence of a vinyl monomer thus leads primarily to a synthesis of graft co-polymers, but some block co-polymer is certainly always present. Irradiation syntheses may be carried out in solution, either in contact with liquid monomer (with or without a diluent) or in contact with monomer in the vapour phase, or in emulsion or suspension. The rubber may be pre-irradiated in the absence of air to produce free radicals for later monomer addition, but the life of these radicals is short as a result of mobility within the rubber matrix. Irradiation at very low temperatures makes it possible to use the trapped radicals technique for a variety of natural and synthetic rubbers. Plastics and polymers with a crystalline phase are more readily pre-irradiated to initiate later grafting by trapped radicals. Irradiation may also be carried out in air to introduce peroxide groupings: These groups can then be used to initiate grafting by any of the methods already discussed. Latex phase grafting is generally favoured for its simplicity; natural rubber grafts with methyl methacrylate styrene, acrylonitrile, and vinyl chloride have been made in this way.

$$CH_2-\overset{CH_3}{\overset{|}{C}}=CH-CH_2\cdot \xrightarrow{O_2} CH_2-\overset{CH_3}{\overset{|}{C}}=CH-CH_2-O-O\cdot$$

$$\text{or}$$

$$CH_2-\overset{CH_3}{\overset{|}{C}}=CH-CH_2-OOH + R\cdot \quad CH_2-\overset{CH_3}{\overset{|}{C}}=CH-CH_2-O-O-CH_2-\overset{CH_3}{\overset{|}{C}}=CH \quad \text{... (32.29)}$$

$$-CH_2-\underset{}{\overset{CH_3}{\overset{|}{C}}}=CH-\dot{C}H \longrightarrow CH_2-\overset{CH_3}{\overset{|}{C}}=CH-\underset{O-O\bullet}{\underset{|}{CH}}$$

$$\downarrow$$

$$CH_2-\overset{CH_3}{\overset{|}{C}}=CH-\underset{OOH}{\underset{|}{CH}} + RH \quad \text{or} \quad CH_2-\overset{CH_3}{\overset{|}{C}}=CH-\underset{\underset{\underset{CH_2-\overset{CH_3}{\overset{|}{C}}=CH-CH}{|}}{\underset{O}{|}}}{\underset{O}{\underset{|}{CH}}} \qquad \text{... (32.30)}$$

The irradiation of mixed lattices for subsequent combination of the ruptured chains is another approach; it has been carried out with natural rubber and poly(vinyl chloride) lattices to prepare graft (and block) co-polymers in fairly high yields without the problem of monomer recovery. The same method has been used to graft polychloroprene onto synthetic polyisoprene dispersions and onto polybutadiene lattices of various compositions.

Photochemical synthesis

Macromolecules containing photosensitive groups which absorb energy from ultraviolet frequencies often degrade by free radical processes. The degradative process as a rule is fairly slow, but by the addition of photosensitisers, such as xanthone, benzyl, benzoin, and 1-chloroanthraquinone, the rate can be speeded up to enable graft co-polymerisation to take place in the presence of methyl methacrylate or other monomers. This can be done in the case of natural rubber in the latex phase with reasonably high yields of graft co-polymer. Natural rubber–*y*-polystyrene and poly(butadiene–*y*-styrene) have both been prepared by ultraviolet irradiation of sensitised latex-monomer dispersions. A combination of photochemical synthesis and redox-type initiation can also be carried out—a process known as one-electron oxidation—to achieve grafting with minimal homopolymer formation.

Bromine atoms on the backbone of a polymer can be liberated readily by ultraviolet irradiation to give free radical sites for grafting reactions. The bromination can be photochemically induced as under:

$$CH_2-\underset{C_6H_{11}}{\underset{|}{CH}}-CH_2-\underset{C_6H_{11}}{\underset{|}{CH}} \xrightarrow[hv]{Br_2} CH_2-\underset{C_6H_{11}}{\underset{|}{\overset{Br}{\overset{|}{C}}}}-CH_2-\underset{C_6H_{11}}{\underset{|}{CH}} + HBR \qquad \text{... (32.31)}$$

or a chain transfer agent such as carbon tetrabromide may be used in the polymerisation step to introduce the labile groups,

$$\sim\sim CH_2-\underset{C_6H_{11}}{\underset{|}{CH\bullet}} + CBr_3 \xrightarrow[hv]{Br_2} \sim\sim CH_2-\underset{C_6H_{11}}{\underset{|}{CHBr}} + \bullet CBr_3 \text{ etc.} \qquad \text{... (32.32)}$$

With the aid of suitable sensitisers, polymers such as brominated butyl rubber, valuable because of their flame retardancy, may act as backbone polymers for a variety of grafting reactions.

An early synthesis of block co-polymers was based on the ultraviolet irradiation of poly(methyl vinyl ketone) in the presence of acrylonitrile. The initial degradative step is.

$$\begin{array}{ccc} \mathrm{H} & \mathrm{H} & \mathrm{H} \\ | & | & | \\ \mathrm{C}- & \mathrm{C}\text{———} & \mathrm{C} \\ | & | & | \\ \mathrm{H} & \mathrm{C{=}O} & \mathrm{H} \\ & | & \\ & \mathrm{CH_3} & \end{array} \longrightarrow \mathrm{CH_2}\bullet + \begin{array}{c} \mathrm{H} \\ | \\ \bullet\mathrm{C}-\mathrm{CH_3}- \\ | \\ \mathrm{C{=}O} \\ | \\ \mathrm{CH_3} \end{array} \quad \ldots (32.33)$$

This degradation reaction, supplemented by various subsequent oxidation steps, has found renewed interest in the form of the introduction of photodegradable plastics as part of the campaign to reduce plastic litter from throwaway packaging. Although as yet there has been no demand for photodegradable rubbers, the incorporation of a small percentage of a vinyl ketone into a rubber co-polymer or homopolymer would open the way to a useful synthesis of block co-polymers.

Many other syntheses of block and graft co-polymers have been reported, but enough has been said to indicate the scope of these reactions and to indicate a potential that has still to be thoroughly explored. Many grafting and block co-polymerisation systems have only been evaluated for plastic materials, but are capable of extension to rubbers.

Metallation using activated organolithium with chelating diamines

Unsaturated elastomers can be readily metallated with activated organolithium compounds in the presence of chelating diamines or alkoxides of potassium or sodium. For example, polyisoprene, polybutadiene, styrene-butadiene co-polymers, and styrene-isoprene co-polymers, can be metallated with *n*-butyllithium. TMEDA complexes (1/1 or 1/2 ratio) to form allylic or benzylic anions. The resulting allylic anion can be employed as an initiator site to grow certain branched or comb polymer species. These polymers can include polystyrene, which would form hard domains, or polybutadiene, which forms soft domains.

Research in this area has resulted in the preparation of several comb polymers. The metallation technique is a useful and versatile method as it can be used with any polymeric material which contains a few double bonds. For example, ethylene-propylene was successfully grafted with norbornene. Similar reactions were performed on polymeric materials which contain aromatic rings, such as polystyrene, poly-α-methyl styrene, and polyphenylene oxide (PPO). In general, polymeric materials that contain either side groups or main-chain allylic or acidic hydrogens can be metallated with organolithium compounds in the presence of chelating diamines. They can also be grafted with ionically polymerisable monomers to produce comblike materials.

Base Polymer Properties

Some properties of the co-polymer particularly germane to this discussion are briefly mentioned here. The properties of, say, natural rubber grafted with poly(methyl methacrylate) cannot be evaluated unless the co-polymer is isolated from either homopolymer species. The methods used are based on fractional precipitation, selective solution, or a combination of these basic techniques. In many cases, though, technologists are concerned with the materials as manufactured, so we consider in this context also the properties of the block and graft co-polymers without homopolymer removed.

The presence of two chemically different polymeric sequences in the same chain causes that macromolecule to act as a soap; i.e., it helps to compatibilise two species of homopolymer in a blend by accumulating at their interface, assisting a more gradual transition from one phase to the other, and thus reducing the interfacial energy. Microphase separation, of course, still occurs, the predominant case in practice, but macrophase separation is thereby usually prevented. In high-impact polystyrene and in A–B–S co-polymers prepared by grafting reactions, the dispersed rubber phase in the glassy matrix and the dispersed glassy phase within the rubbery particles are both prevented from forming separate phases by the graft co-polymer chains, which on a molecular scale have their rubbery segments associated with the rubber particles and their plastic segments with the glass phase. In this respect there is little difference in properties between a graft and a linear block co-polymer—the essential feature is the presence of the two types of sequences in the same macromolecule.

The block co-polymeric thermoplastic polymers owe their properties to this very structure, whereby the polystyrene end blocks (along with any homopolymeric polystyrene) form the microphase, which is dispersed within a continuous phase of polybutadiene formed from the polybutadiene segments of the central sequences in S–B–S-type block co-polymers. (For this to happen, the total volume of the polybutadiene segments must exceed the total volume of the polystyrene segments. When the reverse is the case, the product exhibits the properties of a high-impact polystyrene. Although the polystyrene 'structures' act as physical cross-links at low temperature, at processing temperatures above the softening temperature of polystyrene both segment types exhibit viscoelasticity, allowing the material to be extruded, injection moulded, etc. On cooling, the polystyrene domains become rigid again and assert their influence on the material properties.

When block and graft co-polymers are dispersed in solvents, the solutions have properties which depend on whether or not the co-polymer is eventually fully solvated. If the solvent is a 'good' solvent for both sequences, [e.g., chloroform in the case of natural rubber graft co-polymerised with poly(methyl methacrylate)], then both segment types are expanded and films cast from dilute solutions will usually be intermediate in properties to the two homopolymers (in this example the properties of a reinforced rubber film). If the solvent is a good solvent for the rubber but a poor solvent or non-solvent for poly(methyl methacrylate), e.g., petroleum ether, then the solutions show the typical turbidity of a block or graft co-polymer and the cast film is highly elastic. When the solvent is acetone, a good solvent for poly(methyl methacrylate) but a non-solvent for rubber, the cast films are plastic with high tear strengths.

When grafting is carried out on a polymer under conditions such that the physical form of the substrate polymer is maintained, then the original properties of the substrate usually predominate, and supplementary properties accrue as a result of the grafting. This is invariably the case when the substrate is in fibrous form, e.g., cellulose, nylons, and terylene grafted with various monomer systems. The nature of the grafting reaction to these fibres is usually such as to form a surface coating over the substrate polymer; the surface characteristics, such as dyeing, are therefore usually those of the grafting system. It is very doubtful that any blends of two polymers, or of chemically different co-polymers, can from a thermodynamic point of view ever be fully compatible. Even most block or graft co-polymers systems therefore show microphase separation which will be typical for the properties of a given system. Chemical modification, as discussed in this chapter, will in general lead to the formation of polymeric single phases, provided the reaction has been carried out homogeneously. The choice need not be restricted, however, to just these two approaches, as chemical modification can be carried out after block or graft co-polymerisation or vice versa. Very little has been published on such a consecutive use of these two physically different ways of modifying polymers chemically, so that there is considerable scope for developing new modifications of long-established rubbers, as well as generally for changing old into new polymers.

CHAPTER 33

Control of Biological Degradation of Polymers by means of Biocides

INTRODUCTION

This chapter deals with the control of biological degradation of polymers by means of biocides.

Many natural and synthetic polymers are subject to attack by biological agents. At some time during the life of a polymer, it may be a possible source of food for living organisms, a barrier to be penetrated in the search for sustenance, a substrate subject to enzymatic or chemical by-product attack, or simply a material for shelter or attachment during the life cycle of certain organisms. This may or may not manifest itself in an attack on the object, depending on the inherent properties, including the susceptibility of the polymer or its component parts to penetration or degradation. The degree of susceptibility, the anticipated end use of the material, cost, and other factors determine whether or not a chemical or other protective treatment should be applied. Active chemical compounds used as preservatives control degradation in several ways; those which produce death are termed biocides. Many chemical agents, however, may inhibit the reproduction and growth of organisms without causing death. These may be exemplified in the two common classes of control agents known as bacteriostats and fungistats. With higher organisms, some of the mammals, for example, the repellent qualities of a chemical agent may minimise damage adequately. These protective agents make the polymer less susceptible to biological attack and prolong the useful life of the polymer material by preservation of its physical properties and esthetic appearance.

The more rigorous definition of a biocide as an agent which kills living organisms precludes the somewhat more common usage of the term, which implies that a biocide is an agent which kills, repels, or inhibits the growth of the biological environment in which the polymer exists.

The subject of natural and synthetic polymers is a broad one covering a multitude of substances and products. Natural polymers include cellulosic materials such as wood and natural fibres. Also included are polysaccharides and their derivatives (industrial gums), waxes, natural rubber, resins, leather, and other proteinaceous materials.

Synthetic polymers, such as polyolefins, polyesters, or vinyl polymers, are considered, for the most part, particularly in relation to the polymeric constituents of the compounded polymers, to be resistant to biological attack. However, this is a maxim which has not been scientifically investigated in depth. As more general use is made of these polymers, they may be subjected to different and exotic biological environments which may affect their resistance to biological attack.

Synthetic polymers are seldom used alone. Other materials are mixed with polymers in amounts from minor to major proportions. These include plasticisers, lubricants, stabilisers, fillers, and other materials which may change the biological resistance of both the polymer and the finished composition. Present indications are that these materials particularly should be reviewed for their vulnerability to biological attack. Considerable work with plasticisers has been reported and the indications are that the improper choice of plasticiser can directly affect the durability of the finished product, specifically by consumption of the plasticiser by bacteria and/or fungi.

The history of biocide use for polymer preservation dates to antiquity. Application of cedar and other essential oils was known and used by embalmers in Egypt, ca 3000 BC, to protect mummies and their cloth wrapping from degradation by micro-organisms. In 1000 BC, *Homer* mentioned the use of sulphur to protect against 'pests'. *Moses* was reported to have instructed Joshua to preserve the books of the Pentateuch with cedar oil. Other examples can be found in which chemical treatment was used against the ravages of insects, rodents, and micro-organisms.

After the fall of the Roman Empire, references to the use or development of biocides in the literature are difficult to uncover. This condition lasted until the start of the Renaissance. The biocides used then were inorganic or natural "extractions" including sulphur, copper, mercury compounds, and arsenical compounds. Bordeaux mixture (slaked lime in a copper sulphate solution), elemental sulphur for mildew control, and arsenical compounds for insect control emerged as important chemical agents.

The investigations of many men including *Pasteur*, *Prevost*, De Bary, and *Semmelweis* were instrumental in establishing the concept of micro-organisms as the causative agents of disease and microbial degradation. Foremost among these distinguished scientists was *Pasteur*. His work with the problems of wine souring, silkworm disease, cholera, anthrax, and rabies are milestones in the annals of science.

The importance of the inorganic chemicals for biological control was successfully challenged by the development of the organic biocides. The discovery of the fungicidal activity of dithiocarbamates and of the insecticidal properties of phenothiazine (9-thia-10-azaanthracene, thiodiphenylamine) initiated investigations into the value of organic compounds as biocides. Since that time, many chemicals have been developed and marketed, including DDT (1,1,1-trichloro-2,2-bis(*p*-chlorophenyl)ethane), DDD (1,1-dichloro-2,2-bis(*p*-chlorophenyl)ethane), aldrin, chlordane, dithiocarbamate salts, phenylmercury compounds, organocopper compounds, organophosphorus compounds, substituted urea compounds, and a multitude of others.

World War II stimulated the study of deterioration and the use of biocides for preservation. A combination of material shortages and microbial degradation of military equipment and clothing triggered a comprehensive search for causative agents and effective treatments. One result of this study was the establishment of the Prevention of Deterioration Centre in Washington, D.C., a unit of the National Research Council.

The literature of biological degradation, including that of polymeric materials, continues to grow as new problems are encountered, old problems are solved or continue to elude solution, chemicals are developed, and the biological environment flourishes. A valuable source of information is the abstracts issued by the Prevention of Deterioration Centre.

BIOLOGICAL ENVIRONMENT

The biological environment, that is, the biological surroundings in which polymers are present, includes the biological agents responsible for the deterioration of polymeric substances; these agents are listed in the following sections.

The Plant Kingdom

Eumycetes, Schizomycetes, Actinomycetes, and algae are considered here.

Eumycetes

Eumycetes or true fungi, are micro-organisms of particular importance in causing the degradation of materials. Fungi are nucleated, spore-forming, non-chlorophyllous organisms, which reproduce both sexually and asexually; most of them possess filamentous, somatic structures, and cell walls of chitin and/or cellulose. More than 80,000 species are known.

The true fungi are present everywhere. Their importance as deteriorative agents stems mainly as a result of the production of enzymes which break down non-living substrates in order to supply nutrient materials for support of fungal growth. These substrates can be polymers or other materials present in polymer compositions. Certain environmental conditions are essential for optimum growth and degradative activity. These include an optimal ambient temperature, presence of nutrient materials, and high humidity. Fungal growth is slow in arctic areas and more restricted or absent in low-humidity areas. The growth of fungi is probably greatest in the tropics, although in relation to economic losses, the problem is significant in temperate zones particularly where high humidities prevail. It is interesting to note that some environmental conditions which appear highly unlikely to support fungal growth do allow fungi to exist. Few would suspect that hides in brine solution would be infected by fungi, but there are halophilic and facultative species living and thriving in this hostile environment.

The eumycetes vary widely in their morphology, ranging from microscopic single-cell yeasts to the more highly evolved macroscopic structures of the mushroom type. Representatives from most fungal genera have been isolated from polymeric materials. Many fungi play an important role in degradation, although others grow on polymer surfaces with no known effect on the substrate. However, adventitious growth, that is, growth on the dust particles, inorganic salts, hair, dead insects, epidermal cells, and similar materials deposited on the surface of polymers, can greatly detract from the esthetic properties and other characteristics of end-use items.

Schizomycetes

Schizomycetes the bacteria, and *Actinomycetes*, a type of organism intermediate between fungi and bacteria, have played an undetermined role in relation to fungi in polymer deterioration. The importance of bacteria in some materials such as water-based paints is usually recognised. Bacteria can be single-cell rods, cocci, or spirilla; others are chainlike or filamentous. Bacteria can be either aerobic, or anaerobic; in contrast, the actinomycetes and fungi are necessarily aerobic. Some bacteria are motile; bacteria are predominantly non-chlorophyllous. Their degradative action also is chiefly a result of enzyme production and resultant breakdown of non-living substrates in order to obtain nutrient materials.

Bacteria

Bacteria present in soil are important agents in material degradation. Particularly affected are cellulosic plant life, wood products, and textiles subject to cellulytic degradation. Some actinomycetes cause discoloration of vinyl films. There are excellent references on the subject of bacteria and related organisms.

Algae

Algae differ from fungi and bacteria in that they are chlorophyllous and capable of photosynthesis of nutrient materials. Species types range from single cells to large multicellular seaweeds and kelps.

The algae are not associated with deterioration of polymers to any large extent. Although they were found on deteriorated tentage material in the tropics, this breakdown may have been caused by other organisms. Cellulolytic activity of algae is limited or non-existent. The chief problem with algae is their unsightly and odorous nature when present on polymeric materials, and the fact that they may constitute a source of food for other organisms. They are present as fouling organisms in salt and fresh-water environments.

Plant life representative of the higher phyla in the plant kingdom are not thought of as degradative agents to polymers.

The Animal kingdom

Insects, birds, and rodents are considered here.

Insects

The most varied and prolific organisms of the biological environment are the insects, which number in the hundreds of thousands of species. Insects of particular interest are those which affect cellulosic and proteinaceous materials, such as termites, carpenter ants, boring beetles, moths, silverfish, and grasshoppers.

Birds

Material damage by bird life is of minor importance. There is interest in bird-repellent chemicals to prevent the secondary effects of fungal and bacterial attack after exposure resulting from initial bird damage.

Rodents

Rodents have played an important role in man's struggle for survival. In particular, the rat has played an important part both as a disease bearer and as an agent of material destruction. Rodent damage to polymers is of importance when polymers are used as barriers or packaging materials impeding access to foodstuffs. Rodents and other members of the animal kingdom have had considerable biocidal research directed toward their control or elimination.

Marine Fouling Organisms

Marine fouling organisms include members of both the plant and the animal kingdoms. Some 2000 species have been reported to cause fouling. Many are unique to marine environments. Representatives include algae, bacteria, fungi, protozoans, barnacles and other arthropods, molluscs, tunicates, hydroids, and annelids. Extensive damage, direct or indirect, to wood pilings, hulls, buoys, other immersed materials and devices, and organic coatings, are a result of the activity of fouling organisms.

SUSCEPTIBILITY OF POLYMERS TO BIOLOGICAL ATTACK

Natural and synthetic polymers, or materials containing them, may be attacked by a variety of species within the entire environmental complex to which they are exposed. The introductory statements briefly presented the obvious and important reasons for direct attack on these substances. The following sections elaborate on these reasons and present a summary of polymer susceptibility to the biological agents of degradation. The discussions of attack by these agents will be divided into attack as nutrients, as attachments, and as barriers.

Nutrients

Extensive and costly damage to polymeric materials results from the attack of organisms in the biological environment in search of nutrients. Damage by insects and insect larvae to the natural polymers of fruits, vegetables, ornamental plants staple crops, and stored grains, as well as damage by rodents and birds in similar areas, is well documented. The most prevalent type of degradation is caused by bacteria and fungi. Many instances of microbial deterioration are encountered in daily experiences with bread mould, cheese fungi, putrid meat, 'athlete's foot,' sour wine or cider, etc., but the extent and variety of attack by micro-organisms is seldom suspected by the layman.

Fermentation, putrefaction, and associated chemical changes are carried on by many species of micro-organisms through catalytic agents called enzymes. The action of enzymes is specific to a substrate (eg., cellulases degrade cellulose, lipases split fats, proteases attack proteins). The susceptibility of a polymer to microbial (or, actually, enzymatic) attack is a function of enzyme availability; co-enzyme presence, if required; availability of a point in the polymer susceptible to enzyme attack; and specificity of enzyme for the type of polymer. The polymer may be protected by altering or blocking the point of attack by the enzyme (eg., acetylation of cellulose, resin impregnation), or the biological agent may be killed or inhibited through the use of biocides.

Attachments

The organisms in a biological environment may adhere to polymers without causing degradation of the substrate. Certain marine fouling organisms attach themselves to ship bottoms and may be considered as doing little direct damage; lichens grow on roofing materials with little effect. However, problems of esthetics, loss of fuel, economy and speed in ships, secondary infestations, or impairment of function may arise. For example, fungal and bacterial slimes may attack papermaker's pulp and cause a physical plugging of the felt which reduces water drainage from paper pulp. Bacterial slimes on ship bottom coatings may initiate the growth of other more degradative agents. In such cases, the use of biocides for protection of the preventative measures is necessary.

Barriers

A polymer may act as a barrier to the organisms in a biological environment in the search for food. Rodents and insects will attempt a penetration through any material to get at food. The producer of packaging materials knows only too well the extensive damage caused by such pests. Crickets have damaged plastic-glass fibre screening. Damage to trees and telephone poles by woodpeckers in search of insects and larval food is also well known. Insect penetration of poly (vinyl chloride) joint seals on clay pipes carrying water has been reported. The use of biocides or repellent chemicals is required here also. Not all polymers will be used or found under circumstances where they are subject to attack by biological agents. Thought and planning are required to determine the potential forces of biological deterioration polymeric materials will encounter in actual use. Preventative treatment should be applied where required, not only to polymers that are sources of microbial nutrients, but also to those functioning as points of attachment or as physical barriers.

BIOLOGICAL DEGRADATION

The following section is a summary of the susceptibility of polymers to biological degradation. Rodent and insect susceptibility is assumed to be general for these materials. Therefore, the main topic covered relates to microbial deterioration except in cases of particular interest.

Natural Polymers

Resins

Dammar, East India, Copals, and miscellaneous resins have no history of microbial degradation. There may be a possibility of superficial fungal growth occurring on these resins without degradative effects.

Lignin

Native lignin, an insoluble polymer of phenolic composition found as a major component of plant cell walls, is somewhat resistant to microbes, although some white-rot fungi and certain soil micro-organisms may attack it. Chemically processed lignin is attacked by bacteria and fungi.

Industrial gums

Plant polysaccharides or their derivatives are referred to as industrial gums. They are usually plant exudates, seaweed extracts, seed extracts, starches, and other naturally occurring polysaccharides.

Natural rubber

Natural rubber is the latex exudate of a group of plants, the most common of which is the rubber tree (*Hevea brasiliensis*). After collection, the latex is subject to microbial attack. Slime formation, darkening, decomposition, and spontaneous coagulation occur partly, if not totally, through bacterial action. Crude rubber is discoloured by microbial pigments, attacked by some actinomycetes, and supports surface mould (*Penicillium glaucum*), developing discoloured, soft, tacky areas. Natural rubber products are completely consumed by soil micro-organisms. Some biocidal protection is given by mercapto-benzothiazole and tetramethylthiuram disulphide, which are used as vulcanisation accelerators. The latter is the more efective biocide when zinc oxide is included in the rubber compounds.

Starch

Starch, amylose, amylopectin, dextrins, and their ethers are degraded by numerous bacteria and fungi, but by few yeasts.

Proteins

Purified native proteins are not substantial nutrients for the growth of bacteria, and are not readily attacked even by strongly proteolytic organisms. A nitrogen source other than native protein must be present for active proteolysis to occur. Few bacteria attack elastin or keratin, although the native keratin of hair and wool is susceptible to attack by certain insects, fungi, streptomycetes, and actinomycetes. Collagen (a native protein not resistant to microbial attack) is subject to hydrolysis. Living bacterial cells and their proteins are not normally hydrolysed by enzymes, although heating seems to stimulate enzyme attack. Denatured proteins, casein, and gelatine are readily attacked as are the peptides (peptones and proteoses).

Leather

Leather is subject to microbial degradation both during processing (tanning) and in service. Both bacteria and fungi are involved. Biocides used include phenylmercuric acetate, coconut amine salt of tetrachlorophenol, sodium *o*-phenylphenoxide, sodium 2,3,4,6-tetrachlorophenoxide, sodium pentachlorophenoxide, *o*-phenylphenol, pentachlorophenol.

Cellulose

Cellulose is the most abundant organic material in nature. It makes up the greater portion of the plant body of wood, seed hair, bast fibre, straw, etc. Cotton contains about 91 per cent cellulose, and wood about 40–50 per cent. Cellulose is found in minute quantities in the animal kingdom. The products manufactured from cellulose and its derivatives are utilised as fibres, textiles, in home and boat construction, in paints as thickening agents, as paper and paper products, and in numerous other ways.

Cellulose and cellulose derivatives are susceptible to attack by biological agents through enzymatic hydrolysis or physical penetration.

Biocides used include copper naphthenate, copper oleate, bis(8-quinolinolato)-copper, zinc naphthenate, 2,2′-dihydroxy-5,5′-dichlorodiphenylmethane, sodium salt of 2-mercaptobenzothiazole, quaternary ammonium carboxylates, coconut amine salt of tetrachlorophenol, sodium *o*-phenylphenoxide, sodium pentachlorophenoxide, chlorinated phenols, sodium 2,4,5-trichlorophenoxide, *o*-phenylphenol. Cotton textiles have been reported to be completely resistant to mildew after impregnation with urea-formaldehyde or melamine-formaldehyde resins.

Synthetic Organic Polymers

Synthetic resins or polymers show an inherent resistance to attack by biological agents. This resistance is a result of their hardness, limited moisture absorption, surface smoothness, the absence of enzymatic systems capable of action on the chemical compositions of some polymers, or the chemical inertness of certain polymers.

Inorganic Polymers

Although many of the phosphonitrilic, siloxane, boron-phosphorus, boron-boron, titanium-nitrogen, and other inorganic polymers and their organic derivatives are not commercially important at the present time, their unusual properties have warranted increasing investigation. Data on biological attack on the inorganic polymers are scanty. The nature of the inorganic bond eliminates the utilisation of this type of polymer for food by the micro-organisms. Microbiological degradation through conventional enzyme systems would not be expected, but a possible attack on the polymer used as a barrier could be anticipated based on the total degradative effect of environmental factors.

Polymer Additives

A product may be highly susceptible to biological attack even though the polymer upon which it is based may itself be inert or only mildly susceptible, since other compounding ingredients may be of nutritive value to pests and form a focal point of growth or attack. Although the early history of polymer degradation related the degree of fungal growth to the type of polymer used in the formulation, actually the plasticiser, lubricant, or filler used was the susceptible moiety. Other examples of susceptible additives are the thickening agents in latex emulsion paints, starch sizings on synthetic fibre textiles, and cellulosic fillers in urea-formaldehyde and other resins.

It must be stated that not all additives render the product or polymer more susceptible to attack by biological agents. Some may increase the resistance of the substance to attack or degradation through an antibiological activity inherent to the additive. For example, a compounded plastic may contain an amount of certain stabilisers sufficient to exert a fungal-resistant effect.

ADDITION OF BIOCIDES TO POLYMERS

Biocides are usually added to the polymers themselves, rather than being incorporated during polymerisation of the monomer. They may be added during processing, such as during mastication of rubber, or during mixing and fusion of poly(vinyl chloride) compounds, or be incorporated by post-addition to finished products such as impregnation of textiles, wood and cordage. Biocides may be applied as undiluted active compounds, as solutions in water or organic solvents, as emulsions or dispersions in volatile solvents, or as dispersions in organic vehicles or binders.

EVALUATION OF BIOCIDES

The ultimate effectiveness of a biocide can be determined only by actual use in full-scale field exposure under a variety of conditions. The record of its performance guides the manufacturer's recommendations for dosage, areas of application, and duration of effectiveness of the biocide. Properties of ideal biocide are shown in Table 33.1.

Table 33.1. Properties of the ideal biocide

A broad spectrum of activity	Stability in use environment
High toxicity to the deleterious biological organisms	Ease of application
	Compatibility with other chemicals (stabilisers, lubricants, antioxidants, etc.)
Low toxicity to the host or neutral biological environment	Permanency, when required
Low toxicity to humans	No negative effects on the use system
Low use cost	No dermatitic manifestations
Stability in storage	

A series of screening tests are applied to a candidate biocide to eliminate those having no activity and to establish a spectrum of effectiveness against the biological agents of interest. The test procedures used during the screening programme are varied and depend on the type of chemical, its potential use, the industry served, and the testing facilities available to the researcher. Many governmental and private agencies, such as the American Society for Testing and Materials, have established test methods to evaluate biocide efficiency. Screening tests follow this basic pattern.

Preliminary Screening (*In Vitro*) Tests

Laboratory tests using high dosage levels or range-finding procedures usually establish an approximate effective dose. These tests are followed by more sensitive or sophisticated methods to establish definitive dose levels. Activity spectra against a variety of biological agents from one or more of the plant or animal species is then developed. A typical preliminary test is the modified paper disk bioassay method.

Secondary Screening (*In Vivo*) Tests

Based on preliminary screening data, the candidate is tested under simulated use conditions. The chemical is applied to a textile, incorporated in a plastic or paint formulation, or subjected to plant-scale tests; functionality under use conditions is established. A typical secondary test is the Soil Burial Method 5762.

Field Exposure

Actual field studies or short-term in-plant studies before or during commercial development yield further information about the condidate biocide.

Table 33.2 lists some of the means of evaluating the suitability and efficacy of biocidal treatment of polymeric materials.

Table 33.2. Some properties measured in evaluation of biocidal treatments of polymers.

Presence of organism growth or damage by organisms estimated by visual observations (microscopic and macroscopic)
Phytotoxicity, either selective or general
Destruction or inhibition of organisms through interference with essential life functions or reproduction odours indicative of spoilage or fermentation
Weight gain or loss by the polymer
Changes in mechanical properties, eg, tensile strength, bursting strength, ultimate elongation, elastic modulus, stiffness, torsional modulus, etc.
Changes in permeability to solvents, plasticisers, moisture, or gases
Changes in electrical properties
Chemical changes in the polymer
Colour changes

Whatever the results of its laboratory tests, a successful biocide must function effectively under the conditions of its actual use, and with the minimum number of negative characteristics, including its cost.

TOXICOLOGY

Biocides are, of course, toxic materials. Ideally, a biocide for use with polymers should have a high degree of toxicity to the organisms responsible for degradation, but should be harmless, or relatively so, to man. In actual practice, many biocides do present varying types and degrees of hazards towards man, and must be handled with due care. The fact that such hazards may exist should not, and does not, preclude the commercial use of these biocides where they are required, or where a less biologically susceptible polymer does not exist for use in a given application. The hazards may exist during the manufacture of the biocide itself, during the process of applying it to a polymer, or during end-use exposure of the treated polymer. They may be minimised by the use of protective clothing and equipment, observation of good habits of cleanliness, judicious choice of dosage, and knowledge of the use to which the polymeric products are to be put. The stigma of toxic hazard does not rest solely on the biocide. Examination of the industrial hygiene literature shows that many chemicals (including the polymers themselves) used in polymer manufacture and processing possess toxicity equal to or greater than that of many biocides. Man may be injured by toxic materials, by contact with skin and mucous membranes, by inhalation, and by ingestion.

It is impractical in a short article to discuss the hazards associated with each individual biocide. It is suggested that when the use of a specific biocide is contemplated, some excellent literature sources be consulted, as well as the supplier of the material in question. Some slight toxic hazard to the consumer of biocidally treated polymeric products may exist, primarily through misapplication or improper choice of a biocide. The greater hazard, however, lies in the handling of biocides during polymer processing. A possible source of exposure by ingestion could be via extraction of biocides by and into foods from packaging materials made from polymers. It is suggested that the section of the Food Additive Regulations pertaining to the intended use, component, article of manufacture, or manufacturing aid be consulted for the prescribed conditions concerning the use of the biocide.

INDUSTRIAL BIOCIDES

In 1959, the damage caused by microbial agents to natural and synthetic polymers was estimated to be in excess of 500 million dollars per year. The bulk of the industrial biocides have been developed to counter fungal and bacterial attack on wood, textiles, and agricultural products (both in the ground and in storage). Many of these materials are useful for preservation of protective coatings, rubber, and other natural and synthetic polymers. Approximately 100 different basic chemicals or combinations are offered for use as biocides by 200 companies under 300 or more different trademarks. Table 33.3 contains a partial list of commercially available biocides.

Table 33. 3. Some industrial biocides

Biocide	*Uses*
l-acetoxymercuric-2-hydroxyethane	Paper-mill slimicide
Acrolein	Paper-mill slimicide
Alkyl benzyldimethylammonium bromides	Textile fungicide
Bis(2-hydroxy-5-chlorophenyl)methane (dichlorophene)	Textile, rope, cordage, and paper fungicide and bactericide
Bis(8-quinolinolato)copper (copper 8-quinolinolate)	Textile, rope, cordage, wood, plastic, paper, adhesive, and paint fungicide
Bis(tri-*n*-butyltin) oxide	Paper, pulp, textile, and plastic preservative
Chloromethoxypropylmereuric chloride	Slimicide, paint emulsion preservative
Creosote (coal tar)	Wood preservative
Cupric oleate (copper oleate)	Textile, rope, and cordage fungicide
3,5-dimetyl-4-chlorophenol (*p*-chloro-*m*-xylenol)	Paint, adhesive, and industrial protein preservative
3,5-dimethyl-4-chlorophenol, derivative of (p-chloro-m-xylenol)	Fungistat and fungicide for vinyl compounds
Dimethyldithiocarbamic acid and 2-mercapto-benzothiazole, sodium and zinc salts	Slime control in pulp paper manufacture, preservativefor cotton fabrics, adhesives
3,5-dimethyl-1,3,5(2H)-tetrahydrothiadiazine 2-thione	Paper-mill slimicide
di(phenylmercuric) dodecenylsuccinate	Paint preservative
p-hydroxybenzoic acid, ester of (paraben)	Vinyl compounds
Laurylpyridinium 5-chloro-mercaptobenzothiazole	Slimicide, textile preservative
2-mercaptobenzothiazole (MBT)	Textile and leather preservative
2-mercaptobenzothiazole, sodium salt of	Adhesives, textiles, paper
Monoethanolammonium salt of 2-mercapto-benzothiazole	Slimicide, textile preservative
Naphthenic acids, copper salt of (copper naphthenate)	Wood, cordage, and textile fungicide
Naphthenic acids, zine salt of (zinc naphthenate)	Wood and textile preservative
Pentachlorophenol	Wood, fibre, textiles, leather cordage, and paint preservative
Phenylmercuric acetate	Water-based paint bactericide and fungicide, used in leather processing
Phenylmercuric dimethyldithiocarbamate	Paper-millslimicide, paper and paperboard fungicide

(Contd ...)

Biocide	*Uses*
Phenylmercuric oleate	Paint preservative
Quaternary ammonium 2-ethylhexoate	Textiles
Quaternary ammonium naphthenate	Vinyl film, vinyl coating, fabric bactericide, fungicide
Salicylanilide	Textiles, varnishes for electrical equipment, leather
Salicylanilide, brominated	Paint and textile fungicide
Sodium *o*-phenylphenoxide (sodium *o*-phenyl plenate)	Textile, leather, adhesive, paint, paper, natural and synthetic gum preservative
Sodium 2,3,4,6-tetrachlorophenoxide	Adhesive, leather, paint, wood and cordage preservative
Tetrachloro-*p*-benzoquinone (chloranil)	Leather
2,3,4,6-tetrachlorophenol	Paint, cordage, wood, leather, textile preservative
Tetrachlorophenol, coconut amine	Adhesives, cordage, textiles, paint, leather preservatice
N-trichloromethylmercapto-4-cyclohexene 1,2-dicarboximide (captan)	Paint, lacquer, paper, plasticiser resins
N-(trichloromethylthio)phthalimide	Paint and vinyl fungicide
2,4,5-trichlorophenol	Paper-mill slimicide, paint, wood, adhesive and textile preservative

CHAPTER 34

Basic Concepts of Rheology

INTRODUCTION

It is anticipated that many readers will have little previous knowledge about rheology but will wish to find out how it can be useful to them in solving practical problems involving the flow of molten elastomers. For this reason, it is our intention to supply sufficient basic information about rheology to enable the reader to understand and make use of the methods described. With this in mind, we begin at the beginning, with a definition of rheology.

Rheology is the science that deals with the way materials deform when forces are applied to them. The term is most commonly applied to the study of liquids and liquid-like materials such as paint, catsup, oil well drilling mud, blood, polymer solutions and molten plastics, i.e., materials that flow, although rheology also includes the study of the deformation of solids such as occurs in metal forming and the stretching of rubber.

The two key words in the above definition of rheology are deformation and force. To learn anything about the rheological properties of a material, we must either measure the deformation resulting from a given force or measure the force required to produce a given deformation. For example, let us say you wish to evaluate the relative merits of several foam rubber pillows. Instinctively, you would squeeze (deform) the various products offered, noting the force required to deform the samples.

A pillow that required a high force to compress would be considered 'hard,' and you probably wouldn't buy it, because it would be painful to sleep on. On the other hand, if it required too little force (too 'soft') it would not provide adequate support for your weary head. Foam rubber is a lightly cross-linked elastomer, and in squeezing it you would be evaluating it primarily on the basis of its resistance to deformation, i.e., its elastic modulus, which is a *rheological property*.

Now let us say that you are looking for a lubricating oil for some household applications, and that your local hardware store has a display in which samples of several oils are contained in glass bottles. You can pick up the bottles and invert them to make the oil flow from one end to the other. The 'light' oils flow very rapidly and splash as they hit the end of the bottle. The 'heavy' oils creep slowly in response to your test and take several seconds to accumulate at the end of the bottle. This is a rheological experiment. You have used the earth's gravitational field to supply the force, and the speed with which the oils flow is a measure of their rates of deformation. Lubricating oils are viscous liquids, and in tipping the bottles you would be evaluating them on the basis of their viscosity, i.e., their resistance to flow, and this is another example of a rheological property.

There are two principal aspects of rheology. One involves the development of quantitative relationships between deformation and force for a material of interest. The information for the development of such a relationship is obtained from experimental measurements. For example, for the foam pillow it might be observed that the force required to compress it a certain distance is proportional to the distance. In the case of the lubricating oil, it might be found that the speed with which it flows through a small hole in the bottom of a can is proportional to the height of the oil remaining in the can.

For a linear elastic material or a Newtonian fluid, such simple observations are sufficient to establish a general equation describing how such a material will respond to any type of deformation. Such an equation is called a 'constitutive equation' or a 'rheological equation of state'. However, for more complex materials such as molten plastics, the development of a constitutive equation is a much more difficult task, which may require the results of many types of experiment.

The second aspect of rheology is the development of relationships that show how rheological behaviour is influenced by the structure and composition of the material and the temperature and pressure. Ideally, one would like to know how these parameters affect the constitutive equation, but this has not been accomplished at the present time, except for very simple materials such as Newtonian fluids. In the case of more complex materials, one can at least develop relationships showing how specific rheological properties such as the viscosity and the relaxation modulus are influenced by molecular structure, composition, temperature and pressure.

Molten plastics are rheologically complex materials that can exhibit both viscous flow and elastic recoil. A truly general constitutive equation has not been developed for these materials, and our present knowledge of their rheological behaviour is largely empirical. This complicates the description and measurement of their rheological properties but makes polymer rheology an interesting and challenging field of study.

IMPORTANCE OF RHEOLOGICAL PROPERTIES

The forces that develop when a lubricant is subjected to a high-speed shearing deformation are obviously of central importance to mechanical engineers. The rheological property of interest in this application is the viscosity. The stiffness of a steel beam used to construct a building is of great importance to civil engineers, and the relevant property here is the modulus of elasticity.

The viscoelastic properties of molten polymers are of importance to plastics engineers, because it is these properties that govern flow behaviour whenever plastics are processed in the molten state. For example, in order to optimise the design of an extruder, the viscosity must be known as a function of temperature and shear rate. In injection moulding, the same information is necessary in order to design the mould in such a way that the melt will completely fill it in every shot. In blow moulding, the processes of parison sag and swell are governed entirely by the rheological properties of the melt.

STRESS AS A MEASURE OF FORCE

It has already been emphasised that force and deformation are the key words in the definition of rheology. In order to describe the rheological behaviour of a material in a quantitative way, i.e., to define rheological material constants (such as the viscosity of a Newtonian fluid) or material functions (such as the relaxation modulus of a rubber), it is necessary to establish clearly defined and quantitative measures of force and deformation. Furthermore, it is necessary to define these measures in such a way that they describe the state of the material of interest without detailed reference to the procedure used to make the rheological measurement.

For example, in the case of the evaluation of the pillows already discussed above, one way to quantify the test results would be to place the pillow on a table, place a flat board on top, and measure the distance between the board and the table both before and after a weight of a certain mass was placed on top of the board, as shown in Fig. 34.1. If our objective is simply to compare several pillows of the same size, it would be sufficient to simply list the amount of compression, in centimetres, caused by a weight having a mass of 1 kg.

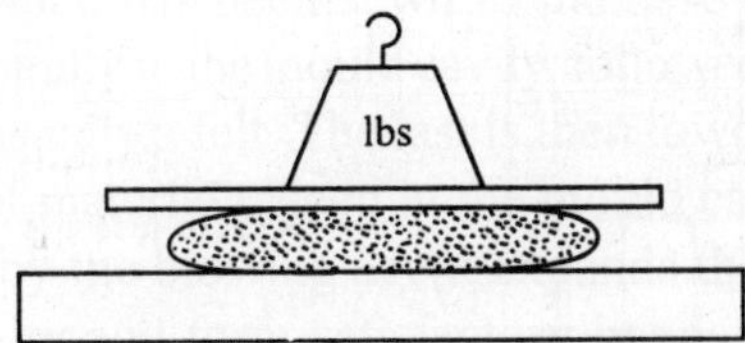

Fig. 34.1. Setup for testing pillows.

However, if our objective is to make a quantitative determination of the elastic properties of the foam rubber, the reporting of the test results is awkward. We must report the size and shape of the sample (the pillow), the mass of the weight applied, and the amount of compression. If one wishes to compare the behaviour of this foam with that of a second foam, the second material must be tested in exactly the same way as the first. It would be advantageous to be able to describe the elastic behaviour of the rubber using physical quantities which are defined so that they describe the state of the material under test, without reference to the details of the test procedure.

First let's look at force. Two types of force can act on a fluid element. A "body force" acts directly on the mass of the element as the result of a force field. Usually only gravity need be considered, but a magnetic field can also generate a body force. A surface force is the result of the contact of a fluid element with a solid wall or with the surrounding fluid elements. It is the surface forces that are of interest in rheology.

The force exerted by a weight sitting on top of a pillow is an example of a surface force. The fact that the ultimate cause of the surface force acting on the pillow is the body force acting on the weight is not of rheological importance, as the compressive force on the sample could equally well be supplied by means of a testing machine, and the observed relationship between force and deformation would be the same.

Placing a weight having a mass of 1 kg on a small pillow will cause more compression than placing it on a large pillow. From the point of view of the material, it is obviously not the total force that is important. In fact, since the deforming force acts on the upper surface of the sample, it is found that if the force is divided by the area of the surface we obtain a quantity suitable for describing the properties of the material. We call this quantity the 'stress'.

In general, then, the stress is calculated by dividing the force by the area over which it acts. In the case of a test like the one with the pillow which involves squeezing, we call this the compressive stress. A more common type of test method for elastic materials involves stretching rather than compression, as shown in Fig. 34.2. Again, the stress is the force divided by the cross-sectional area of the sample, and in this case we call it a "tensile stress". We will use the symbol σ_E for this quantity.

$$\sigma_E = \frac{\text{Stretching force}}{\text{Cross-sectional area}} \qquad \text{... (34.1)}$$

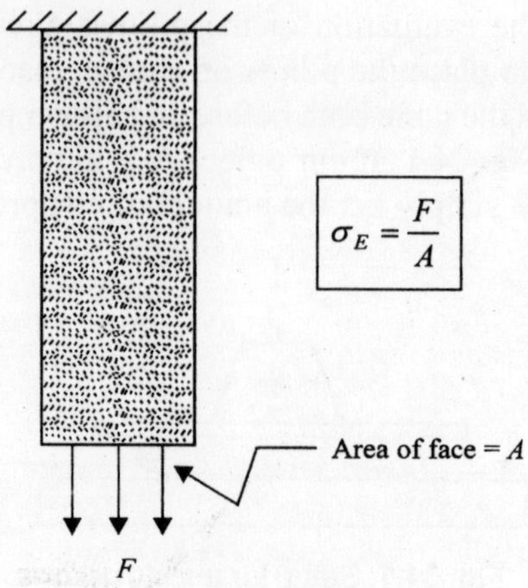

Fig. 34.2. Uniaxial (simple) extension.

Compressive and tensile stresses are the two types of 'normal stress,' so called because the force is normal (perpendicular) to the surface on which it acts. In addition to normal stresses, we can have a "shear stress"; in this case, the direction of the force is tangential to the surface on which it acts, as shown in Fig. 34.3.

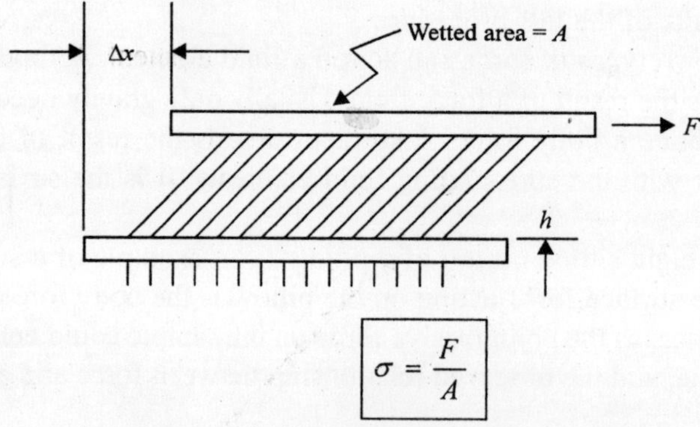

Fig. 34.3. Simple shear (2 plates, gap = *h*).

Fig. 34.3 shows the deformation called "simple shear", in which the sample is contained between two flat plates with a fixed spacing, *h*, between them. The upper plate moves in a direction parallel to itself while the lower plate is stationary. The shear stress is the shear force divided by the tangential area. We will used the symbol σ, with no subscript, to refer to the shear stress in a simple shear deformation.

$$\sigma = \frac{\text{Shear force}}{\text{Tangential area}} \quad \text{... (34.2.)}$$

STRAIN AS A MEASURE FOR DEFORMATION

In the previous section, we stated that shear stresses and normal stresses are useful measures of the forces that act to deform a material. Now we need a quantitative measure of deformation that is rheologically significant. The description of deformation in terms of strain is more complex than the description of force in terms of stress, and there are many alternatives, rheologically significant, measures of strain. We will define here only those measures that are useful in the description of deformations commonly used to make rheological measurements, namely simple shear and simple extension. Later in this section a strain measure for small deformations that is not restricted to describing simple shear or simple extension will be defined. The thing that complicates the definition of a measure of strain is that it is necessary to refer to two states of a material element. In other words, it is not possible to specify the strain of a material element unless we specify at the same time the reference state relative to which the strain is measured. In the case of an elastic material that cannot flow, for example a cross-linked rubber, this is straightforward, because there is a unique, easily identifiable, unstrained state that a material element will always return to when ever deforming stresses are not acting.

For materials that flow, i.e., fluids, such as unique reference state does not exist. In the case of a well-controlled experiment, however, in which a simple homogeneous deformation is imposed on a sample initially at rest and free of all deforming stresses, this initial condition provides a meaningful reference state with respect to which strain can be defined. This can be used to define strain measures for simple extension and simple shear.

Strain Measures for Simple Extension

Consider the simple extension test illustrated in Fig. 34.4. Let L_o be the length of the sample prior to the application of a tensile stress and L the length after deformation has occurred. A simple measure of the deformation is the quantity $(L - L_o)$. However, this quantity is meaningful only in terms of a specific sample, whereas we desire a measure of deformation that describes the state of a material element. We can easily form such a quantity by dividing this length difference by the initial length to obtain the "linear strain" for simple extension.

$$S = (L - L_o)/L_o \qquad \text{... (34.3)}$$

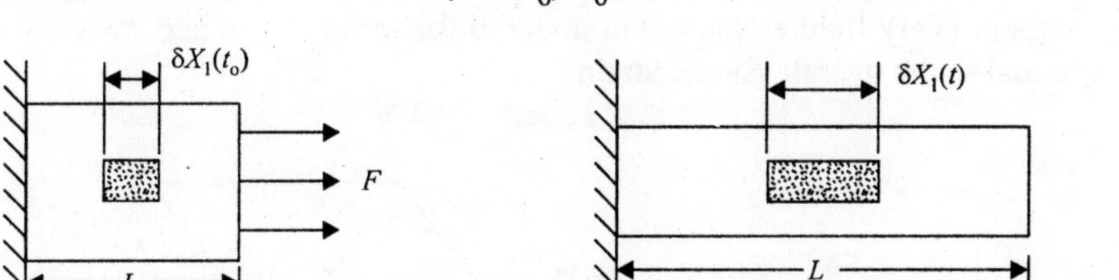

Fig.34.4. Quantities used to describe simple extension.

For a uniform deformation, every material element of the sample experiences this same strain. For example, if the initial length at time t_o of a material element measured in the direction of stretching is $\delta X_1(t_o)$, and the length at a later time, t, after deformation has occurred is $\delta X_1(t)$, the linear strain of the material element is:

$$S = [\delta X_1(t) - \delta X_1(t_o)]/\delta X_1(t_o) \qquad \text{... (34.4)}$$

This measure of deformation has some convenient features. It is independent of sample size, and it is zero in the unstressed, initial state.

However, it is not the only measure of deformation that has these properties. Another is the Hencky strain, which is defined as follows in terms of the length of a material element.

$$\varepsilon = \ln[\delta X_1(t)/\delta X_1(t_o)] \qquad \text{... (34.5)}$$

For a sample with initial length L_o undergoing uniform strain the Hencky strain can also be expressed as:

$$\varepsilon = \ln(L/L_o) \quad \text{... (34.6)}$$

For materials that flow, e.g., molten plastics, this quantity is more useful than the linear strain. In fact, the linear and Hencky strains become equivalent in the limit of very small strains. This can be demonstrated for simple extension by noting that $\varepsilon = \ln(1 + S)$ and that the first term of the series expansion of $\ln(1 + S)$ is S.

The Hencky strain rate is also a useful quantity for describing rheological phenomena in simple extension. This is defined in equation 34.7 in terms of the length, L, of the sample.

$$\varepsilon = d \ln(L)/dt \quad \text{... (34.7)}$$

We note that the initial length does not enter into the Hencky strain rate but does enter into the linear strain rate dS/dt.

Shear Strain

Now consider simple shear, which is the type of deformation most often used to make rheological measurements on fluids. Referring to Fig. 34.3, an obvious choice of a strain measure is the displacement of the moving plate, ΔX, divided by the distance between the plates, h.

$$\gamma = \Delta X/h \quad \text{... (34.8)}$$

Referring to the two material particles shown in Fig. 34.5 rather than to the entire sample we can define the shear strain, γ, for the fluid element located at (x_1, x_2, x_3) as

$$\gamma(x_1, x_2, x_3) = \Delta X_1/\Delta X_2 \quad \text{... (34.9)}$$

where ΔX_1 is the distance, measured in the x_1 direction, between two neighbouring material particles that are separated by a distance ΔX_2 in the x_2 direction. In the absence of edge effects, i.e., for a uniform deformation, every fluid element will undergo the same strain, and the local shear strain will everywhere be equal to the overall sample strain:

$$\gamma(x_1, x_2, x_3) = \Delta X/h \quad \text{...34.10)}$$

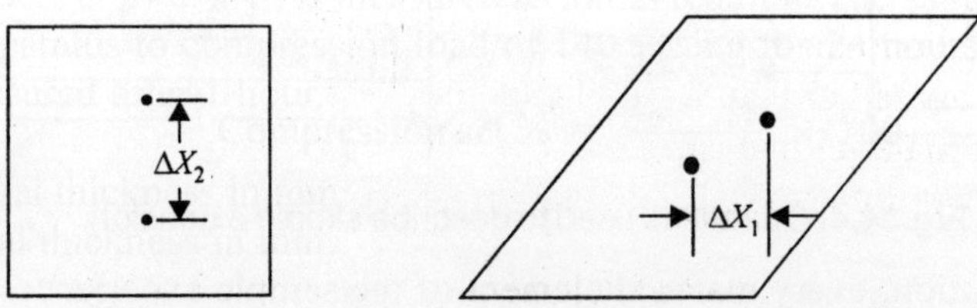

Fig. 34.5. Two material particles in simple shear.

And the shear rate is simply the rate of change of the shear strain with time:

$$\dot{\gamma} = \frac{1}{h}\frac{dX}{dt} = \frac{V}{h} \quad \text{... (34.11)}$$

where V is the velocity of the moving plate.

RHEOLOGICAL PHENOMENA

In this section we will examine the general types of rheological behaviour that can be exhibited by materials. These are elasticity, viscosity, viscoelasticity, structural time dependency, and plasticity. Although we will use simple extension and simple shear behaviour in this section to describe rheological phenomena, it is important to remember that for rheologically complex materials such as polymeric liquids, the behaviour observed in these simple deformations does not tell the whole rheological story.

Elasticity: Hooke's Law

Elasticity is a type of behaviour in which a deformed material returns to its original shape whenever a deforming stress is removed. This implies that a deforming stress is necessary to produce and maintain any deviation in shape from the original (unstressed) shape, i.e., to produce strain. The simplest type of elastic behaviour is that in which the stress required to produce a given amount of deformation is directly proportional to the strain associated with that deformation. For example, in simple extension this can be expressed as:

$$\sigma_E = E\,(L - L_o)/L_o = ES \qquad \text{... (34.12)}$$

The constant of proportionality, E, is called Young's modulus. This relationship is called Hooke's law. The corresponding form of Hooke's law for simple shear is:

$$\sigma = G\gamma \qquad \text{... (34.13)}$$

where G is the shear modulus or modulus of rigidity. We note that in a purely elastic material like this, all the work done to deform the material is stored as elastic energy and can be recovered when the material is permitted to return to its equilibrium configuration.

Another way of describing elastic behaviour is to specify the strain that results from the application of a specific stress. For a Hookean material we have, for simple shear:

$$\gamma = J\sigma \qquad \text{... (34.14)}$$

where J is the shear compliance. Obviously, for a material following Hooke's law:

$$J = 1/G \qquad \text{... (34.15)}$$

Viscosity

Viscosity is a property of a material that involves resistance to continuous deformation. Unlike elasticity, the stress is not related to the amount of deformation but to the rate of deformation. Thus it is a property peculiar to materials that flow rather than to solid materials. We will consider first the simplest type of rheological behaviour for a material that can flow. For simple shear this type of behaviour is described by a linear relationship between the shear stress and the shear rate:

$$\sigma = \eta\dot{\gamma} \qquad \text{... (34.16)}$$

where η is the viscosity. A material that behaves in this way is called a Newtonian fluid.

For a Newtonian fluid, the viscosity is a "material constant", in that it does not depend on the rate or amount of strain. Single phase liquids containing only low molecular weight compounds are Newtonian for all practical purposes. For multiphase systems, for example suspensions and emulsions, and for polymeric liquids, the relationship between stress and strain rate is no longer linear and cannot be described in terms of a single constant. It is still convenient, however, to present the results of a steady simple shear experiment in terms of a viscosity function $\eta(\dot{\gamma})$ defined as follows:

$$\eta(\dot{\gamma}) \equiv \sigma / \dot{\gamma} \qquad \text{... (34.17)}$$

where σ is the shear stress and $\dot{\gamma}$ is the shear rate. A typical viscosity-shear rate curve for a molten polymer is shown in Fig. 34.6. The important features of this curve are listed below.

1. At sufficiently low shear rates, the viscosity approaches a limiting constant value η_o called the zero shear viscosity.
2. The viscosity decreases monotonically as the shear rate is increased. This type of behaviour is called shear thinning. (An older terminology is "pseudoplastic").
3. At sufficiently high shear rates the viscosity might be expected to level off again, although a high-shear-rate limiting value is not observed in melts, because viscous heating and polymer degradation usually make it impossible to carry out experiments at sufficiently high shear rates.

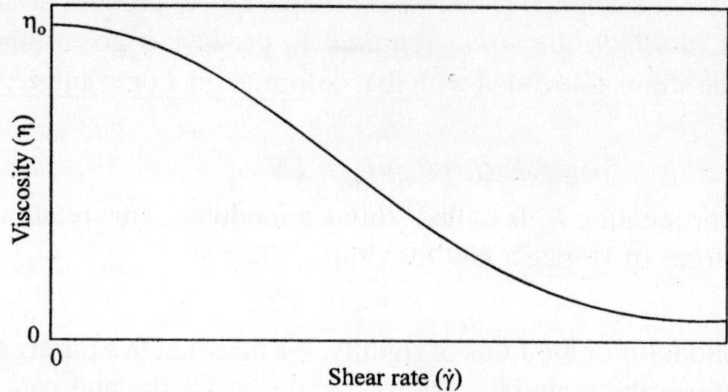

Fig. 34.6. Viscosity-shear rate curve for a typical molten polymer.

Viscoelasticity

Polymeric materials, including solutions, melts, and cross-linked elastomers, exhibit both viscous resistance to deformation and elasticity. In the case of a vulcanised (cross-linked) rubber, flow is not possible, and the material has a unique configuration that it will return to in the absence of deformating stresses. However, the viscous resistance to deformation makes itself felt by delaying the response of the rubber to a change in stress. To illustrate this point, consider the phenomenological analog of a viscoelastic rubber shown in Fig. 34.7. This mechanical assembly consists of a spring in parallel with a dashpot.

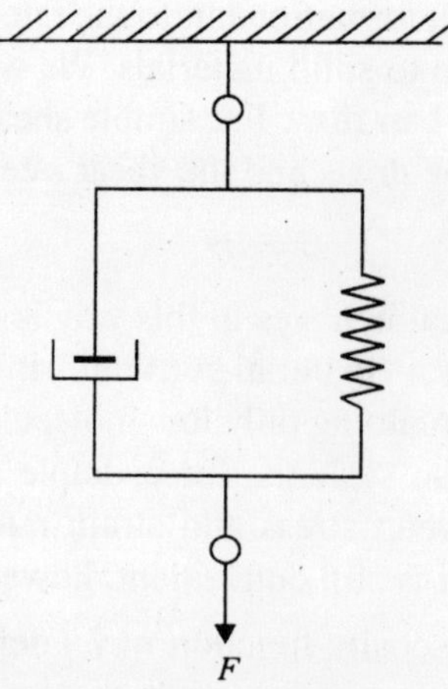

Fig. 34.7. Voigt body analog of a viscoelastic solid.

The force in the spring is assumed to be proportional to its elongation, and the force in the dashpot is assumed to be proportional to its rate of elongation. Thus, the spring is a linear elastic element, in which the force is proportional to the extension, X, and the dashpot is a linear viscous element in which the force is proportion to the rate of change of X. Note that this assembly will always return to a unique length, the rest length of the spring, when no force is acting on it. This assembly, called a Voigt body, is not intended to be a physical or quantitative model for a rubber.

However, the qualitative characteristics of its response to changes in force are similar in some ways to those exhibited by rubbers.

Consider how this assembly would respond to the sudden application of a tensile load, F. This is called a "creep test". The force, F, is the sum of the force in the spring, $K_e X$, and that in the dashpot, $K_v(dX/dt)$. Thus.

$$F = K_e X + K_v(dX/dt) \quad \text{... (34.18)}$$

We note that some of the work put into the assembly to deform it is dissipated in the dashpot, while the remainder is stored elastically in the spring. If X is initially zero, and the force F is suddenly applied at time $t = 0$, this differential equation can be solved to yield:

$$X(t) = F/K_e\ [1 - \exp(-K_e t/K_v)] \quad \text{... (34.19)}$$

The important point to note is that the viscous resistance to elongation introduces a time dependency into the response of the assembly, and that this time dependency is governed by the ratio (K_v/K_e), which has units of time. If we take the force to be analogous to the deforming stress in a viscoelastic material, and the elongation to be analogous to strain, we see that a viscoelastic rubber has time constant and cannot respond instantaneously to changes in stress. This is called a "retarded" elastic response. As the time constant approaches zero, the behaviour becomes purely elastic.

Now we turn to the case of an elastic liquid. To illustrate certain qualitative features of the rheological behaviour of such a material, consider the mechanical assembly shown in Fig. 34.8. This assembly, called a Maxwell element, consists of a linear spring in series with a linear dashpot. Note first that unlike the Voigt body, this assembly has no unique reference length and will deform indefinitely under the influence of an applied force, assuming the dashpot is infinite in length. This is analogous to the behaviour of an uncross-linked polymeric material above its glass transition and melting temperatures. Such a material will flow indefinitely when subjected to deforming stresses.

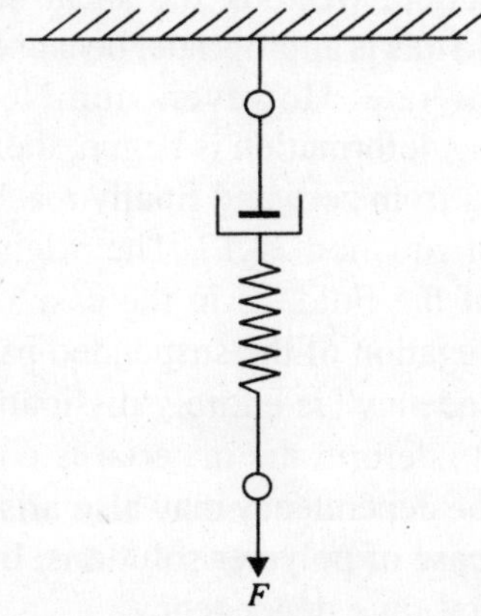

Fig. 34.8. Maxwell element analog of a viscoelastic liquid.

Now we examine the force on the Maxwell element when it is subjected to a sudden stretching by an amount X_o. The force, but not the displacement, is the same in both the spring and dashpot. Thus:

$$F = K_e X_e = K_v (dX_v/dt) \qquad \text{... (34.20)}$$

Again we note that some of the work done is dissipated in the dashpot and the remainder is stored in the spring. The total displacement of the assembly, X_o, is the sum of X_e and X_v:

$$X_e + X_v = X_o \qquad \text{... (34.21)}$$

Thus:

$$F(t) = K_e (X_o - X_v) = K_v (dX_v/dt) \qquad \text{... (34.22)}$$

This ordinary differential equation can be solved to yield

$$F(t) = K_e X_o [\exp(-K_e t/K_v)] \qquad \text{... (34.23)}$$

Note that (K_v/K_e) is a time constant. The force thus decays or relaxes exponentially. If we take the force to be analogous to the deforming stress in an elastic liquid and the elongation to be analogous to the strain, this process is analogous to a stress relaxation experiment. As in the case of the viscoelastic rubber, we note that the combination of viscous and elastic properties endows the material with a characteristic time and makes its response time dependent. As this "relaxation time" becomes shorter and shorter, however, it becomes more and more difficult to devise an experiment that will reveal the elastic nature of the liquid, and its behaviour appears more and more like that of a purely viscous material.

When we examine the rheological behaviour of actual polymeric materials, we find creep and relaxation behaviour that is qualitatively like those described above. In particular, the response to a sudden change in stress or strain is always time dependent, never instantaneous, and there is both elastic storage of energy and viscous dissipation. On the other hand, the creep and relaxation curves cannot be described by a single exponential function involving a single characteristic time. However, practical use can still be made of the concept of a relaxation time by describing the viscoelastic behaviour of real materials in terms of a spectrum of relaxation times.

Structural Time Dependency

In our discussion of the viscosity function, we took the shear stress to be independent of time at constant shear rate. For a Newtonian fluid this is appropriate, because the stress responds instantaneously to the imposition of a constant shear rate. However, non-Newtonian fluids may not respond instantaneously so that when the shearing deformation is begun, there is a transient period during which the shear stress varies with time, starting from zero and finally reaching a steady state value that can be used to calculate the viscosity by use of equation 34.17. The origin of this time dependency may be a flow-induced change in the structure of the fluid, as in the case of concentrated suspension of solid particles. For example, the state of aggregation of the suspended particles can be changed significantly by shearing. This "structural time dependency" is entirely dissipative, that is to say there is no elastic energy storage, and all the work done to deform the material is converted immediately into 'heat,' or more precisely, into internal energy. Time dependency may also arise from changes in the conformation and orientation of molecules, as in the case of polymer solutions, but this type of time dependency is a viscoelastic effect rather than a structural time dependency.

Restricting attention here to inelastic materials such as suspensions of rigid particles in Newtonian fluids, in which no elastic energy storage is possible, the most common type of structural time dependency is that shown in Fig. 34.9, where the shear stress decreases with time after the start up of steady shear flow. This type of behaviour, in which the viscosity decreases with time, is called 'thixotropy'. If the shearing is stopped, the viscosity will gradually increase as the particles respond to inter-particle forces in the absence of a shear stress, but the time required for the complete reformation of the equilibrium structure can be very long. There are some materials, such as corn starch suspensions, that exhibit the opposite effect, viz., the stress increases with time after the start up of steady shear. This type of behaviour is called 'rheopexy' or 'antithixotropy'. In the case of thixotropy, the shear tends to break down structure that has been established in the unsheared material, while in the case of a rheopectic material, the shearing promotes the buildup of structure.

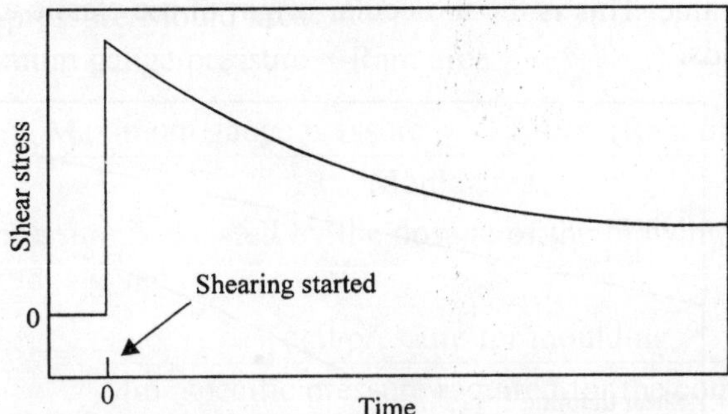

Fig. 34.9. Shear stress versus time for a material with structural time dependency.

Plasticity and Yield Stress

The structure in a concentrated suspension can be sufficiently rigid that it permits the material to withstand a certain level of deforming stress without flow is called the 'yield stress', and this type of behaviour is called 'plasticity'. Metals generally exhibit plasticity, as do semi-crystalline polymers at temperatures between their melting and glass transition temperatures. Highly filled melts are also thought to have a yield stress, although precise measurement of this property is difficult. The simplest type of plastic behaviour is that in which the excess stress, above the yield stress, is proportional to the shear rate. For simple shear flow, this type of behaviour is described by equation 34.24.

$$\sigma = \sigma_o + \eta_p \dot{\gamma} \qquad (\sigma \geq \sigma_o) \qquad \text{... (34.24)}$$

Here, σ_o is the yield stress and η_p is the plastic viscosity. A material that behaves in this way is called a 'Bingham plastic'. This is an idealised type of behaviour that is not precisely followed by any real material, but it is sometimes a useful approximation to real behaviour. Note that Equation 34.24 is not a constitutive equation, as it does not describe all the components of the stress tensor in any type of a deformation but only the shear stress in simple shear. Fig. 34.10 compares shear stress versus shear rate curves for a Bingham plastic, a Newtonian fluid and shear thinning fluid.

POLYMERIC LIQUIDS ARE NON-NEWTONIAN

It is important in applied polymer science to be able to relate physical properties, including rheological properties, to molecular structure. We will mention here only the general mechanisms by which polymeric molecules endow liquids with complex rheological behaviour.

First, we should examine the question of why polymer molecules are elastic. Our physical picture of a polymer molecule is that of a long chain with many joints allowing relative rotation of adjacent links. The presence of this large number of joints makes the molecule quite flexible and allows many different configurations of the molecule. At temperatures above the glass transition temperature a molecule will continually change its configuration due to Brownian motion, but we can describe the state of a large number of molecules in terms of statistical averages. For example, at a given temperature there will be unique average value of the end-to-end distance, *R*, for the molecules of a polymeric liquid that has been at rest for a sufficient length of time that it is in its equilibrium state. Deforming the liquid will alter this average length, but if the deformation is stopped, Brownian motion will tend to return the average value of *R* to its equilibrium value. This is the molecular origin of the elastic and relaxation phenomena that occur in polymeric liquids.

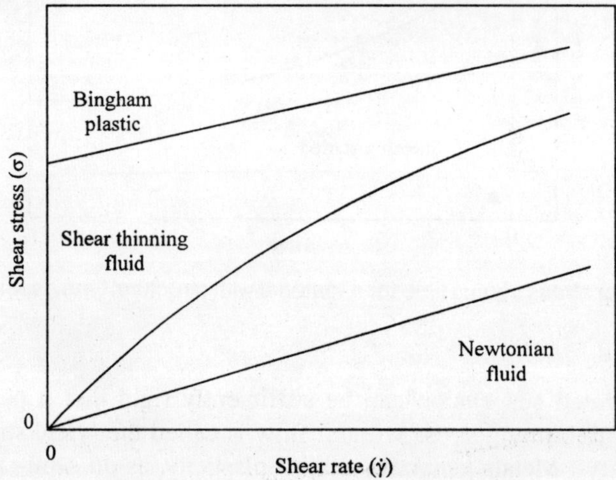

Fig. 34.10. Shear stress versus shear rate for a Bingham plastic, a Newtonian fluid, and a shear thinning fluid.

Polymer Solutions

We consider first the behaviour of a dilute solution in which the forces acting on the polymer molecule are primarily those due to the flow of the solvent. This situation is much simpler than that existing in concentrated solutions and melts, where the rheological behaviours is governed by interactions between polymer molecules. In the absence of a velocity gradient in the solvent, the polymer molecules will have an average end-to-end distance that is small, and a distribution of orientations that is random. When the solution undergoes deformation, dissolved molecule will be dragged along with it but the polymer molecule is so large that different parts of it will be exposed to different solvent velocities. In other words, the solvent moves relative to the segments of the polymer molecule and exerts forces on these segments that change the shape (conformation) of the molecule, increasing the average end-to-end distance and promoting some degree of orientation in a preferred direction. For example, in simple extension, the preferred direction is the stretch direction.

Some types of deformation have a greater ability to increase *R* and orient molecules thus others. Deformations that can generate a high degree of stretching and orientation are said to be ‘strong flows’, while those that cannot are said to be ‘weak’.

We have seen how the deformation of a polymeric liquid can alter the configuration and orientation of the molecules. Thus, the physical nature of the fluid is altered by the deformation, and this, in turn, alters the mechanical (rheological) properties of the fluid. This explains, for example, why the viscosity of a polymer solution depends to some extent on shear rate. At high shear rates, the shape of the polymer molecules is different from that at low shear rates, and this alters the solution's resistance to flow and thus its viscosity. Furthermore, the partial orientation of the molecules, i.e., the shift of the average orientation of the R vectors away from zero, introduces a strain-induced anisotropy that is responsible for differences in the normal stresses in steady simple shear. Finally, the tendency of Brownian motion to return the system of molecules to its most probable, i.e., equilibrium, configuration distribution explains why these fluids are elastic and have relaxation times.

Molten Plastics

In the case of concentrated polymeric liquids, including melts, the transmission of deformation stresses to a molecule is primarily (in the case of a concentrated solution) or totally (in the case of a melt) due to interactions between polymer molecules rather than interactions between the solvent and the molecules. The detailed nature of these interactions is not completely understood at this time, but it is observed that the short term response to rapid deformations of high molecular weight molten polymers is very similar to that of a cross-linked rubber. This has inspired the concept of a 'temporary network' that exists in the melt and acts like a rubbery network at shorter times but whose 'junctions' can slip over longer periods of time to permit flow. The network is sometimes said to arise from 'entanglements' in the melt. However, the modern view is that the rubbery behaviour of melts is due not to an actual looping or knotting of molecules around each other but simply to the constraints on their motion resulting from the fact that molecules cannot cut through each other.

Entanglements occur because of the high degree of spatial overlap of the molecules. The existence of overlap is readily demonstrated by considering the measured size of the polymer coils. One measure of molecular size is the 'radius of gyration', R_g. For linear polyethylene, R_g depends on the molecular weight, M as follows:

$$R_g \text{ (cm)} = 4 \times 10^{-9} \times M^{1/2}$$

For a polyethylene with a molecular weight of 10^6 g/mol, the volume of the sphere occupied by one molecule is therefore about 2.6×10^{-16} cm^3. The mass of this coil is 10^6 divided by Avogadro's number, or 1.7×10^{-18} g. The density of the coil in its occupied volume is thus less than 0.01 g/cm^3. The observed melt density of about 0.77 g/cm^3 can only be accounted for if parts of many other coils are present in the volume occupied by the coil.

A similar calculation for a polymer with a molecular weight of 10^4 g/mol gives a density of 0.1 g/cm^3, which is considerably closer to the measured bulk density. This shows that the degree of coil overlap, and therefore the entanglement density increases sharply with molecular weight.

Rubbery behaviour occurs in a melt when the molecular weight is above some critical value that arise from one polymer to another. Above this molecular weigh the number of entanglements becomes sufficient to produce strong rubber like effects.

The macroscopic effects of the strong interactions between polymer molecules in a melt include high viscosity and high elastic recoil, especially just above the melting point or, in the case of an amorphous polymer, just above the glass transition temperature. At the same time, the nature of this strong interaction can be altered temporarily by deformation so that high molecular weight melts have highly non-linear properties; for example, the viscosity is a very strong function of the shear rate.

Polymeric materials are said to have a 'memory', in that when deformation stresses are eliminated, they tend to return to a previous configuration. Cross-linked polymers have a 'perfect memory' in the sense that since their network is based on permanent chemical cross-links they always return to a unique equilibrium configuration, whereas molten polymers are said to have a 'fading memory', since the entanglement network is not permanent and is altered by flow and relaxation processes.

A WORD ABOUT TENSORS

For those readers who have had little if any experience in the use of tensor notation, the very word 'tensor' probably suggests a mathematical system of impenetrable mystery. However, such readers should have no fear. There is no mystery! While we do not claim to offer here a complete course in tensor analysis, we do present in the next two brief sections everything you will need to know about tensors in order to describe the rheological properties of polymeric liquids. After a careful reading of these sections, you too can impress the uninitiated with your ability to use tensor notation to describe rheological phenomena.

The concept of a tensor was introduced into physics, and thus into rheology, because it is useful; without it, the quantitative description of many physical phenomena would be hopelessly clumsy and tedious. Because of this usefulness, most of the literature on viscoelastic behaviour makes some use of tensor notation. This literature will be inaccessible to a reader having no familiarity with tensor quantities.

We have already stated that rheology involves the relationship between deformation (strain) and force (stress) for a material. It is in the quantitative description of the quantities strain and stress that tensor notation is virtually indispensable. However, before demonstrating this, it will be useful to review briefly the concept of a vector as this is central to an understanding of tensors.

Vectors

Certain physical quantities, such as force and velocity, are best specified in terms of vectors, because a vector has a magnitude and a direction. One example is the velocity vector, v. A vector can be specified by giving its components, v_1, v_2, and v_3, referring to the velocities in the three directions, x_1, x_2, and x_3. Generally, we can refer to a typical velocity component as v_i, where i can be 1, 2, or 3.

Note that while the magnitude is a physical attribute of a vector that does not depend on the choice of a particular co-ordinate system, the components of the vector do depend on the co-ordinate system selected to describe the flow. There is a simple rule that tells how to use the components of a vector in one co-ordinate system to calculate the components of that vector in second co-ordinate system that is rotate with respect to the first.

If vectors are adequate to describe the velocity of a body and the force acting on it, why are they not sufficient for describing rheological phenomena? The answer is that rheology deals not with motion *per se*, but with deformation, and specifically with the relationship between the deformation of a fluid element and the surface forces exerted in this element by the surrounding fluid. Tensors are very useful in specifying these two types of quantities, and the specific tensors that are used to represent these quantities are the strain tensor and the stress tensor.

What is a Tensor?

Like a vector, a tensor can be represented in terms of its components, and the values of these components depend on the choice of the co-ordinate system used. Furthermore, there is a rule for using the components of a tensor in one co-ordinate system to calculate the components of the tensor in another co-ordinate system, rotated with respect to the first. The existence of this rule shows that a tensor has a basic physical significance that transcends the arbitrary choice of the co-ordinate system. However, unlike a vector, the physical significance of a tensor cannot be described in terms of a directed line segment, i.e., in terms of a magnitude and a direction. In describing rheological behaviour associated with a particular type of deformation, we generally select a co-ordinate system that gives the components a physical significance that is easily understood. For example, in describing the stretching of a rod ("uniaxial extension") we take the x_1 direction to be the direction of stretching. Then the σ_{11} component of the stress tensor is simply the tensile stress in the sample.

Whereas a vector has three components, the tensors we will use have nine. A typical component can be written using two indices, for example σ_{ij}, where each index can take on one of the values 1, 2, or 3 corresponding to the three co-ordinate directions. To present the values of all the components of a tensor, matrix notation can be used.

$$\sigma_{ij} = \begin{bmatrix} \sigma_{11} & \sigma_{12} & \sigma_{13} \\ \sigma_{21} & \sigma_{22} & \sigma_{23} \\ \sigma_{31} & \sigma_{32} & \sigma_{33} \end{bmatrix} \qquad \text{... (34.25)}$$

Can any nine numbers form the components of a tensor? No. These numbers have a specific mathematical significance, which we will find particularly suited to the description of deformation and stress. Specifically, these nine components contain all the information necessary to transform one vector into another one that has a certain prescribed relationship with the first. In mathematical language we say that the tensor 'operates on' one vector to yield a second vector, which contains information taken from both the original vector and the tensor. For example, we will see that the strain tensor, i.e., the nine components of the strain tensor, can be used to operate on the components of the vector describing the relative position of fluid particles within an undeformed fluid element, to yield the corresponding position vectors after deformation. Likewise, the stress tensor can be used to operate on the unit normal vector defining the orientation of a surface of a fluid element to yield the surface force vector acting on that element. Since the vector operate on in both cases is an arbitrarily selected one, we see that the strain tensor actually contains a complete description of the deformation that a fluid element undergoes during some flow process, while the stress tensor contains a complete description of the state of stress acting at a point in the fluid at a particular time. The objective here is not to solve flow problems but only to describe rheological phenomena. Thus, tensor calculus will not be required, and the reader need learn no new mathematics but only the definitions of a few quantities.

With regard to notation, we will use a bold face symbol to indicate that it is a vector. For example, the velocity vector will be represented as v. The components of a vector will be indicated by means of a subscript, for example, v_i. For the components of a tensor we will use two subscripts. For example, the components of the stress tensor will be represented by σ_{ij}. In order to minimise the number of new symbols and rules that need be learned, we will not use dyadic notation or the Einstein summation convention. These are methods of notation that simplify the writing of equations involving tensors, and are described here.

THE STRESS TENSOR

The deformations that occur in the processing and use of materials are generally more complicated than simple extension and simple shear and involve a combination of these two types of deformation. For example, consider the deflection of a rubber tyre under load or the flow of a molten plastic into a mould. First, the deformation is not uniform but varies from one place to another within the material. It thus becomes a 'field variable', i.e., a quantity that varies from one point to another and is thus a function of position. Secondly, the stress is not purely tensile, compressive or shear.

The quantitative specification of the forces acting on a solid body as a result of contact with another body is straightforward; one need only give the components of the force vector acting at the interface. However, the specification of the forces acting on the surface of a fluid element is less obvious, since the orientation of the surface is arbitrary, i.e., it depends on how one defines a fluid element. It would appear that in order to completely specify the state of stress on a fluid element one would have to give the components of the stress vector for every possible orientation of the surface. Fortunately, this is not the case, and we will see that by specifying only the components of the stress tensor, the state of stress at a point in a fluid can be completely described.

For a given arbitrary, choice of a fluid element, the surface stress vector is $t^{(n)}$, where n is the unit normal vector for the surface and specifies its orientation. The (n) superscript on the surface stress vector indicates that the components of the vector depend on the orientation of the surface. This vector does not at first seem to be a useful tool for describing the state of stress at a point in a fluid because of the arbitrariness of the choice of the surface orientation.

It is possible to show, however, that if the stress vectors acting on each of three mutually perpendicular planes passing through a point in a fluid are specified, the stress vectors for any other choice of planes can be calculated by means of a simple transformation rule. It is convenient to let these planes be perpendicular to the co-ordinate axes. Thus, the unit normal vector for a surface becomes equal to one of the unit normal vectors for the co-ordinate system:

$$n = e_i$$

and the components of the force vector are given by:

$$t_j^{(e_i)} = \sigma_{ij}$$

where σ_{ij} is the stress tensor.

To understand the physical significance of the nine components of the stress tensor, consider the small cubical element of material shown in Fig. 34.11. The second subscript indicates the direction of the force and corresponds to the co-ordinate axis direction. For example, the stress component shown in Fig. 34.11 acts in the x_1 direction, and the second subscript of this component is thus1. The first subscript indicates the face on which the component acts, and this is specified by reference to the co-ordinate direction normal to this face. Thus, the force shown acts on the face normal to the x_2 direction, and the first subscript of this component of the stress is thus2. The stress component shown is thus σ_{21}.

To complete our definition of the components of σ_{ij}, we need a sign convention. Here, we will use the convention generally used in mechanics, although the reader should be aware that the opposite convention is used by some rheologists. We will take the stress to be positive when it acts in the positive x_j direction, on a face having the higher value of x_i, i.e., the face further from the origin in the x_i direction.

For example, the stress component shown in Fig. 34.11 is positive if the force acts in the direction of the arrow. This is because it acts in the positive x_i direction on a face having the higher value of x_2. Fig. 34.12 shows several additional components of the stress.

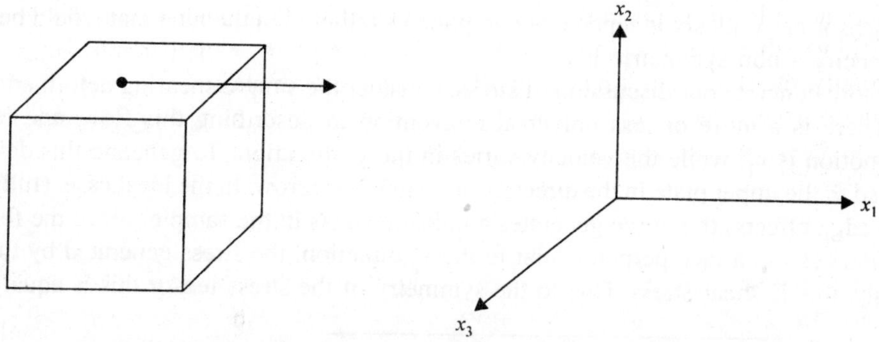

Fig. 34.11. Cubical material element with a typical stress component.

The set of nine components that is needed to specify completely the state of stress at a point in a deformable material is an example of a 'second order tensor', and the members of the set are said to be the 'components' of the tensor.

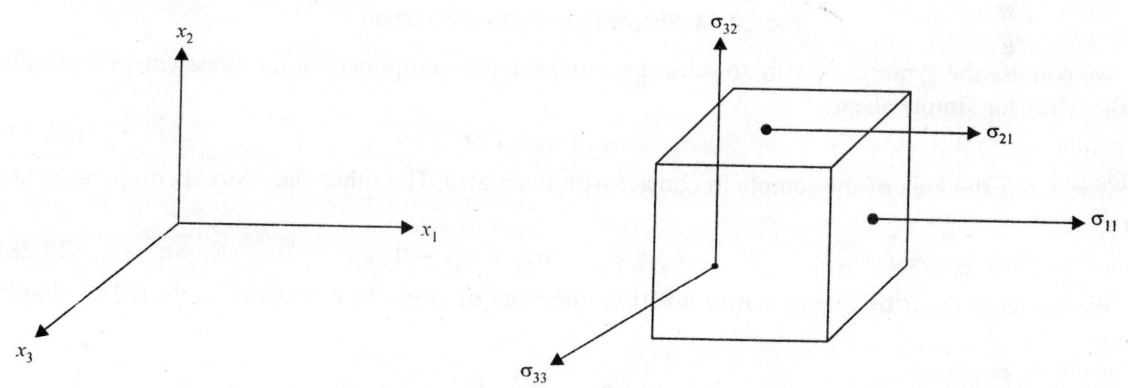

Fig. 34.12. Several additional stress components.

Since the components of the stress tensor describe the state of stress at a point in the material, the cubical element shown in Fig. 34.11 must be shrunk to an infinitesimal size. Thus, the two force vectors shown in Fig. 34.13 are acting in opposite directions at the same point. From Newton's law of action and reaction, these two forces must be equal in magnitude. They are thus both manifestations of the same component of the stress tensor, σ_{11}, and both have positive values if they act in the directions indicated. Thus, according to our sign convention, a tensile stress has a positive value.

Fig. 34.13. Equal; opposite forces at a point are represented by the same component of the stress tensor.

The principle of conservation of angular momentum can be applied to the infinitesimal material element we have been considering to show that the stress tensor has the following property:

$$\sigma_{ij} = \sigma_{ji} \qquad \text{... (34.26)}$$

Thus, any two components that have the same subscripts or indexes, but in reversed order, have the same value. A tensor that has this property is said to be 'symmetric'. One result of this property is that a symmetric tensor has only six independent components rather than the nine that would be required to completely specify a non-symmetric tensor.

To make more concrete our discussion of stress, consider the simple shearing deformation shown in Fig. 34.14. There is a more or less universal convention in describing this flow, and it is that the direction of motion is x_1, while the velocity varies in the x_2 direction. To generate this deformation, a force is applied to the upper plate in the direction shown by the arrow. In the ideal case, (fully developed flow with no edge effects) this force generates a uniform stress in the sample. Since the force is in the x_1 direction and acts on a face perpendicular to the x_2 direction, the stress generated by the force F is σ_{21}. Obviously, this is shear stress. Due to the symmetry of the stress tensor this is equal to σ_{12}.

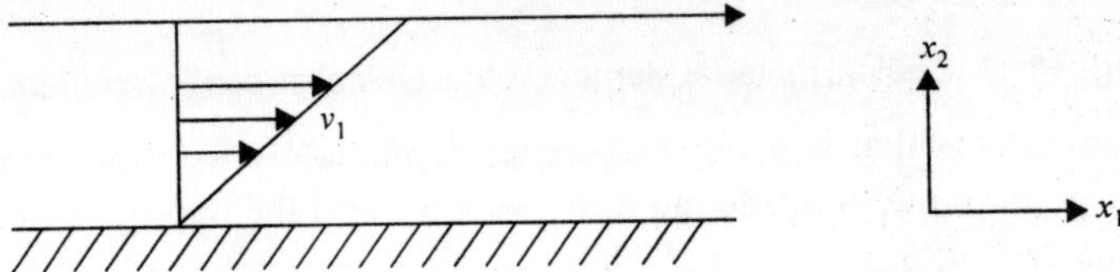

Fig. 34.14. Simple shear index convention.

We will use the symbol, σ, with no subscripts, to mean this component of the stress tensor in simple shear. Thus for simple shear,

$$\sigma_{21} = \sigma_{12} \equiv \sigma = F/A \quad \text{... (34.27)}$$

where A is the area of the sample in contact with the plates. The other shear stress components are zero:

$$\sigma_{13} = \sigma_{31} = \sigma_{23} = \sigma_{32} = 0 \quad \text{... (34.28)}$$

We can now describe, using matrix notation, the state of stress in a material subjected to simple shear:

$$\sigma_{ij} = \begin{bmatrix} \sigma_{11} & \sigma & 0 \\ \sigma & \sigma_{22} & 0 \\ 0 & 0 & \sigma_{33} \end{bmatrix} \quad \text{... (34.29)}$$

Another example of a test that is of practical interest in rheology is simple or uniaxial extension. This test is illustrated in Fig. 34.2. If we let x_1 be the direction of the applied force, the stress component resulting from this force will be σ_{11}, which is a normal stress. If it acts in the direction shown, it is a tensile stress. There are no shear stress components in this case, and the components of the stress tensor are as shown below:

$$\sigma_{ij} = \begin{bmatrix} \sigma_{11} & 0 & 0 \\ 0 & \sigma_{22} & 0 \\ 0 & 0 & \sigma_{33} \end{bmatrix} \quad \text{... (34.30)}$$

There is an additional point regarding normal stresses that should be mentioned here. While all materials are compressible to some extent, in the case of molten plastics, quite high pressures are required to produce a significant change in the volume of a sample. For this reason, for many purposes these materials can be considered to be incompressible. Now consider what happens when we subject an incompressible material to a compressive or tensile stress that is equal in all directions, i.e., an isotropic stress or 'pressure'.

The components of the stress tensor in this situation are as shown below:

$$\sigma_{ij} = \begin{bmatrix} -P & 0 & 0 \\ 0 & -P & 0 \\ 0 & 0 & -P \end{bmatrix} \qquad \text{... (34.31)}$$

A stress of this type is said to be 'isotropic'. The minus signs results from the fact that pressure is considered positive when it acts to compress a material, whereas according to our sign convention compressive stress is a negative quantity. In this situation, the sample will be totally unaffected by the forces associated with this pressure, i.e., it will not change its size or shape. Thus, such an isotropic stress field is of no rheological significance. Only when there are shear stresses acting, as in simple shear flow, or when the normal stress components are different from each other will deformation occur in an incompressible material. Another way of saying this is that if a rheological measurement on an incompressible material is repeated at several different ambient pressures, for example by placing the rheometer in a hyperbaric chamber, the measurements at various pressures will yield exactly the same values of all rheological properties.

This means that for an incompressible material a normal component of stress has no absolute rheological significance. Only differences between two normal components are of rheological significance. For example, in the case of simple shear, it is customary to describe the state of stress from a rheological point of view by specifying the shear stress, σ, and the 'first and second normal stress differences'.

$$N_1 \equiv \sigma_{11} - \sigma_{22} \qquad \text{... (34.32)}$$

$$N_2 \equiv \sigma_{22} - \sigma_{33} \qquad \text{... (34.33)}$$

For Newtonian fluids these two quantities are zero in simple shear, but in polymeric liquids they generally have nonzero values. One manifestation of the first normal stress difference is observed when a liquid is sheared by placing it between two flat parallel disks and rotating one of the disks. It is found that an elastic liquid exerts a normal thrust tending to separate the plates, while a Newtonian fluid exerts no normal thrust on the plates.

In the case of simple extension, there is only one rheologically significant feature of the stress field, because there are no shear stresses acting, and there is axial symmetry, so that $\sigma_{22} = \sigma_{33}$. This is the principal stretching stress, σ_E, defined in equation 34.34.

$$\sigma_E = \sigma_{11} - \sigma_{22} = \sigma_{11} - \sigma_{33} \qquad \text{... (34.34)}$$

Because the magnitude of a single normal stress component is not rheologically by itself, it is customary in rheology to define an 'extra' or 'viscous' stress as follows:

$$\tau_{ij} \equiv \sigma_{ij} + P \; (i = j) \qquad \text{... (34.35a)}$$

$$\tau_{ij} \equiv \sigma_{ij} \qquad (i \neq j) \qquad \text{... (34.35b)}$$

We see from equation 34.31 that all the components of the extra stress tensor are zero in a fluid that has been at rest for a long time. However, for materials undergoing deformation, the stress is no longer isotropic, and the 'P' in the above equation no longer has direct relationship to the pressure as it is usually defined. However, this will not be a problem for us, as we will work only with shear stresses and normal stress differences. For these quantities, the complete specification of P is not essential, and it matters not whether we write our equations in terms of σ_{ij} or τ_{ij}.

A STRAIN TENSOR FOR INFINITESIMAL DEFORMATIONS

In order to describe rheological phenomena in quantitative terms, it is necessary to make use of a quantitative measure of deformation. Such a measure is called strain, and we have defined in the previous sections measures of strain suitable for specific types of deformation, namely, simple shear and simple extension. However, it is advantageous to establish a general definition of strain that is valid for all types of flow kinematics. For the present, we will define only the infinitesimal strain tensor, which will be useful in the presentation of the theory of linear viscoelasticity.

The simplest description of deformation of a fluid element is one based on the behaviour of the position vectors for the fluid particles within this element. Fig. 34.15 shows the position vector, $x_A(t_1)$, for a typical particle 'A' at a time, t_1, as well as the position vector, $x_A(t_2)$, for the same fluid particle at another time, t_2. The use of two times, or two states of the fluid element, is unavoidable in discussing strain, as strain is a measure of the difference between two states. We will take t_1 to represent a time when the fluid is in a reference configuration, with which the configuration at any other time, t_2, will be compared. Thus, we will say that a material element is in a strained state at time t_2 if it undergoes deformation in the time interval between t_1 and t_2.

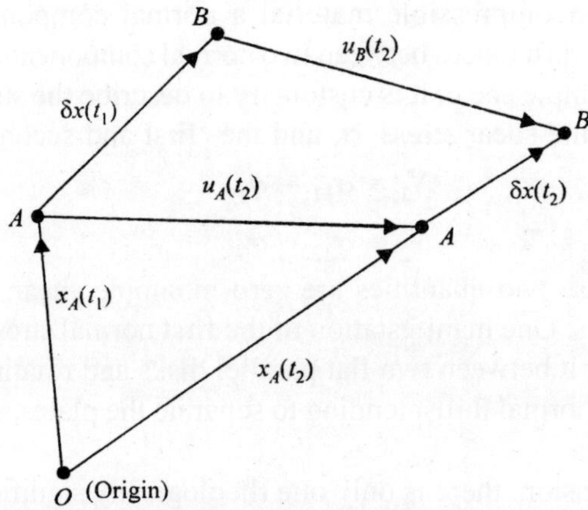

Fig. 34.15. The position vector of a fluid particle (A) at times t_1 and t_2, and the displacement vectors at time t_2 for two neighbouring fluid particles, A and B.

To describe deformation it is necessary to look at two fluid particles and the change that occurs in their relative positions between time t_1 and t_2. In Fig. 34.15, two such particles are shown, with the position of the second particle with respect to the first given by the vector $\delta x(t_1)$ at time t_1 and by $\delta x(t_2)$ at time t_2. The displacement vector, $u(t_2)$ is the position of a particles at time t_2 relative to its position at time t_1. The relative variation of the $u(t_2)$ vectors for the two fluid particles obviously provides information about the deformation of the fluid element containing these particles. For example, in a rigid displacement of the fluid element, without rotation, as shown in Fig. 34.16, the displacement vectors for the two particles are the same, and this is true for any two particles in the fluid element. This suggests that the gradient of u may be useful as a quantitative measure of strain. This quantity has nine components:

$$\frac{\partial u_i(t_2)}{\partial x_j(t_1)} \qquad \text{... (34.36)}$$

These nine quantities are the components of a tensor, and they clearly contain all the information necessary to describe the relative displacements of the two neighbouring particles. This tensor has some characteristics that make it unsuitable as a general measure of strain, but if we limit our attention to flows in which this gradient is always very small, a useful measure of strain can be defined as follows:

$$\gamma_{ij}(t_1,t_2) \equiv \frac{\partial u_i(t_2)}{\partial x_j(t_1)} + \frac{\partial u_j(t_2)}{\partial x_i(t_1)} \quad \text{... (34.37)}$$

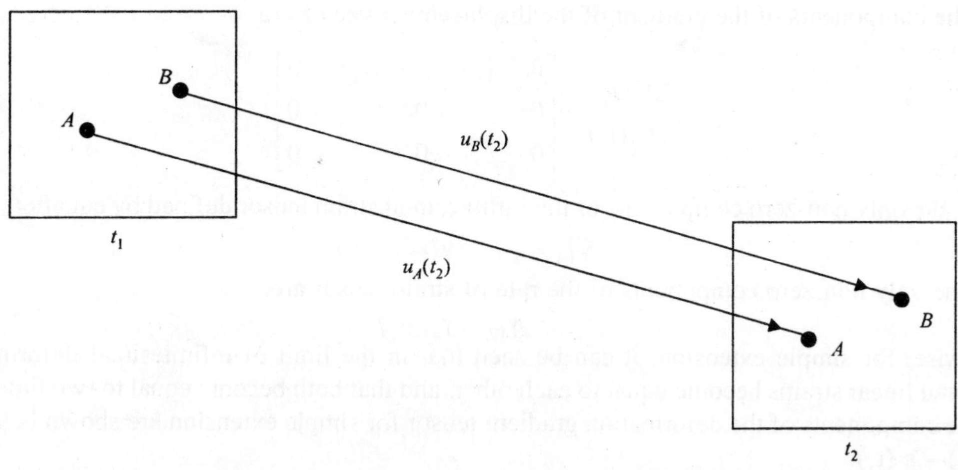

Fig. 34.16. Displacement vectors for particles *A* and *B*, of a fluid element at time t_2 and after a rigid translation of the element.

where $\gamma_{ij}(t_1, t_2)$ is the infinitesimal strain tensor at time t_2, relative to the configuration of the same fluid element at time t_1. Like the stress, this is a symmetric tensor, because if the order of the indexes *i* and *j* is reversed, the value of each component remains unchanged.

The "strain rate tensor", also called the "rate of deformation tensor", is defined by equation 34.38.

$$\dot{\gamma}_{ij} \equiv d\gamma_{ij}/dt \quad \text{... (34.38)}$$

Since the components v_i of the velocity vector are given by:

$$v_i = du_i/dt \quad \text{... (34.39)}$$

Equation 34.38 can be expanded as shown in equation 34.40.

$$\dot{\gamma}_{ij} = \frac{\partial v_j}{\partial x_i} + \frac{\partial v_i}{\partial x_j} \quad \text{... (34.40)}$$

This definition of a rate of deformation is not limited to use in describing deformations that are infinitesimal in total magnitude, as its definition requires only a measure of strain valid during an infinitesimal interval of time.

As an example of the use of the infinitesimal strain tensor, we note that for simple shear of magnitude γ;

$$\gamma_{12} = \gamma_{21} = \frac{\partial u_1}{\partial x_2} + \frac{\partial u_2}{\partial x_1} \quad \text{... (34.41)}$$

The variation with time of the components of the position vector for simple shear are:

$$x_1(t_2) = x_1(t_1) + x_2(t_1)\,[\gamma(t_2) - \gamma(t_1)] \quad \text{... (34.42a)}$$

$$x_2(t_2) = x_2(t_1) \quad \text{... (34.42b)}$$

$$x_2(t_2) = x_3(t_1) \quad \text{... (34.42c)}$$

The only non-zero component of the displacement vector is u_1:

$$u_1(t_2) = x_1(t_2) - x_1(t_1) = x_2(t_1)\,[\gamma(t_2) - \gamma(t_1)]$$

and the components of the gradient of the displacement vector are:

$$\frac{\partial u_i(t_2)}{\partial x_j(t_1)} = \begin{bmatrix} 0 & [\gamma(t_2) - \gamma(t_1) & 0 \\ 0 & 0 & 0 \\ 0 & 0 & 0 \end{bmatrix} \quad \text{... (34.43)}$$

Thus, the only non-zero components of the infinitesimal strain tensor defined by equation 34.37 are:

$$\dot{\gamma}_{12} = \dot{\gamma}_{21} = \gamma(t_2) - \gamma(t_1) \equiv \dot{\gamma} \quad \text{... (34.44)}$$

and the only non-zero components of the rate of strain tensor are:

$$\gamma_{12} = \gamma_{21} \equiv \gamma \quad \text{... (34.45)}$$

Likewise, for simple extension, it can be seen that in the limit of infinitesimal deformation, the Hencky and linear strains become equal to each other, and that both become equal to two times $\partial u_1/\partial x_1$. All of the components of the deformation gradient tensor for simple extension are shown below, where $\varepsilon \equiv \varepsilon\,(t_2) - \varepsilon\,(t_1)$.

$$\frac{\partial u_i(t_2)}{\partial x_j(t_1)} = \begin{bmatrix} \varepsilon & 0 & 0 \\ 0 & -\varepsilon/2 & 0 \\ 0 & 0 & -\varepsilon/2 \end{bmatrix} \quad \text{... (34.46)}$$

Thus, the components of the infinitesimal strain tensor are:

$$\gamma_{ij} = \begin{bmatrix} 2\varepsilon & 0 & 0 \\ 0 & -\varepsilon & 0 \\ 0 & 0 & -\varepsilon \end{bmatrix} \quad \text{... (34.47)}$$

For simple extension, the components of the strain rate tensor are:

$$\dot{\gamma}_{ij} = \begin{bmatrix} 2\dot{\varepsilon} & 0 & 0 \\ 0 & -\dot{\varepsilon} & 0 \\ 0 & 0 & -\dot{\varepsilon} \end{bmatrix} \quad \text{... (34.48)}$$

NEWTONIAN FLUID

We can make use of the tensors defined above to write out the general constitutive equation for the Newtonian fluid, which is valid for low molecular weight, single phase liquids:

$$\tau_{ij} = \eta\dot{\gamma}_{ij} \quad \text{... (34.49)}$$

whereas equation 34.16 gives only the shear stress of a Newtonian fluid undergoing steady simple shear, equation 34.49 provides information about all the components of the stress tensor for any type of deformation.

For example, we can use it to calculate all the stresses in steady simple shear. The velocity distribution for this deformation can be written as follows:

$$v_1 = \dot{\gamma} x_2 \quad \text{... (34.50a)}$$

$$v_2 = v_3 = 0 \quad \text{... (34.50b)}$$

and the components of the rate of deformation tensor are:

$$\dot{\gamma}_{ij} = \begin{bmatrix} 0 & \dot{\gamma} & 0 \\ \dot{\gamma} & 0 & 0 \\ 0 & 0 & 0 \end{bmatrix} \quad \text{... (34.51)}$$

Thus, the components of the extra stress tensor for a Newtonian fluid are, from equation 34.49.

$$\tau_{ij} = \begin{bmatrix} 0 & \eta\dot{\gamma} & 0 \\ \eta\dot{\gamma} & 0 & 0 \\ 0 & 0 & 0 \end{bmatrix} \quad \text{... (34.52)}$$

We note that the normal components of the extra stress are all zero in the Newtonian fluid. This means that:

$$\sigma_{11} = \sigma_{22} = \sigma_{33} \quad \text{... (34.53)}$$

and that the normal stress differences are both zero.

$$N_1 = N_2 = 0 \quad \text{... (34.54)}$$

It is important to remember that these equations for the components of the stress tensor are only valid for steady simple shear of a Newtonian fluid.

Another example of the use of the constitutive equation for a Newtonian fluid is the calculation of the stresses in a deformation other than simple shear, namely simple extension. Combining equation 34.48 to 34.49 we have:

$$\tau_{ij} = \eta \begin{bmatrix} 2\dot{\varepsilon} & 0 & 0 \\ 0 & -\dot{\varepsilon} & 0 \\ 0 & 0 & -\dot{\varepsilon} \end{bmatrix} \quad \text{... (34.55)}$$

For a Newtonian fluid, we have seen that it is possible to measure the viscosity in a simple shear experiment and then use it in the general constitutive equation (34.49) to determine all the components of the extra stress tensor for any type of flow. For non-Newtonian fluids this cannot be done, i.e., it is not possible to predict a fluid's response to arbitrary deformations on the basis of rheological information obtained using only simple shear experiments.

BASIC EQUATIONS OF FLUID MECHANICS

This is the procedure that is necessary, for example, to predict the detailed flow pattern and velocity distribution for melt flow in a die or in a mould. However, in order to understand the role played by rheology in solving such problems, we will simply present the basic equations of continuum mechanics and make a few comments about them. These basic equations are the continuity equation, which is derived from the principle of conservation of mass, and Cauchy's equation, which is a generalisation of Newton's second law of motion, (F = *ma*) for the case of a deformable material.

In deriving these equations, which govern the mechanical behaviour of deformable materials, it is assumed that every point in space (specified by x_1, x_2, x_3) within a material has associated with it values for all the components of the velocity vectors, the strain tensor and the stress tensor. It is further assumed that these 'field variables' are continuous functions of position. In reality, we know that this is not precisely true, because at a sub-microscopic level we find mostly empty space punctuated by regions of concentrated mass, which are the atoms of the molecules of which the material is composed.

However, the scale of our macroscopic observation and of practical flow phenomena is so large compared to the scale at which this inhomogeneity appears, that for all practical purposes we can assume the material to be a 'continuum' and make use of the strain, velocity and stress fields to describe the deformation and flow properties of the material. This is called the 'continuum assumption'. Thus, we assume that the collection of molecules that make up the material of interest can be represented, for purposes of describing its flow and deformation, by a mathematical continuum. The equations based on this assumption are thus called the equations of continuum mechanics.

Continuity Equation

The principle of conservation of mass, when applied to a deformable continuum, tells us that the net flow of mass into any fixed volume element contained within the continuum during some interval of time is equal to the accumulation of mass within the element during that time interval. This principle is summarised by equation 34.56.

$$[\text{Net flow in}] = [\text{Flow in}] - [\text{Flow out}] = [\text{Accumulation}] \quad \text{... (34.56)}$$

For an incompressible material, the accumulation of mass within a volume element is zero. We will not present a derivation here, but in this case it can be shown that the principle of conservation of mass can be written as shown in equation 34.57.

$$\frac{\partial v_1}{\partial x_1} + \frac{\partial v_2}{\partial x_2} + \frac{\partial v_3}{\partial x_3} = 0 \quad \text{... (34.57)}$$

This is the "continuity equation" for an incompressible material.

Cauchy's Equation

The second basic equation of continuum mechanics is based on Newton's second law of motion (F = *ma*). This basic principle describes the relationship between the applied force and the change of momentum of a rigid body. It was later put into a form suitable for describing the deformation of a continuum in response to deforming stresses, and the result is Cauchy's equation.

$$\rho \frac{Dv_i}{Dt} = -\rho g \frac{\partial h}{\partial x_i} + \frac{\partial \sigma_{1i}}{\partial x_1} + \frac{\partial \sigma_{2i}}{\partial x_2} + \frac{\partial \sigma_{3i}}{\partial x_3} \quad \text{... (34.58)}$$

The quantity on the left is the density times the "substantial derivative", which is defined as follows:

$$\frac{Dv_i}{Dt} \equiv \frac{\partial v_i}{\partial t} + v_1 \frac{\partial v_i}{\partial x_1} + v_2 \frac{\partial v_i}{\partial x_2} + v_3 \frac{\partial v_i}{\partial x_3} \quad \text{... (34.59)}$$

The substantial derivative is the quantity that tells how the velocity of a fluid element changes with time as it moves through the flow field.

Equations 34.58 and 34.59 are vector equations, and since i can have any one of the three values 1, 2 or 3, each can also be written as a set of three scalar equations.

Making use of the definition of the extra stress (Eq. 34.35), Cauchy's equation can be re-written in a form involving the pressure.

$$\rho \frac{Dv_i}{Dt} = -\rho g \frac{\partial h}{\partial x_i} - \frac{\partial P}{\partial x_i} + \frac{\partial \tau_{1i}}{\partial x_1} + \frac{\partial \tau_{2i}}{\partial x_2} + \frac{\partial \tau_{3i}}{\partial x_3} \quad \text{... (34.60)}$$

The left hand side describes the acceleration of a material element, while the right hand side accounts for the applied forces. The first term on the right represents the force due to the earth's gravitational field, which is a "body force" and acts directly on the mass of the material element. The second term describes the net force on a material element due to pressure gradients, and the last three terms involve the extra stress tensor and result from the deformation of the fluid. It is through these last three terms that the rheological properties of the material become involved in its mechanical behaviour.

For molten polymers, the very high resistance to deformation resulting from the strong interactions between molecules makes the last three terms very large compared to the acceleration terms on the left hand side. For this reason, in flow situations involving molten plastics, the left hand side can be neglected. This is called the 'creeping flow' approximation.

Taking account of the fact that Cauchy's equation represents three scalar equations, and adding in the continuity equation, we have four equations. The number of dependent variables is much greater than this, however, as it includes the three components of the velocity vector plus the six independent components of the stress tensor. Thus, no problem in continuum mechanics can be solved without an additional equation. This additional equation is the rheological constitutive equation, which relates the components of the stress tensor to the deformation.

Navier-Stokes Equation

The constitutive equation for a Newtonian fluid is given by equation 34.49. When this is combined with Cauchy's equation (34.60) and the continuity equation is used to simplify the terms, the result is the Navier-Stokes equation, which governs the flow of Newtonian fluids. If the viscosity is assumed to be a constant, i.e., independent of position, the Navier-Stokes equation can be written as follows:

$$\rho \frac{Dv_i}{Dt} = -\rho g \frac{\partial h}{\partial x_i} - \frac{\partial P}{\partial x_i} + \eta \left(\frac{\partial^2 v_i}{\partial x_1^2} + \frac{\partial^2 v_i}{\partial x_2^2} + \frac{\partial^2 v_i}{\partial x_3^2} \right) \quad \text{... (34.61)}$$

Note that this is vector equation and that by setting i equal to 1, 2 and 3, we can re-write it as three scalar equations.

In the analysis of flows involving Newtonian fluids, a dimensionless group of central importance is the Reynolds number. This is defined as follows:

$$\text{Re} \equiv \frac{DV\rho}{\eta} \quad \text{... (34.62)}$$

where D and V are a characteristic length and velocity respectively. This is a measure of the relative importance of viscous and inertial (acceleration) effects. In other words, it is a measure of the relative importance of the third term on the right of equation 34.61 as compared with the term on the left. For fluids with very high viscosity, the Reynolds number is generally very low, and the inertia or acceleration term on the left hand side of equation 34.61 can be neglected. This is often referred to as the 'creeping flow' approximation. While molten polymers are not Newtonian, and equation 34.61 is therefore not appropriate, it is generally accepted that for these materials, the creeping glow approximation is valid.

ON-LINE MEASUREMENT OF RHEOLOGICAL PROPERTIES

There are three types of application of on-line rheometers: (i) process monitoring; (ii) quality control; and (iii) automatic process control. The first type arises primarily in the development of a new material or process, when it is useful to be able to monitor the effects of changes in formulation or operating parameters on product characteristics. In a quality control application, a slow and labour-intensive laboratory test procedure is replaced by a direct indication of product quality. The most sophisticated type of application is in automatic process control, where the rheometer is used as a sensor providing an input signal to the controller.

The rheological properties of an in-process material may be of interest for its own sake, or as a measure of some other attribute such as molecular weight. For some resins, such as 'controlled rheology' types or high-flow injection moulding materials, the viscosity is the primary property that distinguishes different grades, so if the resin manufacturing process could be controlled to produce a preset viscosity, quality control would be automatic. More often, the rheological property is of interest as a measure of some other attribute. For example, the storage and loss moduli might serve as a measure of the molecular weight distribution, or the viscosity might be used as a measure of composition or the state of dispersion of a compound or blend.

Types of On-Line Rheometers for Melts

It is common practice to distinguish between 'on-line' and 'in-line' instruments. An on-line rheometer makes use of a side stream and a gear pump to carry a sample of melt to the actual sensor, whereas an 'in-line' rheometer is one that senses the properties of the melt as it passes a particular point in the process. An 'off-line' measurement is one made in a laboratory.

It is more difficult to design a rheometer for in-line use, but such an instrument has the advantage of a short signal delay, and it avoids the use of gear pumps. On the other hand, the sample cannot be brought to a prescribed temperature prior to the measurement. Since fluctations in temperature are a normal part of any industrial process, this means that a temperature compensation procedure must be used to reduce the measured property to a standard temperature. If the material being processed obeys time-temperature superposition, this is quite straightforward, once the shift factor is known as a function of temperature.

On-line capillary and slit rheometers for melts

As is the case in the laboratory, the simplest and most common type of process rheometer is the capillary rheometer. A side stream is taken from the main process line and forced through the capillary by a gear pump. The capillary is thermostated to control the melt temperature. The flow rate is calculated from the speed of the gear pump, and the pressure drop is measured directly by means of pressure transducers. By cycling between several pump speeds, the apparent viscosity at several shear rates can be determined.

If the rheometer exits to the atmosphere, only a single vented pressure transducer is required. However, in this type of installation, all the material that enters the rheometer flows out as scrap at a rate of about 0.5 to 1 kg/h. If this is to be avoided, a second gear pump is required to return the capillary efflux back to the main process stream. A second pressure transducer is then required, and the speed of the second gear pump must be carefully controlled to maintain a reasonable and constant level of pressure in the capillary. Clearly it is not possible to carry out all the measurements necessary to determine the entrance corrections and the Rabinowitch correction, and therefore on-line capillary rheometers are generally used to indicate either the apparent viscosity or to simulate a melt index measurement. The melt index test is simulated as follows. The pressure drop is set at 298.2 kPa, which is the nominal value for the melt index test, and a die having the prescribed L/D is used. Then the motor driving the gear pump is regulated to maintain the pressure drop at the set level, and the resulting pump speed is noted. The simulated melt index value is calculated as follows:

$$MI = 10KSV\rho(D_m/D_c)^3 \quad ...(34.63)$$

where: MI = melt index in g/10 min
V = pump displacement, cc/rev.
S = pump speed, rpm
K = empirical calibration factor to account for all the differences between a melt indexer and the capillary rheometer
ρ = melt density
D_m = standard die diameter for melt index test
D_c = capillary diameter

In practice, the product $KV\rho$ is determined experimentally for the particular resin of interest. On-line rheometers based on slit flow have also been used. Flow in a slit and the equations necessary to calculate quantities of rheological significance are presented there. The advantage of the slit over the capillary is that it is possible to mount pressure transducers flush with the wall. This has been of special interest because, it has been proposed that the first normal stress difference can be determined from wall pressures measured in a slit. Both the hole pressure method and the exit pressure method are described there, and commercial on-line sensors based on both these methods are available. However, as explained questions remain about the reliability of these two methods, and there have been no published reports of their successful use for the on-line measurement of N_1. Another pressure-flow device that can, in principle, be used to measure N_1 is the curved slit rheometer developed by *Geiger*. As in the case of the previously mentioned slit rheometers, however, it is not clear whether this device can form the basis of a useful industrial sensor.

The Rheometrics Melt Flow Monitor is a commercial slit rheometer that was developed not to measure N_1 but to eliminate certain disadvantages of other on-line instruments.

The problems addressed were:

1. Long signal delay due to long side stream.
2. Waste stream produced by rheometer efflux.
3. The need to mount major components near processing machinery.

This rheometer is equipped with two gear pumps, each driven by its own stepper motor, with the speed of the downstream motor regulated to provide a preset pressure at the exit of the capillary. The entire unit is designed to be installed just above a special adapter plate that can be placed at the end of an extruder or incorporated into the barrel assembly so that melt is sampled before it reaches the end of the extruder.

Rotational On-line Rheometers for Melts

In a rotational on-line rheometer, a side stream from a process line carries melt into the rheometer test chamber in which a rotating element is mounted, and means are provided to measure the torque on either the rotating or stationary element. The flow through the rheometer can be continuous, although the through-flow will have an effect on the measured torque, and more reliable results are obtained by periodically stopping the flow to make a measurement. Although the response of such an instrument is relatively slow, it represents a vast improvement over periodic laboratory testing. One challenge in the design of a rotational process rheometer is to provide a method for measuring the torque without interference from a static or dynamic seal. This problem is usually solved by use of a torsion tube, which transmits torque while simultaneously providing a static seal.

In the Rheoprocessor developed by *Kepes* the rotating member is a bicone, and the speed or displacement arising from the application of a constant or time-varying torque is monitored. A seal is not necessary as the rheometer chamber drains to the atmosphere by gravity and is thus not under pressure. A four-way valve allows the periodic stopping of the flow for measurement or the use of a calibration standard.

The Rheometrics On-line Rheometer is a concentric cylinder device designed for the measurement of linear viscoelastic properties. A gear pump provides a controlled flow of melt into the rheometer. The outer cylinder is made to oscillate sinusoidally, and the torque on the inner cylinder is measured by means of a torque tube assembly that provides a perfect seal. The dynamic seal used for the shaft driving the outer cylinder introduces no error, since the torque is not measured on this shaft. *Starita* and *Rohn* have described this rheometer and *Zeichner* and *Macosko* have discussed its application for monitoring the molecular weight distribution of polypropylene.

In another concentric cylinder process rheometer described by *Heinz* the concentric cylinder sensor is a self-contained unit that is inserted into the melt stream, either in an on-line or an in-line installation. The sensor consists of three concentric, thin-walled cylinders.

The second (middle) cylinder is driven in an oscillatory mode, and a shear stress is generated on both sides due to the presence of the two stationary cylinders. The torque on the drive shaft is transmitted to an external displacement sensor by means of a torque tube. Bellows are used to provide seals for the moving parts.

Khachatryan have described a concentric cylinder rheometer for use with fibre-forming resins. Two thin-walled concentric cylinders are suspended in a chamber through which the polymer is made to flow. The inner cylinder is rotated by a motor, and the torque transmitted to the outer cylinder is monitored. The published report does not explain how the outer cylinder is suspended nor how the torque arm is sealed where it passes through the wall of the chamber. Another concentric cylinder on-line melt rheometer has been described by *Menges*.

In-Line Melt Rheometers

All of the on-line instruments described above require the use of a side stream and at least one gear pump. In some installations, where it is important to minimise the signal delay or when the use of a gear pump is not practical, an in-line installation is advantageous a "shear stress transducer" that has been developed for use in a sliding plate rheometer. It has been proposed that such a device could be used as the basis for an in-line rheometer. One possible configuration is shown in Fig. 34.17. A rotating drum (A) generates a nearly uniform shearing deformation in the small gap (B), and the resulting shear stress is monitored by the shear stress transducer (C). The motion of the rotor also generates a drag flow that draws melt into the gap to aid the process of sample renewal. A rheometer of this type has been used to control reactive extrusion processes.

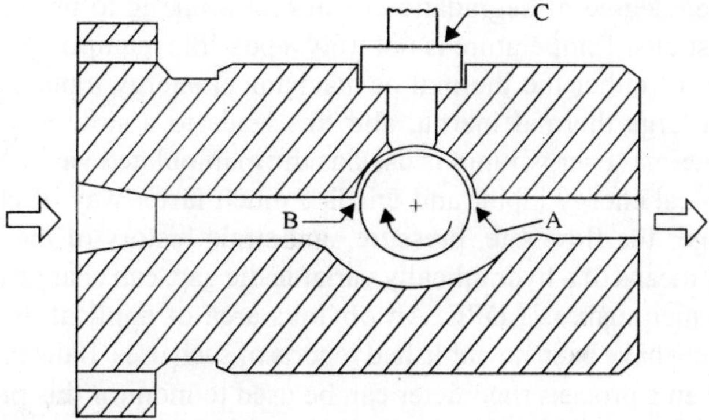

Fig. 34.17. An in-line rotational rheometer that makes use of a shear stress transducer. A = rotary drum driven by consta ıt speed motor or servo-motor; B = gap in which concentric-cylinder drag flow is generated; C = shear stress transdı cer that senses the shear stress at the outer wall of the concentric-cylinder drag flow zone.

A single 'flight' machined onto the drum could be used to ensure a clean sweep of the shearing gap as often as desired to ensure complete sample renewal and to avoid the build-up of a degraded layer of material at the outer wall. Using a servo-motor to drive the drum, any desired shear history can be generated, including steady shear, oscillatory shear and step strain. Thus, not only can the viscosity be monitored but also the storage and loss moduli and even non-linear viscoelastic properties. By using a cam-like drum having two or more diameters over various sections of its perimeter, the viscosity at two or more shear rates could be measured once each revolution by simply driving the drum at a constant rotational speed.

Another possible application of a shear stress transducer for in-line measurements is the measurement of the shear stress on the wall of an extruder barrel. Because the flow field is neither simple nor under complete control, this could not provide information about any well-defined rheological property. However, if a special 'rheometric module' were inserted into the screw, and if the screw speed were accurately monitored, it would be possible to make a measurement of the viscosity.

Revesz has described a device for measuring the shear stress exerted by the melt on the tip of an extruder screw. He suggests a method for estimating the viscosity of the melt as it enters the die, although the principal purpose of the device is to provide information about melt flow rate.

SPECIFIC APPLICATIONS OF PROCESS RHEOMETERS

Some types of processes in which process rheometers have potential applications are: (i) polymerisation; (ii) extrusion and melt forming; (iii) compounding; (iv) blending; and (v) reactive extrusion. The advantages to be gained by the use of such sensors include improvements in product quality, reduction of waste, and increases in production rates.

Possible manipulated variables in process control applications are:

1. Feed rates of reactants or additives.
2. Process temperature.
3. Extruder speed.
4. Geometry of die or other mechanical components.

Feed rate can be controlled by a liquid pump or by a loss-in-weight solids feeder, and this is the obvious choice of manipulated variable in reactive extrusion operations. For example, it may be desirable to accomplish a controlled degree of degradation or of cross-linking to produce an extrudate having a specified viscosity or elasticity. Temperature is not truly a possible manipulated variable, as it is actually changed by manipulation of either the thermal or mechanical energy input to the process. Processing machinery usually has a large thermal inertia, and this leads to a slow response to a change in the temperature set point if thermal energy input is used as the manipulated variable. Changing screw speed alters the rate of mechanical energy input, and this is a much faster way of changing the temperature. However, this also changes the flow rate, pressure, and strain history of the melt. Changing machine geometry, for example by means of a hydraulically variable die gap, can change the rheological properties of a shear sensitive polymer such as LDPE. An obvious area of application is in the polymerisation process itself, although there have been no published reports of such uses. If the product of a polymerisation process is itself a melt, then a process rheometer can be used to monitor this product directly. However, several modern polymerisation processes produce a powder rather than a melt, and in order to use rheology as a tool for process control it is necessary to provide a mechanism for automatically sampling this stream and converting powder to melt. The Rheometrics ROR-QC addresses this problem by use of pneumatic transport. *Starita* and *Macosko* have discussed some ways in which variations in resin characteristics can cause problems in melt forming operations such as blow moulding. If a process rheometer could provide continuous information regarding the nature of the raw material, operating conditions could be altered to compensate for such changes.

Variations in viscosity can cause serious difficulties in an injection moulding operation, especially when large parts are being produced, and it would be useful to be able to monitor melt consistency on a continuous basis. At present, nozzle pressure is sometimes used as a rough indication of viscosity.

The use of compounding operations to produce sophisticated multicomponent materials is growing rapidly, and there is a widely recognised need for methods to ensure that the product has the correct formulation and is properly mixed. Process rheometers show promise in this area, specifically to monitor the state of dispersion of a masterbatch, a compound or a blend.

Considerable attention has been focused on the use of process rheometers for the closed-loop control of reactive extrusion processes. If a rheological property is sensitive to the extent of reaction, then a process rheometer can be used to provide a signal for the control of the feed rate of one of the reactants. Several reports have been published on the use of rheometers to control the peroxide-initiated degradation of polypropylene. This is a process that is used to produce a range of lower molecular weight resin grades starting from a single high-molecular weight grade using an extruder. Both rotational and capillary on-line rheometers have been evaluated for this application. In addition, *Pabedinskas* used the die pressure drop as a measure of the amount of degradation.

CHAPTER 35

Rubber-Resin Pressure Sensitive Adhesive Tape Technology

INTRODUCTION

The subject of pressure sensitive adhesives and tapes comprises a large body of science. This chapter provides an overview of rubber-resin pressure sensitive adhesive tape technology in sufficient depth to be valuable to those unfamiliar with the subject as well as those versed in the technology. Pressure sensitive adhesives (PSAs) are viscoelastic materials that are aggressively and permanently tacky at room temperatures at 100% solids and firmly adhere to myriad surfaces upon contact with finger or hand pressure. PSAs do not require activation by heat or solvent; they can be readily handled; and they afford a convenient way of sealing, holding, reinforcing, insulating, masking, or protecting surfaces and articles, both temporarily and permanently. Although a wide variety of pressure sensitive tapes comprise the overall market, the largest usages of natural rubber based adhesives are in paper tapes, reinforced tapes, polypropylene tapes, duct tapes, industrial cloth tapes, and medical tapes.

PRESSURE SENSITIVE ADHESIVE COMPONENTS

Some materials are inherently pressure sensitive (e.g., certain acrylics and styrene butadiene polymers), but the majority of elastomers used in the industry require key materials to modify their viscoelastic behaviour. The components used in natural rubber elastomers are discussed below.

Tackifying Resins

Tackifying resins are the key materials used to convert elastomers into PSAs. The modifiers of major use are obtained by the polymerisation of petroleum or terpene. The terpene-and petroleum-derived resins are of low molecular weight (typically 1200–2500) and manifest a broad melting point range (typically 85–135°C). The materials based on rosin are of considerably lower molecular weight (1000 or less) and cover similar softening point ranges.

Plasticisers

Plasticisers are often used to lower the elastomer content of an adhesive, thus softening the adhesive and reducing the overall viscosity of the final adhesive. Such modifiers include phthalates, lanolin, naphthenic and aromatic oils obtained from the refining industry, and liquid resins derived from petroleum products as well as from natural products.

Curing Agents

Curing agents help to increase the internal or cohesive strength of an adhesive and are typically used if an adhesive will be exposed to service temperatures above 60°C for extended periods. Commonly used curing agents are organic sulphur compounds containing at least two catenated sulphur atoms, metal salts of organic sulphur compounds, methylol-or halomethyl-containing phenolics, and isocyanates.

Antioxidants

Antioxidants improve the oxidative stability of an adhesive, especially the components bearing carbon-carbon double bonds. It is generally agreed that these modifiers function in unsaturated materials to terminate hydroperoxide free radicals. Commonly used antioxidants are substituted phenols, hindered aromatic amines, hydroquinones, and zinc salts of thiocarbamates.

Fillers

Fillers are used largely to lower cost and to modify the cohesive strength of the adhesive. Often they function as cure accelerators. Typical fillers are zinc oxide, clays, and talcs.

SOLVENT-BASED NATURAL RUBBER ADHESIVES

Typically, because of the high molecular weight, gel content, and high branching characteristics of the natural rubber molecule, it is necessary to subject the elastomer to mechanical shear energy to make it amenable to further processing into a coatable pressure sensitive adhesive solution or dispersion. Accordingly, the elastomer's rheology is usually modified by passing it repeatedly between two steel rolls moving in opposite directions and at different speed or in a Banbury-type mixer, where the compressed rubber is subjected to high shear forces of counterotating rotors.

These mechanical processes cause breakage of the rubber chains, preferentially of the highest molecular weight molecules, thereby reducing the overall molecular weight, the gel content, and the rubber's elasticity, rendering the material soluble in both aromatic and aliphatic solvents. The effects of milling time on natural rubbers at various temperatures, has been studied and a schematic representation of how milling affects the rubber's molecular weight is shown in Fig. 35.1.

In practice, it is recognised that significant variability can exist in the final molecular weight distribution of this processed natural rubber mill base. However, by carefully selecting rubber grades and utilising statistical process control practices, sufficient uniformity of the breakdown process can be attained.

To complete the preparation of the adhesive, the milled elastomer is combined with appropriate solvents and tackifier resins in a low speed mixer commonly referred to as a churn mixer, or in a variety of high speed, heavy-duty mixers, to reduce the time required for dissolution. Solvent selection is dependent on the adhesive formulation ingredients, solvent evaporation rates, and drying times.

The method employed to coat the solvent-based adhesive is usually determined by three factors: (i) the adhesive viscosity; (ii) the coating weight desired; and (iii) the speed of manufacture. Table 35.1 summarises the operational characteristics of various coating methods.

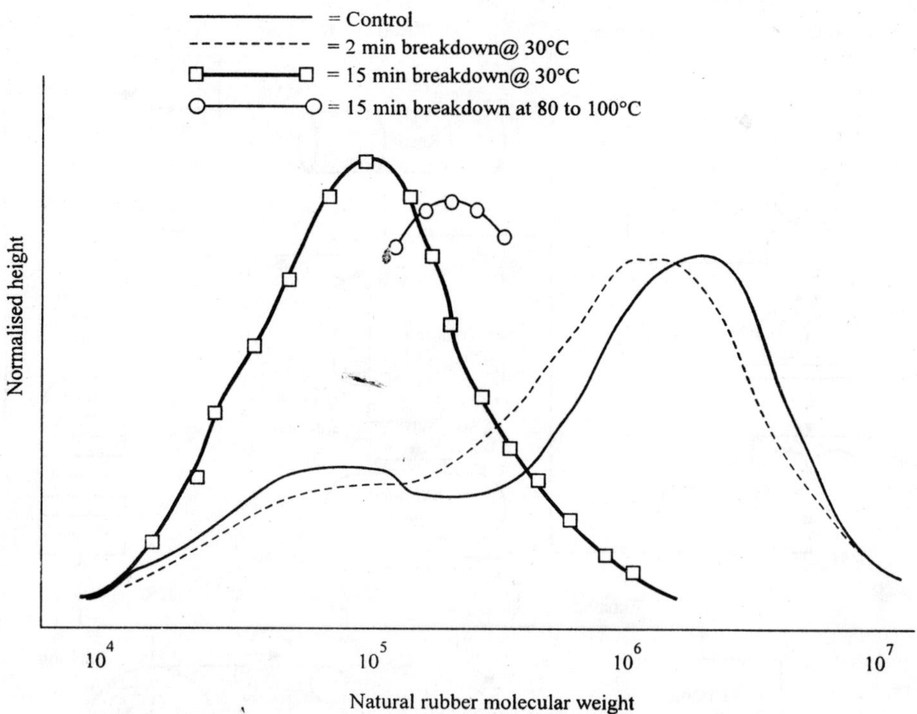

Fig. 35.1. Milling time and temperature effects on NR molecular weight.

Table 35.1. Summary of typical operating characteristics of seven coating systems.

	Viscosity range (cps)		*Solids range (%)*		*Wet coating thickness (mils)*		*Coating speed (m/min)*	
Coating system	*Min*	*Max*	*Min*	*Max*	*Min*	*Max*	*Min*	*Max*
Wire rod	100	1,000	1	Coating rheology limited	0.25	9	3	250
Air knife	5	5,000	3	100	0.1	1.5	1	250
Knife-over-roll	200	25,000	10	100	1	30	2	100
Two-roll squeeze coating	20	1,000	1	Coating rheology limited	0.25	9	3	150
Reversed-roll	100	20,000	10	100	0.3	20	3	150
Gravure	100	1,000	2	100	0.1	3	1	500
Slot orifice (hot melt)	2,000	30,000	100	100	0.5	200	3	200

A typical flowchart of solvent-borne pressure sensitive tape manufacture is shown in Fig. 35.2.

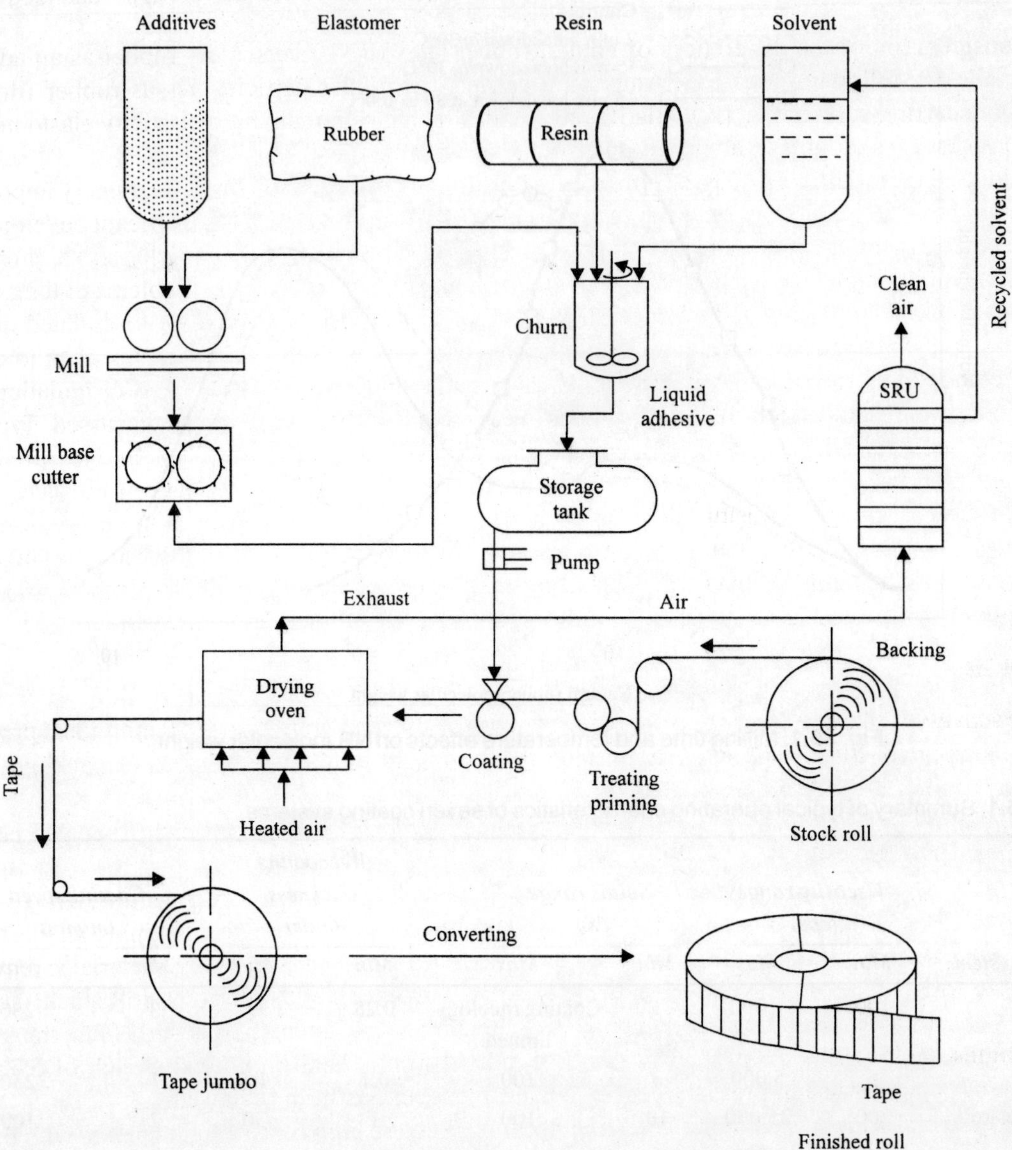

Fig. 35.2. Schematic of solvent PSA tape manufacture.

LATEX-BASED NATURAL RUBBER ADHESIVES

It has been mentioned that natural rubber latex based pressure sensitive adhesives comprise a small portion of the PSA market and afford an alternative to solvent solutions of natural rubber. Four types of natural rubber latex of interest to the PSA industry are described by the American Society for Testing and Materials (ASTM). Natural rubber latex is a colloidal system in which the rubber particles are protected by naturally occurring proteins and lipids that are, in large part, bound to the surface of the rubber particles. Centrifuged natural latices have been shown by light microscopy to have a number average particle diameter in the range of 0.2–0.3 μm.

A consideration of the differences of natural rubber latex (RNL) versus dry rubber as an adhesive base entails the following areas: (i) water as a carrier; (ii) rubber particles versus rubber film; (iii) naturally occurring surfactants; (iv) volatile fatty acids; (v) high strength, high elasticity elastomer; and (vi) no processing available to alter elastomer mechanical properties.

In addition, the modifiers discussed above must be provided in dispersion form. Of primary importance is the availability and quality of tackifier resin dispersions. Here, as a result of significant developments by the tackifying resin industries a wide range of tackifiers are available, with particle sizes approaching those of the rubber molecule itself. This development negates many of the early problems of long drying time at high temperatures needed to effect tackification. The rubber latices can be formulated directly into PSAs, a distinct process advantage compared to solve-borne PSAs. Care must be taken to ensure that the blended latices possess compatible pH ranges and emulsifier types to avoid coagulation. The compounding and stabilisation of natural rubber latex into a PSA has already been discussed. Typically formulated latex adhesives will display coating viscosities from 100 to 2000 cps as measured on a Brookfield LVF viscometer with a number 1 spindle at 60 rpm. These materials can be successfully coated at low coating speeds using Meyer rods and knife-over roll coaters and, at higher speeds, with reversed-roll coating heads. Finally, strong consideration must be given to the role of the migratory characteristics of the naturally occurring surfactants and soaps, as well as the choice of those added via the elastomer modifiers.

BACKINGS

Pressure sensitive adhesive tapes cover a broad spectrum of use, and more than 2000 individual products have been reported. Such wide application requires the use of multiple substrates, or backings, onto which the compounded PSAs are coated.

Three major classifications of paper backings can be listed: (i) saturated papers; (ii) coated papers; and (iii) uncoated papers.

Saturated papers for tapes typically weigh from 60–100 g/m^2. These backings are derived from passing lighter basis weight raw paper (40–80 g/m^2) through a saturation bath containing an elastomer or an elastomer–resin mixture, which impregnates or saturates the sheet. The excess material is removed via squeeze rolls and/or doctor knives, and the wet paper is dried and wound into roll stock. Typical properties of saturated tape base stock are machine–direction tensile strength of 26–50 N/cm, transverse tear strengths of 45–100 g/16 ply, a thickness of around 5 mils, and a minimal stretch of 5%. The saturation of paper for use in pressure-sensitive adhesive tapes has been reviewed.

Coated papers, widely used in the label and decal market, are high basis weight papers, with 65–70 g/m^2 commonly used. These papers are referred to as high or low gloss: the gloss is a function of the clay–binder system employed and whether the papers are direct-coated or cast-coated.

Uncoated papers are common in the lower priced label market and in note pads. A common basis weight is 55–60 g/m^2.

Film backings provide the tape user an opportunity to match such properties and strength, clarity, temperature resistance, resistivity, moisture and air permeability, flexibility, and tear resistance to an application. In practice, biaxially oriented polypropylene backing represents the single largest use for film backing. Other commonly used substrates include, in order of volume, polyethylene, polyester, polyvinyl chloride (PVC), cellulosics, and specialty films (e.g., polyimides, fluoropolymers). Typical applications include packaging tapes, diaper tapes, electrical tapes, protective tapes, medical tapes, and household tapes.

A critical component of essentially all tape products is the release coating, that is, the material that is applied to the side opposite the PSA. These materials function by providing a low energy surface, to which the PSA has a difficult time wetting and establishing strong bond formation. Forces required to separate the adhesive from the release coating are tailored to the specific end use. Accordingly, applications requiring low forces to separate the adhesive from the backing (e.g., transfer tapes) will commonly employ silicone polymers, while applications requiring hand removal of tape from a roll will typically use materials containing long aliphatic hydrocarbon constituents derived by polymerisation or co-polymerisation, graft polymers, especially those containing dimethyl silicone grafts, polymers formulated with controlled, migrating surface-active agents, and some fluoroacrylates. These materials are applied in very thin coatings to prevent cohesive failure and transfer failure and transfer to the adhesive surface.

DYNAMIC PROPERTIES OF RUBBER-RESIN MIXTURES

It is well established that the adhesion behaviour of PSAs is highly dependent on the viscoelastic properties of the materials. Correlations between tack and peel adhesion to viscoelastic parameters have been determined. Values of viscoelastic parameters as obtained by making measurements at different temperatures and frequencies can, by accepted standard procedures, be reduced to a reference temperature T_0 and shifted to generate master curves of the viscoelastic parameter as a function of temperature or frequency. This analytical approach allows for a convenient and quantitative route to assess the effects of adhesive components on critical viscoelastic parameters and for further mathematical treatment of the data to gain additional material insights.

A typical master curve of modulus versus reduced frequency is shown in Fig. 35.3.

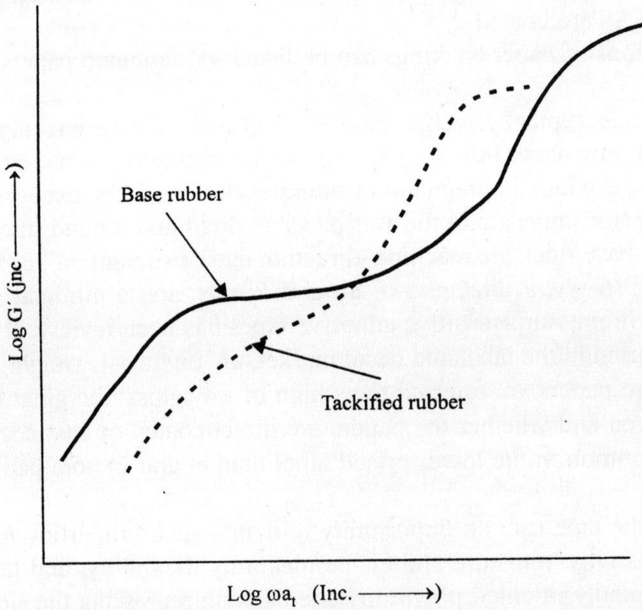

Fig. 35.3. Schematic master curve of shear modulus G´ and reduced ωa_t for NR and tackified NR.

In general, a PSA must flow sufficiently at low rates of deformation to enable a bond to the adherend to be formed, yet to be relatively stiff at high rates of deformation, thereby allowing for higher amounts of energy absorption before bond failure occurs. Such response requires some fractions of the polymer chains in the PSA to have sufficiently high mobility at the temperature or frequency of application to permit wetting of the adherend. Accordingly, PSAs must have low glass transition temperatures T_g, typically –10 to –40°C. As noted earlier, and as evident from Fig. 35.4, which shows how the various adhesive components affect the magnitude and position of the master curve, in the absence of other modifiers, the tackifying resin concentration dominates this response.

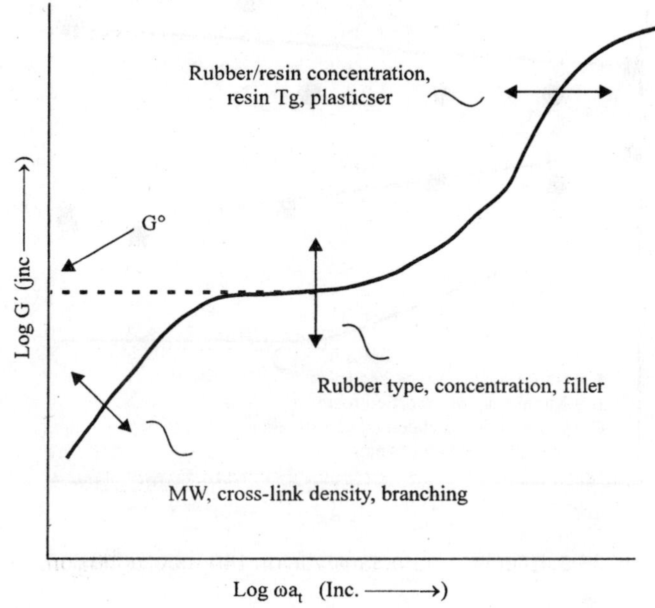

Fig. 35.4. Schematic of master curve response to adhesive formulation variables.

Although it is highly desirable to formulate a PSA to have a T_g high enough to maximise the energy of absorption before bond failure, it is equally important that bond failure be devoid of adhesive residue. This characteristic is a function of the cohesive strength of the adhesive. The Pressure Sensitive Tape Council defines cohesive strength as 'the ability of the adhesive to resist splitting'. Cohesive strength can be related to the viscous component of the viscoelastic response of the pressure-sensitive adhesive and can be shown to relate to the steady state flow viscosity. This response is dominated by the overall system molecular weight and its cross-link density. It has been reported that gel fraction of 30–50% is suitable for a good balance of properties in high shear adhesives. Although proper viscoelasticity is a prerequisite for good adhesion to a substrate, it is not the only factor that affects adhesion. To maximise bond strength, the adhesive must wet and spread over the adherend surface to achieve intimate contact. This phenomenon is largely determined by the relative surface energies of the bonding materials. It is possible to measure contact angles of PSAs, but the data generated have been reported to be unreliable because of solubilisation of low polarity contact angle liquids into the substrate, while the use of high polarity, hydrogen-bonding liquids leads to significant error.

Surface energies of PSAs can be measured, indirectly, however. To assign surface energies to solids, *Zisman* introduced the concept of critical surface tension γ_c.

It has been shown for pressure sensitive adhesives that if psa γ equals or exceeds adherend γ the adhesion will be dependent on the critical surface energy of the adherend. If, however, psa γ is less than adherend γ, then the adhesion will be independent of the critical surface energy of the adherend. The latter condition provides for an estimate of psa γ. *Sherrif* and *Krawiec* have reported on the surface energy effects on tack and peel. These properties are maximised when the surface energies of the materials approximate each other.

Other surface effects, such as surfactant type in compounded NRL pressure sensitive tapes, were found to have a strong negative affect on peel adhesion, as shown in Fig. 35.5.

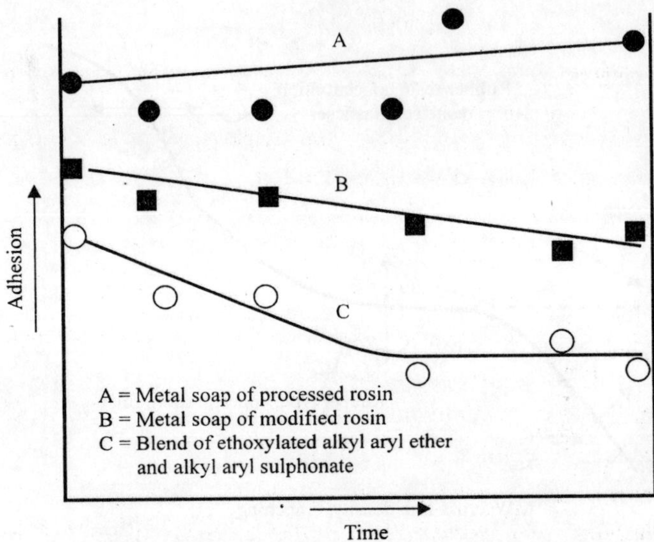

Fig. 35.5. Effects of surfactant type on 180° peel adhesion.

TEST METHODS

The Pressure Sensitive Tape Council has developed a series of test methods for the measurement of adhesion, shear, and tack properties of PSAs. Among the most commonly used are 180° peel adhesion, 90° quick stick, rolling ball tack, and both room temperature and elevated temperature shear adhesion.

Peel adhesion is the force required to remove a PSA tape from a specified test surface at a designated removal angle and rate. The tape is first applied to the test surface using a predetermined force to ensure intimate contact with the surface and a predetermined bonding time to ensure that maximum wetting has occurred. Quick stick is the force required to remove a PSA tape from a specified surface in which bond formation is achieved using only the applied pressure of the weight of the tape backing. The test strip is peeled immediately after application. Rolling ball tack is a value indicated by the distance between the initial contact of a rolling ball onto the surface of the adhesive tape to the point at which it stops. The ball, of a specified diameter and weight, is rolled down an inclined plane onto the adhesive surface. The test measures the resistance to detachment after a short period of contact with the adhesive surface.

Probe tack is the force required to remove a 0.5 cm diameter probe of a specified surface composition after it has been contacted with the adhesive surface for a specified time and application pressure.

Shear adhesion is a measure of force required to remove a PSA tape from a standard test panel in a direction parallel to that surface. When cohesive failure of the adhesive occurs, the time to failure is a measure of the cohesive strength of the adhesive. The test is run under constant force.

References

Brydson, J.A., *Rubbery Materials and their Compounds,* Elsevier Applied Science, London.
Brydson, J.A., *Rubber Chemisty,* Applied Science, London.
Bunn, C.W., *Polymer Science and Technology,* Newnes–Butterworth, London
Treloar, L.R., *Physics of Rubber Elasticity,* Clarendon Press, London.
Gardon, J.L., *Encyclopaedia of Polymer Technology,* Interscience, New York.
Bateman, L., *Chemistry and Physics of Rubber like Substances,* Applied Science, London.
Alliger, G., *Polymer Chemistry of Synthetic Elastomers,* Interscience, New York.
Stannett, V., *Diffusion in Polymers,* Academic Press, London.
Wall, L.A., *Fluoropolymers,* Interscience, New York.
Noll, W., *Chemistry and Technology of the Silicones,* Academic Press, New York.
Blow, C.M., *Rubber Technology and Manufacture,* Butterworth Scientific, London.
Hofmann, W., *Vulcanisation and Vulcanising Agents,* Applied Science Publishers, London.
Smith, C.K., *Polymer Mixing Technology,* Elsevier, New York.
Georgeodian, K.M., *Principles of Polymerisation,* John Wiley and Sons, New York.
Holden, G., *Chemistry and Technology of Rubber,* Reinhold, New York.
Lewis, F.W., *High Polymer Series,* John Wiley and Sons, New York.
Brown, R.P., *Physical Testing of Rubber,* Applied Science Publishers, London
Morton, M., *Rubber Technology,* Van Nostrand Reinhold, New York.
Craig, A.S., *Dictionary of Rubber Technology,* Philosophical Library, New York.
Dorian, A.F., *Dictionary of Plastics and Rubber Technology,* Iliffe Books, London.
Blackley, D.C., *Synthetic Rubbers and their Chemistry,* Applied Science Publishers, London.
Saunders, J.H., *Polyurethanes–Chemistry and Technology,* Interscience, New York.
Wright, P., *Synthetic Rubbers and their Technology,* Maclaren, London.
Alliger, G., *Vulcanisation of Elastomers,* Reinhold, New York.
Scott, G., *Atmospheric Oxidation and Antioxidants,* Elsevier, Amsterdam.
Bruins, P.F., *Plasticiser Technology,* Reinhold, New York.
Kraus, G., *Reinforcement of Elastomers,* Interscience, New York.
Donnet, J.B., *Carbon Black,* Marcel Dekker, New York.

Nourry, A., *Reclaimed Rubber*, Maclaren, London.
Allen, P.W., *Use of Rubber in Engineering,* Maclaren, London.
Anon, W., *Radiation Processing for Rubber and Plastics,* Applied Science, London.
Collyer, H.J., *Environmental Aspects of Polymer and Rubber,* Interscience, New York.
Khromov, M.K., *Mechanical Testing of Unvulcanised and Vulcanised Rubber,* Maclaren, London.
Mooney, M., *Rubber Rheology,* Academic Press, New York.
Norman, R.H., *Conductive Rubber,* Maclaren, London.
Penn, W.S., *Injection Moulding of Rubbers,* Maclaren, London.
Scott, J.R., *Physical Testing of Rubber,* Maclaren, London.
Stamberger, P., *The Colloid Chemistry of Rubber,* Oxford University Press, Oxford.
Stern, H.J., *Rubber: Natural and Synthetic,* Maclaren, London.
Thompson, D.C., *Encyclopaedia of Polymer Science*, John Wiley and Sons, New York.
Sprague, G.R., *The Vanderbilt Rubber Hand Book,* Vanderbilt, New York.
Wheelans, M.A., *Injection Moulding of Elastomers,* Maclaren, London.
White, J.L., *Science and Technology of Rubber,* Academic Press, New York.
Nauntron, W.J., *The Applied Science of Rubber,* Arnold, London.
Kuryla, W.C., *Flame Retardancy of Polymeric Materials,* Maclaren, London.
Killeffer, D.H., *Banbury—The Master Mixer,* Palmerton, New York.
Cooper, W., *Developments in Polymerisation,* Applied Science Publishers, London.
Brennan, J.J., *Toxicity and Safe Handling of Rubber Chemicals,* Academic Press, New York.
Allen, N.S., *Degradation and Stabilisation of Polyolefins,* Elsevier, USA.
Bark, L.S., and **Allen, N.S.,** *Analysis of Polymer Systems*, Elsevier, USA.
Alfrey, T., *Testing of Polymer,* Interscience, New York, USA.
Bernhardt, E.C., *Computer Aided Engineering for Injection moulding,* Pergamon, UK.
Bernhardt, E.C., *Processing of Thermoplastics,* Van Nostrand Reinhold, USA.
Bruins, P.F., *Polyblends and Composites,* Wiley Interscience, New York.
Ceresa, R.J., *Block and Graft Copolymerisation,* Wiley Interscience, New York.
Ferry, J.D., *Viscoelastic Properties of Polymers,* Wiley Interscience, New York.
Harper, C.A., *Handbook of Plastic and Elastomers,* McGraw-Hill, New York.
Iwen, E.D., *Degradation and Stabilisation of PVC,* Elsevier, USA.
MacDermott, C., *Selection Rubber for Engineering Applications,* Marcle Dekker, Reinhold, USA.
Matthews, F.G., *Twin Screw Extruders: A Basic Understanding,* Van Nostrand Reinhold, USA.
McKelvey, J.M., *Polymer Processing,* Wiley Interscience, New York.

Index